91 Springer Series in Solid-State Sciences

Edited by Hans-Joachim Queisser

Springer Series in Solid-State Sciences

Editors: M. Cardona P. Fulde K. von Klitzing H.-J. Queisser

Managing Editor: H. K. V. Lotsch

H. Kuzmany M. Mehring
S. Roth (Eds.)

Electronic Properties of Conjugated Polymers III

Basic Models and Applications

Proceedings of an International Winter School,
Kirchberg, Tirol, March 11–18, 1989

With 251 Figures

Springer-Verlag Berlin Heidelberg New York
London Paris Tokyo Hong Kong

Professor Dr. Hans Kuzmany

Institut für Festkörperphysik, Universität Wien, Strudlhofgasse 4,
A-1090 Wien, Austria
and Ludwig Boltzmann Institut für Festkörperphysik, Kopernikusgasse 15,
A-1090 Wien, Austria

Professor Dr. Michael Mehring

Physikalisches Institut II, Universität Stuttgart, Pfaffenwaldring 57,
D-7000 Stuttgart 80, Fed. Rep. of Germany

Dr. Siegmar Roth

Max-Planck-Institut für Festkörperforschung, Heisenbergstr. 1,
D-7000 Stuttgart 80, Fed. Rep. of Germany

Series Editors:

Professor Dr., Dres. h. c. Manuel Cardona
Professor Dr., Dr. h. c. Peter Fulde
Professor Dr., Dr. h. c. Klaus von Klitzing
Professor Dr. Hans-Joachim Queisser

Max-Planck-Institut für Festkörperforschung, Heisenbergstrasse 1,
D-7000 Stuttgart 80, Fed. Rep. of Germany

Managing Editor:

Dr. Helmut K. V. Lotsch

Springer-Verlag, Tiergartenstrasse 17, D-6900 Heidelberg, Fed. Rep. of Germany

ISBN-13: 978-3-642-83835-4 e-ISBN-13: 978-3-642-83833-0
DOI: 10.1007/ 978-3-642-83833-0

2154/3150-543210 – Printed on acid-free paper

Preface

The International Winter School on Electronic Properties of Polymers and Related Compounds, held March 11–18, 1989, in Kirchberg (Austria), was a sequel to two meetings on similar subjects held there two and four years before. The 1989 winter school was again organized in cooperation with the "Bundesministerium für Wissenschaft und Forschung" in Austria, and with the "Bundesministerium für Forschung und Technologie" in the Federal Republic of Germany. The basic idea of the meeting was to provide an opportunity for experienced scientists from universities and industry to discuss their most recent results and for students and young scientists to become familiar with the present status of research and applications in the field.

Like the previous winter schools, this one concentrated on the electronic structure and application potentialities of polymers. A purely tutorial contribution on high-temperature supercondutors was included. Special attention was paid this time to our present theoretical understanding of the electronic structure of polymers in general, the role of disorder and the nature of the doping process in particular. Accordingly, two discussion meetings were held on the stability of quasiparticles in conjugated polymers and on the thermochromic phase transition in alkylated thiophenes. As a general result, disorder and conjugation length were recognized as fundamental properties, not only for the stability of the quasiparticles but also for linear and nonlinear optics, transport, phototransport, magnetic properties, lattice dynamics, etc.

With respect to the materials studied, the presentations were dominated by polyanilines (18 contributions), polyacetylene (12 contributions) and thiophenes (9 contributions).

Concerning the transport properties, the very high conductivity of the NT-polyacetylene was confirmed and anisotropic conduction in many oriented polymers was reported. The understanding and the application of vibrational spectroscopy have been improved by introducing an effective coordinate model and by using oriented polymer samples. Many more details of the polyaniline system have been revealed, particularly concerning the crystal structure and the nature of the two oxidation processes at 300 mV and 600 mV vs SCE. Magnetic resonance and infrared absorption experiments indicate cluster rather than random doping, but this question is not yet settled. Nonlinear optics has become a most important subject of study and second-order nonlinearity seems to be close to finding applications, particularly in cases which do not rely on extremely low optical damping. Several contributions dealt with molecular electronics, which definitely helped to make the nature of the problems more precise at this stage. The subject was also extended by including biochips and biocomputers. With respect to the family of thiophenes, a detailed study of the thermochromic effect was presented and the ground state for long chains of anellated polythiophenes was determined from experiment and from quantum chemical calculation to be quinoid.

The commercialization of polyaniline as electrodes in secondary electrochemical cells is now well established. Polymers as electrodes for detection of biomolecules and as nonlinear optical elements seem to be close to exploitation. In general, great progress has been made in the development of soluble and processable conducting polymers so that large-scale applications now look much more promising than they did two or four years ago.

This book summarizes tutorial and research lectures presented at the winter school in Kirchberg. We acknowledge all the authors for their contributions and all the discussion speakers at the winter school for their stimulating remarks, which were essential to make the winter school an exciting and informative event.

We are in particular grateful to the "Bundesministerium für Wissenschaft und Forschung" in Austria and to the "Bundesministerium für Forschung und Technologie" in the Federal Republic of Germany, as well as to the sponsors from industry, for their financial support. This support was more than just a great help; it was, in fact, indispensable for the goal of the meeting to be attained.

Finally we thank the manager of the Hotel Sonnalp, Herr Jurgeit, and his staff for their continuous support and for their patience with the many special arrangements required during the meeting.

Vienna, Stuttgart
May 1989

H. Kuzmany
M. Mehring
S. Roth

Contents

Part II Theory

Part III Optical Excitation

Part VI Magnetic Resonance

Part VII Nonlinear Optics

Part VIII Molecular Electronics

Part IX Polyanilines II

Part X Polythiophenes

Part XIII Applications

Part I

Transport, Structure

Electronic Transport in Conducting Polymers

A.B. Kaiser

Physics Dept., Victoria University of Wellington,
P.O. Box 600, Wellington, New Zealand

Abstract. We compile and interpret thermopower and conductivity data for conducting polymers. A striking correlation is seen between the magnitude of thermopower and conductivity. This correlation, together with the thermopower temperature dependence, is consistent with a substantial contribution from variable-range hopping (or similar mechanism) in moderately-doped polymers. In highly-doped samples, the apparent contradiction between the *nonmetallic* temperature dependence of conductivity and the *metallic* behaviour of thermopower can be resolved by a heterogeneous polymer model. The observation of linear thermopower in some 'metallic' samples (without the usual nonlinearity due to the electron-phonon interaction) is an indication of extremely high intrinsic conductivities.

1. Conductivity

The conductivity of moderately-doped conducting polymers typically follows the behaviour associated with variable-range hopping or similar models [e.g. 1]:

$$\sigma = \sigma_0 \exp[-(T_0/T)^\gamma] \tag{1}$$

where σ_0 is weakly temperature dependent, T_0 is related to the localization length, and $\gamma \sim 0.25 - 0.5$. An example is shown by the lower crosses in Fig. 1. This figure also shows the dramatic change that can develop as the doping level increases above about 5 mol. %: the conductivity remains finite in the zero-temperature limit. We use this as the criterion for 'metallic' conduction.

Although conducting polymers have now been made with values of conductivity similar to those of metals [2], the temperature derivative of the conductivity remains of opposite sign to that characteristic of metals (at least at lower temperatures), as shown in Fig. 1. One possibility is that this positive sign has the same origin as that already well-known in amorphous metals, and ascribed to 'weak localization' and electron-electron interaction effects [3]. The latter effect gives a characteristic $T^{1/2}$ term in the conductivity at helium temperatures. There is some indication of a $T^{1/2}$-like term at lower temperatures in some polymer samples near the metal-semiconductor transition (see Fig. 1), but in those samples with less temperature variation the conductivity appears to show the opposite curvature at low temperatures. In these highly-conducting polymers, therefore, in contrast to amorphous

Springer Series in Solid-State Sciences, Vol. 91
Electronic Properties of Conjugated Polymers III
Editors: H. Kuzmany · M. Mehring · S. Roth

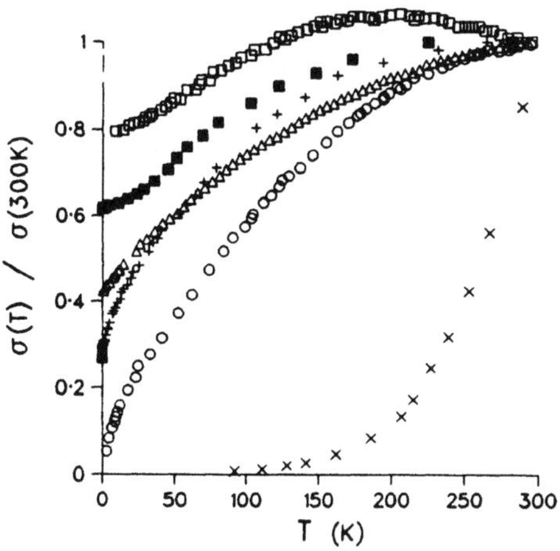

Fig. 1: Conductivity $\sigma(T)$ normalized by its value $\sigma(300K)$ at 300K for several conducting polymers as a function of temperature T. See [3,4] for sources of the data

metals, we are led to consider a *heterogeneous* model to account for the conductivity. The above behaviour is qualitatively what would be expected from fluctuation-induced tunnelling between metallic islands [e.g. 5], but for the high conductivities now obtained the barriers would have to be remarkably thin and uniform.

2. Heterogeneous Polymer Model

Several authors [e.g. 5] have suggested that conduction in highly-conducting polymers occurs in 'metallic' islands separated by low-conductivity 'barriers'. If we model this situation by two different types of material of conductivities σ_1 and σ_2 (with $\sigma_1 \gg \sigma_2$) in series (Fig. 2), we find that the total conductivity is [6]

$$\sigma \approx f\,(L / L_2)\,\sigma_2 \qquad (2)$$

where f is a factor of order unity and L_2/L is the fraction of the path consisting of the 'barrier' material. Hence if the barriers are thin, the conductivity can approach metallic magnitude but *still* show the temperature dependence of the barrier region (or tunnelling) conductivity σ_2.

If the thermopowers of the two materials are S_1 and S_2, the total thermopower is

$$S \approx (W_1 / W)\,S_1 + (W_2 / W)\,S_2 \qquad (3)$$

where W_1 and W_2 are the thermal resistances for the path lengths in each type of material and W is the total thermal resistance [6]. If the barriers are thin and conduct

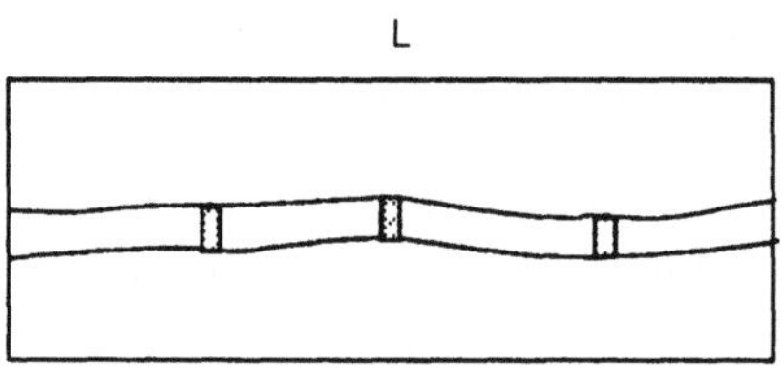

Fig. 2: Fibrillar conduction path in a highly-doped polymer, modelled as metallic regions in series with thin low-conductivity regions

heat better than electricity, we have $W_1 \gg W_2$ and $S \approx S_1$. In this case, thermopower would show metallic behaviour in both size and temperature dependence, in contrast to conductivity. A heterogeneous model can therefore account for the different character of the temperature dependence of conductivity and thermopower seen in highly-doped polymers [6,5].

3. Thermopower

Measurements of thermopower give an indication of the thermal energy of the charge carriers, i.e. the energy at which conduction occurs. In homogeneous material

$$S \sim (k/e)(E_{av}/kT) \tag{4}$$

where e is the charge of the carriers, k is Boltzmann's constant, and E_{av} is the average thermal energy of the charge carriers weighted by their contribution to the conductivity; (k/e) has the value $\pm$ 87 μV/K for holes (positive value) or electrons.

In a metal, conduction occurs mostly within the 'Fermi window' of width ~kT about the Fermi level E_F. The thermopower results from the slight imbalance between the heat carried by electrons and holes above and below the Fermi level. This imbalance increases as kT increases, giving a small approximately linear thermopower XT, but the effective electron-phonon enhancement $\lambda_e(T)$ gives a characteristic nonlinearity [3] at low temperatures (Fig. 3a):

$$S_m = [1 + \lambda_e(T)] X T. \tag{5}$$

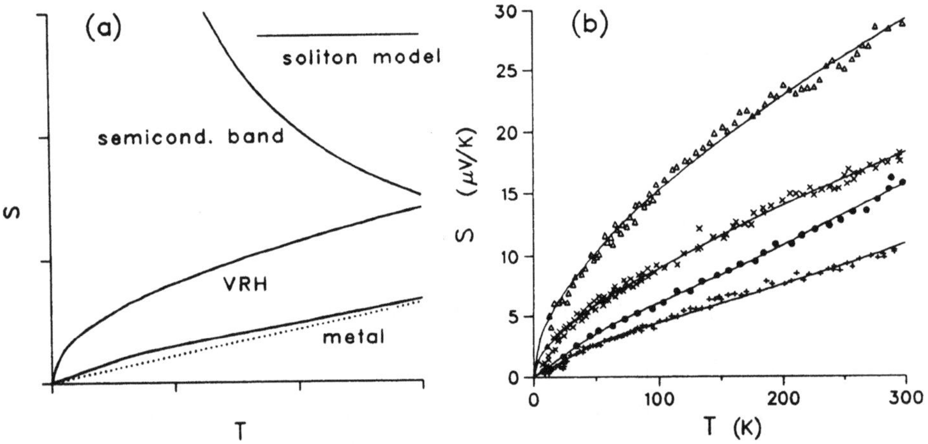

Fig. 3: (a) Schematic picture of thermopower for different conduction mechanisms: metallic conduction (including the electron-phonon nonlinearity), semiconductor band conduction, variable-range hopping (VRH), and inter-soliton hopping [7]; (b) Observed temperature dependence of thermopower for $FeCl_4$-doped polyacetylene, fitted to the variable-range hopping $T^{1/2}$ law with additional linear term for moderately-doped samples (upper two curves), and to the metallic expression (5) for samples with finite zero-temperature conductivity (lower two curves) [6,8]

In a crystalline semiconductor in which conduction occurs at the bottom of the conduction band (energy E_c), the thermal energy per carrier is approximately $(E_c - E_F)$. Hence from (4) the thermopower is large and varies as 1/T. For disordered semiconductors in which variable-range hopping dominates, the thermopower is expected to show behaviour generally intermediate between that for crystalline semiconductors and metals [9]:

$$S_h = (k^2 / 2e) (d \ln N / dE) (T_0 T)^{1/2} \qquad (6)$$

where the energy derivative of the density of states N is taken at the Fermi level.

For moderately-doped polymers (1 to 5 mol. %), the fits to the measured thermopower in Fig. 3b are consistent with a substantial $T^{1/2}$ term, as expected from variable-range hopping, although usually the fit indicates an additional linear term (possibly due to heterogeneity).

Compiling thermopower magnitudes in doped polyacetylene, we find a striking inverse correlation between the sizes of thermopower and conductivity (Fig. 4). Since, according to (6), variable-range hopping thermopower is expected to vary as $T_0^{1/2}$, we also show the variation of this parameter as a function of conductivity, as determined [1] from conductivity temperature dependence. The similarity of the trends (except perhaps for the most highly-conducting polymers) supports the view that variable-range hopping (or a similar mechanism) makes a substantial contribution to conduction in moderately-doped polymers. The value of (dlnN/dE) deduced from the trends in Fig. 4 is about 2 eV^{-1}, which seems reasonable - the values of T_0 are of course sensitive to the value used for γ in (1).

For highly-doped samples in which the conductivity remains finite in the zero-temperature limit, there is a subtle change of thermopower shape [8], a 'knee'

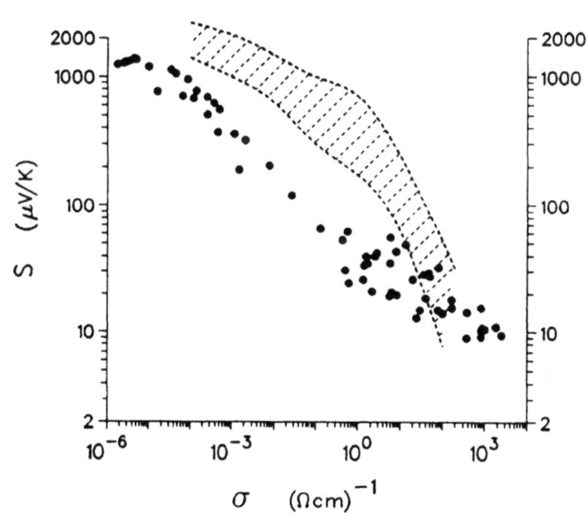

Fig. 4: Compilation of thermopower versus conductivity magnitudes at room temperature (dots) for doped polyacetylene samples, compared to values of $T_0^{1/2}$ (shaded region) obtained [1] by fitting conductivity to equation (1) with $\gamma = 0.33$

just like that expected from the metallic electron-phonon effect fitting the data better than a hopping term in the $FeCl_4$-doped samples (Fig. 3b).

In other metallic samples, the thermopower is essentially proportional to temperature [e.g. 4]. Such behaviour has almost never been seen in metals because of the electron-phonon nonlinearity, even when phonon drag is suppressed. It indicates a very small value of the electron-phonon coupling λ_e for conduction electrons, i.e. little scattering of conduction electrons by phonons and so very high intrinsic conductivities for the metallic fibrils.

4. Conclusion

For further progress in understanding the transition to metallic behaviour in highly-doped polymers, measurements of thermopower and thermal conductivity as well as electrical conductivity are desirable, in particular accurate data at low temperatures. Using a simple heterogeneous polymer model, we have shown how different properties such as conductivity and thermopower can reflect characteristics of *different* regions of the material. For moderately-doped polymers, any detailed theory must account for the correlation we have found between thermopower and conductivity, which appears consistent with dominance by hopping conduction.

References

[1] D. Schäfer-Siebert, C. Budrowski, H. Kuzmany and S. Roth, in *Electronic properties of conjugated polymers*, edited by H. Kuzmany, M. Mehring and S. Roth (Springer, Berlin, 1987), p. 38.

[2] H. Naarmann and N. Theophilou, Synth. Met. **22**,1 (1987).

[3] A.B. Kaiser, in Ref. 1, p. 2.

[4] Y.W. Park, A.J. Heeger, M.A. Druy and A.G. MacDiarmid, J. Chem. Phys. **73**, 946 (1980).

[5] K. Ehinger and S. Roth, Phil. Mag. B **53**, 301 (1986).

[6] A.B. Kaiser, to be published.

[7] S. Kivelson, Phys. Rev. Lett. **46**, 1344 (1981).

[8] Y.W. Park, W.K. Han, C.H. Choi and H. Shirakawa, Phys. Rev. B **30**, 5847 (1984).

[9] N.F. Mott and E.A. Davis, Electronic processes in non-crystalline materials, 2nd ed. (Clarendon Press, Oxford, 1979), p.53.

The Effects of Intrachain Defects on the Electrical Anisotropy of Conducting Polymers

R.H. Baughman and L.W. Shacklette

Allied-Signal Inc., Morristown, NJ 07960, USA

Abstract. Theoretical predictions are presented for the dependence of electrical anisotropy upon conjugation length for conducting polymers. These results, which are consistent with experimental data, provide a method for deriving the average conjugation length and the conductivity of an infinite-conjugation-length polymer from conductivity measurements on polymers having conjugation lengths which limit transport.

I. Introduction

Available experimental and theoretical results suggest that interchain hopping dominates electronic conductivity in all directions for many conducting polymers. Such results include (1) the typically observed temperature independence of electrical anisotropy for chain-oriented polymers [1-10] and (2) the agreement between calculated and observed conductivities as a function of the combined effects of defect concentration and temperature [11-14].

When interchain hopping dominates conductivities in all directions, fundamental relationships exist between electrical anisotropy and chain defect concentration. We herein summarize theoretical predictions of these relationships for chain-oriented polymers in which intercrystallite resistances are negligible. All results presented are for heavily doped polymers for which electronic conductivity is dominated by transport within or between crystallographically equivalent chains. Chain defects are assumed to have dimensions which are short compared with uninterrupted conjugation lengths and to be randomly located along the polymer chain. Also, the key chain imperfections are assumed to provide effectively infinite barriers for chain-direction transport. This assumption will be valid as long as the volumetric frequency of interchain hopping is much larger than the volumetric frequency of hopping across these defects, which is always the case unless the concentration of such defects is very high.

Springer Series in Solid-State Sciences, Vol. 91
Electronic Properties of Conjugated Polymers III
Editors: H. Kuzmany · M. Mehring · S. Roth

II. Predicted Electrical Anisotropy

When both the chain-direction conductivity (σ_1) and the conductivity in an orthogonal direction (σ_2) are entirely limited by interchain hopping then

$$\sigma_1/\sigma_2 = L^2/(6Fd^2), \qquad (1)$$

where L is the average localization length, d is the separation between hopped-from and hopped-to chains, and F is a geometrical factor which is equal to unity for the important case where hopping is only between chains which are in the same plane as the applied electric field [11-14]. More generally, F is the average value of $\cos^2\theta$, where θ is the angle between the electric field normal to the chain-axis direction and the interchain vector from a hopped-from chain to a hopped-to chain. Hence, F is 0.5 for tetragonal phases, such as polyacetylene fully doped with heavy alkali metals [15], and hexagonal phases. An L^2 dependence of electrical anisotropy has been derived by Baughman and Shacklette [11-14], Kahlert [16], Ovchinnikov et al. [17], and Balagurov et al. [18].

Since conducting polymers can contain a variety of defects which interrupt both conjugation and charge transport to different extents, the relationship between the average localization length L, the average separation of defects, and the average conjugation length is easiest to understand when only weak defects and strong defects are present in the polymer. Strong defects are those which terminate conjugation length about as effectively as a chain end, as measured by the effect of these defects on optical bandgap. Weak defects are those which provide much less severe effects on optical properties. The strong defects are assumed to be sufficiently well separated that the volumetric frequency of interchain hopping across these defects is much less than the volumetric frequency of interchain hopping. The weak defects will not effect L as long as the volumetric frequency of interchain hopping across these weak defects is large compared with the volumetric frequency of interchain hopping. Hence, the average localization length, L, corresponds to the average separation of strong defects, which is also approximated throughout this discussion as the average conjugation length. It is convenient to define the conjugation length of a polymer chain segment bounded by strong defects as the length of an oligomer containing the same concentration of weak defects which has the same optical bandgap. If the strong defects have the same effect on optical bandgap as chain ends, then this conjugation length exactly corresponds to the separation between strong defects.

When conjugation lengths are long enough that the interchain hopping frequency is independent of conjugation length, then the average conjugation length appropriate for use in Eqn. 1 is $(L_w L_z)^{1/2}$ where L_w and L_z are the weight-average and z-average conjugation lengths, respectively [14]. Correspondingly, if M is the length of a monomer unit in the chain-axis direction, then $L_w = Mu_w$ and $L_z = Mu_z$, where

$$u_w = \sum_{u=1}^{\infty} uW(u) \text{ and} \tag{2}$$

$$u_z = \sum_{u=1}^{\infty} u^2W(u) / \sum_{u=1}^{\infty} uW(u), \tag{3}$$

and where W(u) is the weight fraction of conjugation lengths containing u monomer units. For narrow conjugation length distributions, such as the Poisson distribution, number-average , weight-average, and z-average conjugation lengths are nearly equal unless conjugation lengths are very short, so the average conjugation length in Eqn. 1 can be replaced by any of these quantities. $(L_w L_z)^{1/2}$ increases relative to L_w with increasing polydispersity of the molecular weight distribution, providing corresponding increased electrical anisotropies for distributions having the same L_w. Nevertheless, the error in replacing $L_w L_z$ for L^2 in Equation 1 by L_w^2 does not exceed about 33% for conjugation length distributions no more polydispersed than the random distribution.

When the conjugation lengths are sufficiently short that interchain hopping frequency depends upon conjugation length, then the appropriate averaging of conjugation lengths results in a value for L which weakly depends upon temperature [14]. The physical origin of this effect is that with decreasing temperatures the fraction of total hops which occur between long conjugation lengths increases (since the barrier for such hops is lowest). Since such hops provide the largest chain-axis displacement per hop, L correspondingly increases with decreasing temperature. Accordingly, using Eqn. 1, the electrical anisotropy increases with decreasing temperature. However, the magnitude of this temperature dependence for electrical anisotropy (σ_1/σ_2) is typically small unless $u_w T$ is very small. Furthermore, this small negative temperature dependence of electrical anisotropy decreases in magnitude with decreasing polydispersity in conjugation length distribution.

Upon increasing conjugation length, electrical resistivity will eventually become dominated by the effective resistivity of uninterrupted conjugated segments. This transition occurs when, on average, the time interval between carrier hops to a conjugated segment becomes comparable with the time for charge equilibration on a conjugated segment after a hop. We can use our previously derived equations [13] to obtain useful predictions for the behavior of electrical anisotropy in the transition region to infinite-conjugation-length properties. These predictions assume that when conjugation lengths are sufficiently long that interchain hopping does not completely dominate chain-direction conductivity, the effective resistance per length of conjugated segment is a fixed quantity - determined, for example, by phonon scattering or scattering due to structural perturbations in a conjugated segment. Also, it is assumed for this transition region that conjugation lengths are long enough that interchain transport across interruptions in conjugation are unimportant. In such case, we have obtained from resistor network theory [13] that

$$\sigma_1/\sigma_1(\infty) = \{1-4b^{-1/2}k^{-1}\{\tanh^{-1}\{b^{1/2}a^{-1}\tanh(k/2)\}\}\} \quad (4)$$

where $a = k\tanh(k/2) + 2$

$b = k^2 + 4k\,\mathrm{ctnh}(k) + 4$, and

$k = (\sigma_2/\sigma_1(\infty))^{1/2}(L/F^{1/2}d)$.

Eqn. 4 provides the dependence of $\sigma_1/\sigma_1(\infty)$ upon $(\sigma_2/\sigma_1(\infty))^{1/2}(L/F^{1/2}d)$, where L is the average conjugation length and $\sigma_1(\infty)$ is the chain-direction electrical conductivity when conjugation length is infinite. Hence, taking the ratio of $\sigma_1/\sigma_1(\infty)$ to k^2, Eqn. 4 provides $(\sigma_1/\sigma_2)(Fd^2/L^2)$ as a function of $\sigma_1/\sigma_1(\infty)$.

Eqn. 4 is important for obtaining a key quantity from experimental data, the electrical conductivity parallel to the chain-axis direction for a polymer having effectively infinite conjugation length ($\sigma_1(\infty)$). Determination of $\sigma_1(\infty)$ is possible using these results if σ_1 and σ_2 have been measured for a polymer which has sufficiently long conjugation lengths that σ_1 is effected by the carrier transport within a conjugated segment. If the polymer is either three-dimensionally oriented or has uniaxial orientation and isotropic conductivity normal to the chain-axis direction, the values of F and d in Eqn. 4 are readily determined from the structure of the polymer. Measurement of average conjugation length (L) then permits calculation of $\sigma_1/\sigma_1(\infty)$ from $(\sigma_1/\sigma_2)(Fd^2/L^2)$, using Eqn. 4.

III. Discussion

Comparison of the present predictions with experimental results is hampered by (1) the general absence of anisotropy measurements on conducting polymers of known conjugation lengths (2) typically unknown deviations from chain alignment, and (3) sample polycrystallinity. Strong anisotropy of conduction orthogonal to the chain-axis direction can result in major deviations from Eqn. 1 for chain-oriented, polycrystalline polymers [13]. However, even in this case, electrical anisotropy remains proportional to L^2.

Electrical anisotropies of about 100 have been reported for iodine-doped polyacetylene obtained by the Durham route [10]. If we neglect crystallite anisotropy normal to the chain-axis direction, as well as deviations from complete chain orientation, and approximate d by the interchain distance in the undoped polymer (~4.1 Å) [12], then this value of σ_1/σ_2 provides, via Eqn. 1, that L_wL_z is 1.0×10^4 Å^2. Making the opposite assumption that interchain hopping is entirely within a plane of polymer chains and that the effective value of conductivity normal to the chain-axis direction is the average value of conductivity components orthogonal to the chain, the calculated L_wL_z is 5.0×10^3 Å^2. Making the reasonable approximation that there is a random distribution of breaks in conjugation length, these estimates correspond to a weight-average conjugation length of 23-33 carbon double bonds. Within the limitations of the above approximations regarding anisotropy orthogonal to the chain-axis direction, this should be viewed as a lower limit estimate - since chain orientation is incomplete in the investigated samples. This predicted average conjugation length is consistent with the number-average and weight-average conjugation lengths (32 and 47 carbon double bonds, respectively) derived from the bimodal conjugation length distribution obtained by Raman spectroscopy [19]. Observations on the weak temperature dependence of electrical anisotropy vary from sample to sample, perhaps due to effects of intercrystallite resistances. Consistent with the predictions of the present theory, a small negative temperature dependence of σ_1/σ_2 is reported by Schimmel et al. [6] and Theophilou et al. [7] for iodine-doped polyacetylene and by Satoh et al. [8] for iodine-doped polythiophene. Negative temperature dependencies were observed only at high pressures by Basescu et al. [9]

and only at low temperatures by Leising et al. [10] for iodine-doped $(CH)_x$.

Eqn. 1 indicates that electrical anisotropy can provide a highly sensitive measure of conjugation length for conducting polymers in which conductivity is determined by interchain hopping. Since interchain hopping can dominate anisotropy until conjugation lengths are quite long, electrical anisotropy provides such information over a range from very short to very long conjugation lengths. In contrast, alternate methods for determining conjugation lengths on solid polymers (such as infrared, Raman, or optical absorption spectroscopies) become insensitive to conjugation length at relatively short conjugation lengths. Furthermore, solution methods for determining conjugation length from properties which depend upon molecular weight are unreliable unless it is known that chain length corresponds to conjugation length. When there is no significant polydispersity in conjugation length or when conjugation lengths are sufficiently long that the interchain hopping frequency is independent of conjugation length, $L_w L_z$ can be directly obtained from the measurement of anisotropy [14]. Even if these conditions are not satisfied, corrections for the effects of the conjugation-length dependence of hopping frequency for polydispersed systems are typically small. If the defects which terminate conjugation length are randomly located, or the functional form of the conjugation length distribution is otherwise obtainable, this $L_w L_z$ product can be used to obtain any desired moment of the conjugation length distribution. This assumption of a particular form for the conjugation length distribution can be checked by comparing theoretical and experimental determinations for the weak temperature dependence of anisotropy, when conjugation lengths are sufficiently short that interchain hopping frequency is a function of conjugation length [14].

This analysis is directly applicable to anisotropy measurements on single crystals of conducting polymer complexes. It is also directly applicable to anisotropy measurements on fully chain-oriented conducting polymer complexes in which interfibril resistances and chain misorientation are negligible and conductivity normal to the chain-axis direction is isotropic. Furthermore, the present single crystal contributions must be included in the interpretation of anisotropy of any conducting polymer complex in which interchain hopping dominates transport in all directions and both σ_1 and σ_2 are not dominated by chain misorientation.

References

1. Y.W. Park, C. Park, Y.S. Lee, C.O. Yoon, H. Shirakawa, and K. Akagi, Solid State Commun. **65**, 147 (1988).
2. P.D. Townsend, C.M. Pereira, D.D.C. Bradley, M.E. Horton, and R.H. Friend, J. Phys. C **18**, L283 (1985).
3. C.R. Fincher, D. Moses, A.J. Heeger, and A.G. MacDiarmid, Synth. Metals **6**, 243 (1983).
4. Y.-C. Chen, K. Akagi, and H. Shirakawa, Synth. Metals **14**, 173 (1986).
5. M. Ogasawara, K. Funahashi, T. Demura, T. Hagiwara, and K.Iwata, Synth. Metals **14**, 61 (1986).
6. Th. Schimmel, G. Denniger, W. Riess, J. Voit, M. Schwoerer, W. Schoepe, and H. Naarmann, Synth. Metals **28**, D11 (1989).
7. N. Theophilou, D.B. Swanson, A.G. MacDiarmid, A. Chakraborty, H.H.S. Javadi, R.P. McCall, S.P. Treat, F. Zuo, and A.J. Epstein, Synth. Metals **28**, D35 (1989).
8. M. Satoh, K. Imanishi, Y. Yasuda, R. Tsushima, H. Yamasaki, and S. Aoki, Synth. Metals **30**, 33 (1989).
9. N. Basescu, Z.-X. Liu, D. Moses, A.J. Heeger, H. Naarman, and N. Theophilou, Nature **327**, 403 (1987).
10. G. Leising, R. Uitz, B. Ankele, W. Ottinger, and F. Stelzer, Mol. Cryst. Liq. Cryst. **117**, 327 (1985).
11. R.H. Baughman and L.W. Shacklette, Syn. Metals **17**, 173 (1987).
12. R.H. Baughman and L.W. Shacklette, Phys. Rev., **39**, 5872 (1989).
13. R.H. Baughman and L.W. Shacklette, J. Chem. Phys. (in press).
14. R.H. Baughman and L.W. Shacklette, preprint.
15. R.H. Baughman, N.S. Murthy, and G.G. Miller, J. Chem. Phys. **79**, 515 (1983).
16. H. Kahlert, unpublished.
17. A.A. Ovchinnikov and K.A. Pronin, Sov. Phys. Solid State **28**, 1666 (1986).
18. B. Ya. Balagurov, G.A. Vinogradov, M.M. El'yashevitch, unpublished.
19. E. Perrin, E. Faulques, S. Lefrant, E. Mulazzi, G. Leising, Synth. Metals **28**, D317 (1989).

Influence of the Catalyst on the Structure and Conductivity of Highly Conducting Polyacetylene

N. Theophilou [1], D.B. Swanson [1], A.G. MacDiarmid [1], J.G. Mantovani [2], B.K. Annis [3], and A.J. Epstein [4]

[1] Department of Chemistry, University of Pennsylvania, Philadelphia, PA 19104-6323, USA
[2] Department of Physics, University of Tennessee, Knoxville, TN 37996, USA and Health and Safety Research Division, Oak Ridge National Laboratory, Oak Ridge, TN 37831, USA
[3] Chemistry Division, Oak Ridge National Laboratory, Oak Ridge, TN 37831-6197, USA
[4] Department of Physics and Department of Chemistry, The Ohio State University Columbus, OH 43210-1106, USA

Factors are discussed which may contribute to the high conductivity of BASF $(CH)_x$ [1,8]. They include: (i) chemical factors such as type of catalyst employed, method of washing, etc. which might influence the amount of catalyst residue remaining in the $(CH)_x$ and hence the conductivity of the resulting I_2-doped polymer and (ii) mechanical factors such as extent of stretch alignment which is also expected to influence the conductivity of the doped polymer.

I. Introduction

Several techniques have been developed during the last decade to polymerize acetylene using various Ziegler-Natta type catalysts [1-4,7-11] or by a precursor route [5-6]. The stretch-alignment of polyacetylene is of considerable interest, due to the appearance of anisotropic optical, magnetic and electrical properties.

Room temperature [1] or thermal [2] mechanical alignment of polyacetylene can be performed. Polyacetylene can also be aligned during synthesis, using an oriented liquid crystal matrix [3,4]. A precursor polymer can also be stretched while being converted by heat treatment to polyacetylene [5,6]. It has been found recently that a thermal treatment (>100°C) of the catalyst system can significantly affect the properties of the catalyst and of the resulting polymer [23].

II. Experimental and Results

The conditions for the synthesis of catalyst, polymerization and doping have been described previously [1,8]. Argon (2 ppm O_2, 8ppm H_2O) was additionally purified by passing first through $CaCl_2$, then through P_2O_5 and then through a molecular sieve (4Å, Aldrich). Silicone oil (linear polydimethylsiloxane SF96/50, Thomas Scientific) having a viscosity of 50 centistokes (25°C) was used. $AlEt_3$ (93%, Aldrich) was used as received. $Ti(OBu)_4$ (Morton Thiokol) was freshly distilled before use.

Springer Series in Solid-State Sciences, Vol. 91
Electronic Properties of Conjugated Polymers III
Editors: H. Kuzmany · M. Mehring · S. Roth

Table 1: Properties of various types of polyacetylene

Properties \ Type	Shirakawa[2]	Liquid [3,4] Crystal	Durham-[5,6] Graz[a]	Non Solvent[7]	New [1,8,11]	ARA [1,8,11]
I. TECHNIQUE						
(i) Catalyst	$Ti(OBu)_4-AlEt_3$			$Ti(OBu)_4-AlEt_3$		
(ii) Solvent	Toluene	Liq. Cryst.		Cumene (150°C)	Silicone oil (120°C)	
(iii) Add. Red. Agents	-----	-----		-----	-----	n-BuLi; $n-Bu_2Mg$
(iv) Polym. T	-78°C	-40-15°C (H=10KG)		-78°C	22°C	22°C
II. MORPHOLOGY						
(i) Fibrils (Å)	100-500 [17]	800-1,000	Non fibrillar	300-500	200	200
(ii) Density (g/cm^3)	0.4	-----	1.05	1.00-1.15	0.85[1]	0.9[1]
III. C, sp^3 Defects						
(i) ^{13}C NMR (%)	1-4[9]	-----	-----	-----	Not det.[1]	Not det.[1]
(ii) Raman	60 = bonds[12] (Conj. length)				60 = bonds[12] (conj. length)	-----
IV. ANISOTROPY						
(i) Optical	-----	4-4.5 [25]	26 [6]	-----	28 [13]	-----
(ii) Mosaic	35° [18]	20° [19]	5° [20]	-----	15° [15]	-----
(iii) Coherence Length (Å)	$L_{\parallel}$=100-130 [21]	-----	80 [22]	-----	160-180 [15]	-----
	$L_{\perp}$= 60-90 [18]	-----	55 [22]	-----	47-53 [15]	-----
(iv) $\sigma_{\parallel} / \sigma_{\perp}$	5-8	2-7 [24,25]	-----	70 [7]	30-100 [8]	30-100 [8]
V. CONDUCTIVITY (I_2 doping)						
(i) Non aligned	200			2,000 [7]	2,000-5,000	
(ii) Aligned	1,600	5,000-16,000 [24,25]	1,200-4,000[20] (AsF_5 doping)	10,000-22,000[7]	20,000-50,000 [1,10]	75,000-147,000 [1,10]

[a] The monomer, 7,8-bis-(trifluoromethyl)-tricyclo-[4.2.2.0]-deca-3,7,9-triene (BTFM-TCDT) is transformed with $WCl_6/(CH_3)_4Sn$ catalyst to poly-(BTFM-TCDT) (precursor polymer) which is converted to polyacetylene by heating at 80°C [20].

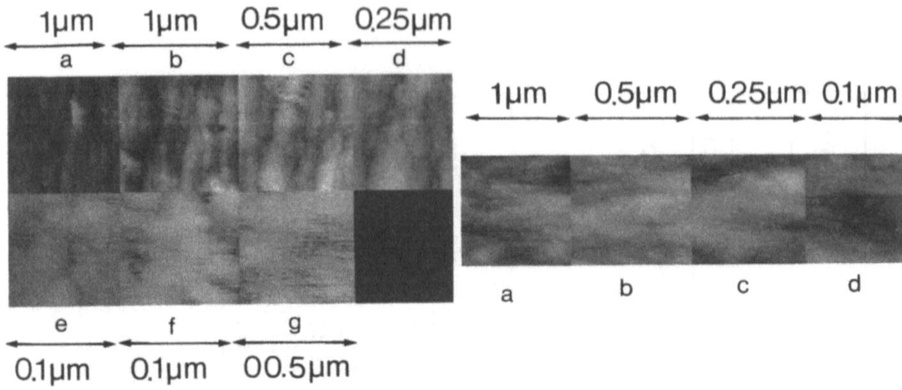

Fig. 1. STM of stretch-aligned (l/l_0=6) New-$(CH)_x$. The stretching direction for Fig. 1(A) is vertical, for Fig. 1(B) horizontal. STM scans are horizontal

Table 2: Relationship between conductivity of unstretched $(CH)_x$ and chemical composition

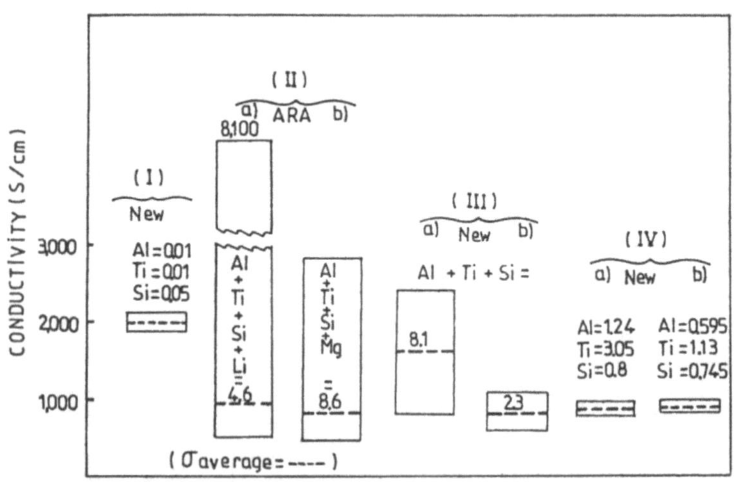

(I) Free-standing New-$(CH)_x$ [1,9], synthesized and washed with toluene followed by HCl-CH_3OH (6%), in a dry box.

(II) Free-standing ARA-$(CH)_x$ with: (a) n-BuLi (b) n-Bu_2Mg, synthesized in a glass reactor and washed with toluene in a dry box [16].

(III) Free-standing New-$(CH)_x$, synthesized in a glass reactor and washed in a dry box, with: (a) toluene and (b) toluene, followed by HCl/CH3OH (6%) [16].

(IV) Free-standing New-$(CH)_x$ synthesized and washed in a dry box with: (a) toluene and (b) toluene/followed by CH_3OH [16].

Selected properties of various types of $(CH)_x$ are given in Table 1. "New"- and "ARA"-$(CH)_x$ have fibril diameters of (~200Å) and high density (0.85-0.9 g/cm^3) [1]. The number of defects by ^{13}C NMR is not detectable for New- and ARA-$(CH)_x$ [8], whereas S-$(CH)_x$ has been reported to have a broad ^{13}C (30-60 ppm) peak corresponding to 1-4% of sp^3 carbon defects [9]. However, other investigators have [14] reported the absence of C sp^3 defects by ^{13}C NMR in S-$(CH)_x$. Raman experiments [12] give a conjugation length of ~60 double bonds for Shirakawa $(CH)_x$ as well as for New-$(CH)_x$. Hence both types of $(CH)_x$ show the presence of some type of defect.

New- and ARA-$(CH)_x$ can be produced with a high degree of alignment (l/l_0=6.5); for example a 5% I_3^--doped New-$(CH)_x$, showed an intensity ratio of oscillator strengths of light polarized parallel and perpendicular to the chain direction of ~28 at 1380 cm^{-1} [13]. These polymers also show a low mosaic distribution (disorientation angle of chain axis of 15°) [15], a value intermediate between that of Shirakawa and Durham-Graz polyacetylene [18,20]. One of the important properties of New-$(CH)_x$ is its higher coherence length (160-180Å), parallel to the stretching direction. This is twice as high [15] as that observed for Durham-Graz-$(CH)_x$ and S-$(CH)_x$. This is consistent with STM measurements on New-$(CH)_x$ aligned samples [Fig. 1(A), (e-g) and Fig. 1(B)(d)] where shiny elliptical areas of ~>200Å are aligned in the direction of stretching on the surface of samples. The anisotropy of conductivity, $\sigma_\parallel/\sigma_\perp$, for the stretch-aligned iodine-doped New- and ARA-$(CH)_x$ was ~30 to 100 corresponding to values of 20,000 to 50,000 S/cm for the New-$(CH)_x$ and 75,000 to 147,000 S/cm for ARA-$(CH)_x$ [1,10].

The chemical purity of the polymers has also been examined in order to understand the role played by the residual catalyst on conductivity. This study was made on non-aligned films in order to eliminate the variation in conductivity due to different extents of stretching.

Table 2 shows the conductivity and the catalytic content of various New- and ARA-$(CH)_x$ samples. From the values obtained, it seems that the average value of conductivity of I_2-doped $(CH)_x$ is insensitive to the amount (≤8.6%) of catalyst. It ranges from ~1,000 to 2,000 S/cm from one apparently identical synthesis to another. We also observe that for the ARA-$(CH)_x$ the conductivity varies greatly from one portion of the film to another. For example, maximum and minimum values of 8,100 and 500/cm were measured within the same sample.

Acknowledgement:

[1,4]Work suported in part by the Defense Advanced Research Projects Agency through a contract monitored by the Office of Naval Research and by NSF Grant No. DMR-86-15475.

[2]Funding was provided by the ONRL Laboratory Director's R&D Fund and by the Office of Health and Environmental Research, U.S. Department of Energy under contract DE-AC05-84OR21400 with the Martin Marietta Energy Systems, Inc.

[3]Research sponsored by the Division of Materials Sciences/Office of Basic Energy Sciences, U.S. Department of Energy, under contract DE-AC05-84OR21400 with the Martin Marietta Energy Systems,Inc.

References

1. H. Naarmann and N. Theophilou, Synth. Met., 22, 1 (1987), Germ. Pat. DE3,617,505 (11/26/87).
2. Y.W. Park, M.A. Druy, C.K. Chiang, A.G. MacDiarmid, A.J. Heeger, H. Shirakawa and S.Ikeda, J. Polym. Sci., Polym. Lett. Ed. 17, 195 (1979).
3. K. Akagi, S. Katayama, H. Shirakawa, K. Araya, A. Mukoh and T. Narahara, Synth. Met., 17, 241 (1987).
4. M. Aldissi, J. Polym. Sci., Polym. Lett. Ed., 23, 167 (1985).
5. J.H. Edwards, W.J. Feast, D.C. Bott, Polymer, 25, 395 (1984).
6. G. Leising, Poly. Bull., 11, 401 (1984); Phys. Rev. B, 38, 15, 10313 (1988).
7. K. Akagi, M. Suezaki, H. Shirakawa, H. Kyotani, M. Shimomura and Y. Tanabe, Synth. Met., 28, D1 (1989).
8. N. Theophilou and H. Naarmann, Makromol. Chem., Macromol. Symp., 24, 115 (1989); N. Basescu, Z.X. Liu, D. Moses, A.J. Heeger, H. Naarmann and N. Theophilou, Nature, 327, 403 (1987).
9. N. Theophilou and H. Naarmann, Conducting Polymers, L. Alcácer (ed.), 65, D. Reidel Publishing Co. (1987).
10. N. Theophilou, D.B. Swanson, A.G. MacDiarmid, A. Chakraborty, H.H.S. Javadi, R.R. McCall, S.P. Treat, F. Zuo and A.J. Epstein, Synth. Met., 28, D35 (1989).
11. T. Schimmel, W. Riess, J. Gmeiner, G. Denninger, M. Schwoerer, H. Naarmann and N. Theophilou, Solid State Commun., 65, 11, 1311 (1988).
12. H. Kuzmany, private communication, (1988).
13. H.W. Woo, D.B. Tanner, N. Theophilou and A.G. MacDiarmid, Bull, APS 34(3), 823 (1989); private communication (to be published).
14. H. Bleier, private communication, (1988).
15. D. Djurado, J. Ma, N. Theophilou and J. Fischer, Synth. Met. in press.
16. N. Theophilou, D.B. Swanson, A.G. MacDiarmid and A.J. Epstein, Bull. APS 34(3) 406 (1989); to be published.
17. A.J. Epstein, H. Rommelmann, R. Fernquist, H.W. Gibson, M.A. Druy, T. Woerner, Polymer 23, 1211 (1982).
18. R. Robin, J.P. Pouget, R. Comes, H.W. Gibson and A.J. Epstein, Phys. Rev. B 27, 6, 3938 (1983).
19. M. Rolland, J.L. Ribet, A. Montaner, M. Galtier, Z. Lakhliai, J.L. Sauvajol, M. Brunet, R. Almairac and P. Bernier, Synth. Met. 24, 1 (1988).
20. G. Leising, Synth. Met. 28, D215 (1989).
21. C.R. Fincher, Jr., C.E. Chen, A.J. Heeger, A.G. MacDiarmid and J.B. Hastings, Phys. Rev. Let. 48, 3, 100 (1982).
22. H. Kahlert, O. Leitner, G. Leising, Synth. Met. 17, 467 (1987).
23. N. Theophilou, R. Aznar, A. Munardi, J. Sledz, F. Schue and H. Naarmann, Synth. Met. 16, 337 (1986).
24. H. Shirakawa, K. Akagi, S. Katayama, K. Araya, A. Mukoh and T. Narahara, J. Macromol. Sci-Chem., A25 (5-7), 643 (1988).
25. H. Shirakawa, K. Akagi, S. Katayama, M. Ito and K. Araya, IUPAC, August 1-6, 1988, Kyoto (Japan).

Chemical n Doping of Polyacetylene with Alkaline and Tetraalkyl Ammonium Cations

B. François, C. Mathis, and F. Ignatious*

Institut Charles Sadron,
6 rue Boussingault, F-67083 Strasbourg Cedex, France

Abstract. Polyacetylene (PA) films, prepared using Shirakawa's procedure, are chemically doped with tetraalkyl ammonium (TAA) cations. The maximum doping level is limited both by the cation size and by a spontaneous deactivation reaction (Hoffmann's type). The d.c. conductivity decreases when the cation size increases. The EPR bands of TAA doped CD_x are specially narrow due to the absence of spin-orbit and hyperfine coupling.

1. Introduction

The electrical conductivity of doped conducting polymers depends on their conjugation length and on the interchain distance. In an attempt to control this interchain distance we describe here n-doping processes allowing the chemical insertion of cations, the size of which can be varied in a very large range.

We have previously reported the necessary conditions to prepare homogeneous n-doped PA by a chemical way with alkaline cations /1/. A set of organoalkaline dopants with suitable redox properties has been proposed. In the present paper we describe the chemical doping of CH_x and CD_x with large TAA cations and some electrical and magnetic properties of these n-doped PA. More detailed data can be found elsewhere /2/.

2. Doping process

The organo-TAA are generally unstable due to the Hoffmann-type elimination reaction /3/ so that, even if prepared in situ by ionic exchange, their lifetime is too short for doping use. Nevertheless we found that radical-ion or dianions of aromatic ketones, with barium or

Springer Series in Solid-State Sciences, Vol. 91
Electronic Properties of Conjugated Polymers III
Editors: H. Kuzmany · M. Mehring · S. Roth

lithium cations, can be mixed with TAA halides without complete ionic exchange. The, so-formed, more or less stable complexes absorb at a wavelength located between the absorption of the initial Ba or Li compounds and the absorption of the pure TAA compounds. Their stability seems to be related to the strongly associated state of Ba or Li aromatic ketone salts, the TAA halide probably being accommodated into aggregates which prevent Ba or Li halide precipitation.

Barium benzophenone or Li fluorenone mixed with TAA bromides were mainly used in this work. The weak solubility of tetramethyl,ethyl and propyl bromides in THF led to very slow doping rates so that the maximum doping levels could be achieved only with tetrabutylammonmium or larger cations.

3. Maximum doping levels

In order to determine unambiguously the doping levels and the nature of the inserted cations, three different determinations were carried out for each sample: the cation content was measured from the weight uptake and from nitrogen elemental analysis ; the number of charges tranferred to the polymer was deduced from the acidimetric titration of a methanolic solution reacted with the doped samples.

Table 1: Maximum doping levels of TAA-doped PA

Cation	Doping level %		
	gravimetry (weight uptake %)	N analysis	Acidimetry
Me_3Dodec N	(123) 7		
But_4 N	(119) 6.4	5.8	6.5
$Pent_4$ N	(174) 7.7	6.5	7
Hex_4 N	(198) 7.3	7.3	6.9
$Hept_4$ N	(243) 7.7	5.9	6.5
Oct_4 N	(247) 6.9	6.4	6.6
$Dodec_4$ N	(249) 4.7	4.4	
Oct dec_4N	(39) 0.5	0.5	

No barium or lithium was found in the polymer and the three analyses are in fairly good agreement as reported in Table 1 for the Ba benzophenone dianion/TAA bromide complex (which allows the highest doping levels).
These maximum doping levels are plotted as a function of cation radius in Fig. 1. The radii were calculated assuming a spherical shape and a density 0.8 g/cm^3.

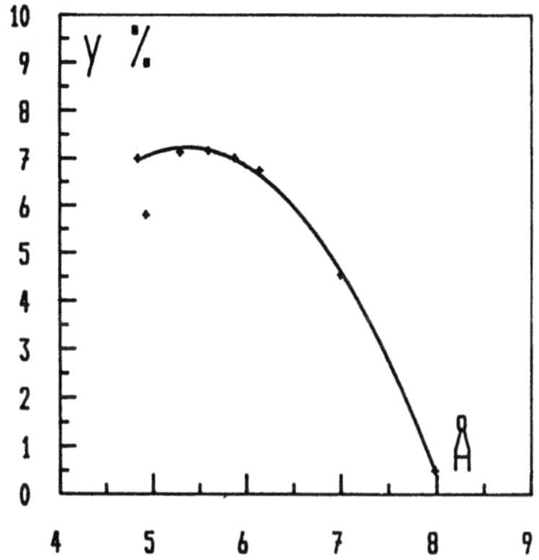

Fig.1:
Maximum doping level
vs
radius of TAA cations

These maximum doping levels have two limitations:

- The first one comes from the size of the cation. The doping level decreases when the radius is larger than 6 Angstroems. The maximum weight uptake corresponds to a cation content of about 71% in weight, or 78 % in volume.

-The second limitation, at 7%, is due to the enhanced chemical reactivity of doped sites above this value.
This last conclusion is based on the following experiment: A lithium-doped PA (y = 0.13) was reacted with Bu_4 N Br in THF. From the analysis of the solution and of the sample after the reaction it can be deduced that

- the doping level of the sample after reaction is 0.08
- the inserted cations are exclusively NBu_4^+
- about 5% of CH units were protonated during exchange
- tributylamine was found in the solution . Its amount corresponds to the decomposition of 0.05 doped sites per CH units.

These data are in perfect agreement with the assumption that above a 0.07-0.08 doping level the doped sites are rapidly decomposed in a Hoffmann-type reaction :

$$= CH - \underset{(N\,Bu_4)^+}{CH} - CH = \longrightarrow\; = CH - CH_2 - CH = \;+\; CH_3 - CH_2 - CH = CH_2 + \underline{N\,Bu_3}$$

4. d.c. electrical conductivities of TAA-doped PA

The d.c. electrical conductivities are systematically lower than those determined with alkaline cations . For a given doping condition theiy vary from sample to sample, depending on the synthesis of the film ,its density ,etc., by a factor of 2 or 3. Therefore the influence of the cation size on the conductivity has to be studied with samples coming from the same synthesis. Despite this relatively poor accuracy a linear correlation can be observed (for a set of samples doped at y = 0.07) between log σ and the cation size, including both alkaline and TAA cations (Fig. 2).

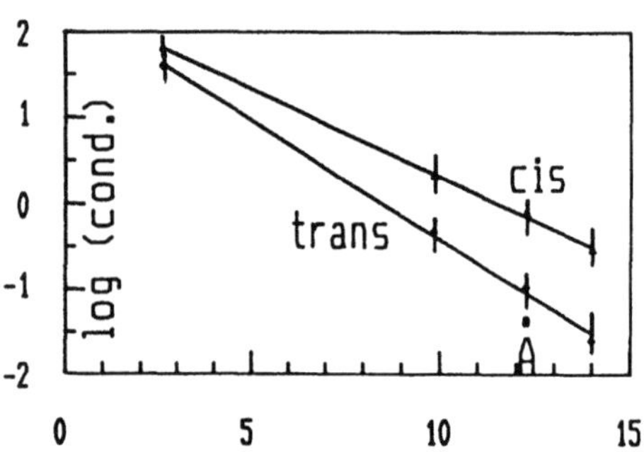

Fig.2:
Variation of log σ
vs
TAA cation diameter

Such a variation can be explained if it is considered that, at room temperature, the transport takes place by a hopping process between nearest neighbours following the law
$\sigma = \sigma_0 \exp(-2\alpha R)\ \exp(-E_a/KT)$ (Mott's model).
This suggests a correlation between the hopping distance R and the radius of cations which control the interchain distance. Up to now no structural data are available for these materials which are probably basically amorphous /4/.The activation energy of this conductivity (Table 2) is higher than observed with alkaline cations, at similar doping level, and does not depend on the TAA size.

Table 2 : Conductivity activation energies (meV)

Cation :	Li	Na	K	Bu_4N	Hex_4N	$Hept_4N$	Oct_4N	$Dodec_4N$
Cis CH_x				57	59	62	55	64
Trans CH_x	36	9	9	62	64	69	55	

5. EPR properties of TAA-doped PA

The E.P.R. band widths measured for TAA doped CH_x and CD_x are reported in Table 3 . They are much narrower than those observed with alkaline cations. The increasing EPR band width with alkaline cation size was attributed /5/ to a spin orbit coupling which apparently does not occur with TAA. The hyperfine coupling then becomes visible in TAA-doped samples and it explains that the band is still narrower (50 mG) in CD_x than in CH_x .

Table 3: EPR band width (Gauss) of 7% n-doped PA

Cation :		Li	Na	K	Bu_4N	Oct_4N	$Dodec_4N$
CH_x :	0.7	4.9	8.3	13.6	0.15	0.15	0.19
CD_x :	0.35	5.8	12	18.2	0.13	0.050	0.094

References

1- C.Mathis ,R.Nuffer ,A.Rudatsikira ,B.François: Makromol. Chem. ,Macromol.Symp. 24 ,99 (1989)
2- F.Ignatious ,B.François ,C.Mathis: papers submitted to Synth.Metals and Makromol.Chem.
3- J. March: Advanced Organic Chemistry (Wiley, New york 1985) p.908
4- L.W.Shacklette ,J.E. Toth: Phys.Rev. 32,9,5892 (1985)
5- F.Rachdi ,P.Bernier: Phys.Rev. B 33(11),7817 (1986)

Super-Molecular Structure of Polyacetylenes and Electrical Conductivity

J. Plocharski

Max-Planck-Institut für Festkörperforschung,
Stuttgart, Fed. Rep. of Germany
(on leave from Warsaw University of Technology, Warszawa, Poland)

Abstract: Polyacetylene samples were synthesized following new procedures applying ageing of catalytic mixtures at elevated temperatures. The foils were aligned by mechanical stretching. The prepared samples were studied using resonant Raman and IR spectroscopies as well as electrical conductivity measurements after doping with iodine. The experimental results obtained for different materials were compared and the role of their morphology was discussed.

1. Introduction

Problems related to the influence of various factors on the electrical conductivity of doped polyacetylene have been studied extensively by many researchers. However, only a few works tended to explain the role of the morphology of the polymer samples in relation to their conductivity. This resulted from significant difficulties in modifying the polymer morphology without changing other parameters like molecular weight or concentration of intrachain defects. Only some kinds of mechanical treatment appear to be safe enough from this respect.

In our previous work [1] we found that by modifying the fibrous morphology of doped polyacetylene by rolling we have increased the density of the samples but we did not change the conductivity calculated per unit mass. This observation suggested that the structure of standard Shirakawa polyacetylene resembles more a sponge than a collection of separate fibrils. The barriers limiting the overall conductivity of the polyacetylene samples are mainly inside the fibers and not between them.

Another way to alter the morphology relies on mechanical stretching which is a common method to improve the properties of traditional polymers. Standard polyacetylene cannot be effectively stretched [2], however some recently described procedures [3,4,5] lead to the synthesis of well-stretchable materials. The conductivity of oriented and doped samples of the new polyacetylenes may reach 100000 S/cm (in comparison to 500 S/cm for the standard material). This great discrepancy suggests the existence of a qualitative difference between these materials. But all efforts to reveal it have failed up to date. This paper tries to explain the influence of mechanical orientation of polyacetylene samples on their conductivity after doping.

Springer Series in Solid-State Sciences, Vol. 91
Electronic Properties of Conjugated Polymers III
Editors: H. Kuzmany · M. Mehring · S. Roth

2. Experimental

Polyacetylene samples were prepared using two different techniques:

"Silicone oil" polyacetylene was synthesized directly following the description published by Naarmann and Theophilou [4]. The catalyst was prepared by dissolving proper amounts of $AlEt_3$ and then $Ti(OBu)_4$ in silicone oil (AV 1000, Th. Goldschmidt AG) and ageing it at 120 °C. Purified acetylene gas was polymerized at 21 °C onto glass substrates covered with a 70 μm layer of the catalyst solution. The obtained foils were removed from the substrates, stretched, washed with toluene and dried.

"Solvent free" polyacetylene was synthesized following the description published by Akagi et al. [5]. The catalyst was prepared by dissolving $AlEt_3$ and $Ti(OBu)_4$ in cumene and keeping this mixture under reflux at ca. 150 °C for a pretty long time. Cumene was removed by pumping prior to the polymerization of acetylene, which was carried out at -78 °C. The obtained foils were washed with toluene, dried and stretched.

Doped polyacetylene was prepared by treatment of cis-rich samples with iodine vapours up to a saturation level.

To get trans-polyacetylene the undoped foils were annealed in vacuum.

Electrical conductivity was measured with the four-probe-technique.

3. Results and Discussion

An exact analysis of the expected influence of stretching on the conductivity is difficult but simple and rather semiquantitative estimates can be done. Considering two extreme cases may help in these estimates:

I. Let us assume that the polymer sample consists of relatively long and highly conductive polymer chains that undergo reorientation during stretching. The overall conductance of the sample is limited by charge carrier transport from one chain to another chain. Stretching of this sample leads to mutual ordering of polymeric chains but rather not to altering the number of barriers limiting the conductance. As a result the total resistance (in ohms) of the sample should remain unchanged during stretching, in contrast to its outer geometry. Simple assumptions of constant resistance and constant mass of a sample upon stretching lead to the following expression:

$$\frac{\sigma_{st}}{\sigma_0} = \left(\frac{l_{st}}{l_0}\right)^2 \cdot \frac{\tau_{st}}{\tau_0}$$

where σ_0, σ_{st} is the conductivity (before and after stretching), l_0, l_{st} the length, and τ_0, τ_{st} the density.

II. If polymeric chains are strongly defected their reorientation would not increase the conductivity. In this situation only the changes of the sample density may affect the final conductivity and hence:

$$\frac{\sigma_{st}}{\sigma_0} = \frac{\tau_{st}}{\tau_0} .$$

Table 1 Properties of polyacetylene prepared during the course of this work

Poly-acetylene	unstretched			
	thickness [μm]	iodine conc. [% at.]	density [g/cm^3]	conductivity [S/cm]
"Silicone oil"	32	13	0.98	250
"Solvent free"	27 3	15 15	0.94 0.95	280 520

Poly-acetylene	stretched			
	thickness [μm]	stretching factor l_{st}/l_0	iodine conc. [% at.]	conductivity $\sigma_{\parallel}$ [S/cm]
"Silicone oil"	15	6	13	2500
"Solvent free"	30 1	6 6	15 15	6000 10000

Experimental results should fall between these two extreme cases under a condition that the procedure of orientation does not modify the character of the barriers limiting conduction. The increase of the conductivity created by stretching should be higher for a material containing a lower amount of chain defects. This effect can be noticed in the case of the samples studied. Some of the obtained results are presented in Table 1.

Fig. 1 presents typical IR spectra of the synthesized polyacetylenes. "Silicone oil" polyacetylene apparently contains residues of silicones (ca. 20 % by weight according to elemental analysis) which could not be removed even by prolonged washing in toluene. These residues make the spectrum difficult to interpret. The IR spectra of "solvent free" polyacetylene are typical for good quality material. More clear conclusions can be extracted from the results of resonant Raman experiments. The respective spectra are depicted in Fig. 2. The low energy peaks characteristic for long conjugated segments of polyene chains are better pronounced in "solvent free" polyacetylene. This is especially the case for curve "b". "Solvent free" polyacetylene apparently contains less intrachain defects than the "silicone oil" one and this is in agreement with effects of stretching. The stretching of the former material by a factor of 6 increases its conductivity about 20 times in contrast to only 10 for the latter material with the same stretching factor. The influence of density changes can be neglected since both materials are very dense even before stretching.

In the course of this study the maximum polyacetylene conductivity reached was 10000 S/cm which was one order of magni-

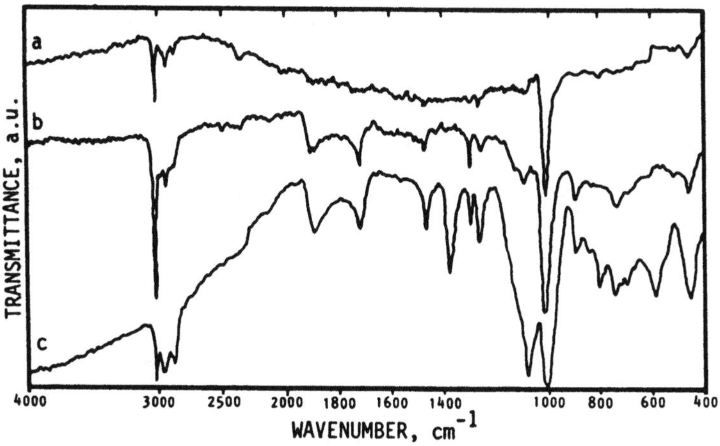

Fig. 1 IR spectra of thermally isomerized polyacetylene
a) "solvent free"-3 μm thick, b) "solvent free"-27 μm thick,
c) "silicone oil" polyacetylene

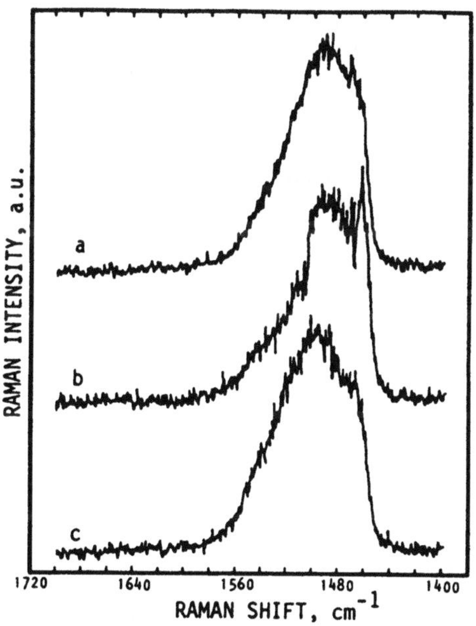

Fig. 2 Resonant Raman spectra of thermally isomerized polyacetylene. Laser wavelength 457.9 nm; a, b, c - see description of Fig. 1

tude below the highest values reported [4]. However, some conclusions concerning the phenomenon of the surprisingly high conductivity of the new polyacetylenes may be drawn. Stretching itself, just by reorientation of the polymer chains, is able to increase the conductivity even by a factor of 50 (for l_{st}/l_0 = 6-8) with no assumed changes in the barriers limiting conductivity. The unstretched material exhibits conductivity value of 500-5000 S/cm. This is not more than 10 times the value of the standard materials. This effect can be attributed to a lower concentration of chain defects or to their different distribution. Such suggestions are present in the literature [6] but they lack experimental confirmation.

Thus, the most amazing feature of the new polyacetylenes is their high stretchability, which property is usually related to highly polymerized materials of linear chains and to a narrow distribution of molecular masses. If this refers also to these new materials the values of molecular masses obtained for standard polyacetylene from radioquenching experiments [7] seem to be overestimated.

Acknowledgements

I would like to thank to Dr. S. Roth for his great help, valuable suggestions and critical reading of the text. I feel also indebted to the Max-Planck-Gesellschaft for their hospitality in Stuttgart. My special thanks are due to Dr. P. Knoll for his help in the Raman measurements.

References

1. J. Plocharski, S. Roth, Synthetic Metals in print
2. H. Shirakawa, S. Ikeda, Synthetic Metals 1 (1979/80) 175
3. G. Lugli, U. Pedretti, G. Perego, J. Polym. Sci.: Polym. Lett. 23 (1985) 129
4. H. Naarmann, N. Theophilou, Synthetic Metals 22 (1987) 1
5. K. Akagi, M. Suezaki, H. Shirakawa, H. Kyotani, M. Shimomura, Y. Tanabe, Synthetic Metals 28 (1989) D1
6. S. Kivelson, A.J. Heeger, Synthetic Metals 22 (1988) 371
7. J.C.W. Chien: "Polyacetylene", Academic Press, Orlando 1984, pp. 50-62

Highly Conducting Polyanilines and Polyacetylene/Polyaniline Composites

N. Theophilou[1], *A.G. MacDiarmid*[1], *D. Djurado*[2], *J.E. Fischer*[2], *and A.J. Epstein*[3]

[1]Department of Chemistry, University of Pennsylvania, Philadelphia, PA 19104-6323, USA
[2]Laboratory for Research on the Structure of Matter, University of Pennsylvania, Philadelphia, PA 19104, USA
[3]Department of Physics and Department of Chemistry, The Ohio State University, Columbus, OH 43210-1106, USA

A technique has been developed to stretch-align polyaniline films, "fibers" and composites, which exhibit after doping, conductivities up to two orders of magnitude greater than previously synthesized non-aligned samples.

I. Introduction

Much effort has been applied in recent years to increase the electrical, magnetic, and optical anisotropic properties of conducting polymer systems by stretch-alignment [1-12,17]. An increase in alignment has been found to give enhancement in conductivity parallel to the stretch direction as compared to the non-aligned polymers.

II. Experimental and Results

The stretch-alignment was performed with the device shown in Fig. 1 in argon. A polymeric stretchable support (5 x 2 x 0.01 cm) such as polypropylene, polyethylene, polybutadiene or polytetrafluoroethylene (Teflon) was clamped in the device of Fig. 1 and an ~2% solution of polyaniline (emeraldine base) in NMP (N-methyl pyrrolidinone) was poured onto the substrate. A heating infrared lamp (250 W; General Electric) was placed ~20 cm from the substrate and the solvent was evaporated at ~60°C over ~20 min. until the semi-solid resulting mixture was just no longer "tacky" to the touch. At this time the temperature was raised to ≥~110°C (see [10]) and stretching was commenced at a rate of ~1 cm/min to stretching ratios up to ~l/l_0=3 (l=final length; l_0=initial length). Transparent films on substrates (thickness <1000 Å), as well as free-standing [13] thick polyaniline films can be stretched by this technique[10]. "Fibers" were obtained by a different method to that previously reported [14] by evaporating the NMP in micro-grooves formed on the substrate by previous stretching of the substrate. Stretch-aligned films and fibers were doped with 1M HCl solution for ~16 hrs. and washed with distilled water, methanol, diethyl ether and pumped for 24 hrs. Maximum conductivities of the film of ~10^2 S/cm (max. $\sigma_\parallel/\sigma_\perp$~10) and of the "fiber" of ~70 S/cm (max. $\sigma_\parallel/\sigma_\perp$~3) were measured (Fig. 5).

Springer Series in Solid-State Sciences, Vol. 91
Electronic Properties of Conjugated Polymers III
Editors: H. Kuzmany · M. Mehring · S. Roth

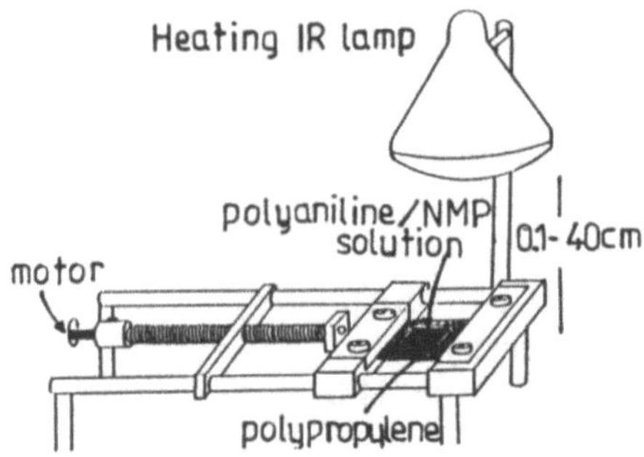

Fig. 1. Stretching device for the alignment of polyaniline

Stretch-aligned (l/l_0=~4) polyacetylene/polyaniline composites were similarly obtained by casting a polyaniline film on both surfaces of a cis-rich polyacetylene film and by stretching during evaporation of the NMP as for the above substrates. Sufficient heat was transferred to the underside of the film to evaporate the NMP from this surface also.

The $(CH)_x$ of the composites was doped for ~1 hr with a saturated I_2/CCl_4 solution, washed twice with CCl_4 and dried under dynamic vacuum for ~30 min. The reaction of emeraldine base with I_2 has not been extensively studied. It is believed that the major increase in conductivity of the composite will be due to doping of the $(CH)_x$. A maximum conductivity of the composite of 1 x 10^4 S/cm (max. $\sigma_{\parallel}/\sigma_{\perp}$~15) was observed (Fig. 5).

Several techniques were used to measure the degree of preferred orientation of aligned emeraldine base films. First, the mosaic distribution in a plane normal to the film was determined from a rocking curve (omega scan) using CuKa at 2θ= 18.4°, corresponding to one of three broad reflections found in a θ-2θ scan (Fig. 2). The film was mounted on an Al frame with the average chain axis in the scattering plane and normal to the scattering vector. The mosaic FWHM is comparable to 2θ, so absorption of the incident or diffracted beam severely distorts the profile. We estimate an OUT-

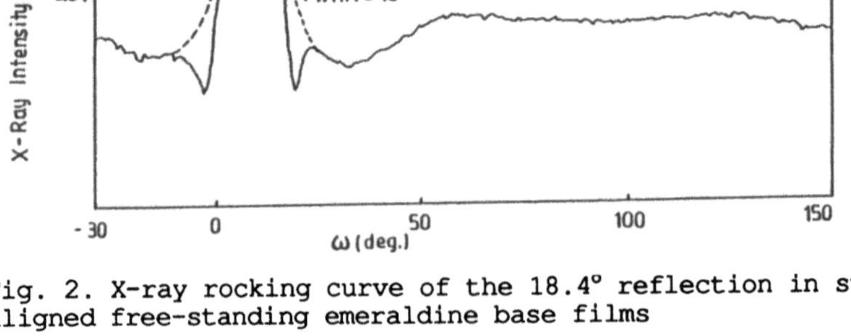

Fig. 2. X-ray rocking curve of the 18.4° reflection in stretch-aligned free-standing emeraldine base films

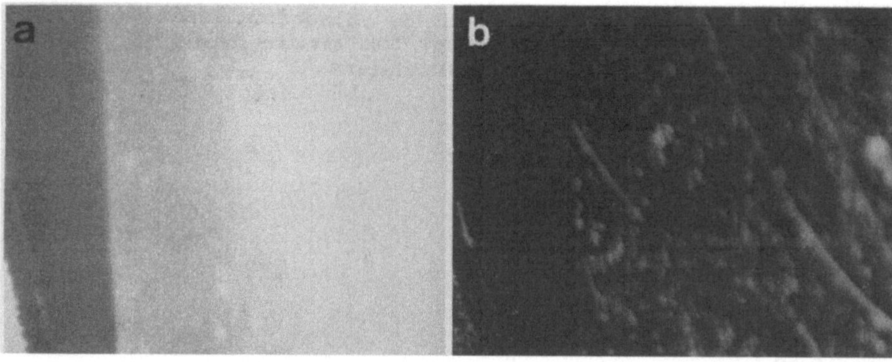

Fig. 3. Polarized transmission optical microscopy of transparent stretch-aligned emeraldine base film supported on polypropylene film. (a) polarized ⊥ to stretch direction (b) polarized ∥ to stretch direction

OF-FILM-PLANE mosaic FWHM of 16-19°, comparable to New-$(CH)_x$ [15]. Second, polarized IR absorption of the 1500 cm^{-1} mode was measured in both free-standing and substrate-supported films [9,17]. The ratio of oscillator strengths parallel and perpendicular to the average chain axis is $\lesssim$3-4, which translates to an IN-PLANE mosaic FWHM in the range 26°-32° using the method of Ref. [18]. This is significantly larger than the X-ray-derived value, indicating that what is often presumed to be a mosaic cone is actually a conical ellipsoid. This is confirmed (at least on the surfaces) by SEM, from which a high degree of local fibril alignment is evident on the sectioned surface (out-of-plane distribution, Fig. 4a), while

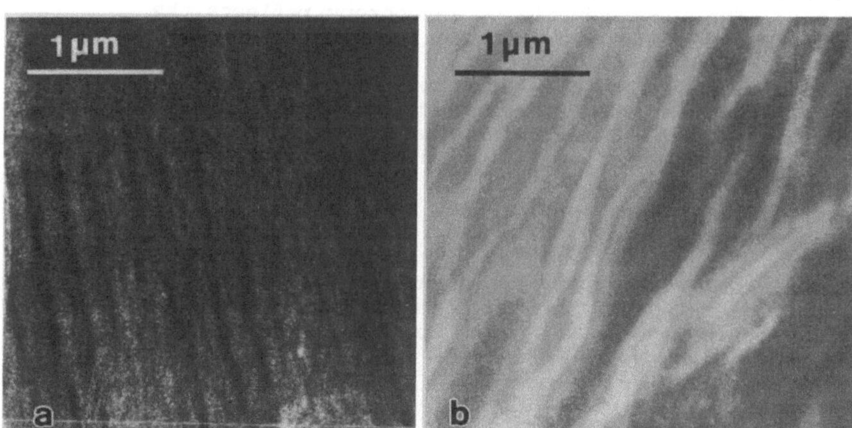

Fig. 4. SEM of (a) cross section and (b) surface of stretch-aligned (l/l_o~3) free-standing emeraldine base film

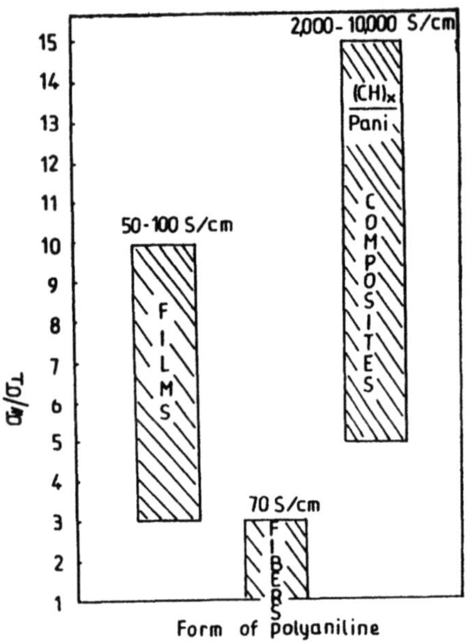

Fig. 5. Electrical anisotropy ($\sigma_{\parallel}/\sigma_{\perp}$) for various forms of polyaniline

the as-grown surface shows substantial fibril axis disorder even after stretch-aligning (in-plane distribution, Fig. 4b).

Our conclusions regarding the texture of stretch-oriented emeraldine base are weakened by the fact that the two principal axes of the mosaic ellipsoid were measured with different techniques (albeit both bulk-sensitive probes) which sample different effective areas in the film plane. They imply however that our understanding of both the characterization of, and the microscopic process involved in the alignment of polymer films is incomplete. The low conductivity anisotropy reflects the relatively broader in-plane distribution but is insensitive to the narrower out-of-plane distribution.

Finally we note that the large constant background in Fig. 2 implies either a large amorphous fraction which diffracts at the same 2θ, or a sizeable fraction of the crystalline component which resists stretch orientation [15]. In either event, the effect is much larger here than in New-$(CH)_x$ [15]. Polarized optical microscopy implies that the IN-PLANE FWHM may be smaller for thinner films (Fig. 3).

Acknowledgement: Work supported in part by the Defense Advanced Research Projects Agency through a contract monitored by the Office of Naval Research and by NSF Grant No. DMR-86-15475, NSF MRL Program under Grant No. DMR-85-19059 (JEF) and by DOE Grant DE-FG02-87ER45254 (DD).

References

1. Y.W. Park, M.A. Druy, C.K. Chiang, A.G. MacDiarmid, A.J. Heeger, H. Shirakawa and S. Ikeda, J. Polym. Sci., Polym. Lett. Ed., 17, 195 (1979).
2. H. Shirakawa, Y.C. Chen and K. Akagi, Synth. Met., 14, 173 (1986).

3. K. Akagi, S. Katayama, H. Shirakawa, K. Araya, A. Mukoh and T. Narahara, Synth. Met., 17, 241 (1987).
4. M. Adissi, J. Polym. Sci., Polym. Lett. Ed., 23, 167 (1985).
5. F.E. Karasz, J.D. Capistran and D.R. Gagnon, Mol. Cryst. Liq. Cryst., 118, 327 (1985).
6. M. Ogasawara, K. Funahashi, T. Dewura, T. Hagiwara and K. Iwata, Synth. Met., 14, 61 (1986).
7. K. Yoshino, S. Nakajima, M. Fuji and R. Sugimoto, Polym. Comm., 28, 309 (1987).
8. H. Naarmann and N. Theophilou, Synth. Met., 22, 1 (1987); Germ. Pat. DE3,617,505 (11/26/87).
9. N. Theophilou, A.G. MacDiarmid, B.K. Annis, R.P. McCall and A.J. Epstein, Bull. A.P.S. 34(3), 823 (1989).
10. K.R. Cromack, M.E. Jozefowicz, J.M. Ginder, R.P. McCall, A.J. Epstein, E. Scherr and A.G. MacDiarmid, Bull. A.P.S. 34(3), 583 (1989).
11. M.E. Jozefowicz, A.J. Epstein, X. Tang, A.G. MacDiarmid and J.P. Pouget, Bull. A.P.S. 34(3), 541 (1989).
12. A. Andreatta, Y. Cao, J.C. Chiang, A. J. Heeger and P. Smith, Synth. Met., 26, 383 (1988); Bull. A.P.S. 34(3), 582 (1989).
13. M. Angelopoulos, G.E. Asturias, S.P. Ermer, A. Ray, E.M. Scherr, A.G. MacDiarmid, M. Akhtar, Z. Kiss and A.J. Epstein, Mol. Cryst. Liq. Cryst., 160, 151 (1988).
14. X. Tang, E. Scherr, A.G. MacDiarmid and A.J. Epstein, Bull. A.P.S., 34, (3) 583 (1989).
15. D. Djurado, J. Ma, N. Theophilou and J.E. Fischer, Bull. A.P.S. 34(3), 406 (1989); Synth. Met. (in press).
16. B.K.Annis, N. Theophilou, A.G. MacDiarmid and A.J. Epstein, to be published.
17. H.S. Woo, D.B. Tanner, N. Theophilou and A.G. MacDiarmid, Bull. A.P.S. 34(3), 823 (1989).
18. H. Shirakawa, K. Akagi, S. Katayama, M. Ito and K. Araya, IUPAC Intern. Congres on Macromolecules, Aug. 1-6, 1988, Kyoto (Japan).

Simulation of the Cooperative Proton Motion in a Model Hydrogen-Bonded Chain

P. Misiak and H. Chojnacki

Institute of Organic and Physical Chemistry,
I-4, Wyb. Wyspianskiego 27, PL-50-370 Wroclaw, Poland

Abstract. The cooperative proton motion in the hydrogen-bonded molecular chain is frequently considered within the framework of the energy and proton transport problems. In the present paper the simple model of the proton motion dynamics has been analyzed with respect to the conditions for propagation of kink-like excitations along the chain. The influence of alternating external electric field on excitation is studied by numerical solving of equations of motion.

1. Introduction

The quasi one-dimensional networks of hydrogen-bonded molecules or molecular units appear in many physical and biological systems, e.g. ice, structured water in narrow channels, protein molecules etc. Especially, the chains of hydrogen bonds play an important role in some biological processes in living systems, like proton or energy transport across molecular membranes [1]. The relatively high electrical conductivity of ice is also due to mobility of protons in hydrogen bonds formed between water molecules [2,3]. The processes mentioned above involve the cooperative behavior of protons in the chain, and their understanding at the molecular level can serve as a starting point for description of more complicated phenomena.

The model chain assumes the heavy molecular units (ions, atoms or atomic groups) X connected by the hydrogen bonds

$$\cdots X{-}H\cdots X{-}H\cdots X{-}H\cdots X{-}H\cdots X{-}H\cdots \quad .$$

The interactions in the system are determined by potential functions denoted as V_{XX} for heavy units interactions, V_{HH} for proton-proton interactions, V_{XH} for coupling between atomic and protonic subsystems, and V_{ext} for the interaction with external field. Heavy molecular units and protons are assumed to have one degree of freedom, i.e. they can move only along the chain. Choosing the specific form of potential functions determines the relevant model.

In real systems the chain of hydrogen bonds is surrounded by molecules and ions which produce some external electric field acting on the protons within the bonds. In general, the field changes from bond to bond and influences the motion of protons.

Springer Series in Solid-State Sciences, Vol. 91
Electronic Properties of Conjugated Polymers III
Editors: H. Kuzmany · M. Mehring · S. Roth

2. Theoretical Outline

In the following we will consider the model in which the heavy molecular units are at rest in their equilibrium positions in equidistant sites, and protons can move between neighboring atoms in the effective double-well potential arising from the interactions between atoms and protons. The potential well has two equivalent minima separated by the potential barrier. The height and width of the barrier depends on the properties of the atomic sublattice. As it was shown by Davydov [1], the assumption of the phonon-like motion of heavy atoms coupled to the protonic motion leads to the effective lowering of the potential barrier in hydrogen bonded systems. Thus, we assume the model hamiltonian in the form

$$H = \Sigma_i T(i) + \Sigma_i V_{loc}(i) + \Sigma_i V_{int}(i,i+1) + \Sigma_i V_{ext}(i) ,$$

where

$$T(i) = 0.5m(du(i)/dt)^2$$

is the kinetic energy of the i-th proton of mass m and position with respect to the middle of the bond u(i),

$$V_{loc}(i) = a(1-u(i)^2/D^2)^2$$

denotes the effective double-well potential with the barrier height a and the distance between the minima 2D,

$$V_{int}(i,i+1) = 0.5k_1(u(i+1)-u(i))^2 + 0.25k_2(u(i+1)-u(i))^4$$

is the anharmonic interaction potential between neighboring protons, and

$$V_{ext}(i) = eEsin(2\pi(iL/\lambda+\phi))u(i)$$

stands for the potential function of interaction between the i-th proton of the charge e and the external field characterized by strength E and parameters λ,ϕ determining variation of the field along the chain; L is the chain constant (distance between heavy atoms). The spatial variation of the electric field we assume in the simple sinusoidal form.

In study of the cooperative proton motion in the system defined by the above hamiltonian the appropriate classical equations of motion for the chain of N=16..32 protons have been solved numerically. The simulations were performed for various combinations of λ and ϕ with other parameters fixed at the following values [5,6]:

$$a=0.2 \text{ eV}, D=0.04 \text{ nm}, E=10^9 \text{ V/m}, L=0.3 \text{ nm},$$
$$k_1=100 \text{ eV/nm}^2, k_2=10^5 \text{ eV/nm}^4.$$

The initial conditions (for the moment t=0) were chosen as follows: u(i)=-D for i=1..N, $du(1)/dt=v_0$, du(i)/dt=0 for i=2..N. The energy injected into the system as kinetic energy of the first proton was set to 0.5 eV, which is the universal biological energy unit i.e. approximately equals to the energy released in the reaction of hydrolysis of ATP [cf. ref.1].

Numerical errors were controlled during the simulation by inspecting the total energy of the system after each time step done. The energy was conserved up to 0.01 %.

3. Results and Conclusions

Simulation of the proton motion within the chain in absence of an external electric field shows formation of the stable kink, which propagates along the chain, carrying most of the energy injected into the system, in the form of the energy hump spread over 3-4 protons. The kink moves with the constant velocity to the end of the chain. The form of the displayed picture does not depend on the length of the chain.

If the external electric field is switched on, it perturbs the kink motion. The effect of the perturbation depends on the initial phase ϕ and wave length λ of the electric field, i.e. on the mode of variation of the field along the chain. Figure 1 shows the dependence of the average kink velocity on λ and ϕ. It can be observed that when the field accelerates the kink motion at the beginning of the chain (ϕ=0.5), the following retardation (kink goes upwards) does not influence the effectiveness of the energy transport along the chain. On the contrary, the dumping of the proton motion in the starting bonds leads to the retardation of the kink motion or - if the dumping extends over several bonds - to stopping (in fact reflecting) the kink.

In conclusion one can state that within the framework of the simple model presented above the external alternating electric field can play an important role in the energy and charge

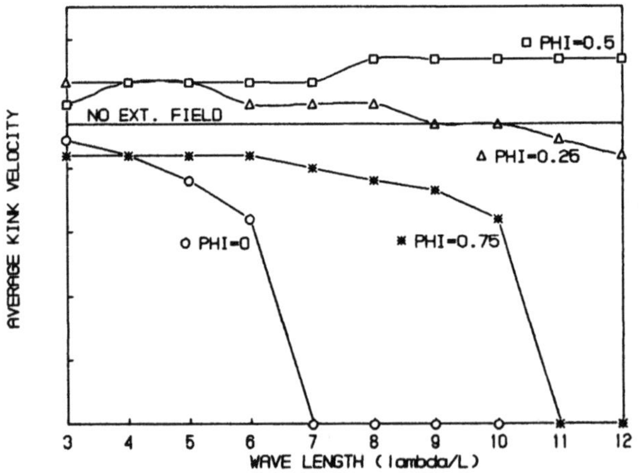

Fig. 1. The average velocity of the kink in the hydrogen-bonded chain versus the wavelength λ/L of alternating external electric field, for the four different values of the initial phase ϕ=0, 0.25, 0.5, 0.75

transfer mechanism in quasi one-dimensional hydrogen-bonded systems. The appropriate configuration of the environment (neighboring molecules etc.) at the beginning of the chain of hydrogen bonds, can create the favorable external conditions for proton motion and in turn significantly increase the effectiveness of the energy and proton transport along the chain.

Acknowledgement

This work has been partially sponsored under the CPBP 3.13 contract.

References

1. A.S. Davydov: Solitons in Molecular Systems (Naukova Dumka, Kiev 1988, in Russian).
2. L. Glasser: Chem. Rev. **75**, 21 (1975).
3. V.Ya. Antonchenko, A.S. Davydov, A.V. Zolotariuk: Phys. Stat. Solidi (b) **115**, 631 (1983).
4. H. Chojnacki: In Quantum Science. Methods and Structure., ed. by J-L. Calais, O. Goscinski, J. Lindenberg and Y. Oehrn (Plenum Press, New York and London 1976) p.561.
5. Y. Kashimori, F. Chien, K. Nishimoto: Chem. Phys. **107**, 389 (1986).
6. J. Halding, P.S. Lomdahl: Phys. Rev. A **37**, 2608 (1988).

Mass Transport in Conducting Polymers: An Experimental and Computer Simulation Study

A. Morin[1], *M. Bénière*[1], *S. Pekker*[1;*], *F. Bénière*[1], *D.A. Morton-Blake*[2], *V.C.A. Hanratty*[2], *and J. Corish*[2]

[1]Laboratoire de Sciences des Matériaux, Université de Rennes I, F-35042 Rennes Cedex, France
[2]Department of Chemistry, Trinity College, Dublin 2, Ireland

ABSTRACT. We have measured the microscopic diffusion coefficients of $FeCl_4^-$ in $(CH)_x$ using a new experimental method. An activation energy of 1.0 eV is derived from the temperature dependence of the diffusion coefficient . Simultaneously, atomic lattice simulation techniques have been developed for conducting polymers. They provide information on the location and migration of the dopant species in the crystalline parts of the polyacetylene.

1. GENERAL DIFFUSION MODEL

The heterogeneous morphology of conducting polymers is essential for the diffusion properties. The importance of interfibrillar spaces and rather regular fibrils have allowed us to develop a model of diffusion from a liquid source, based upon the following assumptions :

- all the fibers are perfect cylinders with an average radius $a = 200$ Å ;
- constant diffusion coefficient ;
- very fast diffusion of the dopant and solvent through the interfibrillar space ;
- constant concentration C_s of the dopant at the surface of the fibrils.

Using the mathematical equation for diffusion in a cylinder [1] :

$$\frac{\partial C}{\partial t} = \frac{1}{r}\frac{\partial}{\partial r}\left(r\, D \frac{\partial C}{\partial t}\right) ,$$

the quantity of diffusing substance M_t is given by

$$\frac{M_t}{M_\infty} = 1 - \sum_{n=1}^{\infty} \frac{4}{a^2\alpha_n^2} \exp\left(-D\alpha_n^2 t\right) \qquad (1)$$

(α_n 's : Bessel roots).

2. EXPERIMENTAL AND RESULTS

We record continuously the weight uptake of a polyacetylene sample immersed in a solution of $FeC\ell_3$ in nitromethane (concentration : 7.10^{-2} M). The experimental data, giving M_t versus time are converted into M_t/M_∞ versus $\sqrt{t}$ for a comparison with the theoretical equation (1).

The absorption curves, recorded at various temperatures in the range - 16 °C up to + 51 °C, show clearly a faster kinetics with increasing temperature (fig. 1).

---

* *Permanent address :* Central Research Institute for Physics, Budapest (Hungary).

Springer Series in Solid-State Sciences, Vol. 91
Electronic Properties of Conjugated Polymers III
Editors: H. Kuzmany · M. Mehring · S. Roth

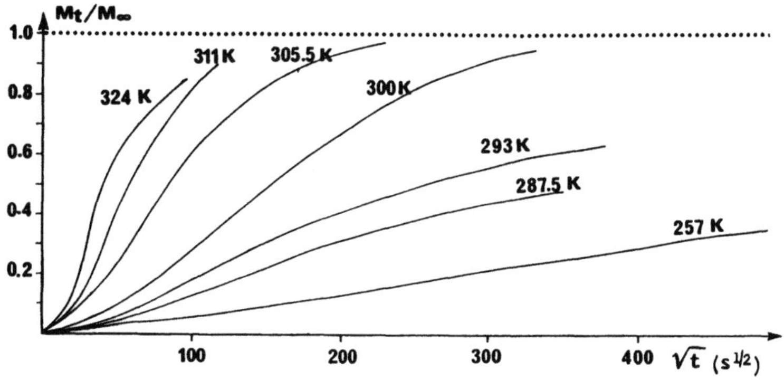

Fig. 1 - Temperature dependence of the absorption kinetics

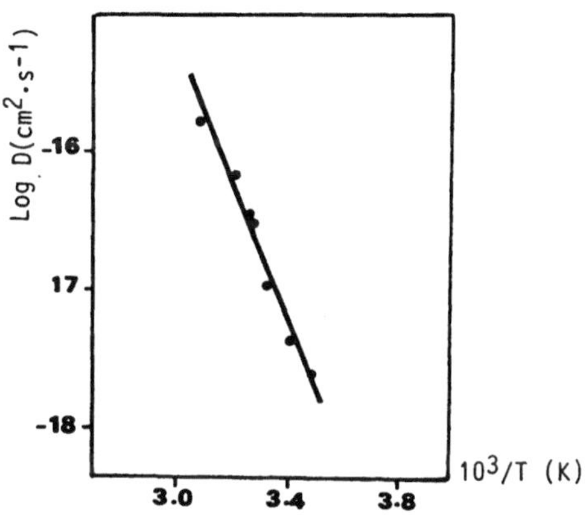

Fig. 2 - Temperature dependence of the diffusion coefficient

The diffusion coefficients are calculated by a least-squares fitting of the theoretical equation (1). The experimental value of D at 300 K is 10^{-17} $cm^2.s^{-1}$. We notice the small magnitude compared to the macroscopic diffusion coefficient which is about 9 orders of magnitude higher when the sample is considered as an homogeneous material. The detailed results are discussed in [2].

The temperature dependence of the diffusion coefficients is indicated in the Arrhenius plot (fig. 2). It shows that, between 14 °C and 50 °C, the diffusion of $FeC\ell_4^-$ in $(CH)_x$ follows an Arrhenius law defined by

$$D = 0.82 \exp\left(-\frac{1.0\ \text{eV}}{kT}\right) cm^2.s^{-1} .$$

3. LATTICE SIMULATION : STRUCTURE INVESTIGATION OF K^+ DOPED POLYACETYLENE

Lattice modelling calculations such as those in alkali metal intercalated graphite [3] are extended to doped polyacetylene lattices. The first requirement in such a calculation is to generate a set of atomic lattice points using data provided by diffraction investigations and secondly to specify a set of "short-range" potentials between declared pairs of bonded and non-bonded atoms. We have developed interatomic potentials which lead to an accurate lattice simulation of K^+-doped polyacetylene [4].

Interatomic potentials :

C--C interaction bonded : $V_{C-C}(r) = 4.5873[1-e^{-1.9169(r-1.396)}]^2$

$V_{C=C}(r) = 5.3060[1-e^{-2.1353(r-1.346)}]^2$

(From bond energy and force constant data)

non-bonded : $V_{CC}(r) = 3627.41\ e^{-r/0.2778} - 24.6367\ r^{-6}$

(Williams IV potential)

C--H interaction bonded : $V_{CH}(r) = 4.3051[1 - e^{-2.1029(r-1.08)}]$

(From bond energy and force constant data)

non-bonded : $V_{CH}(r) = 380.2212\ e^{-r/0.2725} - 5.4118\ r^{-6}$

(Williams IV potential)

K^+--K^+ interaction : $V_{KK}(r) = 3796.9\ e^{-r/0.2603} - 285.9\ r^{-6}$

(Catlow et al. 1976)

C--K^+ interaction : $V_{KC}(r) = 18851.5\ e^{-r/0.2690} + 186.9\ r^{-6}$

(Fitting to experimental structure)

Bond angle bonding potential : $V(C-C-C) = 0.5 \times 12.42 \times (q - 125.3027)^2$

$V(H-C-C) = 0.5 \times 5.2431 \times (q - 117.348)^2$

The bond lengths and bond angles of the relaxed chains in the lattice containing a K^+ vacancy (fig. 3b) show very little distortion from those in the perfect lattice (fig. 3a): less than 0.022 Å in the C-C bond lengths and less than 1.1° in the bond angles. The effective negative charge on the vacancy attracts the adjacent K^+ ions in the same channel. They are displaced by 2.1 Å toward the position of the missing K^+ ion.

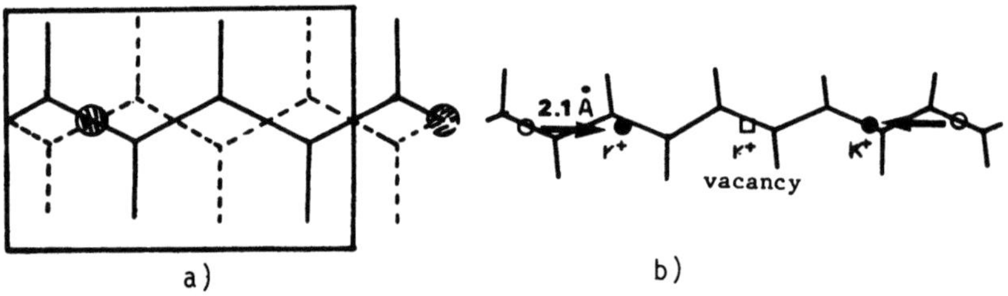

Fig. 3 - $(CHK_{0.125})_x$ lattice projected parallel to the plane of the polymer chain. a : perfect lattice ; b : lattice containing a K^+ vacancy

4. CONCLUSION

We have derived the effective diffusion coefficient for the motion of the dopant into the fibrils and from its temperature dependence, an activation energy of 1.0 eV. This energy usually contains a defect formation term and a migration enthalpy. We have not enough experimental data to separate clearly both contributions. However, we are extending the computer simulation studies to the calculation of the different terms in order to give a clear description of the diffusion process in conducting polymers.

ACKNOWLEDGEMENT. This work is supported by the "Commission of the European Communities" under a Stimulation Grant.

REFERENCES

1. Peter R H 1968 in "Diffusion in Polymers" ed. J Crank and G S Park (Academic Press)
2. Morin A, Bénière M and Bénière F, accepted for publication in J. Phys. Chem. Solids
3. Morton-Blake D A, Corish J and Bénière F 1988 Phys. Rev. B 37 4180
4. Baughman R H, Murthy N S and Miller G G 1983 J. Chem. Phys. 79 515

On the Structure of Lithium-Doped trans-Polyacetylene

O. Leitner[1], *G. Leising*[1], *H. Kahlert*[1], *J. Fink*[2], *and H. Fritsche*[2]

[1]Technische Universität Graz, Institut für Festkörperphysik,
Petersgasse 16, A-8010 Graz, Austria
[2]Kernforschungszentrum Karlsruhe,
Institut für Nukleare Festkörperphysik,
Postfach 36 40, D-7500 Karlsruhe, Fed. Rep. of Germany

We present high energy electron diffraction (HEED) studies on undoped and lithium-doped highly oriented trans - polyacetylene prepared via the Durham-Graz route. The doping was performed under ultra high vacuum conditions by an evaporation and diffusion technique. New sharp diffraction lines develop in the HEED-scans of lithium-doped samples with intensities increasing continuously with the incorporation of the dopant species.

Charge transfer to the $(CH)_x$ chains increases the lattice parameter c gradually upon doping. Based on structural model calculations the unit cell of lithium doped $(CH)_x$ is obtained.

INTRODUCTION

Alkali-doping of trans-polyacetylene has been subject of numerous investigations, but most of them are dealing with Shirakawa-type polyacetylene, and the doping was performed in solution or by chemical or electrochemical means [1]. These facts introduce some uncertainties whether the new arising intensities in the diffraction scans are due to the insertion of the dopant species or solvant molecules.

For our investigation of the structural changes of polyacetylene upon doping we used very thin films of oriented highly crystalline trans-$(CH)_x$ prepared via the Durham-Graz route [2]. All the steps of the preparation of the undoped polymer were done either in argon atmosphere or under dynamic high vacuum. The samples were sealed in quartz glass tubes under dynamic vacuum for the transport into the electron spectrometer, there these tubes were opened in order to perform the doping. The samples therefore were never exposed to air prior to the measurement. Details of the experimental setup are described by J. Fink et al.[3]. The doping was done with a combined evaporation and diffusion technique. Metallic lithium films of varying thickness were deposited onto the polymer surface by evaporation (in order to achieve different doping levels). The diffusion of lithium into the bulk of the polyacetylene films was activated by raising the temperature up to 100°C - 200°C for several hours. At liquid nitrogen temperature the diffraction patterns were obtained by measuring the number of elastically scattered 170keV electrons with a counter as a function of the momentum transfer.

Springer Series in Solid-State Sciences, Vol. 91
Electronic Properties of Conjugated Polymers III
Editors: H. Kuzmany · M. Mehring · S. Roth

RESULTS AND DISCUSSION

In Fig. 1, hk0 and 001 diffraction scans are presented, which show the evolution of new diffraction lines. The numbers labelled at the different curves denote the thicknessses of the deposited lithium layers obtained with an Inficon thickness measuring device. Because of the necessarily limited size of our samples (area about 1mm², thickness 10 to 100 nm), which is necessary to avoid multiple scattering and to achieve a homogeneous doping throughout the sample, it is not possible to determine the doping concentration by simply measuring the weight uptake or by elemental analysis. Therefore the integrated intensity of the first most prominent new peak at about .95Å is taken as a rough measure of the doping concentration rather than the thickness of the deposited lithium layers. In Figure 2 the integrated intensity of the first diffraction line is shown as a function of the thickness of the lithium layer. The curve does not saturate indicating that the highest possible doping level is not yet achieved. With increasing doping concentration, the first line shifts to lower k-vectors. This means that the spacing between corresponding crystallographic planes increases as pointed out in Fig. 2. The largest observed diffraction spacing provides a useful guide for the evaluation of model structures. This reflection line is taken to be the 010 reflection defining the unit cell parameter b. The third diffraction line can be attributed to the second order reflection of the b-spacing. For the further calculation the following assumptions are considered. The second peak in the diffraction scan is taken as the 110 reflection for two reasons: Firstly,

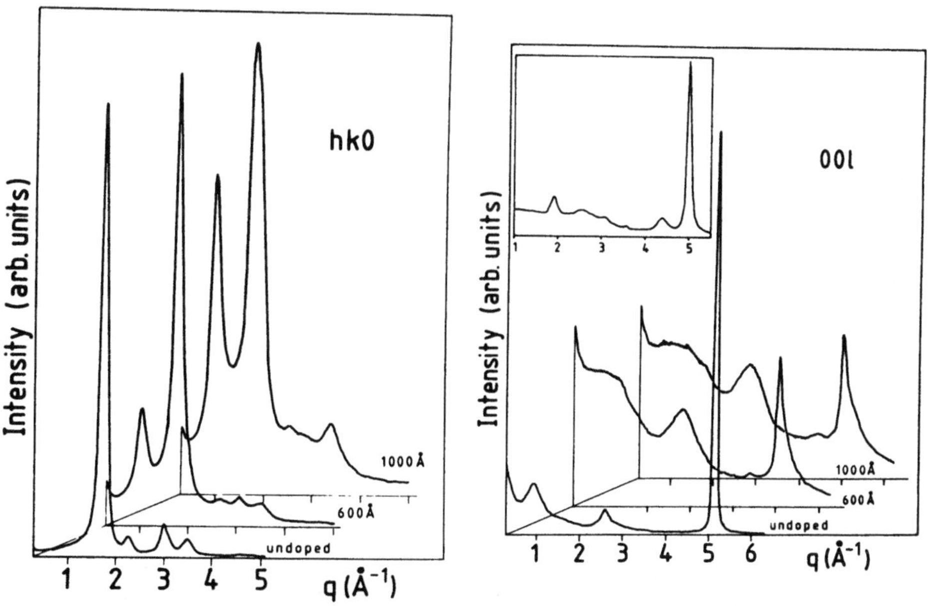

Fig.1: Sequence of equatorial (hk0) and meridional (001) HEED-scans of undoped and lithium doped trans polyacetylene. The degree of doping increases from the bottom to the top

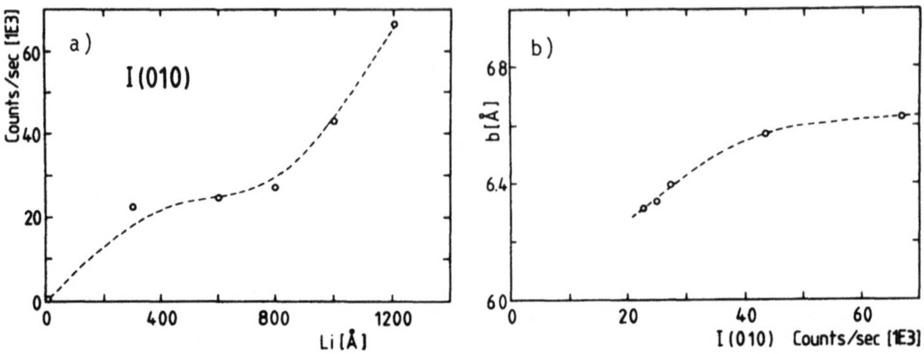

Fig.2:a) Increase of the integrated intensity of the doping induced single diffraction line as a function of the thickness of the lithium layer. b) Variation of the unit cell parameter b with increasing doping concentration

if one assumes that lithium ions are too small to distort the host lattice to a larger extent, the doped structure should have a and b axes at least almost rectangular. Secondly, taking the second reflection as resulting from the 100 plane, the area of the a-b plane would be much less than that of the undoped polymer ruling out this choice. With these considerations, the lattice parameter a can be derived and the shorter spacings observed in the experiment can be checked for consistency. All lines can be indexed. First calculations were made with a doubled b-spacing in order to simulate diffraction scans at moderate doping levels. The positions of the two lithium ions in the unit cell (0.25/0) and (0.75/0.5) gave the highest possible intensity for the first reflection (in this case indexed by 020). The calculated intensity cannot explain the much

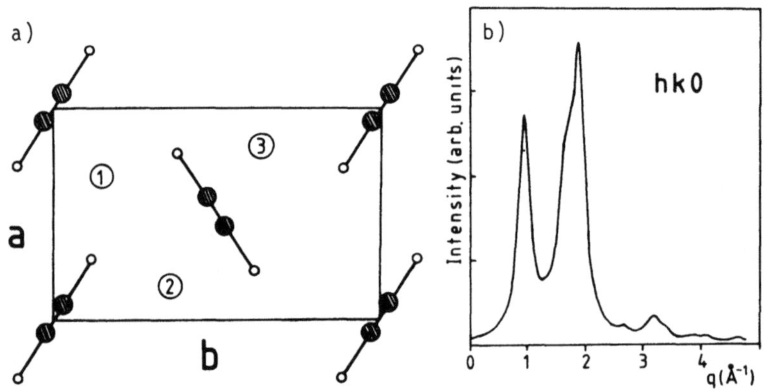

Fig.3: a) Unit cell proposed for the lithium doped polyacetylene complex. The little circles represent hydrogen, the dashed ones carbon and the large circles discussed possible lithium positions. b) Calculated diffraction spectrum for lithium doped $(CH)_x$, with lithium ions in positions 1 and 2

higher diffraction intensities observed for higher doping concentrations. The unit cell chosen for this case is sketched in Fig. 3.
Circles labelled with numbers are the possible positions of lithium ions discusssed in this work. With lithium at the positions 1 and 3 an almost equidistant distribution of lithium in the a-b plane can be achieved. But to fit the high intensity of the 010 reflection one has to move the lithium ions apart from each other or closer to each other. This results in carbon-lithium interatomic distances too small to be explained without a strong covalent bond. As the lithium ions are not in commensurate positions along the chain this can be excluded. Only the positions 1 and 2 for the lithium ions are suitable to be varied to fit the experimental results without this restriction. Figure 3 shows a calculated spectrum obtained for the second choice discussed.

In Fig. 4 the variation of the lattice parameter in chain axis direction is plotted. Due to the charge transfer onto the $(CH)_x$ chains the cell dimension parallel to the chains is elongated. In contrast to the theory [4], the dependence upon the doping concentration is not linear which is consistent with experimental results obtained by M.J. Winokur et al.[5]. The inset of figure 1 shows additional doping induced peaks which appear in samples of high quality. The spacing between neighbouring lithium ions, deduced is about 3.3 Å and thus incommensurate to the lattice parameter c.

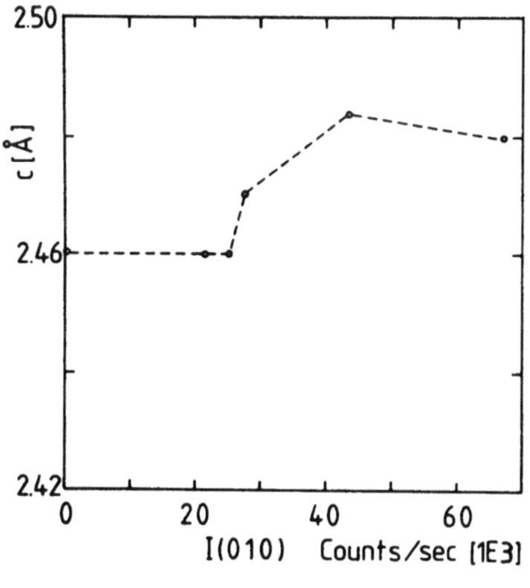

Fig. 4: Elongation of the lattice parameter c versus the intensity of the doping induced 010 peak taken as an approximate measure for the lithium incorporated into the $(CH)_x$ host lattice. The dashed line serves as a guideline for the eye

REFERENCES

1 R.H.Baughman, N.S. Murthy, and G.G. Miller, J. Chem. Phys. 79, 515 (1983).
2 G. Leising, Polymer Bulletin 11,401 (1984).
3 J. Fink, N. Nücker, B. Scheerer, W. Czerwinski, A. Litzelmann, and A. vom Felde, Springer Series in Solid-State Sciences 76

4 M. Kertesz, F. Vonderviszt, and S. Pekker, Chem. Phys. Lett., 90,430 (1982).
5 M.J. Winokur, Y.B. Moon, A.J. Heeger, J. Barker, and D.C. Bott, submitted to Phys. Rev. B, Rapid Commun.

Part II

Theory

3-D Structures and Intrinsic Defects in trans-Polyacetylene

P. Vogl[1] *and D.K. Campbell*[2]

[1]Institut für Theoretische Physik, Universität Graz, A-8010 Graz, Austria

[2]Center for Nonlinear Studies, Los Alamos National Laboratory, Los Alamos, NM 87545, USA

Abstract. We report first-principles local-density functional pseudopotential calculations of static and dynamic electronic structure properties of crystalline 3-D *trans*–$(CH)_x$. We find a broken symmetry ground state of $P2_1/a$ symmetry with in-phase dimerizations. Using a Green's function technique, we show that the 3-D character of the electronic band edge states strongly suppresses self-trapping, destabilizing polarons and possibly bipolarons as well in perfectly ordered 3-D *trans*–$(CH)_x$.

1. Introduction

Within strictly one-dimensional (1-D) models of conducting polymers, such as *trans*-polyacetylene ($(CH)_x$), one expects a variety of localized, "self-trapped", nonlinear excitations – kink solitons, polarons, and bipolarons – with possible significant effects on optical and transport properties [1]. Present conducting polymers, on the other hand, are typically inherently three-dimensional (3-D), partially crystalline films, in which interactions between polymer chains play an important role [1]. Indeed, "solid state screening" effects, both intra- and inter-chain, are frequently invoked to explain the apparent differences between the excited states observed in *trans*–$(CH)_x$ and those expected from extrapolations of both measurements and calculations on long but finite polyenes in the gas phase.

Although several studies [2,3] based on simplified theoretical models have suggested that 3-D interchain coupling can have strong effects on the 1-D excitations, previous *ab-initio* theories [4,5,6] of the 3-D structure of both *trans*- and *cis*-$(CH)_x$ have left for further investigation the critical question of whether the interchain couplings are sufficiently strong to destabilize the carrier self-trapping associated with intrinsic defects in 1-D systems.

In this paper we report the results of extensive first-principles studies of 3-D crystalline *trans*–$(CH)_x$ and intrinsic defects in this material. For the electronic structure of crystalline 3-D *trans*–$(CH)_x$, we employed parameter-free local density functional theory (LDA) [7] with *ab-initio* norm-conserving pseudopotentials for carbon [8] and the Coulomb potential for hydrogen in a momentum-space representation [9]. The exchange-correlation energy is evaluated with the functional

Springer Series in Solid-State Sciences, Vol. 91
Electronic Properties of Conjugated Polymers III
Editors: H. Kuzmany · M. Mehring · S. Roth

given by Painter [10]. The Kohn-Sham equations [9] are solved in a plane-waves basis with up to 1400 plane waves corresponding to a maximum kinetic energy of 30 Rydbergs. All total energies are converged to 1 meV.

2. Crystal and electronic structure

Since the crystal structure of 3-D *trans*–$(CH)_x$ in its ground state is still not firmly established, [11,12] we have determined it from self-consistent minimizations of the total energy within LDA, augmented by a calculation of the Hellman-Feynman forces [9] on the ions.

In order to characterize fully either of the proposed monoclinic structures, 10 structural constants need to be calculated [12]. With the exception of the dimerization amplitude $u_C = \pm 2u_0$ of the C-C bonds in the chain-direction, our total energy and force calculations yield all structural parameters in excellent agreement with the experimental results [12]. They will be given in detail elsewhere [13]. The calculation of the Hellman-Feynman forces proved crucial for an accurate assessment of u_C since they are an order of magnitude more sensitive to u_C than the total energy itself; because of this, previous total energy studies [6] were not able to obtain unambiguous results for u_C. Our calculations give $u_C = 0.01$ Å. Thus, we find the LDA to correctly predict a broken symmetry ground state but –consistent with earlier studies, [5,6] – to quantitatively underestimate the experimental value [12] which is $u_C = 0.05$ Å.

Irrespective of the precise value of u_C, our results show that the dimerized P21/a structure is ***lower*** in total energy than the dimerized $P2_1/n$ structure. For the experimental value for u_C, the energy difference is 0.01 ± 0.003 eV per $(C_2H_2)_2$ unit [14]. This difference is consistent with the similarity of the two structures, and is of the order of optical phonon energies in solids.

We have analyzed the electronic valence charge density of both monoclinic structures and find, as expected [15], that the band states above and below the energy gap in crystalline *trans*–$(CH)_x$ consist primarily of molecular carbon-π-bands. In addition, however, we find a non-negligible admixture of hydrogen p-type states. This H-admixture is larger for the top valence band states than for the bottom conduction band states, leading to a particle-hole asymmetry [16] which shows up also in the band structure. Each H-atom on one chain approximately points towards a p-orbital of a C-atom on one of the neighboring chains. The C-H distance across the chains is 2.9 Å in the $P2_1/a$ structure versus 3.1 Å in the $P2_1/n$ structure. This difference provides a crucial additional bonding energy that favors the $P2_1/a$ structure.

In the band structure for the dimerized $P2_1/a$ phase of crystalline *trans*-$(CH)_x$ we find that the pair of the two lowest conduction and highest valence bands near the k-point B = (0, 0, 1/2) are split by 0.53 eV and 0.46 eV, respectively. This splitting is caused by the interaction of electronic carbon and hydrogen states on adjacent chains. In the band structure of the $P2_1/n$ phase – which has been studied previously [4,5] as well as in our present work – the splittings occur near the **k**-point A = (-1/2, 0, 1/2) and are 0.23 eV for each of the band edge states; these smaller splittings reflect the larger distance between the neighboring C and H atoms.

3. Intrinsic Defects

In a strictly 1-D deformable chain and invoking adiabatic phonons, a charge carrier will always self-trap due to the electron-phonon interaction; this is in contrast to the behavior in 3-D systems, [5] where self-trapping is not guaranteed and requires a strong electron-phonon coupling. Since our results show that the interchain interaction crucially alters the effective dimensionality (as well as the symmetry) of the electronic band states close to the minimum gap, it is clearly essential to study the influence of the interchain interaction on the intrinsic, nonlinear defects predicted by the 1-D theories. To carry out this study, we have calculated the electronic structure of 3-D crystalline *trans*–$(CH)_x$ containing frozen-in intrinsic defects and focused on the formation of bound states in the energy gap.

We have mapped our first-principles calculations for the perfect *trans*–$(CH)_x$ crystal onto an accurate, multi-orbital, three dimensional tight-binding model [13]. The inter-site Hamiltonian matrix elements have been fitted to explicit functions of the distance between the atomic sites. The model reproduces both the energy bands, the atomic character of the bands as obtained from integrated charge densities in each valence band, and the electron-acoustic phonon deformation potentials for the band edge states. Details will be given elsewhere [13]. Due to the theoretical uncertainty of u_C, we have performed these calculations for several values of u_C.

Employing a multi-orbital Koster-Slater Green function method [17], we have calculated the electronic structure of 3-D *trans*–$(CH)_x$ in the $P2_1/a$ structure with a localized polaron defect [18] in an otherwise perfectly 3-D dimerized crystal.

Significantly, we find that the 3-D character of the band edge states prevents the formation of self-trapping intragap states; this destabilizes the one-dimensional polaron excitation [3] in crystalline *trans*-polyacetylene.When we artificially set all interchain matrix elements equal to zero, the polaron bound states appear in the gap for any lattice distortion, in complete agreement with known 1-D results [18,19].

We have also computed the bound states due to a bipolaronic displacement pattern – corresponding to a separated kink (S) and anti-kink ($\bar{S}$) soliton. Again, the interchain coupling strongly impedes the formation of bound states in the energy gap of three-dimensional *trans*–$(CH)_x$. Only for separations exceeding 6 lattice constants (14 carbon sites in chain direction) between the kink- and antikink-soliton ($S - \bar{S}$), a pair of shallow bound states appears in the energy gap.

If we simply extrapolate the numerical results to larger $S - \bar{S}$ distances, the binding energies barely exceed room temperature thermal energies for distances $S - \bar{S}$ smaller than 8-10 lattice constants. There are two additional factors which limit the stability of a bipolaron. First, as has long been recognized, the formation of a separated kink-antikink pair costs interchain bonding energy due to misalignment of the dimerized phases on adjacent chains in the region between the $S - \bar{S}$ pair. We estimate from our total energy calculations that the maximum kink separation attainable in a perfectly ordered crystal is of the order of 50 lattice constants in chain direction. Since the electronic wave function in a kink-bound state is expected to extend over 5-10 lattice constants, [16] this interchain confinement to 50 lattice constants strongly limits the formation of a bipolaron. Second, if two electrons are captured in the $S - \bar{S}$ bound state, the Coulomb repulsion will further reduce the total effective binding energy. These qualitative arguments, being backed by our Green function calculations, make it plausible that bipolarons are also destablized in perfectly ordered crystalline *trans*–$(CH)_x$.

To conclude, we discuss the limitations, both theoretical and in application to presently existing conducting polymers, of our results and mention some important open issues. First, the LDA calculations capture only part of the correlation energy due to electron-electron many-body interactions [7]. We are presently applying the so-called self-interaction corrected LDA [20] to single chains of $(CH)_x$; we find that this method tends to give a substantially larger dimerization than LDA. For the ground states of many semiconductors, insulators, as well as van der Waals crystals, [21] however, the LDA methods are known to be accurate, providing correct energies, pressure dependences, phonons and phase transition information [9]. Second, the calculated gap in the LDA effective single particle spectrum is *not* the optical gap but differs from it by a self-energy, Σ_{el} [22,23]. Third, even though the force-free value of the dimerization in our study is too small, the critical band-edge splittings are unaffected both by the self-energy correction and – as our calculations show – by the precise value of the dimerization. Fourth, previous studies [1] suggest that for localized intrinsic defects Coulomb effects typically drive the intragap levels towards the band edges, further reducing their depth. We therefore expect that none of these theoretical limitations will invalidate our ground state calculations of the interchain coupling strengths and their effects on the localized intrinsic defects.

With regard to applications of results to presently existing conducting polymers, we stress that our calculations have been made for perfectly ordered crystalline 3-D *trans*–$(CH)_x$. Any *extrinsic* defect is likely to produce gap states. At present, the chain lengths even in highly oriented, crystalline Durham/Graz *trans*–$(CH)_x$ are believed to be smaller than 40 C-H units, [12,24] and there is a high density of broken bonds. In the morphologically complex Shirakawa material, many chains are in a disordered environment near the fibril surface [25]; it is thus conceivable that optical excitations in particular in this material exhibit properties which are more characteristic of single-chains of *trans*–$(CH)_x$ than of the 3-D crystal. Our results strongly suggest that a number of experimental issues, including the interpretation of optical absorption spectra, require extensive further study. In this regard, recent discussions [26] of the non-solitonic and extrinsic nature of the picosecond photoconductivity in *trans*–$(CH)_x$ may prove illuminating.

We thank the Centers for Materials Science and Nonlinear Studies for computational support at Los Alamos National Laboratory and the Computational Sciences Division of the US DOE for computational support at the NMFECC at Livermore. This work has been supported also by the Fonds zur Förderung der wissenschaftlichen Forschung in Österreich.

References

[1] Excellent review articles may be found in *Handbook of Conducting Polymers*, edited by T.A. Skotheim (Dekker, New York, 1986).
[2] Yu. N. Gartstein and A. A. Zakhidov, Sol. State. Commun. **60**, 105 (1986).
[3] D. Emin, Phys. Rev. B **33**, 3973 (1986).
[4] P. M. Grant and I. P. Batra, J. Physique **44**, C3-437 (1983); Synth. Metals **1**, 193 (1979).
[5] J. Ashkenazi, E. Ehrenfreund, Z. Vardeny, and O. Brafman, Mol. Cryst. Liq. Cryst. **117**, 193 (1985).
[6] J. Ashkenazi, W. E. Pickett, B. M. Klein, H. Krakauer, and C. S. Wang, Synth. Metals **21**, 301 (1987).
[7] For a review, see J. Callaway and N. H. March, Solid State Physics **38**, (1984) 136.
[8] R. W. Jansen, O. F. Sankey, Phys. Rev. B **36**, 6520 (1987).
[9] J. Ihm, A. Zunger, M.L. Cohen, J. Phys. C **12**, 4409 (1979) .
[10] G. S. Painter, Phys. Rev. B **24**, 4264 (1981).
[11] Y.B. Moon, M. Winokur, A. J. Heeger, J. Barker, and D. C. Bott, Macromolecules **20**, 2457 (1987).
[12] H. Kahlert, O. Leitner, and G. Leising, Synth. Metals **17**, 467 (1987).
[13] P. Vogl and D. K. Campbell, Phys. Rev. Letters **62**, 2012 (1989), and to be published.
[14] We note that the optimization procedure yields the same structural parameters for both monoclinic structures.
[15] W.-P. Su, J. R. Schrieffer, and A. J. Heeger, Phys. Rev. B **22**, 2099 (1980); Phys. Rev. B **29**, 2309 (1984).
[16] B. Horovitz, Phys. Rev B **12**, 3174 (1975); D. Baeriswyl and K.Maki, Phys. Rev. B **28**, 2068 (1983).
[17] G. F. Koster and J. C. Slater, Phys. Rev. **95**, 1167 (1954) and *ibid.*, 96, 1208 (1954).
[18] D. K. Campbell and A. R. Bishop, Phys. Rev. B **24**, 4859 (1981).
[19] S. Phillpot, D. Baeriswyl, A. R. Bishop, P. Lomdahl, Phys. Rev. B **35**, 7533 (1987).

[20] A. Svane and O. Gunnarson, Phys. Rev. B **37**, 9919 (1988).
[21] S. B. Trickey, F.R. Green, Jr., and F.W. Averill, Phys. Rev. B **8**, 4822 (1973).
[22] We find this self energy (the "scissors operator" [23]) to be given by $\Sigma_{el} = 0.74$ eV for *trans*-$(CH)_x$. It does not enter the ground state calculations of intrinsic defects in this work, however; for a discussion of these subtleties, see Reference 23.
[23] R. W. Godby, M. Schlüter, L. J. Sham, Phys. Rev. B **37**, 10159 (1988); M. S. Hybertsen, S. G. Louie, Phys. Rev. B **34**, 5390 (1986).
[24] B. Ankele, G. Leising, and H. Kahlert, Solid State Commun. **62**, 245 (1987).
[25] T. Ito, H. Shirakawa, S. Ikeda, J. Polymer Sci. Polymer Ed. **12**, 11 (1974).
[26] H. Bleier, K. Donovan, R.H. Friend, S. Roth, L. Rothberg, R. Tubino, Z. Vardeny, and G. Wilson, Synthetic Metals **28** (1988).

Polarons and Bipolarons in Conducting Polymers: The Role of Disorder and Coulomb Interaction

D. Baeriswyl [1]

Centre d'Etudes Nucléaires, Département de Recherche Fondamentale/ Service de Physique/Groupe DSPE, 85X, F-38041 Grenoble Cedex, France

Abstract. It is shown that in any single-particle theory bipolarons (provided that they exist) are more stable than polarons. Correspondingly, the Coulomb interaction has to be taken into account for obtaining a Curie susceptibility, at least in equilibrium. On the other hand, disorder appears to be essential for stabilizing polarons and bipolarons as self-trapping is likely to be suppressed in a perfect three-dimensional crystal. A simple model including both disorder and Coulomb interaction yields a magnetic susceptibility consisting both of a Curie and a Pauli term.

1. Introduction

It has become routine to interpret experimental data on conducting polymers in terms of polarons and bipolarons. (Pairs of topological solitons in an idealized trans-polyacetylene chain can be considered as bipolarons with vanishing confinement ; we don't consider here the possible role of individual solitons). Typically, weakly doped polymers are found to have charge carriers with spin ("polarons"), whereas results on heavier doped systems often require the coexistence of "polarons" and spinless charge carriers ("bipolarons").

The models used for interpreting optical and magnetic experiments are usually effective single-particle Hamiltonians such as the Su, Schrieffer, Heeger model (SSH) [1] or Valence Effective Hamiltonians (VEH) [2]. It is the purpose of this paper to show that these models are too limited to be applicable to real systems and that the combined effects of disorder and Coulomb interaction are essential for understanding qualitatively the evolution of electrical and magnetic properties upon doping.

2. The relative stability of polarons and bipolarons in single-particle theories

Suppose that the ground state of the undoped system is a semiconductor (or insulator), i.e. the filled valence band is separated from the empty conduction band by a gap 2Δ. In a single-particle theory, which we discuss in this

[1] Permanent address : Institut für Theoretische Physik, ETH-Hönggerberg, CH-8093 Zürich

Springer Series in Solid-State Sciences, Vol. 91
Electronic Properties of Conjugated Polymers III
Editors: H. Kuzmany · M. Mehring · S. Roth

section, the gap may arise as a simple band-structure effect (as, e.g., in the Hückel model for polyparaphenylene), or from the Peierls mechanism (as in the SSH model of trans-polyacetylene), or due to both effects (generalized SSH model for cis-polyacetylene [3]). The addition of a single electron to the system can result in a self-trapped state or polaron, where the electron occupies a level shifted into the gap as a result of lattice relaxation. Similarly the addition of two electrons can yield a self-trapped pair state or bipolaron where both electrons occupy the same level in the gap. We show now that in any single-particle theory the bipolaron energy is lower than the energy of two polarons.

We notice first that any polarisation of the lattice together with the resulting shifts of the electronic energy levels require a positive energy δE, otherwise the system would not have been in its ground state originally. Adding an electron to the undistorted system, i.e. to the bottom of the conduction band, costs an energy Δ (we take the zero of energy at midgap). This additional electron can provoke a lattice polarisation which in turn lowers the level of the added electron to $\epsilon < \Delta$. This self-trapped state or polaron requires the energy

$$E_p = \epsilon + \delta E, \qquad (\delta E > 0) \ . \tag{1}$$

We repeat now the same steps for a second electron. First, we add it to the lowest available level, ϵ, keeping the lattice fixed. Therefore the total energy required for adding the two electrons is

$$E_2 = 2\epsilon + \delta E \ . \tag{2}$$

Since any subsequent relaxation can only lower the total energy, the formation of a bipolaron requires an energy

$$E_{bp} \leqslant 2\epsilon + \delta E < 2E_p \ . \tag{3}$$

This completes the proof that the bipolaron is the preferred state of any single-particle theory as compared to two polarons.

3. The role of disorder and Coulomb interaction

The proof for the higher stability of bipolarons given above does not depend on the crystallinity of the system or any type of disorder as long as the model is described by a single-particle Hamiltonian. In fact, it has been recently recognized that ideal crystallinity is not very favorable for self-trapping [4,5]. The reason is that the coherent three-dimensional delocalization of the added charge is likely to yield a lower energy state than a polaron on a single chain which does not profit from interchain bonding. Disorder easily destroys the transverse coherence and therefore stabilizes polarons and bipolarons. Impurities thus act as pinning or even nucleation centers for polaronic defects [6]. To a certain degree it is a matter of taste whether such a defect should be classified as a pinned polaron or a relaxed impurity state.

The arguments of Section 2 are however no more valid if Coulomb interactions are explicity included (and not simply in terms of renormalized single-particle parameters). The most important modification is an energy U representing the Coulomb interaction between two electrons occupying the same level. This energy depends on the extent ℓ of the localized wavefunction and thus has to be distinguished from the Hubbard parameter U_0 describing the interaction between two π-electrons on the same carbon atom. A simple estimate yields $U(\ell) = U_0/\ell$. In polyacetylene the effective U for the localized wavefunction associated with a soliton has been estimated to be of the order of 0.5 eV [7] which is certainly not negligible on the scale of typical binding energies (of the order of 0.1 eV). Adding U to Eq. (2) we realize that a delicate balance exists between lattice relaxation favoring bipolarons and Coulomb repulsion favoring polarons. The lattice relaxation effects are sometimes included within the effective U which can then assume either sign, positive U favoring polarons, negative U bipolarons.

4. A simple model for the magnetic susceptibility

The general problem of interacting electrons in a random potential has been studied extensively in recent years [8,9]. Here we limit ourselves to a very simple model which requires the bandwidth to be determined mostly by the random potential (in contrast to the band formation due to the overlap of wavefunctions) and the Coulomb interaction to be short-ranged. The important quantities of the model are then the single-particle levels ϵ_ν and the Coulomb energies U_ν required for putting two electrons into the ν-th level where $\nu = 1, 2, \ldots, N_a$. Neglecting all other terms, in particular exchange interactions induced by the overlap of wavefunctions or Coulomb interactions between electrons occupying different levels, we are left with a simple statistical problem. All possible configurations can be represented in terms of the four possible occupancies of the individual levels $(n_{\nu\uparrow}, n_{\nu\downarrow}) = (0,0)$, $(1,0)$, $(0,1)$, $(1,1)$. In a magnetic field H the partition function for the grand canonical ensemble is given by

$$Z = \prod_\nu \left\{1 + \exp[-\beta(\epsilon_\nu-\mu-\mu_B H)] + \exp[-\beta(\epsilon_\nu-\mu+\mu_B H)] + \exp\left[-2\beta\left(\epsilon_\nu-\mu+\frac{1}{2}U_\nu\right)\right]\right\}, \quad (4)$$

where μ is the chemical potential and $\beta = 1/(k_B T)$. A similar model has been discussed some time ago by Kaplan et al. in the context of amorphous semiconductors [10]. For $T \longrightarrow 0$ the largest of the four weights in Eq. (4) survives. In the absence of a magnetic field the ground state configuration corresponds to doubly occupied levels for $\epsilon_\nu < \mu - U_\nu$, singly occupied levels for $\mu - U_\nu < \epsilon_\nu < \mu$ and empty levels for $\epsilon_\nu > \mu$ (we have assumed that the interaction terms U_ν are positive). The application of a magnetic field has a twofold effect. On the one hand the region of singly occupied levels is increased to $\mu - U_\nu - \mu_B H < \epsilon_\nu < \mu + \mu_B H$ at the expense of the region of doubly occupied levels which is decreased to $\epsilon_\nu < \mu - U_\nu - \mu_B H$. On the other hand the spins of electrons in singly occupied levels are aligned parallel to the field. Therefore we expect both a Pauli-type and a Curie-type contribution to the magnetic susceptibility. This is indeed the case as will be shown now.

The magnetic susceptibility is obtained from the partition function (4) through

$$\chi = k_B T\, \partial^2 (\ln Z)/\partial H^2$$
$$= 2\beta\mu_B^2 \sum_\nu \{\exp[\beta(\epsilon_\nu - \mu)] + 2 + \exp[-\beta(\epsilon_\nu - \mu + U_\nu)]\}^{-1} \qquad (5)$$

$(H = 0)$. For the special case $U_\nu = 0$ this expression reduces to the Pauli susceptibility,

$$\chi_p = -2\mu_B^2 \sum_\nu (\partial/\partial\epsilon_\nu)\{\exp[\beta(\epsilon_\nu - \mu)] + 1\}^{-1}$$
$$= -2 N_a \mu_B^2 \int d\epsilon\; g(\epsilon)\, (\partial/\partial\epsilon)\{\exp[\beta(\epsilon-\mu)] + 1\}^{-1}$$
$$\longrightarrow 2 N_a \mu_B^2 g(\mu) \quad \text{for } T \longrightarrow 0, \qquad (6)$$

where $g(\epsilon)$ is the density of states. This shows that the Pauli susceptibility does not require extended states. It also shows that in the absence of (explicit) Coulomb interactions there is no Curie susceptibility, as has been emphasized already by Kaplan et al. [10]. This is of course only true if equilibrium can be attained after the application of the field. It is in fact conceivable that the low mobility of electrons prevents them from redistributing completely. In this case some non-equilibrium Curie spins can appear.

We consider now a more general situation with $U_\nu = U > 0$ at low temperatures ($k_B T \ll U$ and $g(\epsilon)$ slowly varying on the scale $k_B T$). Proceeding as above one finds

$$\chi = \frac{N_s \mu_B^2}{k_B T} + N_a \ln 2\, \mu_B^2\, [g(\mu) + g(\mu - U)], \qquad (7)$$

where

$$N_s = N_a \int_{\mu-U}^{\mu} d\epsilon\; g(\epsilon) \qquad (8)$$

is simply the number of singly occupied states. The susceptibility consists of a Curie term and a Pauli contribution, as anticipated.

In this derivation it has been assumed that the chemical potential μ is not altered by the application of the magnetic field. This will not always correspond to the experimental situation ; sometimes it can be more appropriate to keep the number of electrons N_e constant. The corresponding transformation is easily performed on the basis of the thermodynamic relation $N_e = k_B T\, \partial \ln Z/\partial\mu$. We have also assumed that the density of states $g(\epsilon)$ is not changed as the chemical potential is varied. This is certainly not always fulfilled in doping experiments. In fact very often the addition of an electron can be associated with the addition of a level in the gap. For weak doping all these levels would be singly occupied (due to the Coulomb repulsion) and there would be neither doubly occupied nor empty states (in the gap). Therefore there would be no Pauli contribution to the susceptibility. In this sense the appearance of a Pauli term signals the overlap between dopant-induced levels and "pre-existing" levels, for instance the merge of an impurity band with the conduction or valence band. A further remark concerns the assumption of a constant U. As the chemical potential is varied, the characteristic localization length within the "active region" and therefore U will change. It follows that also the number of Curie spins, Eq. (8), will change. Finally it is obvious that a negative U will completely suppress the singly occupied states at low temperature and yield a vanishing susceptibility for $T \rightarrow 0$, as is easily confirmed on the basis of Eq. (5).

5. Concluding remarks

In this note we have pointed out that in equilibrium the existence of Curie spins or, loosely speaking, "polarons", is intimately related to the effects of Coulomb repulsion. A simple model including both disorder and Coulomb interaction has been analyzed and shown to yield a susceptibility consisting of a Curie- and a Pauli-term. Recent experiments on polythiophene have indeed revealed a coexistence of Curie- and Pauli-spins with a gradual change from Curie- to Pauli-behavior upon doping [11]. Similar results have been observed for other conducting polymers, and, interestingly, also close to the metal-insulator transition in phosphorus-doped silicon [12]. The picture developed here is to a certain extent phenomenological and a more microscopic approach will be required for justifying (or disproving) the underlying assumptions. In the meantime it can serve as a valuable tool for analyzing magnetic susceptibility measurements. An adaptation of the model to other experimental situations, in particular transport and optical absorption, could yield additional insight, as the evolution of optical or transport data is often closely connected to that of the magnetic susceptibility.

Acknowledgment. I have benefitted from numerous discussions with the members of the DSPE group, especially with M. Nechtschein, J.P. Travers, C. Odin and C. Menardo.

References

1. For a recent review see A.J. Heeger, S. Kivelson, J.R. Schrieffer and W.-P. Su : Rev. Mod. Phys. 60, 781 (1988)
2. For a review see J.L. Brédas : In Handbook of Conducting Polymers, Vol. 2, p. 859, ed. by T. Skotheim (Marcel Dekker, New York 1986)
3. S.A. Brazovskii and N.N. Kirova : Pis'ma Zh. Eksp. Teor. Fiz. 33, 6 (1981) [Sov. Phys. JETP Lett. 33, 4 (1981)]

4. Yu. N. Gartstein and A.A. Zakhidov : Solid State Commun. 62, 213 (1987); Erratum ibidem 65, ii (1987)
5. D. Baeriswyl and K. Maki : Synth. Met. 28, D507 (1989)
6. S.R. Phillpot, D. Baeriswyl, A.R. Bishop and P.S. Lomdahl : Phys. Rev. B 35, 7533 (1987)
7. D. Baeriswyl, D.K. Campbell and S. Mazumdar : Phys. Rev. Lett. 56, 1509 (1986)
8. B.L. Altshuler and P.A. Lee : Physics Today 41, No.12, p. 36
9. M.J. Hirsch : Comments Cond. Mat. Phys. 13, 249 (1987)
10. T.A. Kaplan, S.D. Mahanti and W.M. Hartmann : Phys. Rev. Lett. 27, 1796 (1971)
11. M. Schärli, H. Kiess, G. Harbeke, W. Berlinger, K.W. Blazey and K.A. Müller : Synth. Met. 22, 317 (1988)
12. M. A. Paalanen, S. Sachdev, R.N. Bhatt and A.E. Ruckenstein : Phys. Rev. Lett. 57, 2061 (1986) ; M.A. Paalanen, J.E. Graebner, R.N. Bhatt and S. Sachdev : Phys. Rev. Lett. 61, 597 (1988)

Superconductivity in Models of Conducting Polymers

J. Voit

Theoretische Physik 1 and BIMF, Universität Bayreuth,
D-8580 Bayreuth, Fed. Rep. of Germany

Abstract. We investigate a one-dimensional model of interacting electrons coupled to either acoustic or optical (intramonomer) phonons of arbitrary frequency at arbitrary band filling and discuss the competition between insulating charge density wave and superconducting ground states. Minimal requirements for the stabilization of superconductivity are: the presence of a high-frequency dispersionless vibration mode, doping of the polymer into the incommensurate regime (i.e. at least five percent), and well screened electronic interactions.

1 Introduction

The competition and cooperation of electron-electron and electron-phonon interaction in models for conducting polymers [1,2] has been a subject of debate over nearly a decade [3]-[7]. Traditional approaches find an enhancement of dimerization by weak electronic interactions within extended Hubbard models [3]. However, dimerization may also be decreased when more complete interaction potentials are considered [4]-[7]. While these results have been derived for adiabatic phonons, other work has emphasized the important role played by the frequency of the phonons involved, in controlling the interplay of electron-phonon and the different electron-electron interaction terms [8].

A thorough understanding of this interplay is of interest, since electron-electron interactions are (i) shown experimentally to be important in polyacetylene [$(CH)_x$] and (ii) believed to play an important role in high-temperature superconductivity and possibly also in organic superconductors. We therefore inquire if and under what conditions these interactions can not only destabilize dimerization but also trigger (singlet) superconductivity (SS) in polymers. Such a programme should, on the one hand, help to clarify the discrepancy between a recent theory suggesting that $(CH)_x$ should be a high-temperature superconductor [9] and experiment [1]; on the other hand, it might provide some guideline to the chemists' attempts to synthesize a superconducting polymer.

2 Model

Our model is given by the following Hamiltonian

$$H = H_0 + H_{int} + H_{SSH} + H_{intra} \ , \tag{1}$$

$$H_0 = -t_0 \sum_{n,s} [c^\dagger_{n+1,s} c_{n,s} + \text{H.c.}] \ , \tag{2}$$

Springer Series in Solid-State Sciences, Vol. 91
Electronic Properties of Conjugated Polymers III
Editors: H. Kuzmany · M. Mehring · S. Roth

$$H_{int} = \sum_{n,m,l,p,s,s'} V(n,m,l,p) c^{\dagger}_{n,s} c^{\dagger}_{m,s'} c_{l,s'} c_{p,s} + \delta\mu \sum_{n,s} c^{\dagger}_{n,s} c_{n,s} \ , \quad (3)$$

$$H_{SSH} = \alpha \sum_{n,s} [u_{n+1} - u_n][c^{\dagger}_{n+1,s} c_{n,s} + \mathrm{H.c.}] + \frac{K}{2} \sum_n (u_{n+1} - u_n)^2 + \sum_n \frac{p_n^2}{2M} \ , \quad (4)$$

$$H_{intra} = g \sum_{n,s} q_n c^{\dagger}_{n,s} c_{n,s} + \sum_n \left(\Pi_n^2 / 2m_r + (f/2) q_n^2 \right) \ . \quad (5)$$

$H_0 + H_{SSH}$ define the standard SSH [2] model where the nearest neighbour hopping integral t_0 is modulated via the coupling α of electrons to acoustic phonons $u_{n+1} - u_n$. K is the spring constant and M the ion mass. (3) is the most general parametrization of the electron-electron interaction which is taken non-zero only for neighbouring sites/bonds, yielding $U = 2V(0,0,0,0)$ and $V = 2V(0,1,1,0)$ of the extended Hubbard model, and two new terms $W = 2V(0,0,1,1)$ and $X = 2V(0,0,1,0)$ involving bond charges [4]. Importantly, definite inequalities exist among these parameters reflecting the screening range of the potential $V(n,m,l,p)$ [6,7]

$$U > V(> X > W) \qquad \text{and} \qquad U > X(> V \geq W) \quad (6)$$

for a badly (well) screened potential respectively (terms in parentheses are eventually negligible [7]). g in (5) is the coupling constant to an intramonomer vibrational (more generally bosonic) degree of freedom q_n characterized by a restoring force f and a reduced mass m_r.

We treat this model by applying a renormalization group (RG) analysis based on a continuum field theory, as outlined earlier [8,10]. The continuum theory correctly describes the asymptotic low energy properties of the original lattice model, thereby restricting to weak coupling; also, RG theory allows to determine accurately low energy properties, consistent with this limitation. The main advantage of the present method is that it allows to treat the interaction of electrons among themselves and with phonons of arbitrary frequencies in (1) on an equal footing. This is essential since it has been shown previously [8,10] that the phonon frequency ω_{ph} dramatically influences the competition between different interaction terms and thus the physical behaviour of the system. The tendency of the system to develop an ordered ground state of type j (e.g. charge density wave [CDW] or SS) is assessed by computing renormalized exponents α_j of correlation functions

$$R_{CDW}(2k_F + q) \sim | v_F q |^{-\alpha_{CDW}} \ , \qquad R_{SS}(q) \sim | v_F q |^{-\alpha_{SS}} \ . \quad (7)$$

Despite the simplicity of the method, solving the complete Hamiltonian (1) is exceedingly difficult since it involves competing interactions associated with three very different energy scales. We shall therefore discuss shortly in the following Section the case with acoustic phonons only (SSH-model) while the coupling of electrons to intramonomer degrees of freedom will be investigated in more detail in Section 4.

3 Acoustic Phonons

If only an acoustic phonon mode is present, and if U, V, W, X respect (6), a half-filled (undoped) system dimerizes for arbitrary ω_{ph} [8]. Doped into an incommensurate (IC) state, dimerization is lost at any $\omega_{ph} > 0$; even then, however, very strong short range bond order survives and SS flucutations are strongly depressed [10]. In contrast to others [9], we find that in the IC limit, electronic interactions (e.g. U) *decrease* SS strongly suggesting that a polymer as simple as $(CH)_x$ *cannot* become a superconductor [10]. Moreover, 3D coupling is inefficient in stabilizing SS in such situations [11]. These conclusions are valid also close to half-filling where a charged soliton lattice is formed: its band is either completely filled or empty and there is no SS instability.

The absence of strong SS fluctuations in models based on electron - acoustic phonon coupling is due to the very inefficient coupling at small wave vectors [12], while coupling at $q \approx 2k_F$ favouring CDW is good. The situation changes dramatically if one considers a dispersionless phonon mode which arises from an intramonomer vibration (or optical phonon). Such modes couple with comparable efficiency at small and large q [12]. Then SS is enhanced considerably and may even overcome CDW.

4 Intramonomer Vibrations

In order to exhibit more clearly the novel physics introduced by coupling of electrons to intramonomer modes (5) we drop (4) from our model in this Section. For a half-filled band, strong umklapp processes practically always yield a CDW ground state. However, in the IC limit umklapp scattering is absent resulting in a delicate competition between CDW and SS fluctuations which sensitively depends on the model parameters.

We first reduce H_{int} (3) to the Hubbard model. Figure 1 shows the evolution with U of the correlation function exponents α_{CDW} and α_{SS} at fixed electron-phonon coupling for different phonon frequencies. While both α_j decrease with increasing U (yielding, for large U, a spin density wave state) α_{SS} generally decreases faster than α_{CDW}. Interestingly, for low enough ω_{ph} (e.g. $\omega_{ph}/E_F = 0.1$) CDW fluctuations are always stronger than SS ones. However, with increasing ω_{ph}, there will be an increasing range of U-values where SS fluctuations dominate CDW ones. The dominant type of fluctuation in the 1D system will order in a 3D array of chains [11]. Notice that SS is rather suppressed than enhanced by repulsive interactions in 1D. This is also supported by independent work starting out from the large-U-limit [13].

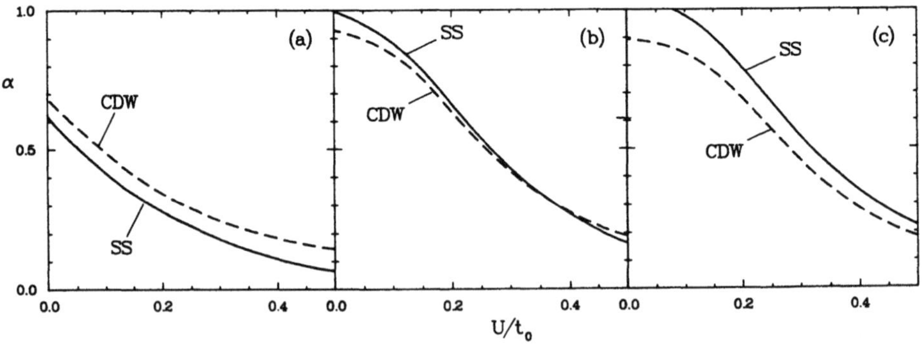

Fig. 1 Exponents α for SS and CDW correlation functions as a function of U. $g^2/\pi t_0 f = 0.25$, $\omega_{ph}/E_F = 0.1$ (a), 0.5 (b), and 1.0 (c)

The Hubbard model is only an approximation to more realistic electron-electron potentials and additional terms have to be taken into account according to (6). For bad screening, V may be important. V however strongly acts in favour of CDWs implying in general that badly screened electronic interactions are unfavourable for SS. For well screened interactions, next important to U is the X-term; it is of the same order as the hopping integral t_0 and must therefore be included in a consistent tight-binding approach [7]. X is not charge conjugation symmetric. It enhances CDWs in a less than half-filled band and SS in a more than half-filled one. This is interesting since both organic and most of the oxide superconductors are hole superconductors.

These results clearly exemplify the dramatic influence of ω_{ph} on the CDW/SS competition. The frequency dependence of α_{CDW} and α_{SS} (the 1D equivalent of the isotope effect) is given in Figure 2. Notice that the coupling constants have been scaled down by a factor 100 with respect to Fig. 1. Interestingly then, the system without electron-electron interaction seems to display SS at all phonon frequencies; possibly, for $\omega_{ph} = 0$ CDW and SS are degenerate. Adding a $U > 0$ increases the tendency towards CDW, in particular at low phonon frequencies. The physical origin of this isotope effect is very different from that of the standard BCS theory where it reflects the energy scale over which the attractive pairing interaction exists. Here, CDW and SS both are due to the interaction of electrons with the same phonons, and the isotope effect is due to the different scale dependence of the $q \approx 2k_F$ and $q \approx 0$ parts of this interaction and their competition in influencing the electronic properties. However, the variation of T_c in an array of coupled chains will not only reflect that of α_j but will depend also on the coupling mechanism involved.

Finally, the doping dependence of these effects, in particular the freezing out of umklapp scattering at light doping levels, can also be studied by the present method. While a value of about 5% is representative for suppression of umklapp scattering, establishment of SS clearly depends on additional parameters.

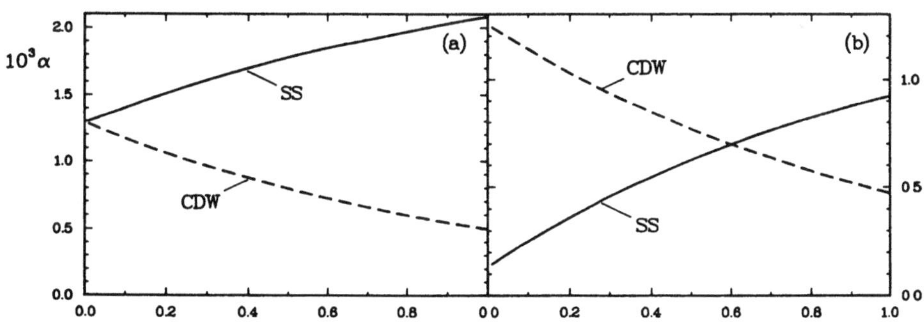

Fig. 2 Exponents α for SS and CDW correlation functions as a function of ω_{ph}/E_F. $g^2/\pi t_0 f = 0.0025$, $U/t_0 = 0$ (a) and 0.005 (b)

5 Conclusions

We have studied the competition between insulating and superconducting ground states a general 1D model of interacting electrons coupled to acoustic phonons or intramonomer vibrations in conducting polymers. We have found that coupling to acoustic phonons only is extremely unfavourable to superconductivity, thereby rationalizing its absence in $(CH)_x$. Coupling to a dispersionless vibrational mode is considerably more efficient; we have discussed the CDW/SS competition in such a model. We have found in particular that superconductivity is strongly favoured by good coupling to high-frequency intramonomer vibrations, heavy doping to the incommensurate limit, and sufficiently well screened electronic interactions so that only the U- and X-interactions are important. Specifically, if X is important, hole superconductors are more likely to be found than electron superconductors. Notice that any dispersionless bosonic excitation (e.g. charge transfer fluctuations) will give similar results.

We now comment on the full problem including both acoustic phonons and intramonomer vibrations. If the effects of both phonon modes were simply additive, the acoustic phonon would certainly destroy most if not all of the SS enhancement brought up by the intramonomer vibrations. This additivity is, however, far from clear. Specifically, their $q \approx 2k_F$ interactions with the electrons lead to a charge pile-up between/on the sites, respectively. Then it is conceivable that they destructively interfere in building up CDWs while SS might be favoured even more strongly.

References

[1] For reviews on recent progress, see e.g. Proceedings of the International Conference on Synthetic Metals (ICSM'88), Santa Fe, 1988, Synth. Met. **27 - 29** (1989).

[2] W.-P. Su, J. R. Schrieffer, and A. J. Heeger, Phys. Rev. Lett. **42**, 1698, (1979).

[3] The most recent work is G. W. Hayden and Z. G. Soos, Phys. Rev. B **38**, 6075, (1988); V. Waas, J. Voit, and H. Büttner, Synth. Met. **27**, A21, (1988); cf. also references therein.

[4] S. Kivelson, W.-P. Su, J. R. Schrieffer, and A. J. Heeger, Phys. Rev. Lett. **58**, 1899, (1987), and **60**, 72, (1988).

[5] C. Wu, X. Sun, and K. Nasu, Phys. Rev. Lett. **59**, 831, (1987).

[6] D. Baeriswyl, P. Horsch, and K. Maki, Phys. Rev. Lett. **60**, 70. (1988), and J. T. Gammel and D. K. Campbell, ibid. 71.

[7] A. Painelli and A. Girlando, Synth. Met. **27**, A15, (1988).

[8] J. Voit, Synth. Met. **27**, A33, (1988), Phys. Rev. Lett. **62**, 1053, (1989),

[9] G. T. Zimanyi, S. A. Kivelson and A. Luther, Phys. Rev. Lett. **60**, 2089, (1988).

[10] J. Voit and H. J. Schulz, Phys. Rev. B **37**, 10068, (1988).

[11] Yu. A. Firsov, V. N. Prigodin, and Chr. Seidel, Phys. Rep. **126**, 245, (1985).

[12] J. Voit and H. J. Schulz, Molec. Cryst. Liq. Cryst. **119**, 449, (1985).

[13] E. Tosatti and Lu Yu, Physica C **153-155**, 1253, (1988).

Investigation of the Electronic Structure of Polymeric Vinylene Derivatives of Pyrene

J.M. Toussaint [1], *F. Wudl* [2], *and J.L. Brédas* [1;2]

[1]Service de Chimie des Matériaux Nouveaux,
Département des Matériaux et Procédés, Université de Mons,
avenue Maistriau 21, B-7000 Mons, Belgium
[2]Institute for Polymers and Organic Solids, University of
California at Santa Barbara, Santa Barbara, CA 93106, USA

Abstract. Theoretical studies are reported on the geometric and electronic structures of two polymeric vinylene derivatives of pyrene: poly(1,6-pyrenylene vinylene) (P1,6PV) and poly(2,7-pyrenylene vinylene) (P2,7PV). The strong bandgap reduction (~1.3 eV) when going from P1,6PV to P2,7PV is shown to be due to the stabilization of quinoid contributions to the electronic structure in P2,7PV.

The Modified Neglect of Diatomic Overlap (MNDO) semiempirical method [1] was used to optimize the geometries of the pyrene molecule and dimers of P1,6PV and P2,7PV. The results presented in figure 1 indicate very similar geometries for the three systems. The influence of the vinylene linkages on the pyrene unit geometry is thus weak (only the adjacent bonds elongate from ~1.40 Å up to ~1.42 Å in P1,6PV and up to ~1.44 Å in P2,7PV).Note that the degree of bond-length alternation is almost identical for the two polymers.

These geometries are then used as input for Valence Effective Hamiltonian [2] (VEH) band structure calculations. The VEH electronic parameters are given in Table 1 for pyrene and for P1,6PV and P2,7PV. Note that the VEH polymeric ionization potential values are scaled down by 2.3 eV to take into account the solid-state polarization energy [3,5].

Table 1. Π-Π* transition and ionization potential of pyrene, P1,6PV, and P2,7PV. All values are in eV. Experimental values for pyrene (in parentheses) are taken from Ref. 4

	Π-Π* transition	IP
Pyrene	3.35 (3.83)	7.91 (7.41)
P1,6PV	2.97	5.38
P2,7PV	1.71	4.77

Springer Series in Solid-State Sciences, Vol. 91
Electronic Properties of Conjugated Polymers III
Editors: H. Kuzmany · M. Mehring · S. Roth

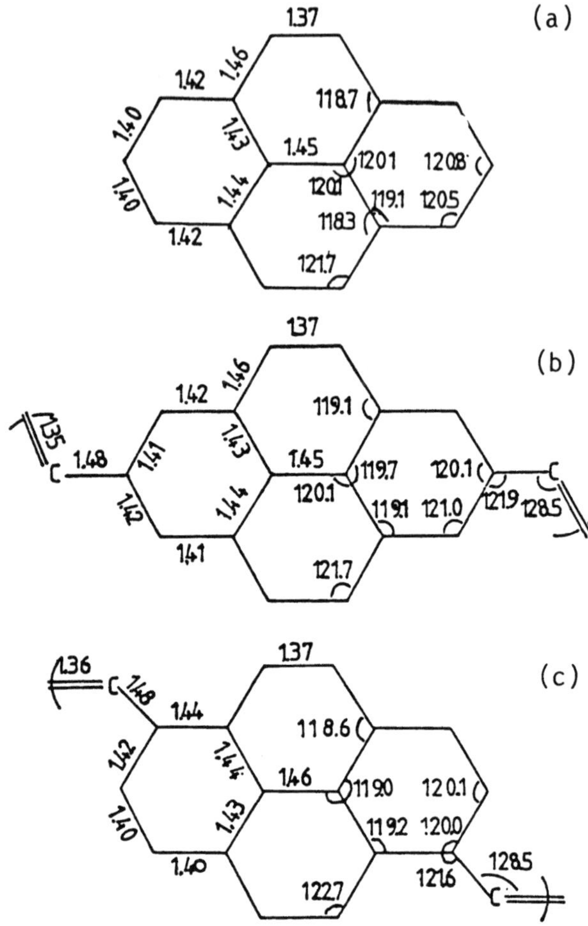

Fig. 1. MNDO-optimized geometric structures of (a) pyrene, (b) central part of the P1,6PV dimer, and (c) central part of the P2,7PV dimer. (Bond lengths in Å and angles in degrees; the systems are fully coplanar).

In order to rationalize the 1.3 eV Eg difference between P1,6PV and P2,7PV, we have examine the bonding-antibonding electronic patterns of the highest two occupied and lowest two unoccupied (LUMO and LUMO+1) electronic levels of pyrene, P1,6PV, and P2,7PV and analyzed in detail the influence of the vinylene linkages on these four levels of pyrene [6].

In the 1-6 mode of connection, the interactions between the pyrene HOMO and LUMO and the vinylene linkages are very weak and lead to dispersionless bands located at the same energy levels as in the molecule. However, strong interactions between the pyrene and vinylene moieties do occur for the HOMO-1 and LUMO+1 levels of pyrene. As a result, the HOMO and LUMO bands of P1,6PV mostly derive from the HOMO-1 and LUMO+1 levels of pyrene:

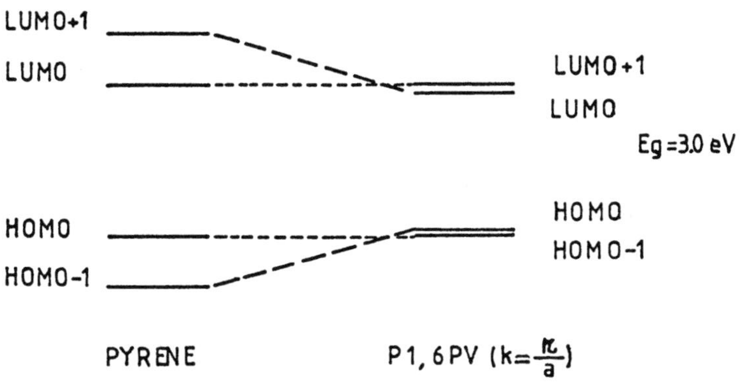

The situation is totally different for the 2-7 mode of connection. The electronic overlap between the vinylene units and the pyrene HOMO and LUMO is here very effective and leads to the P2,7PV HOMO and LUMO bands. (It is the interaction with the pyrene HOMO-1 and LUMO+1 levels which is here very weak, resulting in the appearance of two flat bands in the polymer). Quinoid contributions to the electronic structure are thus favored in the 2-7 compound [6] since the pyrene HOMO is destabilized and the pyrene LUMO is stabilized to give rise to the P2,7PV HOMO and LUMO levels, respectively:

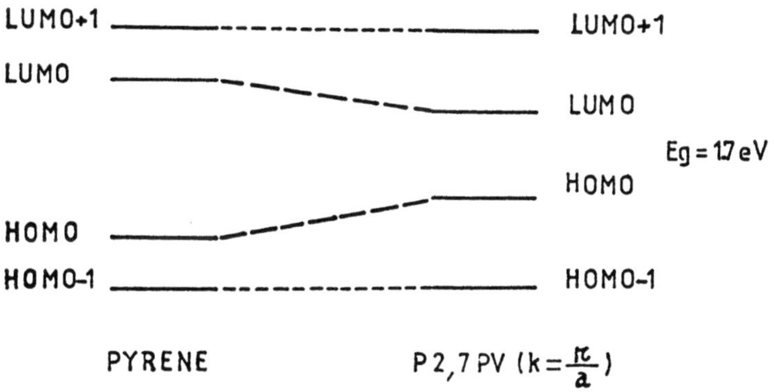

Acknowledgements

One of us (JLB) is most grateful to Prof. A.J. Heeger for making possible his stay at the UCSB Institute for Polymers and Organic Solids. The authors thank the University of Mons Computer Center for the use of the CCI facility and the Belgian Fund for Scientific Research (FNRS), IBM Belgium, and Facultés Universitaires Notre-Dame de la Paix for the use of the Namur SCF facility. This collaboration work has been sponsored by the NATO Division of Scientific Affairs under Research Grant 407/84.

References

[1] M.J.S. Dewar and W. Thiel, J. Am. Chem. Soc. 99, 4899 (1977); 99, 4907 (1977).
[2] J.M. André, L.A. Burke, J. Delhalle, G. Nicolas, and Ph. Durand, Int. J. Quantum Chem. Symp. 13, 283 (1979).
[3] J.L. Brédas, R.R. Chance, R. Silbey, G. Nicolas, and Ph. Durand, J. Chem. Phys. 75, 255 (1981); J.L. Brédas, R.L. Elsenbaumer, R.R. Chance, and R. Silbey, J. Chem. Phys. 78, 5656 (1983).
[4] E. Clar and W. Schmidt, Tetrahedron 35, 1027 (1979).
[5] N. Sato, K. Seki, and H. Inokuchi, J. Chem. Soc. Faraday Trans. 77, 1621 (1981).
[6] J.M. Toussaint, F. Wudl, and J.L. Brédas, J. Chem. Phys., to be published.

Quinoid–Aromatic Transition in Polythiophene-like Systems

J. Kürti [1] *and P.R. Surján* [2]

[1] Department of Atomic Physics, R.Eötvös University Budapest, H-1088 Budapest, Puskin u. 5-7, Hungary
[2] Chinoin Research Centre, H-1325 Budapest, P.O. Box 110, Hungary

Hückel calculations with geometry relaxation on long polythiophene (PT), polyisothianaphthene (PITN) and polyisonaphthothiophene (PINT) chains were performed. Results on energetics and geometry are presented.

1. Introduction

Considerable interest has been paid recently to the problem of conjugated polymers with small bandgaps [1-8]. It has been pointed out that design of such systems is possible by decreasing the bond length alternation of a polyacetylene-like backbone by means of heteroatoms and/or cyclic sidegroups [1,2,5,6]. In other words, a promising way to find a small bandgap polymer is to change systematically the structure of an initially aromatic polymer towards a quinoidal one or vice versa. An example of this is the PT - PITN - PINT series. PT itself is not a small bandgap polymer, E_g=2.1 eV experimentally [9]. Attaching one (PITN) or two (PINT) benzene rings to the heterocyclic backbone leads to a decrease of the gap. The experimental values are: 1.0 eV and 1.4 eV for PITN and PINT, respectively [3,7].

The question of the actual geometry of the above systems is not yet solved, however. To our knowledge there are no experimental data for the bond lengths of these polymers. Theoretically, contradictory results have been reported in the literature. For PITN, Brédas et al. predicted an aromatic structure [4] whereas Lee and Kertesz obtained a quinoidal geometry [5].

2. Method

We used a tight-binding Hückel Hamiltonian, augmented by a σ-core potential, accounting for geometry relaxation, according to Longuett-Higgins and Salem [10-13]:

$$H = \sum_{i,s} \alpha_i a^+_{i,s} a_{i,s} + \sum_{i<j,s} \beta_{ij}(a^+_{j,s} a_{i,s} + a^+_{i,s} a_{j,s}) + H_\sigma . \qquad (1)$$

The second summation runs over bonds. This Hamiltonian is diagonalized for finite long chains numerically. Electron-phonon coupling is introduced by the dependence of the resonance integral (β_{ij}) on bond length (r_{ij}):

$$\beta_{ij} = A * exp(-r_{ij}/B), \qquad (2)$$

Springer Series in Solid-State Sciences, Vol. 91
Electronic Properties of Conjugated Polymers III
Editors: H. Kuzmany · M. Mehring · S. Roth

and Coulson's linear relationship between bond length and bond order (P_{ij}):

$$r_{ij} = R_1 - (R_1 - R_2) * P_{ij}, \quad \text{where} \quad P_{ij} = \sum_n N_n * C^*_{ni} C_{nj}. \tag{3}$$

(N_n = 0,1 or 2 is the occupation number of the nth MO, C_{ij}-s are the LCAO coefficients.)

The above model should be solved self-consistently by relaxing the geometry. The single and double bond lengths (R_1 and R_2) are standardized values: 1.54Å / 1.33Å for C-C and 1.82Å / 1.71Å for C-S bonds. The A and B values in Eq.(2) were optimized for C-C to obtain 1.5 eV energy gap and 1.36Å / 1.45Å bond length alternation for a long polyacetylene chain (200 C atoms). The B value for the C-C bond obtained in the above manner is in agreement with the standardized orbital exponent of Hehre et al. [14]. This means that the resonance integral is in close connection with the corresponding overlap integral. On this ground the B value for C-S bond was determined using the standard exponent for S from [14]. All other parameters correspond to Streitwieser values [15]. The applied parameters are collected in Table 1.

Table 1. Parameters used. (Distances in Å , energies in eV.)

	R_1	R_2	$\beta(R_1)$	$\beta(R_2)$	α
C-C (C)	1.54	1.33	-1.626	-3.220	0.000
C-S (S)	1.82	1.71	-1.674	-2.564	-3.846

3. Results

Figure 1 shows the calculated energy band structure for the three polymers studied. For PITN and PINT a pair of localized midgap states (see below) are found. Ignoring these levels in calculating the band-band transition the following gap values are obtained: 1.97 eV (PT), 1.00 eV (PITN) and 1.37 eV (PINT). These values are in a good agreement with the experimental ones, cited previously.

Having used a model of finite chains we can monitor the changes in geometry along the oligomer. Apart from a few unit cells at the ends, the geometry of the monomers turned out to be uniform. This structure is aromatic for PT, while it is quinoidal for PITN and PINT (see Fig.2) in agreement with the results of Lee and Kertesz [5]. PITN has the smallest bond length alternation in the middle of the polyacetylenic backbone, as can be seen from Fig.2 . This explains why it has the smallest bandgap among these polymers.

In the quinoid case, if all the bonds between monomers were short ("double bonds"), two localized free radicals would appear at the ends of the chain. As a matter of fact, the distance between unit cells is long ("single bond") at the chain ends. It gradually shortens over a few monomers until it reaches the uniform value shown on Fig.2 . This variation in the bond length can be seen from Table 2. There is no such change in the case of PT. The aromatic-quinoid

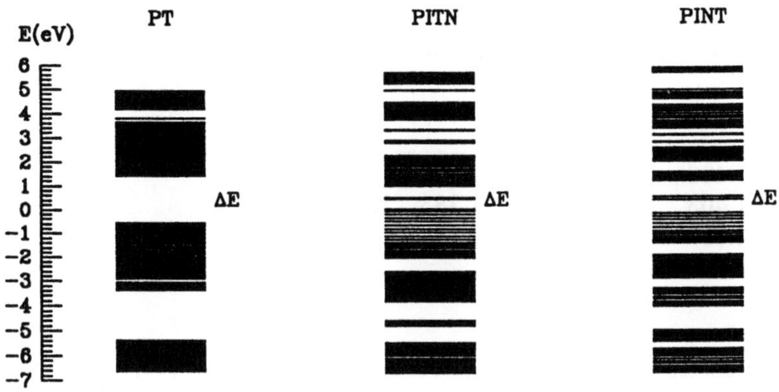

Fig. 1 Orbital energy levels (band structure) of thiophene, isothianaphthene and isonaphthothiophene oligomers, consisting of 40, 22 and 15 monomers, respectively

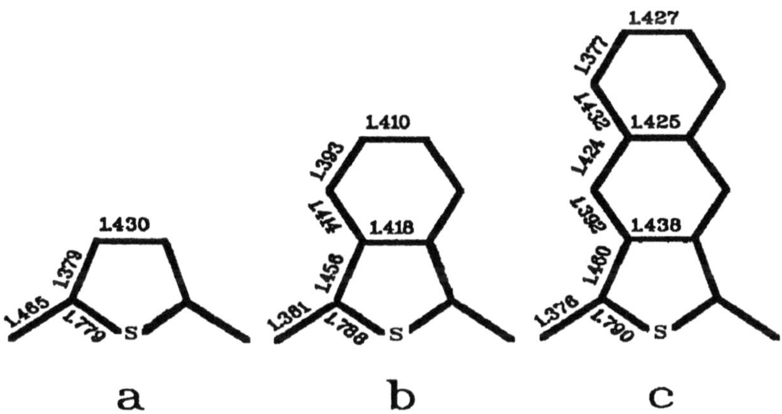

Fig. 2 Equilibrium geometry far from chain ends in PT(a), PITN(b) and PINT(c)

switch-over in PITN and PINT leads to some spreading out of the free radicals at the chain ends. Exactly these are the two midgap states as can be verified on the basis of the LCAO-coefficients from our calculations. It should be mentioned that this aromatic-quinoid transition within a PITN or PINT chain resembles the butatriene-acetylene structural change in the case of polydiacetylene [16].

If the chain is shorter than about 8-10 monomers, there is no possibility for developing the quinoid structure. Thus, the effect of the chain ends is to stabilize the aromatic geometry for short ITN and INT oligomers.

In summary, we have shown by Hückel calculations with geometry relaxation that the structures of PITN and PINT strongly depend on the chain length. There is an aromatic to quinoid switch-over in these materials if the oligomer is longer than ≈10 monomers. We think there is no contradiction between the results of refs. [4] and [5]. Lee and Kertesz calculated the geometry of PITN with periodic boundary conditions, whereas Brédas et al. performed the

Table 2. Variation in the bond lengths between monomers (in Å) along the chain for the three polymers. The numbering begins at the chain end

	1–2	2–3	3–4	4–5	5–6	6–7
PT	1.468	1.466	1.465	1.465	1.465	1.465
PITN	1.427	1.403	1.391	1.386	1.383	1.382
PINT	1.399	1.381	1.378	1.377	1.376	1.376

calculations for clusters not longer than tetramers [17]. Furthermore, our results are in agreement with the predictions of the Raman measurements of Wallnöfer et al. on PITN [18].

Acknowledgements

The authors thank W.Wallnöfer and J.L.Brédas for valuable discussions. This work was supported by the Hungarian Academy of Sciences under the project number MTA OTKA 64/1987.

References

1 J.L.Brédas, Springer Series in Solid-State Sciences **63**, 166, (1985)
2 J.L.Brédas, J.Chem.Phys. **82**, 3808, (1985)
3 F.Wudl, M.Kobayashi, N.Colaneri, M.Boysel and A.J.Heeger, Mol.Cryst.Liq.Cryst. **118**, 195, (1985)
4 J.L.Brédas, A.J.Heeger and F.Wudl, J.Chem.Phys. **85**, 4673, (1986)
5 Y.S.Lee and M.Kertesz, J.Chem.Phys. **88**, 2609, (1988)
6 M.Kertesz and Y.S.Lee, Synthetic Metals **28**, C545, (1989)
7 A.O.Patil, A.J.Heeger and F.Wudl, Chem.Rev. **88**, 183, (1988)
8 J.M.Toussaint, F.Wudl and J.L.Brédas, to be published in J.Chem.Phys.
9 T.C.Chung,J.H.Kaufman,A.J.Heeger and F.Wudl, Phys.Rev**B30**, 702,1984
10 H.C.Longuett-Higgins and L.Salem, Proc.R.Soc. London, **A251**, 172, 1959
11 M.Kertesz and P.R.Surján, Solid State Commun. **39**, 611, (1981)
12 P.R.Surján and H.Kuzmany, Phys.Rev. **B33**, 2615, (1986)
13 J.Kürti and H.Kuzmany, Springer Series in Solid-State Sciences **76**, 43, 1987
14 W.J.Hehre, R.Ditchfield, R.F.Stewart and J.A.Pople, J.Chem.Phys. **52**, 2769, (1970)
15 A.Streitwieser: Molecular Orbital Theory for Organic Chemists, Wiley, New York, (1961)
16 H.Sixl, Springer Series in Solid-State Sciences **63**, 240, (1985)
17 J.L.Brédas private communication
18 W.Wallnöfer, H.Kuzmany and E.Falques, this volume

The Effect of Helical Conformation on the Electronic Structures of σ- and π-Conjugated Polymers

C.X. Cui and M. Kertesz

Department of Chemistry, Georgetown University,
Washington, DC 20057, USA

Abstract. Many inorganic and organic polymers possess stable helical conformations. Torsional potential curves of polyethylene, polysilane, polygermane and polythiophene are calculated. The flatness of the torsional potential curves of σ -conjugated polysilane and polygermane could explain their thermochromic behavior. Three minima have been located on the energy surface of polyethylene: the most stable one is an all-trans conformation, and the other two are gauche. The effect of deviation from planarity of polythiophene on its band gap is discussed. The effects of side groups and doping on the conformation are analyzed as well.

1. INTRODUCTION AND METHOD

Great many polymers have a helical ground state, including some conducting polymers [1-7]. The concept of band theory has been extended by several groups to the case of a screw axis of symmetry which is present in every helix [8-13]. The effect of deviation from planarity is expected to be the most crucial for π-conjugated polymers, but generally the effect of conformational change on the electronic structure, and thereby on physical properties such as IR, Raman and UV-vis spectra and conductivity are expected to be important.

Recently we have developed calculational tools [13] to predict geometrical and electronic structures of helical polymers based on the MNDO (Modified Neglect of Diatomic Overlap) band theory approach [14]. The derivatives of the total energy of the helical polymer with respect to bond lengths, bond and dihedral angles as well as the translational vector can be calculated by this new version of MNDO solid state program, which is based on Stewart's MOSOL [15].

Therefore, it is now possible to perform full geometrical optimizations for helical polymers for the first time. In this communication, we present recent results of our calculations of geometrical and electronic structures of σ- and π-conjugated polymers.

It is well known that the MNDO approach produces reasonable geometries of organic and inorganic molecules and polymers without using vast amounts of computer time [14,16]. The parameters to describe the geometry of a helical polymer are illustrated in Fig.1. Because all operators of a screw axis form an Abelian group, there are only one-dimensional irreducible representations, which are complex roots of unity. In this way

Springer Series in Solid-State Sciences, Vol. 91
Electronic Properties of Conjugated Polymers III
Editors: H. Kuzmany · M. Mehring · S. Roth

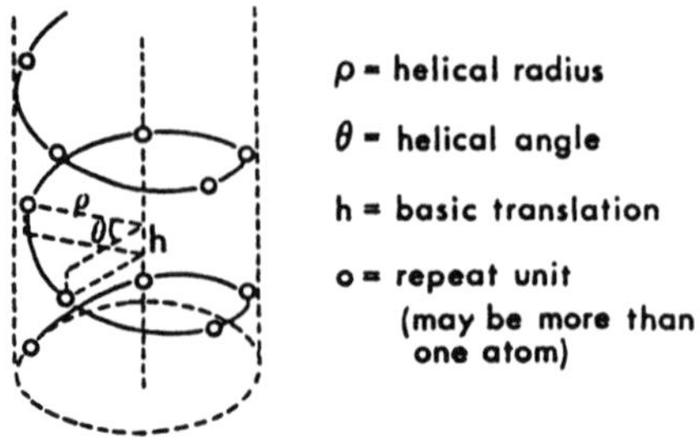

Fig. 1. Parameters to describe a helical polymer

the symmetry adapted linear combinations of atomic orbitals for helical polymer are formally very similar to usual Bloch functions [17]. Therefore, the Hartree-Fock equation and the formulas for the calculations of derivatives of the total energy of a helical polymer are similar to those for the translational symmetry. Further calculational details can be found elsewhere [13].

2. RESULTS AND DISCUSSION

Polyethylene has been extensively studied by various quantum chemical methods [11,18] such as ab initio and semi-empirical approaches due to the relative simplicity of its structure. Most of those studies focus on the all-trans conformation, and some of them are directed to the relative stability of the all-trans vs. gauche conformations. Fig. 2 shows the heat of formation of polyethylene per repeat unit as a function of the helical angle. Three minima have been located on the energy surface. The lowest one corresponds to the all-trans conformation, the other two to gauche conformations. The all-trans conformation is more stable than the gauche one by 0.83 kcal/mol, which is close to the value of 0.87 kcal/mol as calculated by ab initio calculation without full geometry optimization [11]. This is also in agreement with other known experimental and theoretical results [1].

Polysilane and Polygermane exhibit very unusual electronic spectra and are considered to be σ-conjugated polymers [19,20]. Experimental work on their derivatives reveals the strong conformational dependence of ultraviolet (UV) absorption. The

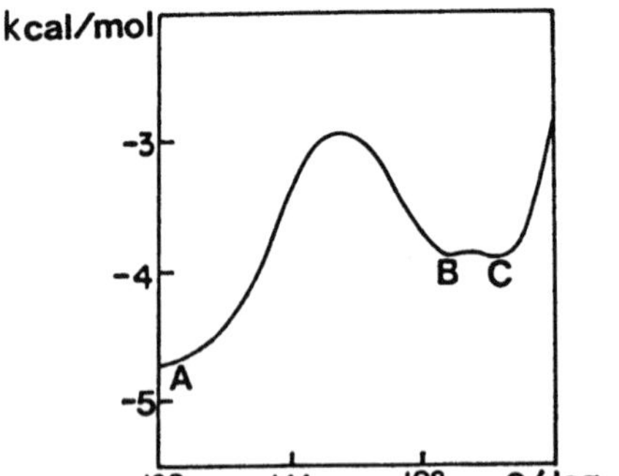

Fig. 2. Torsional potential curve of polyethylene

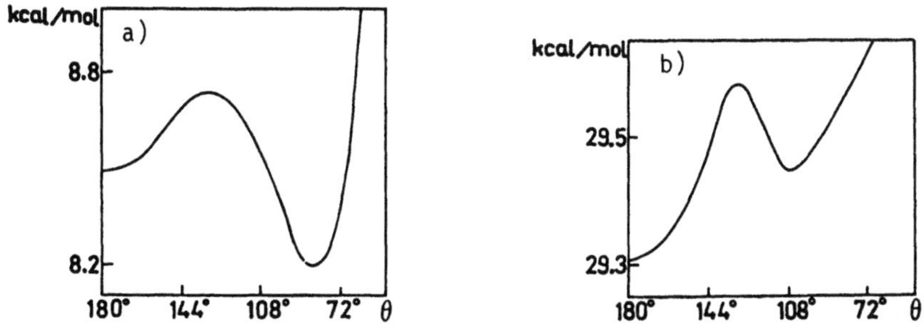

Fig. 3. Torsional potential curve of polysilane (a) and polygermane (b)

wavelength of maximum absorption of a range of disubstituted polysilanes and polygermanes increases with increasing length of the silicon backbone and with increasing substituent size. These materials have been subject to theoretical studies [21,22,23], all of which agree that the torsional potential is flat. Figure 3 presents the calculated torsional potential curves of polysilane and polygermane. For polysilane the gauche conformation is more stable than the all-trans one, the reverse is true for polygermane. Ab-initio calculations predicted that all-trans polysilane is more stable than the gauche one based on the rigid rotation model, wherein the total energy of the gauche conformation is calculated using the geometrical parameters of all-trans polysilane [21]. The most important feature of the torsional potential curves of Fig. 3. is their flatness. This could rationalize the dependence of their UV absorption on temperature. As the temperature changes, the ratio of all-trans and gauche conformations varies as well. Due to the difference between the band gaps of all-trans and gauche conformations, the UV absorption shows temperature dependence [20,21].

Polythiophene is one of the most important conducting polymers which have been synthesized up to now [24]. Torsional potential curve of polythiophene is shown in Fig. 4, indicating that it is a very flexible chain and the most stable form is a helix. This is in agreement with the behavior of polythiophene in solution [2] and electron diffraction studies on bithiophene

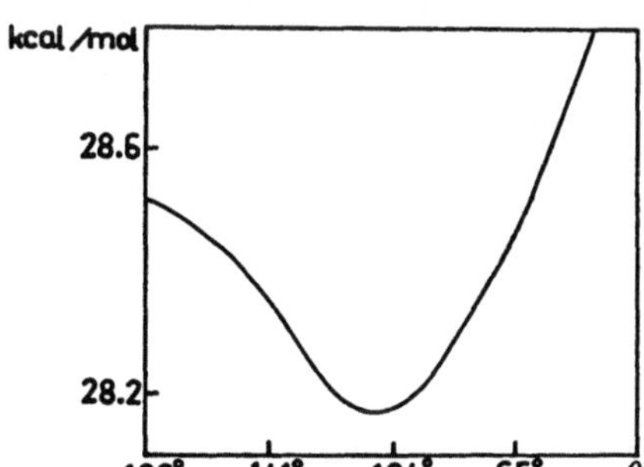

Fig. 4. Torsional potential curve of polythiophene

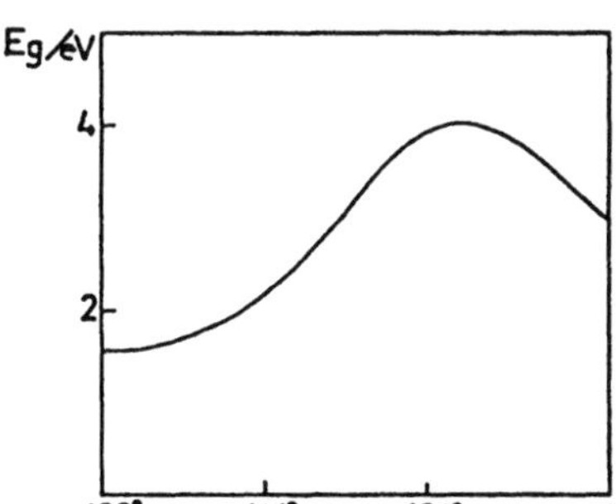

Fig. 5. The band gap as a function of dihedral angle (τ)

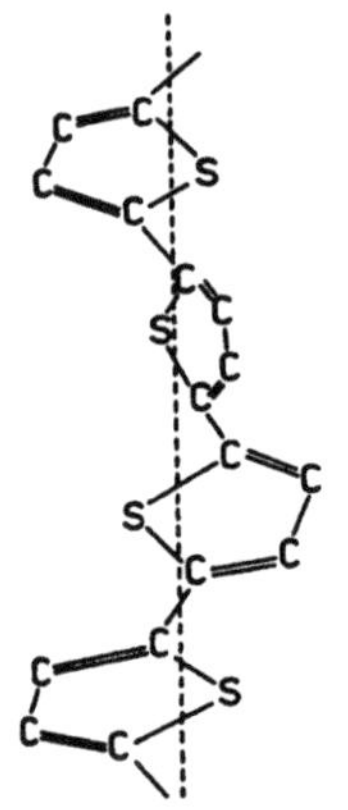

Fig. 6. The optimized geometry of polythiophene

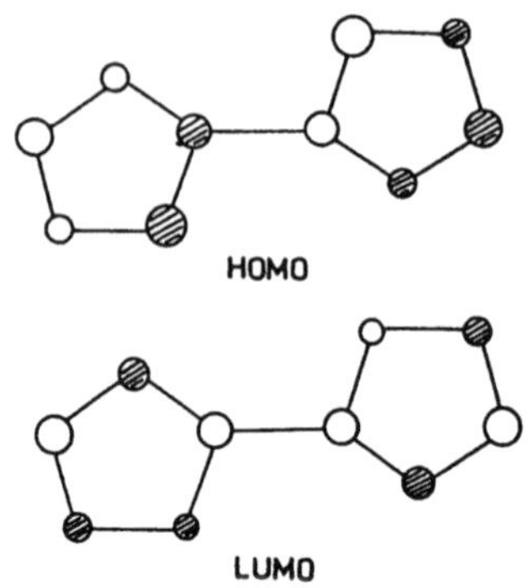

Fig. 7. The HOMO and LUMO of bithiophene

[25]. The existence of helical forms of conducting polymers has been known for some times. For example, the helical form of polyacetylene has been identified [6,7]. The effect of deviation from planarity of polythiophene on its band gap is shown in Fig. 5, indicating that small deviation form planarity will not lead to significant increase of the band gap. Fig. 6 shows the optimized geometry [26].

The effect of side group on the geometry of polythiophene is analyzed based on the calculations of bithiophene derivatives. One and two methyl groups attached to bithiophene will lead to non-planar structure more stable than plane ones by 3.75 and 8.90 kcal/mol, respectively. This indicates that the substituted polythiophene is more likely to be found in helical conformation [26].

The effect of doping on the geometry of polythiophene is discussed by charging bithiophene and its derivatives. For the unsubstituted bithiophene, upon charging (+2 or −2) it becomes planar. It is easy to understand this by inspecting the HOMO and LUMO as shown in Fig. 7. It can be seen that the orbitals of the HOMO on the atoms linking one ring to another is anti-bonding. If one takes out an electron from bithiophene, it will increases the interring bonding, and the planar geometry becomes favorable. Addition of electrons to the LUMO will also strengthen the bonding between the two rings, improving the relative stability of the planar form. Similar results have been obtained for one methyl group substituted bithiophene. For the two methyl group substituted bithiophene, the dihedral angle at the minimum energy shifts from about 95.0° for uncharged to about 160.0° for the charged one, indicating that even though the planar form becomes more favaroble upon doping, the most stable form is still helical due to side group to sulfur repulsions.

ACKNOWLEDGMENT: This work was supported by the National Science Foundation through a grant (DMR-8702148), and the Camille and Henry Dreyfus Foundation. We thank Dr. A. Karpfen for his coordinate transformation program.

REFERENCES

1. H. Tadokoro, Structure of Crystalline Polymers, (Clarendon: Oxford University, 1984).
2. J.P. Aime, F. Bargain, M. Schott, H. Eckhardt, G.G. Miller, and R.L. Elsenbaumer, Phys. Rev. Lett. **62**, 55(1989).
3. F. Garnier, G. Tourillon, J.Y. Barraud, and H. Dexpert, J. Mater. Sci. **20**, 2687(1985).
4. W.R. Salaneck, R. Erlandsson, J. Prejza, I. Lundstrom, and O. Inganas, Synth. Met. **5**, 125(1983).
5 R. Yang, K.M. Dalsin, D.F. Evans, L. Christensen, and W.A. Hendrickson, J. Phys. Chem. **65**, 23(1989).
6. F.S. Bates and G.L. Baker, Macromolecules, **16**, 1013(1983).
7. M.L. Elert and C.T. White, Phys. Rev. **B28**, 7387(1983).
8. A. Imamura and H. Fujita, J. Chem. Phys. **61**, 115(1974).
9. I.I. Ukrainskii, Theor. Chim. Acta **38**, 139(1975).
10. A. Blumen and C. Merkel, Phys. Stat. Sol. **B83**, 425(1977).
11. A. Karpfen and A. Beyer, J. Comput. Chem. **5**, 11, 19(1984).
12. J.M. Andre, D.P. Vercauteren, V.P. Bodart, and J.G. Fripiat, J. Comput. Chem. **5**, 535(1984).
13. C.X. Cui and M. Kertesz, J. Am. Chem. Soc. 1989 (in press).
14. M.J.S. Dewar and W. Thiel, J. Am. Chem. Soc. **99**, 4907(1977)
15. J.J.P. Stewart, QCPE BULL. **5**, 62(1985).
16. Y.S. Lee and M. Kertesz, J. Chem. Phys. **88**, 2609(1988).
17. F. Bloch, Z. Phys. **52**, 555(1928).
18. M. Kertesz, Adv. Quantum Chem. **15**, 161(1982).
19. R. West, J. Organomet. Chem. **300**, 327(1986).
20. R.D. Miller and R. Sooriyakumaran, J. Polym. Sci. Polym. Chem. Ed. **25A**, 111(1987).
21. H. Termae and K. Takeda, J. Am. Chem. Soc. **111**, 1281(1989).
22. J.V. Ortiz and J.W. Mintmire, J. Am. Chem. Soc. **110**, 4522(1988).
23. J.R. Damewood, Jr., and R. West, Macromolecules, **18**, 159(1985).
24. G. Tourillon, in Handbook of Condunctinq Polymer T.A. Skotheim Eds. (Marcel Dekker, Inc. New York, 1986)
25. A. Almenningen, O. Bastiansen, and P. Svendsas, Acta Chem. Scand. **2**, 1671(1958).
26. Further details see: C.X. Cui and M. Kertesz, to be published.

Part III

Optical Excitation

Phototransport in trans-Polyacetylene – A Short Review

H. Bleier

Max-Planck-Institut für Festkörperforschung,
Heisenbergstr. 1, D-7000 Stuttgart 80, Fed. Rep. of Germany

Abstract: In this paper a short review on the photoconductivity properties of trans-polyacetylene is given. The transport properties are discussed in detail and the possible excitation and recombination pathways are presented. I will show that the picosecond photoresponse observed in this material can be attributed to a transport, which is limited by impurities present in the samples, and is carried by non-solitonic carriers, which are either free electrons and holes or polarons.

1. Introduction

Electroactive conjugated polymers have been extensively investigated during the last decade, since the discovery that their conductivity can be changed by more than ten orders of magnitude upon doping [1]. Especially polyacetylene, $(CH)_x$, has become the prototype of this class of quasi-one-dimensional electron systems. Highly doped polyacetylene reaches metallic conductivity up to about 10^5 S/cm [2], whereas undoped trans-$(CH)_x$ is a semiconductor with a bandgap of ≈1.7 eV. Photoexcitation across this gap leads to the generation of charged carriers, as in conventional semiconductors. The relatively strong electron-phonon-coupling in 1d-systems, however, leads to a distortion of the lattice around these carriers and therefore they are often regarded as new quasi-particles with solitonlike or polaronlike properties. Recently some review articles on this topic have appeared [3,4].

In this paper, I want to concentrate on the transport properties and the recombination mechanism of the photogenerated charge carriers.

2. Transport Properties of Photogenerated Charge

When illuminating trans-$(CH)_x$ with a chopped cw-light source at a wavelength corresponding to above gap-excitation, a very slow response of the photocurrent (PC) can be observed, demonstrating that photoexcited charge is living up to seconds [5]. Illumination with short (25 ps) laser pulses, however, results in a photosignal, which relaxes on a time scale of $\tau_{1/2}$≈100 ps (Fig. 1) [6]. These two components, the so-called slow and the fast component behave in a different way, when looking on their temperature- and intensity-dependence. As shown in Fig. 2, the fast picosecond component is temperature independent from room temperature down 10 K. Moreover it is exactly proportional to

Springer Series in Solid-State Sciences, Vol. 91
Electronic Properties of Conjugated Polymers III
Editors: H. Kuzmany · M. Mehring · S. Roth

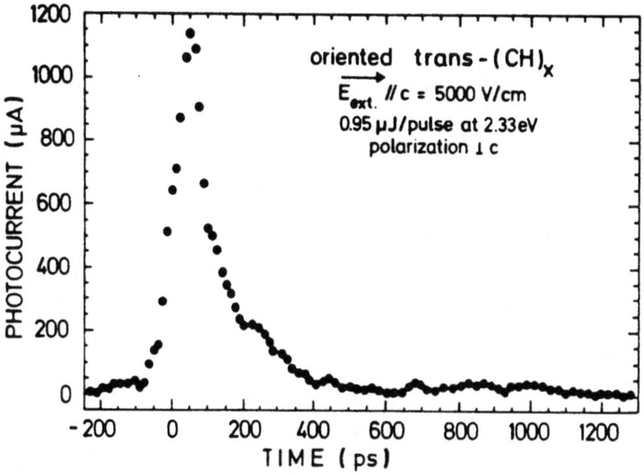

Fig 1 Fast component of the photocurrent in oriented trans-polyacetylene excited with 25 ps laser pulses at 2.3 eV with the applied electric field parallel to the chain orientation at T=300 K

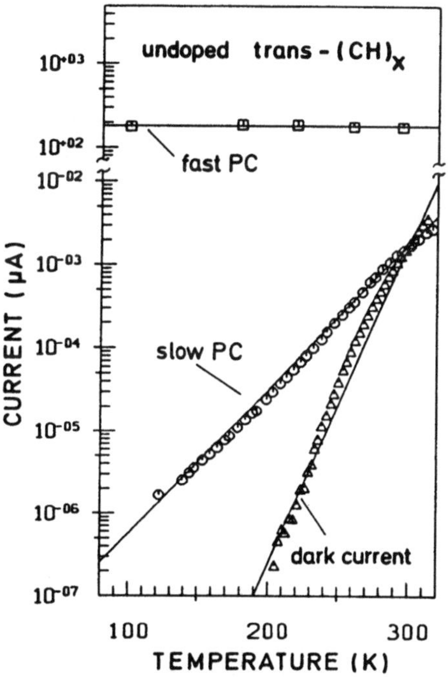

Fig. 2 Temperature dependence of the fast and slow PC-component and of the dark current. The fast component is temperature independent down to about 10K

the incident light intensity from 0.2 to 20 µJ/pulse (corresponding to an energy density of 0.5-50 mJ/cm² or a photon fluence of 10^{15}-10^{17} photons/cm²). The slow component shows a strong temperature dependence (see Fig. 2) and goes sublinear with the light intensity, with a clear tendency to saturation at high intensities.

For oriented samples two different types of anisotropy can be observed, an anisotropy with respect to the applied electric field and with respect to the polarization of the incident light. The anisotropy of the fast as well as the slow PC-component with respect to the applied electric field is found to be around 50 in the highly oriented Durham-samples [7] and somewhat lower for other types of oriented material [8] (depending on the degree of chain orientation). This clearly demonstrates that the charge transport is primarily along the chain direction and not perpendicular to it. Moreover, the temperature independent picosecond component of the photocurrent tells us that the photogenerated charge is supposed to move initially relatively free along the conjugated chain segments. In the hypothetical case of nearly defect-free infinitely long chains, this fast component would mainly be limited by phonons, resulting in an increasing photoresponse when lowering the temperature. From the observed experimental results we infer that the transport in the first hundred picoseconds is limited by the temperature independent defects on the chains and not by phonons. As shown in detail in Refs. 5 and 6 the migration distance l and the mobility μ of this fast component can be estimated to l=400 Å at 15000 V/cm and μ=2 cm^2/Vs for the oriented Durham-samples with applied electric field parallel to the chain direction.

When reaching the chain ends the carriers loose their mobility, the further motion can be explained by hopping transport in analogy to the dark conductivity, where several hopping models for undoped and lightly doped samples have been discussed [9,10]. The crossover from the high to the low mobility regime, can directly be seen in Fig. 3, by following the PC-decay over eight orders of magnitude in time.

At short times up to 10^{-6} s and at room temperature (solid line) the fast component dominates the photocurrent. For times longer than 10^{-5} s the slow component becomes more important.

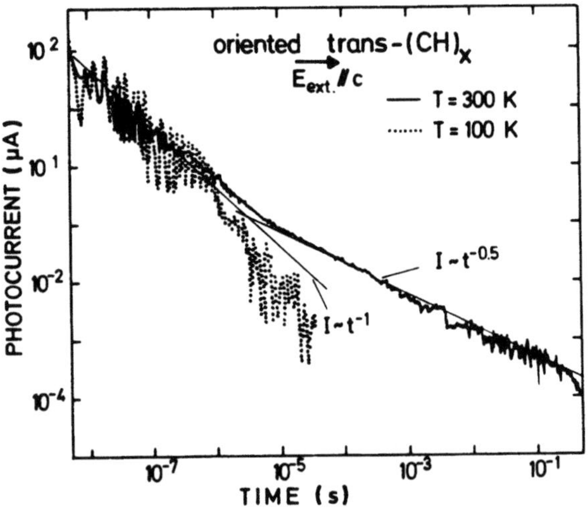

Fig. 3 Decay of the photocurrent after 500 ps excitation investigated from nanoseconds to seconds at 300 K (solid line) and at 100 K (dotted line)

At a lower temperature of 100 K (dashed line) the slow component is frozen out and only the temperature independent fast component remains. In the recently synthesized highly conducting material [11], with the above cited conductivity up to 10^5 S/cm for doped samples, the slow component already dominates at much shorter times of some hundred picoseconds [12]. This result suggests that the higher DC-conductivity in this material has its origin mainly in a more efficient hopping transport and not in a better motion of the free carriers (in the fast component), which would mean in a longer conjugation length.

3. Excitation and Recombination Mechanism

As briefly mentioned in the previous paragraph the PC in highly oriented samples is also anisotropic with respect to the polarization of the photoexciting light. Fig. 4 shows this anisotropy for a 1.7 μm thick sample at an excitation energy of 2.6 eV. Due to the anisotropic optical absorption and reflection [14], the measured data have to be corrected to a constant number of absorbed (and not incident) photons, which lowers the anisotropy. Nevertheless the photocurrent is still by a factor of 1.7 higher for light polarized perpendicular to the chain orientation. The reason why this anisotropy is relatively weak is discussed in detail in Refs. 5 and 13. The higher photoresponse for perpendicular polarization can be explained by an increased probability to separate the photoexcited electron hole pairs to different chains. Therefore we can already conclude that interchain separation is an important process for carriers contributing to the photoconductivity. In a second step we can compare the PC-data with results of photoinduced absorption (PA) experiments at the so-called low-energy (LE) peak at 0.45 eV, which can be attributed to charged excitations [15], i.e. charged solitons. This "midgap"-peak has been studied by time resolved IR-spectroscopy [16]. What turns out is that this excitation is formed almost instantaneously within the experi-

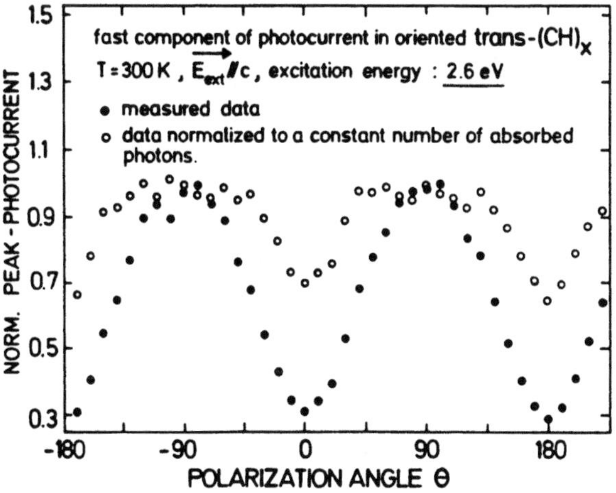

Fig. 4 Anisotropy of the fast PC-component with respect to the polarization of the incident light. θ is the angle between the polarization and the chain direction. Excitation energy is 2.6 eV

mental resolution of 0.4 ps. Most of these photoexcited solitons decay on a time scale nearly as fast as their creation of 0.5 to 1 ps. A relatively weak tail remains for times longer than 300 ps if the pump beam is polarized perpendicular to the chain direction. If the carriers giving rise to the photocurrent are the same carriers, i.e. the solitons, which are seen spectroscopically, they should behave in the same way. As shown in Ref. 13 the amplitude of the tail in the PA-signal at 0.45 eV goes strongly sublinear with the intensity of the incident pump beam. In contrast we have shown that the picosecond photoresponse (fast component) remains linear with intensity up to a pump fluence 20 times as high. Therefore it can be concluded that the carriers responsible for the fast PC-component are not the solitons seen in the PA-experiments. Since they are not solitons, we believe that they are all interchain-carriers, which have to be free electrons and holes or polarons. These carriers can then undergo one (or all) of the steps drawn on the right side of Fig. 5. Electrons and (or) holes may form like-charged pairs of solitons, the same can happen with the polarons. Another possibility is that neutral solitons and charged polarons can convert into isolated charged solitons. After such a conversion the carriers are trapped and contribute only to slow PC-component. All intrachain-solitons, formed by the Su-Schrieffer-mechanism [17] and giving rise to the superfast PA-response, decay geminately or convert into bound neutral soliton pairs as proposed by Kivelson and Wu [18] (left side of Fig. 5) and do not contribute to the photocurrent.

Some more information about the different possibilities of charge conversion, shown on the right part of Fig. 5, can hope-

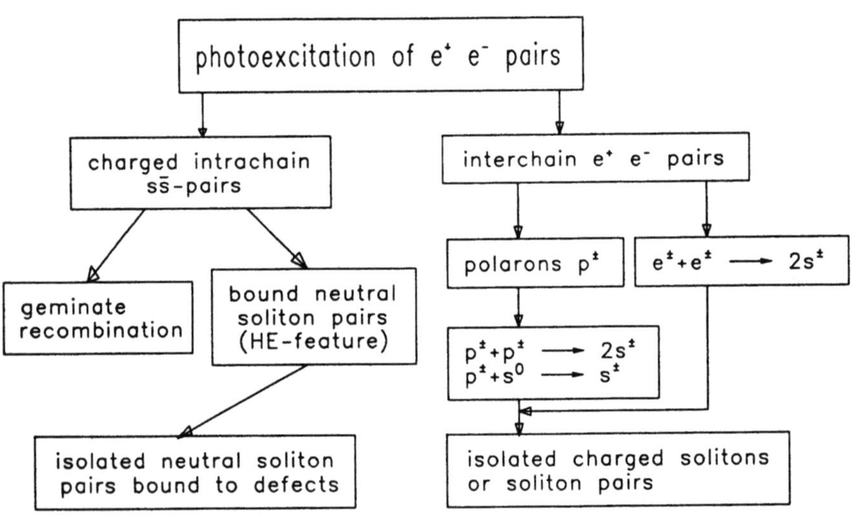

Fig. 5 Possible scenario for the recombination of photoexcited electron-hole pairs in trans-$(CH)_x$. The left part (neutral excitations) has been proposed by Kivelson and Wu [18]. The right side shows several possibilities for the conversion of interchain-excited e^+-e^--pairs

fully be obtained by transient photocurrent investigations with an increased time resolution down to ≈1 ps.

Acknowledgement

I want to thank all colleagues from Polymer Hill for their help with the manuscript and valuable discussions. G. Leising I want to thank for providing the highly oriented samples. Financial support from the Volkswagenstiftung and the DFG (SFB 329) is gratefully acknowledged.

References

1. C.K. Chiang, C.R. Fincher, Y.W. Park, A.J. Heeger, H. Shirakawa, E.J. Louis, S.C. Gau, A.G. MacDiarmid, Phys. Rev. Lett. 39, 1098 (1977)
2. Th. Schimmel, W. Rieß, J. Gmeiner, G. Denninger, M. Schwörer, H. Naarmann, N. Theophilou, Solid State Commun. 65, 1311 (1988)
3. S. Roth, H. Bleier, Advances in Physics 36, 385 (1987)
4. A.J. Heeger, S. Kivelson, J.R. Schrieffer, W.P. Su, Rev. Mod. Phys. 60, 781 (1988)
5. H. Bleier, S. Roth, Y.Q. Shen, D. Schäfer-Siebert, G. Leising, Phys. Rev. B38, 6031 (1988), and references therein
6. H. Bleier, S. Roth, H. Lobentanzer, G. Leising, Europhys. Lett. 4, 1397 (1987)
7. J.H. Edwards, W.J. Feast, Polymer 21, 595 (1980)
8. R. Tubino, R. Dorsinville, A. Walser, A. Seas, R.R. Alfano, Synthetic Metals 28, D175 (1989)
9. S. Kivelson, Phys. Rev. Lett. 46, 1344 (1981)
10. K. Ehinger, S. Roth, Phil. Mag. B53, 301 (1986)
11. H. Naarmann, N. Theophilou, Synthetic Metals 22, 1 (1987)
12. S.D. Phillips, A.J. Heeger, Phys. Rev. B38, 6211 (1988)
13. H. Bleier, K. Donovan, R.H. Friend, S. Roth, L. Rothberg, R. Tubino, Z. Vardeny, G. Wilson, Synthetic Metals 28, D189 (1989)
14. G. Leising, Phys. Rev. B38, 10313 (1988)
15. Z. Vardeny, J. Orenstein, G.L.Baker, Phys. Rev. Lett. 50, 2032 (1982)
16. L. Rothberg, T.M. Jedju, S. Etemad, G.L. Baker, IEEE QE-24, 311 (1988)
17. W.P. Su, J.R. Schrieffer, Proc. Natl. Acad. Sci. 77, 5626 (1980)
18. S. Kivelson, W.K. Wu, Phys. Rev. B34, 5423 (1986)

Photomodulation and Photoluminescence Spectroscopy of Poly(3-methylthiophene)

J. Poplawski and E. Ehrenfreund

Department of Physics and Solid State Institute,
Technion–Israel Institute of Technology, Haifa 32000, Israel

Abstract. In this paper we discuss optical data of photoexcited neutral and doped poly(3-methylthiophene) (PMT). These data include: photoinduced absorption (PA) of intra-gap electronic and infrared active vibrational transitions characteristic of charged defects and photoluminescence (PL) due, probably, to an exciton.

1. Introduction

PMT belongs to the class of conjugated polymers with a non-degenerate ground state symmetry [1, 2]: the aromatic form has a lower total energy than its quinoid counterpart [3]. The low lying electronic excitations consist, therefore, of relatively short regions of quinoid-like structure in an otherwise perfectly aromatic polymer [4]. Depending upon the charge of the defect, the formation of self localized polarons and bipolarons is expected upon doping or photoexcitation [2]. Doping induced polarons were observed in a number of non-degenerate ground state conjugated polymers. They are characterized by three subgap electronic absorption bands, a set of infrared active vibration (IRAV) lines and strong electron spin resonance (ESR) absorption. Bipolarons induced by doping showed only two subgap electronic transitions and no ESR absorption [5]. Photogenerated defects in these polymers also show subgap absorption bands and IRAV. In poly(thiophene), for instance, two PA bands were observed, indicating photogenerated bipolarons [6].

Another feature of these polymers is the strong PL, below the absorption threshold [4, 7]. The PL is accompanied by strong phonon sidebands whose frequency is similar to the strongest A_g Raman active intrachain vibrations, indicating thus an intrachain exciton.

Springer Series in Solid-State Sciences, Vol. 91
Electronic Properties of Conjugated Polymers III
Editors: H. Kuzmany · M. Mehring · S. Roth

2. Photoinduced Absorption

Fig. 1a shows the PA spectrum of neutral PMT [7]. There are two well defined asymmetric PA bands, with peaks at 0.4 (LE) and 1.3 eV(HE), which are assigned to transitions from band states to the two gap states (inset, Fig. 1) associated with the photoinduced defect. The HE band features a shoulder at 1.6–1.8 eV. The zero crossing at 1.9 eV is coincident with the zero phonon PL line (see below). Below 0.25 eV a set of photogenerated IRAV lines appear [7] showing that the defect thus produced is charged. To establish the connection between the different features of the PA, we show in Fig. 2 the dependence of each of the bands upon the pump intensity, I_L. It is seen that a sublinear dependence I_L^α with $\alpha \simeq 0.6$, is observed for all the PA bands. The temperature dependence of the PA shows that all the bands decrease by approximately the same factor between 15 K and 270 K, but the HE band decreases faster at low temperatures.

Fig. 1b shows the PA of heavily doped PMT. Comparing this spectrum to the undoped case, we see that the PA is substantially reduced, featuring now very broad PA bands, one around $\simeq$ 0.6 eV and one extending from $\simeq$ 1.2 eV to $\simeq$ 1.9 eV, which may perhaps be the sum

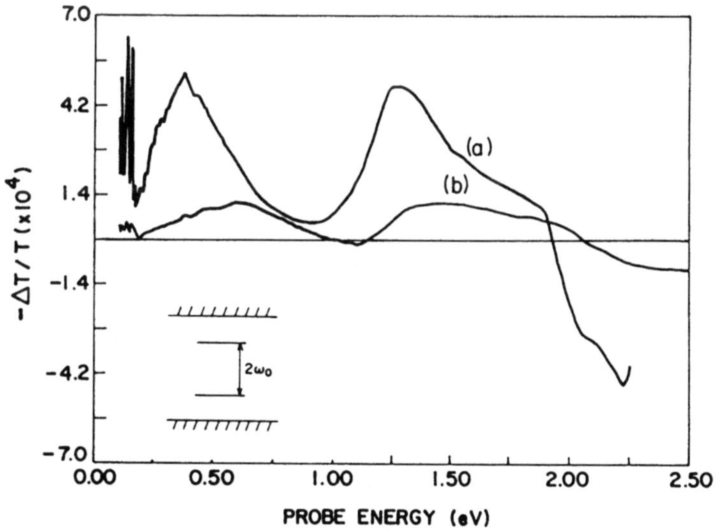

Fig. 1. The PA spectra of PMT at 15 K: *a*, neutral; and *b*, with 33 *at*% ClO_4^- doping. The inset shows the energy level scheme for a polaron or a bipolaron excitation

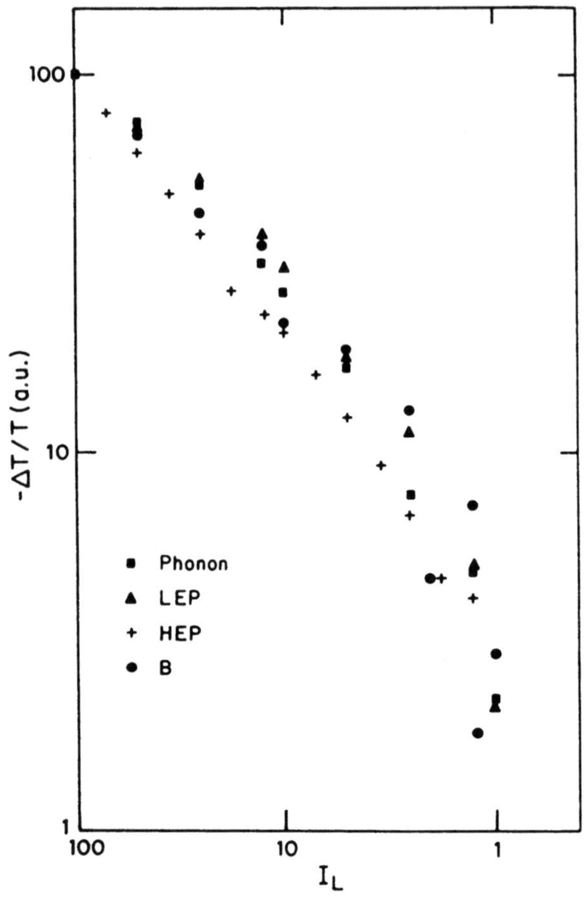

Fig. 2 A log–log plot of $(-\Delta T/T)$ vs. the pump intensity I_L

of two unresolved bands at 1.4 eV and 1.7–1.8 eV. The zero crossing is shifted to 2.02 eV, indicating higher band edge.

An independent RRS and IRAV analysis [7], within the amplitude mode formalism, gave the value $\gamma \simeq 0.27$ for the confinement parameter $\gamma = \Delta_e/\lambda\Delta$ (λ is the dimensionless electron phonon coupling parameter and $2\Delta_e$ is the extrinsic contribution to the gap 2Δ). Within the Fesser–Bishop–Campbell model [2], this γ corresponds to $\omega_0 = 0.4\Delta$ ($2\omega_0$ is the interlevel spacing, Fig. 1(inset)) if the charged defect is a bipolaron and $\omega_0 = 0.8\Delta$ if it is a polaron. Thus, with $2\Delta \simeq 2.2$ eV for PMT [7], a strong absorption at $2\omega_0 \simeq 1.8$ eV should have been observed for photoinduced polarons. This is not the case for neutral PMT, for which the appearance of two well defined PA peaks suggests that bipolarons are the energetically favored charged excitations.

This assignment is supported by the value of the ratio $\omega_0/\Delta = (1.3 - 0.4)/2.2 = 0.4$, as expected for bipolarons in PMT. We should remark, however, that the final assignment can be done using LESR, as was previously done for poly(thiophene) [6].

3. Photoluminescence

The PL spectrum of PMT at 15 K is shown in Fig. 3. The emission spectrum shows well resolved phonon sidebands having energy spacing of $\simeq 175\ meV$, with the zero phonon line at 1.90 eV. The sharp lines at 2.420 and 2.238 eV are the RRS first and second order lines of the strong stretching vibration of the carbon backbone (1469 cm^{-1}). The observation of well resolved phonon sidebands with energy spacing which is nearly identical to the intrachain stretching vibration indicates an intrachain exciton strongly coupled to the carbon backbone.

Treating the vibronic structure in the weak coupling limit [8], we write the total emission strength as

$$F(E) = \sum_{\nu} exp(-S)(S^{\nu}/\nu!)F_0(E - \nu\hbar\omega) \ , \qquad (1)$$

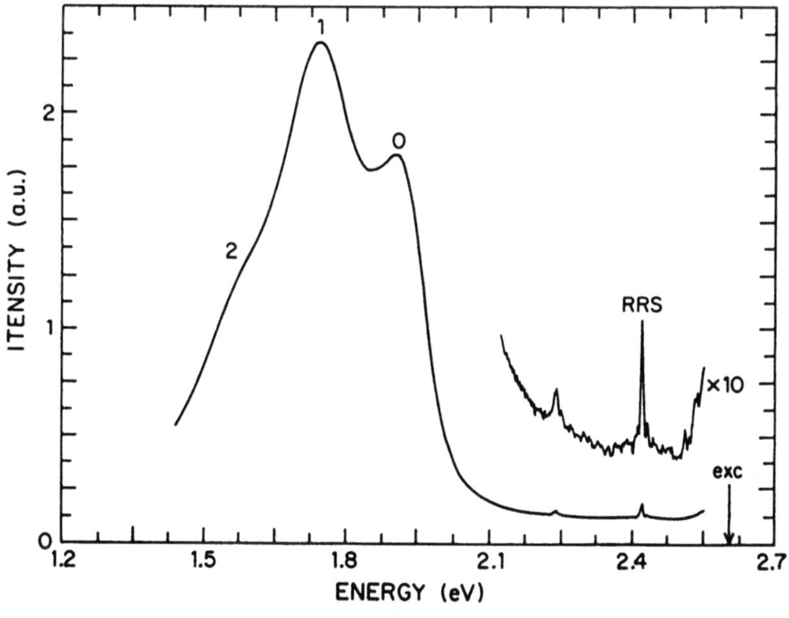

Fig. 3 The PL spectrum of PMT excited with the 4765Å Ar^+ line

where $F_0(E)$ is the functional form of the emission strength in the absence of electron–lattice vibration interaction, $\hbar\omega$ is the phonon energy and $\nu = 0, 1, 2....$ We analyzed the PL spectrum (Fig. 3) using eq. 1, and found $S \simeq 1.1$. This small value of S together with the resolved vibronic structure justifies the use of the weak coupling formula. The dimensionless coupling strength parameter S is a measure of the effective electron–vibration interaction A_{eff}. Writing, $S^2 = \bar{q}A_{eff}/\hbar\omega$ ($\bar{q}$ is the vibrational zero point amplitude), we find $A_{eff} \simeq 2\ eV/\mathring{A}$. This value is comparable to the electron–phonon interaction in polyacetylene. We thus conclude that the excited electron–hole pair interacts strongly with the lattice, possibly forming a confined "polaron–exciton" [4].

Acknowledgments. This work was supported by the Fund for Basic Research–Administered by the Israel Academy of Sciences and Humanities, and by the Technion V.P.R Fund–P. and E. Nathan Research Fund.

References

1 S. Brazovskii and N. Kirova, JETP Letters, 33, 4 (1981).

2 K. Fesser, A.R. Bishop and C. Campbell, Phys. Rev. B27, 4804 (1983).

3 M. Kertesz and Y–S Lee, Synthetic Metals 28, C545 (1989).

4 R.H. Friend, D.D.C. Bradley and P.D. Townsend, J. Phys. D: Appl. Phys. 20, 1367 (1987).

5 T.-C. Chung, J.H. Kaufman, A.J. Heeger and F. Wudl, Phys. Rev. B30, 702 (1984)

6 Z. Vardeny, E. Ehrenfreund, O. Brafman, M. Nowak, H. Schaffer, A.J. Heeger and F. Wudl, Phys. Rev. Lett. 56, 671 (1986).

7 J. Poplawski, E. Ehrenfreund, S. Glenis and A.J. Frank, Synthetic Metals 28, C335 (1989).

8 S. Nakajima, Y. Toyozawa and R. Abe, "The Physics of Elementary Excitations", Springer Series in Solid State Sciences, Vol. 12 (1980).

Photoexcited Triplet States in Poly(p-phenylene vinylene) and Model Oligomers

N.F. Colaneri [1], *D.D.C. Bradley* [1], *R.H. Friend* [1], *and C.W. Spangler* [2]

[1]Cavendish Laboratory, Madingley Road, Cambridge CB3 0HE, UK
[2]Department of Chemistry, Northern Illinois University, DeKalb, IL 60115, USA

Abstract. We compare the generation and decay of charged and neutral photoexcitations in poly(*p*-phenylene vinylene) and short model oligomers. The radiative recombination of intramolecular polaron-excitons is detected through luminescence measurements, while interchain photogeneration of charged bipolarons in the polymer leads to photoinduced absorption. An additional photoinduced absorption, present in the polymer and model oligomers, is assigned to triplet excitons.

1. Introduction

Some of the unusual self-localized electronic excitations of conjugated polymers also play a role in the electronic structure of shorter oligomers. Hence, investigation of the electronic properties of the oligomers is useful for obtaining a full understanding of the properties of the corresponding polymers. In the case of poly(*p*-phenylene vinylene) (PPV), single photon absorption experiments on model oligomers have previously been used to interpret the optical absorption spectrum of the polymer [1,2]. In addition, the photoinduced absorption spectra of the polymer and oligomers show related features. A photoexcitation tentatively identified as a triplet exciton has been observed in photoinduced absorption measurements of PPV [3]. This assignment is made more plausible by a comparison with the well-known triplet-triplet absorption of the dye molecule *trans*-stilbene [4], which can be viewed as the shortest oligomeric fragment of PPV. In this contribution we report further spectroscopic and magnetic investigations of the lowest triplet excitation in PPV and a three-ring oligomeric analog, *trans,trans*-1,4-distyrylbenzene.

2. Experimental Details

For photoluminescence and photoinduced absorption measurements, samples were mounted in an Oxford instruments optical access helium-flow cryostat. To probe the spectrally resolved absorption of the samples, broadband light from a tungsten halogen lamp was passed through the sample, dispersed with a monochromator, and measured with a photomultiplier or appropriate solid state detectors. Photoexcitation was provided either by radiation from a Coherent Innova 100 argon ion laser, lasing

Springer Series in Solid-State Sciences, Vol. 91
Electronic Properties of Conjugated Polymers III
Editors: H. Kuzmany · M. Mehring · S. Roth

all-lines in the ultraviolet (laser emission at 363.8 nm, 351.1 nm and a trio of lines centered near 335 nm), or by the frequency-tripled line of a Nd:YAG laser (355 nm).

Electron spin resonance measurements were performed with a Varian E109-E commercial ESR spectrometer operating at a frequency of 9.14 GHz. For low temperature measurements, the spectrometer was fitted with a system for flowing cold nitrogen gas through the microwave cavity. Photoexcitation was provided by a 150 W xenon arc lamp, which illuminated the sample through an optical access in the cavity. Quartz focussing optics were used to concentrate the light on the sample.

3. Results

Unlike intrachain photoexcitation of electron-hole pairs in PPV, which leads to the production of neutral polaron-excitons [5], interchain excitation can result in the generation of doubly charged bipolarons. These non-linear electronic excitations can be observed through measurements of the polymer's photoinduced absorption. Figure 1 shows the photoinduced absorption signal due to bipolarons, measured at 40 K, when PPV is excited with the Ar^+ laser, chopped at 67 Hz. At this relatively low chopping frequency the bipolaron absorption signal appears in near quadrature with the pump beam, implying that their lifetime is greater than 15 ms. The magnitude and zero of the high energy bipolaron feature, seen at 1.6 eV in figure 1, have been adjusted to compensate for the presence of an additional induced absorption

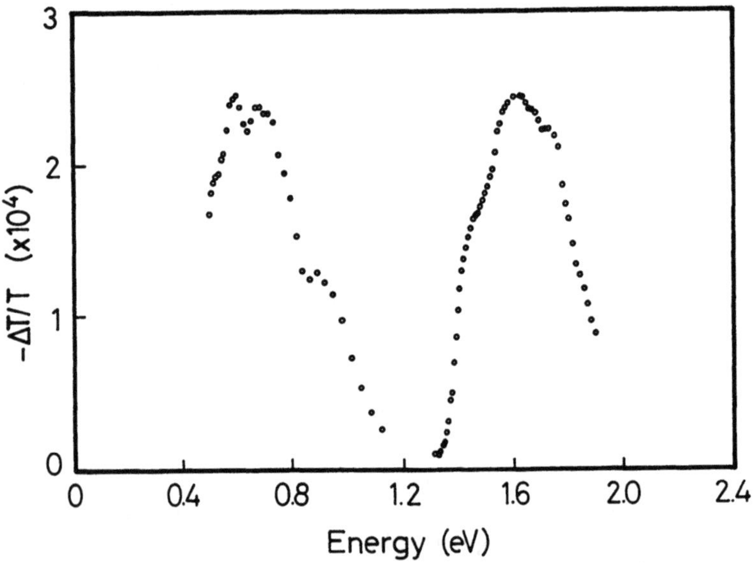

Figure 1. Photoinduced absorption spectrum of PPV at 40 K. Excitation was provided by UV radiation from an Ar^+ ion laser, chopped at 67 Hz. The data show the absorption signal in quadrature (90° phase lag) with the modulated excitation source

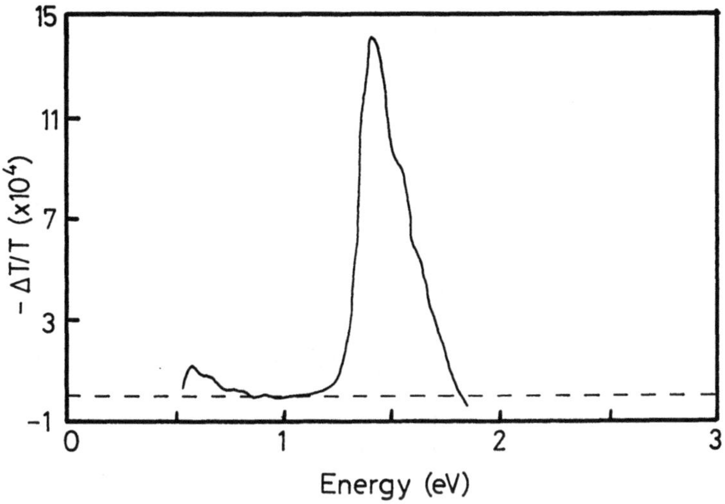

Figure 2. Photoinduced absorption spectrum of PPV at 40 K. The figure displays the same data as in figure 1, with the phase adjusted to show the absorption signal in phase with the modulated excitation

which appears nearly in phase with the pump at this temperature and chopping frequency. This new feature, which peaks near 1.45 eV, is shown in figure 2.

In reference 3 we suggested a possible assignment of the 1.45 eV photoinduced absorption in PPV to a triplet excitation. This identification is made more plausible by comparison with transient absorption measurements on model oligomers of PPV. The lowest triplet state of *trans*-stilbene (the chemical structure of *trans*-stilbene is shown in figure 3a) has been extensively investigated by both spectroscopic [4] and magnetic [6] techniques over the past 20 years. The non-phosphorescent nature of the triplet's decay indicates that the decay kinetics are dominated by non-radiative mechanisms [7]. In glassy medium, the triplet lifetime is of order 2 ms at 77 K. The zero-field splitting of the triplet sublevels, inferred from light-induced ESR measurements [6], has been used to argue that the triplet electronic excitation of *trans*-stilbene extends along the whole molecule.

We have also examined the lowest triplet state in *t,t*-DSB, a slightly longer oligomeric analog of PPV, whose chemical structure is shown in figure 3b. The onset of the π-π^* transition in *t,t*-DSB occurs near 3 eV [8]. During excitation with the Ar^+ ion laser, its absorption spectrum exhibits a new feature which we assign to

(a) 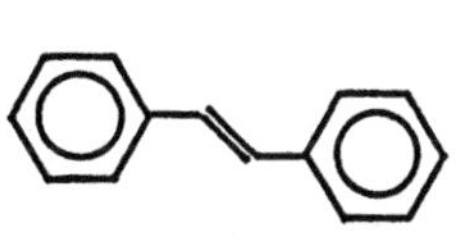 (b) 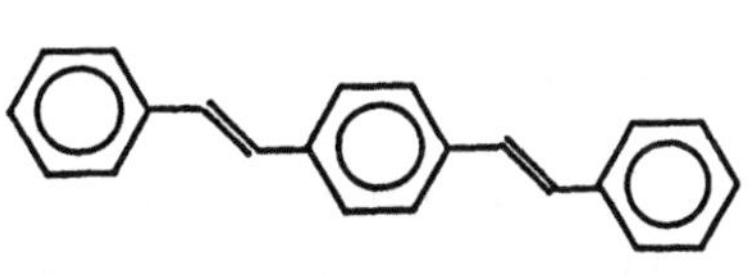

Figure 3. Chemical structure of two short oligomeric analogs of PPV: (a) *trans*-stilbene, and (b) *trans,trans*-1,4-distyrylbenzene (*t,t*-DSB)

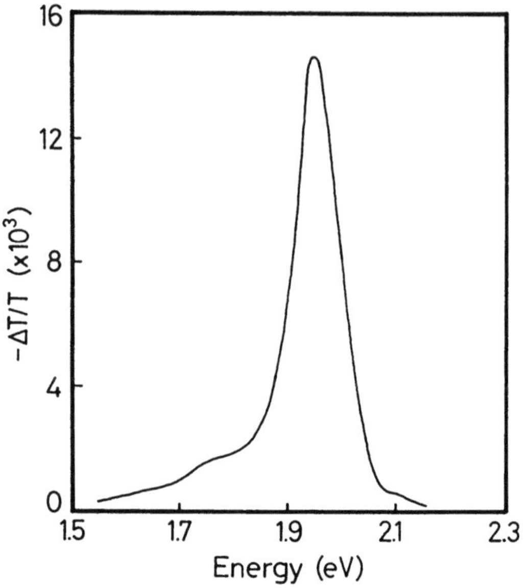

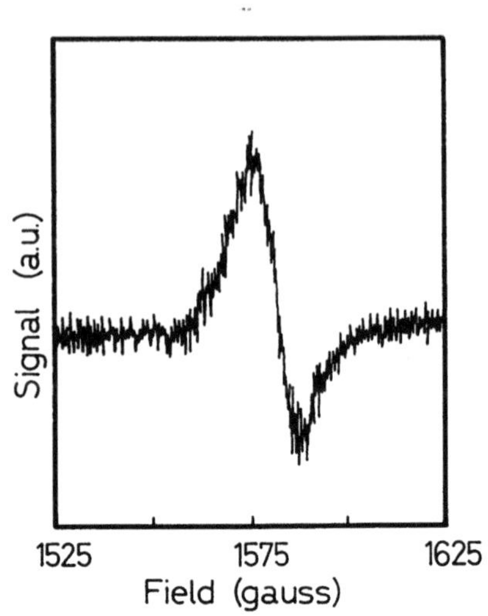

Figure 4. Triplet-triplet absorption in *t,t*-DSB at 80 K, during UV laser irradiation

Figure 5. Light induced ESR absorption signal of *t,t*-DSB at 80 K for an incident microwave frequency of 9.14 GHz. The asymmetric "half field" line arises from the forbidden $\Delta m_S=2$ transition between triplet sublevels

a triplet-triplet transition. The spectrum of this photoinduced absorption, which peaks near 1.95 eV, is shown in figure 4.

The extent of the triplet excitation in *t,t*-DSB will affect the zero-field splitting of the triplet sublevels. This can, in turn, be inferred from the Zeeman splitting of the sublevels in the presence of a moderate magnetic field. The light-induced ESR signal of *t,t*-DSB at 77 K consists of a single weak asymmetric line of width $\Delta H_{pp} \approx 12$ G, centered near 1580 G, as shown in figure 5. We assign this feature to the $\Delta m_S=2$ transition. The position of the minimum field at $H_{min} \approx 1575$ G corresponds to a zero-field splitting parameter $D^* \approx 0.07$ cm^{-1}, significantly smaller than the value reported for the lowest triplet in *trans*-stilbene [6]. This implies that the triplet in *t,t*-DSB is of greater extent than that in *trans*-stilbene.

4. Discussion

The relative importance of electron-electron interactions in a correct description of the electronic properties of conjugated polymers has not been resolved. While one electron models adequately describe many ground state properties, a variety of theoretical investigations suggest that even moderately strong interactions need not substantially alter this description. A clearer distinction can be made in the case of excited states. We have previously argued [3] that differences between the spectroscopic signatures of polaron-excitons and bipolarons in PPV, identical excitations from the point of view of a one electron model [9], indicate that Coulomb interactions are important in this material. This conclusion can be supported by measurements of the triplet excited state. If the excitation responsible for luminescence quenching in partially converted samples of PPV [10] is a higher excited triplet, then the observed triplet-triplet transition at 1.45 eV implies a singlet-triplet splitting of order 1 eV. This splitting is a direct measure of the strength of the Coulomb interaction. Confirmation of this conclusion requires direct measurement of singlet-triplet absorption, which is more difficult.

References

1. H-H. Horhold, *Z. Chem.* **12**,41 (1972); G. Drefahl, R. Kuhmstedt, H. Oswald, and H-H. Horhold, *Makromol. Chem.* **131**, 89 (1970)
2. J. Obrzut and F. E. Karasz, *J. Chem. Phys.* **87**, 2349 (1987)
3. D. D. C. Bradley, N. F. Colaneri, and R. H. Friend, *Synth. Met.* **29**, E121 (1989)
4. W. G. Herkstroeter and D. S. McClure, *J. Am. Chem. Soc.* **90**, 4522 (1968)
5. R. H. Friend, D. D. C. Bradley, and P. D. Townsend, *J. Phys.* **D20**, 1367 (1987)
6. M. Yagi, *Chem. Phys. Lett.* **124**, 459 (1986)
7. T. Ikeyama and T. Azumi, *J. Phys. Chem.* **89**, 5332 (1985)
8. E. A. Andreyeshchev, V. S. Viktorova, S. F. Kilin, Yu. P. Kushakevich, and I. M. Rozman, *Opt. i Spektrosk.* **24**, 387 (1968)
9. K. Fesser, A. R. Bishop, and D. K. Campbell, *Phys. Rev.* **B27**, 4804 (1983)
10. D. D. C. Bradley and R. H. Friend, *to be published*

Transient Photoconductivity in PPV – Influence of the Conjugation Length on Transport and Recombination

Y.Q. Shen, H. Lindenberger, H. Bleier, and S. Roth

Max-Planck-Institut für Festkörperforschung,
D-7000 Stuttgart 80, Fed. Rep. of Germany

Abstract Transient photoconductivity measurements in highly oriented polyphenylene-vinylene (PPV) are reported. The PPV films show a fast picosecond component of the photocurrent as well as a contribution of long-living carriers. The conjugation length of PPV can be easily controlled by varying the heating temperature and time during the thermal conversion from the precursor polymer. The values of the intrinsic mobility which have been extrapolated from the fast component increase strongly with the conjugation length. The highest value of the intrinsic mobility is found to be ca. $2cm^2/Vs$. Bimolecular recombination has been observed in PPV films with long conjugation length and the value of the recombination rate γ is found to be 0.2 - $0.9s^{-1}$.

1 Introduction

Photocurrent investigations are an important tool to study the creation and the transport of charged defects (i.e. solitons, polarons and bipolarons). Bleier et al. have carried out transient photocurrent investigations in polyacetylene films [1]. They find that the photocurrent consists of two components i.e. a fast component and a slow component. The fast component, which dominates the photocurrent on a time scale less than some nanoseconds, reflects the migration of the charge carriers on polymer chains where they have been created (i.e. the intrinsic transport properties). On a longer time scale the charge carriers will reach the end of the conjugated segments of the chain and therefore the transport is dominated by hopping across defects just as in the case of the dark conductivity [2]. The slow component of the photocurrent is due to this type of charge transport. The fast component of the photocurrent is four orders of magnitude larger than the dark current and the intrinsic mobility of polyacetylene is estimated from the fast component to be ca. $2cm^2/Vs$ [1].

In this paper we report on transient photocurrent investigations in precursor-route PPV films [3]. The dependence of the intrinsic mobility of PPV films on the conjugation length, as well as the recombination mechanisms in PPV films, will be discussed.

2 Conjugation length in PPV films

Precursor-route PPV is well-known for its simple chemical preparation, processability and stability and for these reasons it is very useful for investigations of different properties of conducting polymers. The conductivity of iodine doped PPV has been reported to be $5.2 \times 10^3 S/cm$ [4]. The water soluble precursor polymer can be converted into PPV by heating at 300°C to eliminate the dialkyl sulphonium halide [3]. Therefore the conversion process increases the conjugation length.

Springer Series in Solid-State Sciences, Vol. 91
Electronic Properties of Conjugated Polymers III
Editors: H. Kuzmany · M. Mehring · S. Roth

We have prepared a group of PPV films under different conversion conditions (i.e. conversion temperature and conversion time). In order to study the chemical structure and to estimate the conjugation lengths in these PPV films, elemental analysis and IR spectroscopy have been carried out. The results of the IR spectroscopies are identical to those reported by Bradley et al. [5]. The results of the elemental analysis show that the content of sulfur decreases as the conversion time and temperature increase. The ratios of $C : H : S$ obtained from elemental analysis can be well fitted by the structure of -[-Ø-C_2H_2-]$_n$-[-Ø-CH_2-$CH(SCH_3)$-]$_m$-, where n is the number of monomers of the PPV and m is the number of monomers of the structure containing thioether. The elemental analysis shows also an excess of Cl (ca. 6%), which is independent from the conversion conditions. If the Cl is assumed to contribute to the saturated structure, it can not fit the results of the elemental analysis. Since the IR spectroscopy does not show any C-Cl stretch mode, we conclude that the conjugation interruption is mainly formed by the thioether structure. When assuming a homogeneous distribution of the conjugation interruptions and taking the length of a PPV monomer to be 6.5Å [6], the conjugation length can be then estimated from the mole content of S. We find that the conjugation length l_c of the PPV films converted at 300°C obeys the relation $l_c \propto t^{0.48}$, where t is the conversion time. The PPV films used in our photocurrent investigations are highly oriented and are converted at 300°C for 2, 24 and 96 hours. Their conjugation lengths are 32, 115 and 260 Å, respectively.

3 Experimental results and discussions

In the transient photocurrent measurements a Dye-laser pumped by a 500ps pulsed N_2-laser is used for optical excitation. The wavelengths of the excitation light have been chosen to be 488nm and 366nm which correspond to the absorption edge and the absorption maximum of the PPV films. Moreover the excitation light is polarized parallel

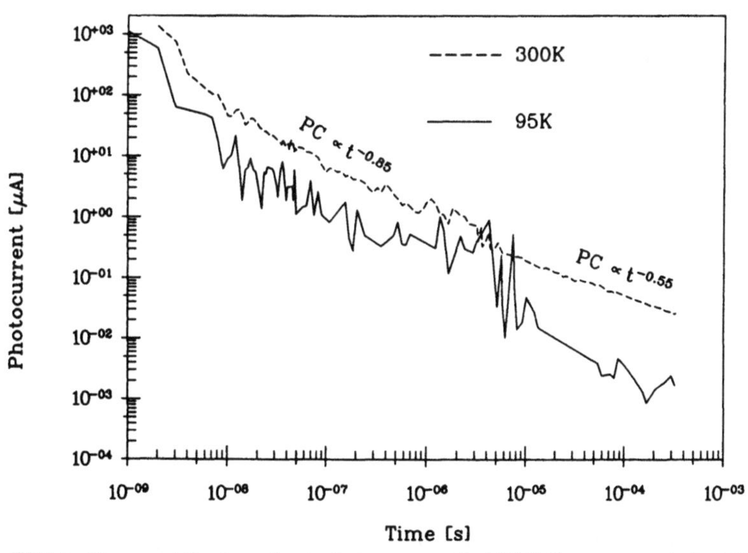

FIG.1: Decay of the transient photocurrent in PPV films converted at 300°C for 95 hours, excited by perpendicular polarized light (488nm), for $E = 20kV/cm$ and $Int.abs = 8\mu J$, at 300K and 95K

or perpendicular to the stretch orientation. The electric field is applied parallel to the stretch orientation so that the photocurrent reflects the transport along the chains. A detailed description of the experimental set-up can be found in reference 1.

The measurements of the field dependent transient photocurrent show that the peak-photocurrent is proportional to the electric field E for all the PPV films. The decay of the transient photocurrent in PPV films converted for 96 hours is shown in FIG.1 for two different temperatures. As in the case of polyacetylene the photocurrent in PPV films contains two components, a fast component in the short time region and a slow component which becomes dominant at $1\mu s$. The amplitude of the transient photocurrent (peak photocurrent) is therefore dominated by the fast component. At room temperature the fast component I decays with time t as $I \propto t^{0.85}$ and the slow component decays as $I \propto t^{0.55}$. At $95K$ the fast component remains and the slow component is too small to be detected. The decay of the fast component is independent of the electric field but depends on the light intensity. The decay becomes faster with increasing intensity, which is a clear evidence for bimolecular recombination.

The bimolecular recombination is also reflected in the intensity dependence of the peak photocurrent. For PPV films converted for 96 hours the peak photocurrent increases linearly with intensity in the low intensity region and increases sublinearly for higher intensities (see FIG.2). For PPV films converted for 2 and 24 hours the peak photocurrent varies linearly with the intensity. The sublinearity of the intensity dependent photocurrent in PPV films converted for 96 hours, is due to bimolecular recombination and not to saturation, since it has only been observed in PPV films with long conjugation length and the photocurrent decays faster for higher light intensities.

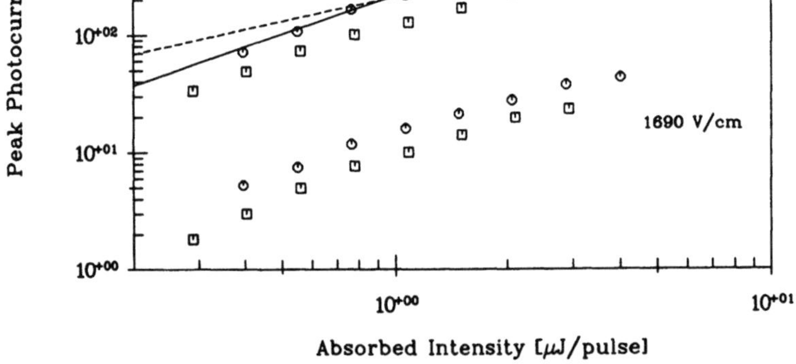

FIG.2: The intensity dependent peak photocurrent in oriented PPV films converted at 300°C for 96 hours, for excitation light (366nm) parallel and perpendicular to the chain orientation. The electric fields are 1.7 and $20kV/cm$, and the gap width is $145\mu m$. The measurement is carried out at $300K$

For the coexistence of linear and bimolecular recombination, the photocurrent decay can be described by the following equation:

$$\frac{dn}{dt} = -\gamma n^2 - \frac{n}{\tau} + \eta\phi n_{ph}$$

where γ is the rate of bimolecular recombination, τ is the lifetime of the charge carriers, η is the quantum efficiency, ϕ is the escape factor for geminate recombination and n_{ph} is the number of incident photons per unit time. For a light pulse with a total number of photons N_{ph} and a pulse duration of t_0, it obeys $n_{ph} = N_{ph}/t_0$ for $0 \leq t \leq t_0$ and $n_{ph} = 0$ for $t > t_0$. This equation can be solved and the integrated charge (in electronic units) N_e can be obtained as a function of N_{ph}. The unknown parameters η, ϕ, γ and the mobility μ occur in this expression of N_e only as two products, i.e. $\eta\phi\gamma$ and $\eta\phi\mu$. We have succeeded to fit the expression of N_e to the measured data of PPV films converted for 96 hours, and the values of $\eta\phi\gamma$ and $\eta\phi\mu$ have been obtained. In absence of bimolecular recombination (i.e. $\gamma = 0$), N_e obeys $N_e = N_{ph}\eta\phi\frac{\mu E\tau}{d}$, where d is the distance between the two electrodes on the sample. The value of $\eta\phi\mu$ for the PPV films converted for 2 and 24 hours can be calculated in terms of this expression. The values of $\eta\phi$ for PPV have not yet been directly determined. We apply therefore the corresponding values of polyacetylene, where $\eta\phi = 0.01$ [1], which results in a value of γ for PPV films converted at 300°C for 96 hours of 0.2 - $0.9s^{-1}$, depending on the wavelengths and the luminosity of the excitation light. The maximum measured value of the intrinsic mobility for the same PPV films is ca. $2cm^2/Vs$, which is comparable to the mobility of polyacetylene, and the mobility decreases as the conjugation length decreases(see Table 1).

Table 1: Comparison of conjugation lengths and mobilities μ in PPV films converted at 300°C for 92, 24 and 2 hours

Conversion time of PPV [hours]	Conjugation length [Å]	$\mu[cm^2/Vs]$
96	260	1.86
24	115	0.39
2	38	0.028

4 References

1. H.Bleier, S.Roth, Y.Q.Shen, D.Schäfer-Siebert and G.Leising: Phys.Rev.B, 38, 6031 (1988).
2. K.Ehinger, S.Summerfield, W.Bauhofer and S.Roth: J.Phys.C, 17, 3753(1984).
3. J.D.Capistran, D.R.Gagnon, S.Anton, R.W.Lenz and E.F.Karasz: A.C.S Polymer Preprints, 25, 282(1984).
4. I.Murase, T.Ohnishi, T.Noguchi and M.Hirooka: Synth.Met., 17, 639(1987).
5. D.D.C.Bradley: J.Phys.D, 20, 1389(1987).
6. T.Hartmann: M.Phil Thesis, University of Cambridge (1986).

Optical Properties of Polyparaphenylene Prepared by Precursor Polymers

G. Leising[1], *K. Pichler*[1], *and F. Stelzer*[2]

[1]Institut für Festkörperphysik, Graz University of Technology, Petersgasse 16, A-8010 Graz, Austria
[2]Cal.Tech., Department of Chemistry, Pasadena, USA

ABSTRACT

We present UV-visible and infrared investigations on Polyparaphenylene (PPP) thin films synthesized by the precursor-polymer route invented by Ballard et al [1]. Cast thin films of the precursor polymers, poly(5,6-dihydroxy-2-cyclohexene-1,4-ylene) with different restgroups (methyl, biphenyl, naphthalene), are converted to polycrystalline PPP at various temperatures under high vacuum conditions.

The optical properties (band gap, band tailing) and the characteristic infrared bands of these PPP samples are compared to those of PPP synthesized by other preparation routes.

INTRODUCTION

PPP is a representative of conjugated conducting polymers of the polyaromatic-type, which can be doped by oxidizing and and reducing agents (acceptor and donor doping) [2]. The electrical conductivity of the pristine PPP can be changed from values typical for insulators (less than 10^{-10} S/cm), by many orders of magnitude, up to values close to those of poor metals (10^{3} S/cm) upon doping. One of the most appealing properties of PPP, concerning possible technical applications, is its thermal stability in air up to temperatures of about 500°C. Major disadvantages of PPP and most of the conjugated polymers are their insolubility and infusability, which are severe restrictions to their processability, again thinking of technical applications. Many of the studies on PPP have been carried out on samples prepared by the Kovacic method [3] and by the Yamamoto method [4]. These preparation routes normally give a powdered material, which is pressed into pellets. A shear-flow technique based on the Kovacic method, allowed the direct polymerization of thin partially oriented PPP films [5]. More recently electrochemical techniques have been developed for the synthesis of PPP films on ITO-coated glass electrodes [6,7] and a vacuum deposition technique has been reported [8] to produce thin films of PPP by evaporation of PPP powder at 680°C.

A common feature of PPP samples prepared by the above mentioned methods is a granular or fibrillar morphology. For the investigation of the intrinsic optical and transport properties, this complex morphology poses serious problems. Therefore, the invention of routes using non-conjugated precursor polymers, which are soluble in suitable solvents, allows the

Springer Series in Solid-State Sciences, Vol. 91
Electronic Properties of Conjugated Polymers III
Editors: H. Kuzmany · M. Mehring · S. Roth

film-casting technique to be applied for the preparation of thin conjugated polymer films of high quality, with a compact non-fibrous morphology. Since the precipitation technique can be used to purify the precusor polymers, conjugated polymers having extremely low catalyst residues can be synthesized. Since PPP is insoluble, conventional molecular weight determination techniques cannot be applied. The average number of repeat units in the polymer chains therefore, could only be estimated from the position of the C-H deformation vibration of the para-disubstituted benzenes in the infrared and from the position of the interband transition in the optical spectra, by extrapolation from the values of the phenylene oligomers. Based on infrared studies an average value of 16 phenyl rings was determined for PPP prepared by the Kovacic method [9].

EXPERIMENTAL

The preparation of PPP via precursor polymers is outlined in Fig. 1 and follows the route invented by Ballard and coworkers [1]. A detailed description of the monomer synthesis and the polymerization conditions for the precursor polymers will be published elesewhere. In the present study naphtalene, methyl and biphenyl species where used as sidegroups in the precursor polymers. Films were cast from solutions (solvents: acetone, chlorobenzene and methylenechloride) of the precursor polymers onto glass or quartz substrates for UV-VIS-NIR (ultraviolet-visible-near-infrared) spectroscopy and KBr crystals for UV-VIS-NIR and infrared spectroscopy. The precursor polymer films are converted to PPP under high vacuum conditions at temperatures up to 620°C for several hours. Both the optical and infrared spectra measured under inert atmosphere (Ar) and under ambient conditions (air) appeared to be identical. The UV-VIS-NIR have been taken with a Perkin-Elmer Lambda-9 spectrophotometer and for the collection of the infrared spectra we used a Perkin-Elmer 684 infrared spectrophotometer.

Doping experiments have been carried out by exposing the PPP-films to vapour of AsF_5 at pressures up to one atmosphere in a stainless steel doping/high-vacuum line. Only an extremely small doping effect could be observed by a slight increase of the optical absorption in the sub-bandgap region and a moderate increase of the conductivity by more than five orders of magnitude, up to values of about 10^{-5} S/cm. However, electrochemical doping procedures appeared to work and those results

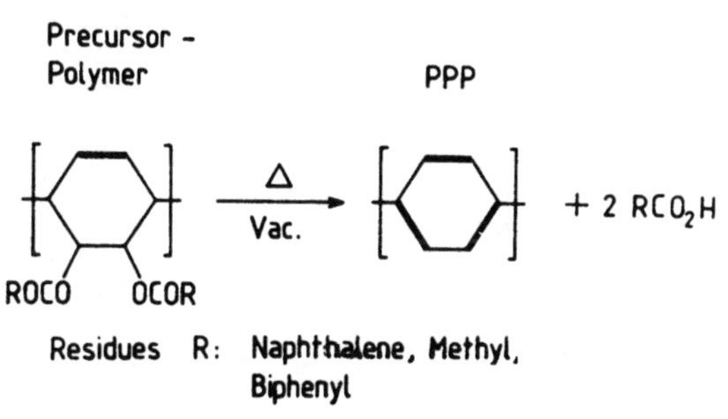

Fig. 1: Preparation route of PPP-films by precursor-polymers; Conversion of the precursor-polymer to PPP at high vacuum and temperatures up to 620°C for several hours

will be reported soon. We attribute this doping behaviour to the dense morphology of the PPP-films prepared by this route, empeding the diffusion (very low diffusion coefficient) of the dopant species in the case of gas phase doping.

RESULTS AND DISCUSSION

The infrared absorption spectrum of a precursor-polymer film with methyl as a restgroup, cast from a chlorobenzene solution onto KBr, is shown in Fig. 2a. This polymer film is converted to PPP at 400°C under high vacuum for 20 hours and in Fig. 2b the infrared absorption spectrum of the PPP film is depicted. The position of the C-H bending-vibration of the para-substituted rings is located at 807 cm^{-1}, which compares to a value of 805 cm^{-1} for Kovacic-PPP [7]. The bands for the C-H bending-vibrations of the mono-substituted rings (end-groups) are found at 765 cm^{-1} and 697 cm^{-1} (the three C-H bending-vibrations are indicated by arrows in Fig. 2b).

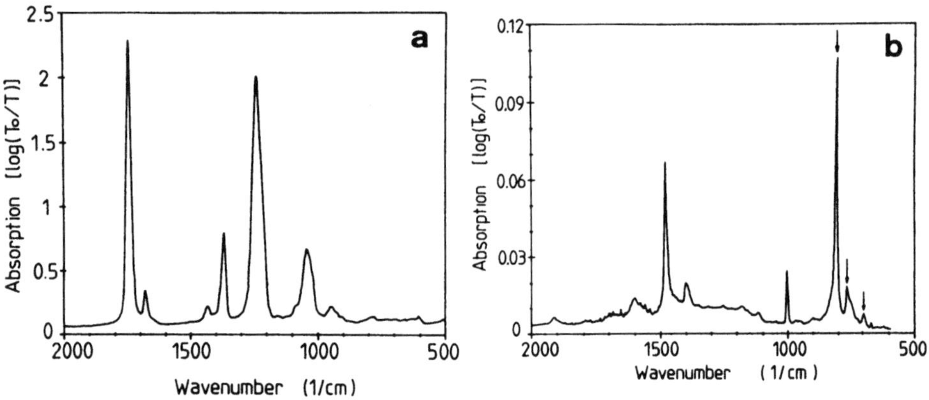

Fig. 2: a) IR-spectrum of the precursor-polymer film with methyl restgroup (on KBr); b) IR spectrum of the PPP-film received by conversion of the precursor-polymer film of a) at 400°C/20 hours (arrows see text)

The UV-VIS-NIR absorption spectrum of a thin PPP film on a quartz substrate and prepared under the same conversion conditions as the sample above is shown in Fig. 3. The absorption maximum appears at 3.65 eV, which is slightly higher than the values of 3.2 ev [7] up to 3.4 eV [6] for PPP prepared by other routes, but the same as for extracted Yamamoto-PPP [10]. The linear extrapolation of the band edge leads to a value of 2.7 eV, which is the same as for the other PPP's [5,6,7,10]. The nearly perfect plane-parallel geometry of the polymer films produces pronounced Fabry-Perot interferences in the infrared region, which have been used for the determination of the film thickness. Therefore, we can give an estimate of the value of the absorption coefficient at the maximum of the interband transition of PPP (neglecting reflectivity contributions) to be $2*10^5$ cm^{-1}. For comparing the optical and

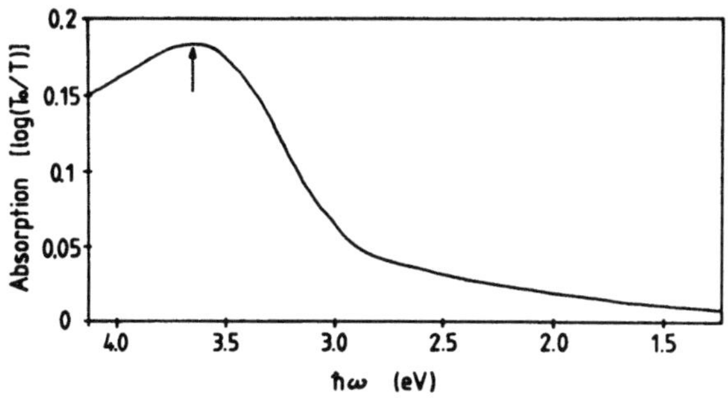

Fig. 3: UV-VIS-NIR absorption spectrum of PPP; the arrow indicates the peak position of the interband transition at 3.65 eV

infrared properties of our PPP samples to that of oligophenyles, we plot in Fig. 4 the peakpositions of the interband absorption (E_n) and the position of the C-H para-sustituted bending vibration (v_n) versus the inverse chain lengths (1/n). Using the relation $E_n = E_\infty + E/n$ [11] for the interband transition the fit gives E_∞ = 3.36 eV, in good agreement with the experimental values of other PPP's [5,6,7]. With our value for the interband peakposition we derive the average chain length for PPP prepared by this route to be about ten phenyl rings.

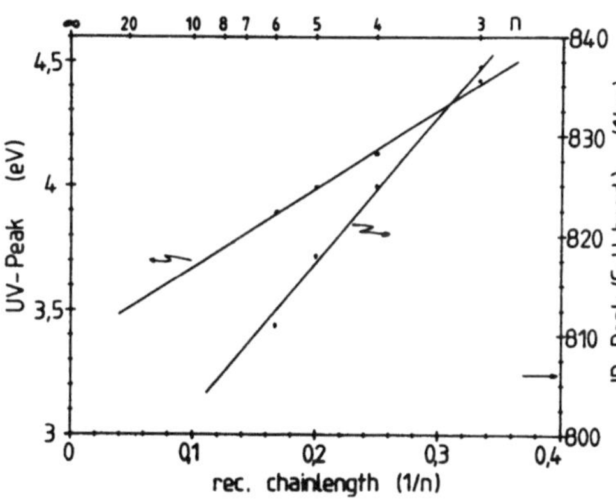

Fig. 4: Dependence of the peakposition of the interband absorption and the C-H para-substituted bending-vibration on the reciprocal chain-length (n) of oligophenyles; The fits (solid lines) are calculated by $E_n = E_\infty + E/n$ for the interband transition and $v_n = v_\infty + v/n$ for the infrared band (the values from the fit are: E_∞=3.36eV, v_∞=786.5cm^{-1})

REFERENCES

1 D.G.H. Ballard, A. Courtis, I.M. Shirley and S.C. Taylor, J. Chem. Soc., Chem. Commun. 954 (1983).
2 D.M. Ivory, G.G. Miller, J.M. Sowa, L.W. Shacklette, R.R. Chance and R.H. Baughman, J.Chem. Phys. 71 1506 (1979); L.W. Shacklette, R.R. Chance, D.M. Ivory, G.G. Miller and R.H. Baughman, Synth. Met. 1 307 (1979).
3 P. Kovacic and A. Kyriakis, J. Am. Chem. Soc. 85 454 (1963).
4 T. Yamamoto and A. Yamamoto, Chemistry Letters 353 (1977).
5 B. Tieke, C. Bubeck and G. Lieser, Macromol.Chem., Rapid Commun. 3 261 (1982).
6 M. Satoh, K. Kaneto and K. Yoshino, J. Chem. Soc., Chem. Commun. 1629 (1985).

7 J.F. Fauvarque, M.A. Petit, A. Digua and G. Froyer, Macromol. Chem. 188 1833 (1987).
8 M. Komakine, T. Namikawa and Y. Yamazaki, Macromol. Chem., Rapid Commun. 7 139 (1986).
9 M.B. Jones, P. Kovacic and D. Lanska, J. Polym. Sci. Polym. Letts. Ed. 19 89 (1981).
10 G.Froyer, J.Y. Goblot, J.L. Guilbert, F.Maurice and Y.Pelous, J. de Physique 44 C3-745 (1983).
11 B.E. Kohler, Springer Series in Solid State Sciences 63 101 (1985).

Part IV

IR and Raman Spectroscopy

Interpretation of Vibrational Spectra of Conducting Polymers by Means of an Effective Coordinate

G. Zerbi[1], *C. Castiglioni*[1], *Bogang Tian*[1], *and M. Gussoni*[1;2]

[1]Dipartimento di Chimica Industriale, Politecnico, Piazza L. da Vinci 32, Milano, Italy
[2]CNR, Istituto Chimica Fisica Materiali, Genova, Italy

Abstract It is shown that the observed vibrational infrared and Raman spectra of pristine and doped polyconjugated systems can be interpreted in lattice dynamics with the definition of an effective conjugation coordinate.

The most relevant <u>common</u> characteristic of those organic materials which are now known as one-dimensional metals or conducting polymers is the occurrence of a one-dimensional backbone of carbon atoms sharing σ- and π-electrons. By one-dimensional backbone we mean chains with large intra-chain and small interchain coupling. The same backbone may sustain different electronic configurations (dimerized or undimerized polyacetylene, arrays of "benzenoid" or "quin-oid" rings, occurrence of solitons or polarons in the chain etc.). The different situations of "conjugation" produce different Raman spectra in the 1500-1000 cm^{-1} region, where CC bond stretching frequencies occur; these frequencies decrease with increasing conjugation length.

Solid state theory stresses that the general characteristic of a one-dimensional metal with a partly filled band is that the regular structure is not the most stable one (1): a Peierls distortion occurs which gives rise to a gap at the edge of the Fermi distribution with energy $E_g = 4\ \partial\beta/\partial u\ u_o$. where $\partial\beta/\partial u$ is the electron-phonon interaction, i.e. the variation of the nearest neighbor resonance integral with the distortion u. In polyacetylene (2, 3) u is the dimerization amplitude projected onto the chain axis; its equilibrium value u_o decreases with increasing conjugation length, and so does also E_g. For $k=k_F$ (namely at the zone boundary of the dimerized structure) and for phonon wavevector q=0 (lattice vibrating with translational symme-

Springer Series in Solid-State Sciences, Vol. 91
Electronic Properties of Conjugated Polymers III
Editors: H. Kuzmany · M. Mehring · S. Roth

try), coupling among phonons and electrons takes place. This coupling is stronger the larger the conjugation length and causes the vibrational frequencies involving the oscillation of u to decrease. This accounts for the dependence of the 1500–1000 cm^{-1} frequencies on the conjugation length.

However, this model reduces all the vibrational degrees of freedom to one, namely u, while in the repeat unit of $(CH)_x$ there are several in-plane vibrational motions which may be coupled to u. The model has been extended by Horowitz to the multidimensional case (4). The energy gap is now thought of as a sum of contributions due to the different distortions which may occur in the chain. Again a unique parameter, $\tilde{\lambda}$, accounting for e-p interaction, can be defined. $\tilde{\lambda}$ may be expressed in terms of the ratio between renormalized and bare frequencies: $2\tilde{\lambda} = \pi_n (\omega_n / \omega^o_n)^2$.

The parameter $\tilde{\lambda}$, obtained from the experimental data ω_n, decreases with increasing conjugation length, thus giving a measure of the latter . This approach has led to very good results in both PA and other polyconjugated polymers. However, it may be questioned by molecular spectroscopists for some reasons: i) $\tilde{\lambda}$ is a relative quantity, which depends on the "bare" frequencies which are chosen in an arbitrary way; ii) $\tilde{\lambda}$ cannot be related in a straightforward way to a variation of force constants, as molecular spectroscopists would like; iii) $\tilde{\lambda}$ refers simultaneously to all degrees of freedom, making the choice of the force constants to be altered even more difficult.

A different approach for justifying the dependence of the frequencies on the conjugation length has been proposed (5-8): the force constants able to correctly represent the vibrations of conjugated chains are thought to depend on the e-p coupling constant $\partial\beta/\partial R$ through the bond-bond polarization coefficients (π_{ij}) which can be computed within the Hückel frame:

$$F_{ij} = F^o_{ij}\,\delta_{ij} + F^\pi_{ij} \quad \text{where} \quad F^\pi_{ij} = \pi_{ij}\;\partial\beta_i/\partial R_i\;\;\partial\beta_j/\partial R_j . \qquad (1)$$

Here R indicates the CC bond stretching coordinates. This approach has led to good results for PA, but it is not easily transferred to other more complex polymers; moreover, also in this model the uniqueness of the distortion which opens the gap has been lost. This happens both when ij in eq.1 refers to vibrational displacements (6) and when it refers to cartesian displacements (7).

The above two methods suggested to us (9,10) to justify the decrease of ω_{CC} with the conjugation length in a way which is more immmediate for molecular spectroscopists . Let us discuss the case of PA. One can define $\mathcal{R}=\frac{1}{2}(R_{C=C}-R_{C-C})$, which is the most effective coordinate in representing the dimerization in PA while $R_+=\frac{1}{2}(R_{C=C}+R_{C-C})$ has practically null projection on the chain axis. The energy gap can then be written as a function of only $\mathcal{R}_o$, i.e. $E_g = 2(\partial\beta/\partial\mathcal{R})\,\mathcal{R}_o$, thus stressing that $\mathcal{R}$ is the dimerization coordinate. One also expects that the only force constant which is affected by the conjugation length is $F_{\mathcal{R}}$. This can be confirmed by a simple calculation: notice that

$$F_{\mathcal{R}}=\frac{1}{2}((K_{C=C}+K_{C-C})-\sum_n^N(4f^n_{C-C/C=C}-2f^n_{C=C/C=C}-2f^n_{C-C/C-C}))\quad(2)$$

$$F_{R+}=\frac{1}{2}((K_{C=C}+K_{C-C})+\sum_n^N(4f^n_{C-C/C=C}+2f^n_{C=C/C=C}+2f^n_{C-C/C-C}))\quad(3)$$

where N extends as far as the conjugation extends. Since, as predicted by any theory on conjugated polyenes, $f^n_{C-C/C-C}$ and $f^n_{C=C/C=C}$ are negative while $f^n_{C-C/C=C}$ is positive, even if the values of the interactions decrease with increasing n(12,13), the sum in $F_{\mathcal{R}}$ is non-negligible and increases with N while the sum in F_{R+} becomes smaller and smaller. Let us give an estimate of both force constants defined in eqs.2,3 using eq.1 and the π_{ij} coefficients reported in (6). One thus obtains, in the conjugation conditions described in (6) and corresponding to E_g=1.6eV, a $F^{\pi}_{\mathcal{R}}$=-4.99 contribution to $F_{\mathcal{R}}$ and a F^{π}_{R}+=0.09 contribution to F_R+, confirming that only the first depends on the conjugation.

We like to stress that the value of $F_{\mathcal{R}}$ depends both on the values of the individual terms of the sum in eq.2 and on the extension of the sum. Molecular orbital calculations on oligoenes have shown that when the length of the polymer increases the distance of interaction increases and the extent of interaction increases (12,13).

In Table 1 we have reported the values of $F_{\mathcal{R}}$ and F_R+ corresponding to various force fields of the literature; it is apparent, for all cases, that F_R+ remains practically unaltered while $F_{\mathcal{R}}$ decreases in going to longer chains. The force fields used to compile Table 1 are both from molecular orbital calculations (a-d) and from refinement on experimental data (e-g): f.f.(a,b) was obtained with "ab initio" calculations on very short polyenes (12,13) while f.f.(c) was obtained with MNDO on longer polyenes (14). Another "ab initio" calculation on longer polyenes, performed with extended bases, has been recently reported in the literature (15); unfortunately the published results cannot be used in this discussion because only a few of the relevant force constants have been published and because there is no comparison with polyenes of different length. f.f.(d) was calculated with eq.3, as reported in (16): the first value corresponds to $\partial\beta/\partial R=0$ (no e-p interaction), namely to a hypothetical case of completely localized π-electrons. The f.f.(e) was fitted to experimental values of a sample of PA and of a sample of β-carotene (17); f.f.(f) was obtained from observed frequencies of IR and Raman spectra (different excitation lines) of various samples of PA(18); f.f.(g) was obtained by us from a large set of experimental data from short and long polyenes. The most recent f.f. derived from experimental data (19) is not reported in Table 1. The comparison of this field with the others of Table 1 requires a detailed discussion which is beyond the purpose of this paper and which will be reported elsewhere.

The variation of the frequencies with the value of $F_{Я}$ is reported in Figure 1: the frequencies clearly decrease with decreasing $F_{Я}$: in this way we can describe in a continuous way all the "conjugational" states, just as Я does. As stated above, $F_{Я}$ may decrease even if the conjugation length N does not increase (eq.6), but simply because the terms in the sum have increased values. This is what happens in the case of doped samples, as confirmed by molecular orbital calculations (20,21). Thus $F_{Я}$ in a doped sample is expected to be much smaller than that of the corresponding pristine material. It can be shown that some of the CC modes of the doped portions of PA or of polyenes may become infrared active (3,10) because the symmetry of the charge distribution is lowered. The corresponding frequencies lay on the same curve where one finds the Raman frequencies of the pristine sample, but at a lower value of $F_{Я}$. The same happens in the Horowitz model; however, in that case, the

Table 1. Dependence of $F_{Я}$ and of F_{R+} on the conjugation length

	$F_{Я}$ (mdyne/A)	F_{R+} (mdyne/A)
a. butadiene (12a)	6.14	7.53
hexatriene (12b)	5.62	7.28
b. butadiene (13)	6.14	7.33
hexatriene (13)	5.61	7.04
octatetraene (13)	4.73	7.40
c. C_4H_6 (14)	6.52	8.93
C_6H_8 (14)	6.23	8.76
C_8H_{10} (14)	6.16	8.85
$C_{10}H_{12}$ (14)	6.11	8.82
$C_{12}H_{14}$ (14)	6.07	8.83
d. polyacetylene, $\partial\beta/\partial R=0$ (16)	8.66	8.66
polyacetylene, $\partial\beta/\partial R=10$ (16)	3.66	8.59
e. β-carotene (17)	4.49	6.97
polyacetylene (17)	3.85	6.97
f. polyacetylene:		
short chains (18)	6.14	7.37
medium chains (18)	5.30	7.27
long chains (18)	4.73	7.85
g. polyacetylene:		
short chains (10)	5.01	7.36
medium chains (10)	4.14	7.36
long chains (10)	3.75	7.36

e-p coupling parameter relative to doped and/or photoexcited samples is called the pinning potential (α_{dop} and α_{PG}).

The same procedure may be adopted i) to describe variation of ω_{CC} in any kind of conjugated macromolecule, ii) to account for variations of ω_{CC} with actual conjugation length (which causes dispersion in the resonance Raman spectrum) and for variations of force constants in eq.6 by great changes in electronic configuration (which occur in doped and photoexcited samples, leading to very intense IR bands). In any case, variations in conjugation can be represented as the transition between different equilibrium geometries: in

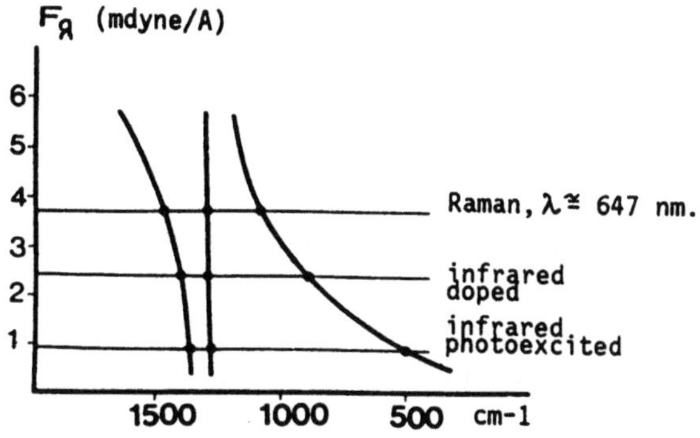

Fig.1 A_g frequencies in pristine (Raman), doped (IR) and photoexcited (IR) polyacetylene as a function of $F_{\mathcal{R}}$

the case of a chain of rings, for instance, the two states are one prevalently "benzenoid" and the other prevalently "quinoid". There is only one $\mathcal{R}$ coordinate, in each case, which describes this variation in the most ef-fective way. In the case of polyparaphenylene $\mathcal{R}$ is (11)

$$\mathcal{R} = -R_1 + R_2 - R_3 - R_4 + R_5 - R_6 + R_7$$

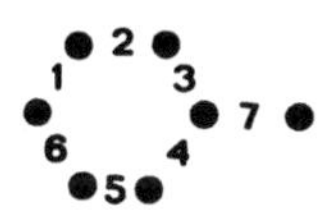

In the case of polypyrrole(22) $\mathcal{R}$ is

$$\mathcal{R} = R_1 - R_2 + R_3 - R_4 + R_5 - R_6 + R_7 - R_8$$

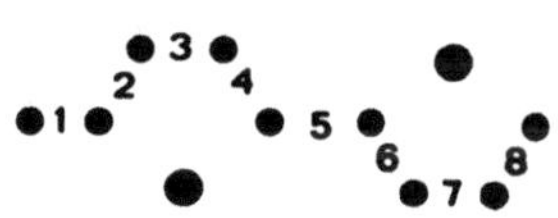

Curves similar to that reported for PA (Figure 1) have been obtained also in these cases and the agreement with experimental data (Raman data of pristine and infrared data of doped and photoexcited species) is reasonable.

References

1. R.Peierls, "Quantum Theory of Solids", Oxford University Press, London, 1955
2. W.Su,J.Shrieffer,A.Heeger,Phys.Rev.Lett.,42,1698(1979)
3. E.J.Mele in "Handbook of Conducting Polymers"(T.A. Skotheim, ed.), Dekker, N.Y., 1986
4. E.Ehrenfreund,Z.Vardeny,O.Brafman,B.Horowitz, Phys.Rev. B 36, 1535 (1987)
5. T.Kakitani, Progr.Theor.Phys., 51, 656 (1974)
6. L.Piseri,R.Tubino,G.Dellepiane, Solid State Commun. 44, 1589 (1982)

7. E.J.Mele,M.J.Rice, Phys.Rev.Letters, 45, 926 (1980)
8. E.Mulazzi,G.P.Brivio,E.Falques,S.Lefrant, Solid State Commun., 46, 851 (1983)
9. C.Castiglioni, M.Gussoni, J.T.LopezNavarrete, G.Zerbi, Solid State Commun., 65, 625 (1988)
10. G.Zerbi, C.Castiglioni, J.T.LopezNavarrete, Tian Bogang, M.Gussoni, Synth.Metals, 28, 359 (1988)
11. C.Castiglioni,M.Gussoni,G.Zerbi,Synth.Met.,28,375(1988)
12. a.P.Pulay,G.Fogarasi,G.Pongor,J.E.Boggs,A.Vargha,J.Am. Chem. Soc., 105, 7037 (1983);
b.G.Fogarasi, P.G.Szalai, P.P.Liescheski, J.E.Boggs,P. Pulay, J.Mol.Structure(Teochem), 151 341 (1987)
13. H.Yoshida,M.Tasumi, J.Chem.Phys., 89, 2803 (1988)
14. J.T.LopezNavarrete,G.Zerbi, Synth.Metals, in press
15. H.Villar, M.Dupuis, J.Watts, G.Hurst, E.Clementi, J.Chem.Phys., 88, 1003 (1988)
16. L.Piseri, R.Tubino, L.Paltrinieri, G.Dellepiane, Solid State Commun., 46, 183 (1983)
17. F.Inagaki,M.Tasumi,T.Myazawa,J.Raman Spectr., 3,335 (1975)
18. G.Zannoni and G.Zerbi, J.Mol.Structure,100 485 (1983)
19. Y.Furukawa,T.Arakawa,H.Takeuchi,I.Harada,H.Shirakawa, J.Chem.Phys. 81, 1964 (1984)
20. H.Villar,M.Dupuis,E.Clementi,J.Chem.Phys.88,2859(1988)
21. J.T.LopezNavarrete,G.Zerbi, unpublished results
22. Bogang Tian,G.Zerbi, in this volume

"Effective Coordinate" and the Interpretation of IR and Raman Spectra of Polypyrrole, Pristine and Doped

Bogang Tian and G. Zerbi

Dipartimento di Chimica Industriale, Politecnico di Milano,
Piazza L. da Vinci 32, Milano, Italy

Abstract We calculate the resonance Raman spectrum of undoped and the IR spectrum of doped polypyrrole using the effective coordinate theory. Agreement with the experimental spectra is good. The extension of the calculations to the alkyl-polypyrrole shows that the spectra are dominated by the motions of the skeleton where π electrons are involved.

1. Introduction

The use of the effective coordinate theory [1] provides a way to calculate the Raman and IR spectra of polyconjugated electroactive polymers. We present here the case of polypyrrole.

2. Dynamical Calculation of Polypyrrole

In our work polypyrrole is taken as a 1-d crystal. Wilson's GF matrix method [2] extended to polymers [3] is employed to calculate the vibrational properties such as vibrational frequencies and vibrational displacements for polypyrrole. In this kind of calculation both the geometry and the force field of polymer are needed. The geometry of polypyrrole is taken from that of middle ring of tripyrrole in the crystalline state [4]. The chemically convenient internal coordinates, such as bond length and bond angle, are used in our calculations.

Two independent force fields are used in this work : i) a valence force field with the force constants transferred from pyrrole [5] and biphenyl [6]; ii) a force field based on a MNDO calculation [7] of tetrapyrrole [8]. Scaling of the force field has been carried out using the method of ref. [9]. This starting force field has been refined and fitted to the experimental Raman data [10] on the basis of "Effective Coordinate Theory" (see next section).

3. Effective Coordinate of Polypyrrole

The so-called "Effective Coordinate Theory" (ECT) has been developed in our group as an extension of Horovitz's "Amplitude Mode Theory" [11]. The central idea of the theory is to handle the skeletal vibrations of the π conjugated polymers

Springer Series in Solid-State Sciences, Vol. 91
Electronic Properties of Conjugated Polymers III
Editors: H. Kuzmany · M. Mehring · S. Roth

as a whole by defining an "effective coordinate" $\mathcal{R}$. For polypyrrole

$$\mathcal{R} = R_1 - R_2 + R_3 - R_4 + R_5 - R_6 + R_7 - R_8$$

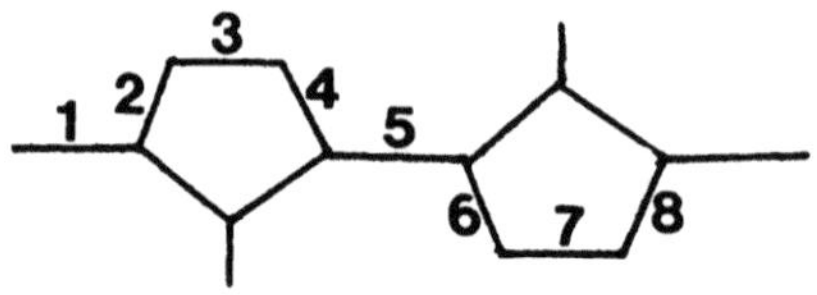

This coordinate describes a totally symmetric (A_g) skeletal vibration, which most effectively involves the conjugation of the π electrons along the polymer chain. The resonance Raman spectrum is mainly contributed by this kind of vibration [12], whereas for the doped polymer the doping induced bands are thought to be the IR activated modes which contain the $\mathcal{R}$ vibration [1].

In our theory, after defining the effective coordinate, the corresponding force constant $f_{\mathcal{R}}$ is taken as the measure of conjugation and the dependency of the vibrational frequencies on $f_{\mathcal{R}}$ provides a description of the π conjugation in the system. The case of polypyrrole is shown in Fig. 1. This figure is calculated based on the force field ii). The experimental Raman frequencies [10] are fitted quite well by the calculated curves with $f_{\mathcal{R}} \cong 5.0$ mdyn/Å.

It has been shown that the intensity enhancement in the resonance Raman spectrum or the intensity of the doping induced bands of the polyconjugated ma-

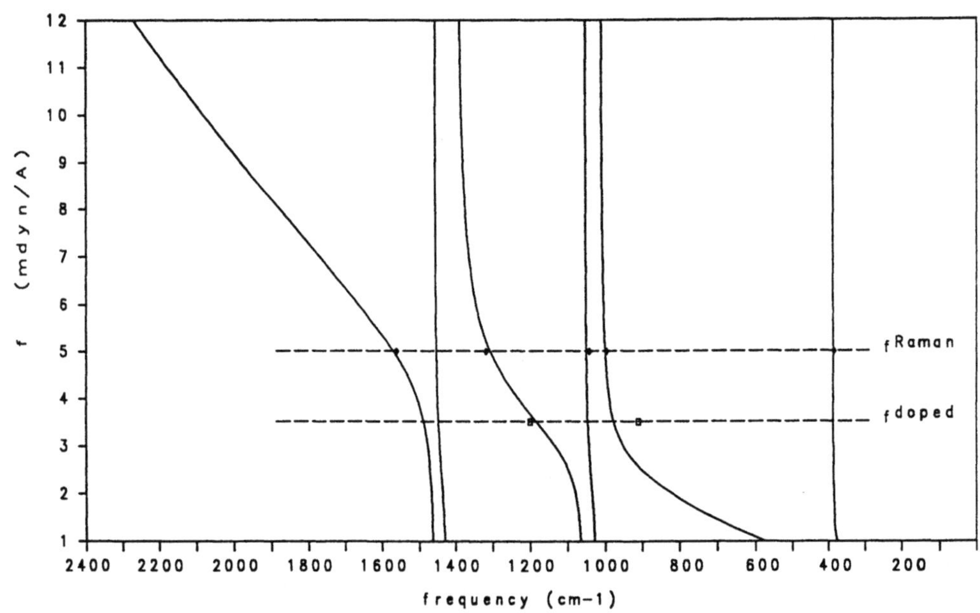

Fig. 1 Dependency of normal frequencies on the effective force constant $f_{\mathcal{R}}$ of the A_g modes in polypyrrole, based on force field ii)

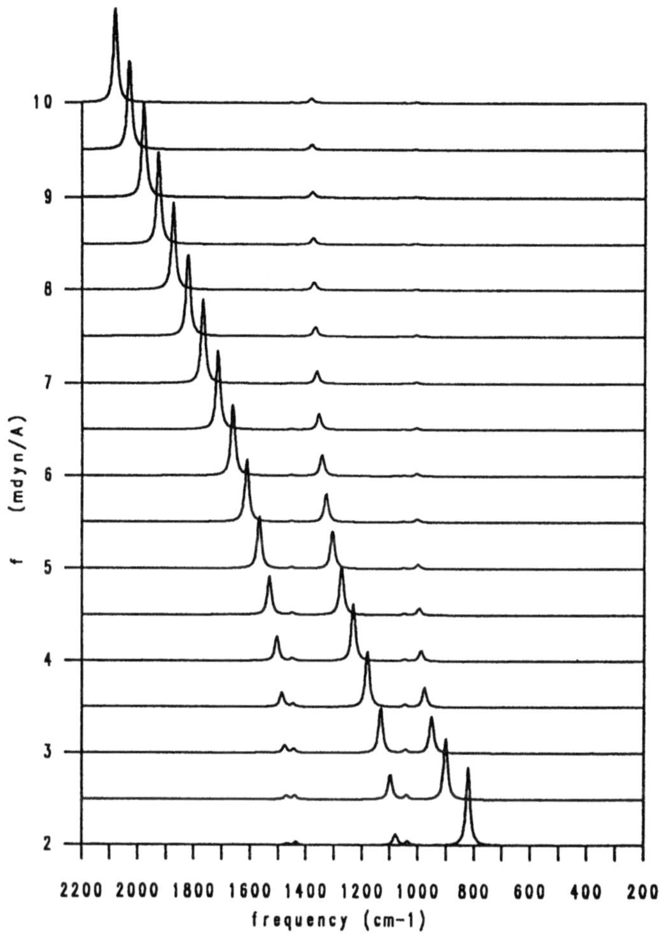

Fig. 2 Dependency of the calculated Raman intensities on $f_{\mathfrak{R}}$ of the A_g modes in polypyrrole, based on force field ii)

terials is propotional to the square of $\mathfrak{R}$ component of the vibrational amplitude (eigenvector) $L_{\mathfrak{R}}$. In Fig. 2 we show the calculated Raman intensities as functions of $f_{\mathfrak{R}}$ to be compared with the experiments. With $f_{\mathfrak{R}} \cong 5.0$ mdyn/Å the experimental Raman spectrum of polypyrrole in ref. [10] is well predicted, with the exception of the medium intensity of C-H deformation at 1050 cm^{-1}.

The doping induced bands can be predicted also from Figs. 1 and 2. This is based on the idea that the strong doping induced bands are the IR activated modes corresponding to the totally symmetric ones [1,11]. In Figs. 1 and 2 the two doping induced bands at $\cong$ 1200 cm^{-1} and $\cong$ 910 cm^{-1} in the IR spectrum of doped polypyrrole [13] are fitted by setting $f_{\mathfrak{R}}$ = 3.0/3.5 mdyn/Å.

Care must be taken in the use of the force field (e.g. case i) above). If not properly chosen it may lead to unreliable results, as will be shown elsewhere.

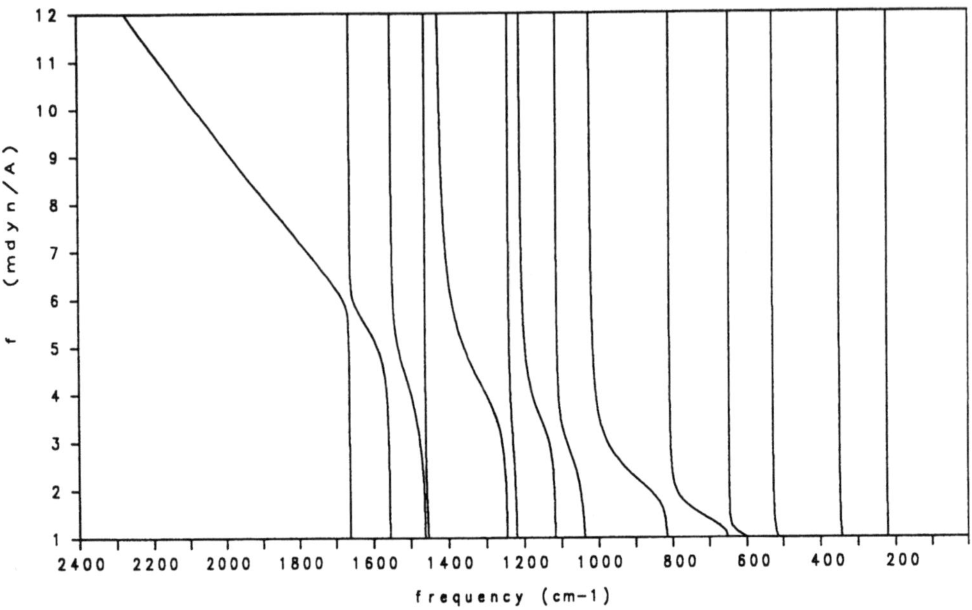

Fig. 3 Dependency of normal frequencies on the effective force constant $f_{\mathcal{A}}$ in alkyl-polypyrrole

Since the IR spectra of alkyl-substituted, doped polypyrrole are available [14] we have applied ECT to this case. We have carried out, so far, calculations on a simplified model of poly 3-methyl pyrrole. Since symmetry is lowered many more modes become totally symmetric and a few of them will couple with $\mathcal{A}$, thus aquiring some infrared (if doped) or Ramn intensity (pristine). Figs. 3 and 4 show the results. The main concept which emerges is that the intensity (IR or Raman) associated with the $\mathcal{A}$ mode is spread over many modes. Thus the IR or Raman spectra of poly alkyl-pyrroles will be mainly dominated by the absorption or scattering of $\mathcal{A}$ due to the skeletal vibrations which mostly affect the network of π electrons. Such predictions are nicely verified by the experiments.

References

1. i) C. Castiglioni, J. T. Lopez Navarrete, M. Gussoni and G. Zerbi, Solid State Comm., 65, 625 (1988); ii) G. Zerbi, C. Castiglioni, J. T. lopez Navarrete, B. Tian and M. Gussoni, Synthetic Metals, 28, D359 (1989).
2. E. B. Wilson, J. C. Decius and P. C. Cross, Molecular vibrations, McGraw-Hill, New York, 1955.
3. L. Piseri and G. Zerbi, J. Chem. Phys., 48 3561 (1968).
4. The X-ray diffraction data on $(Py)_2$ and $(Py)_3$ were kindly provided by Dr. G. B. Street as a private communication; see also ref. 13.

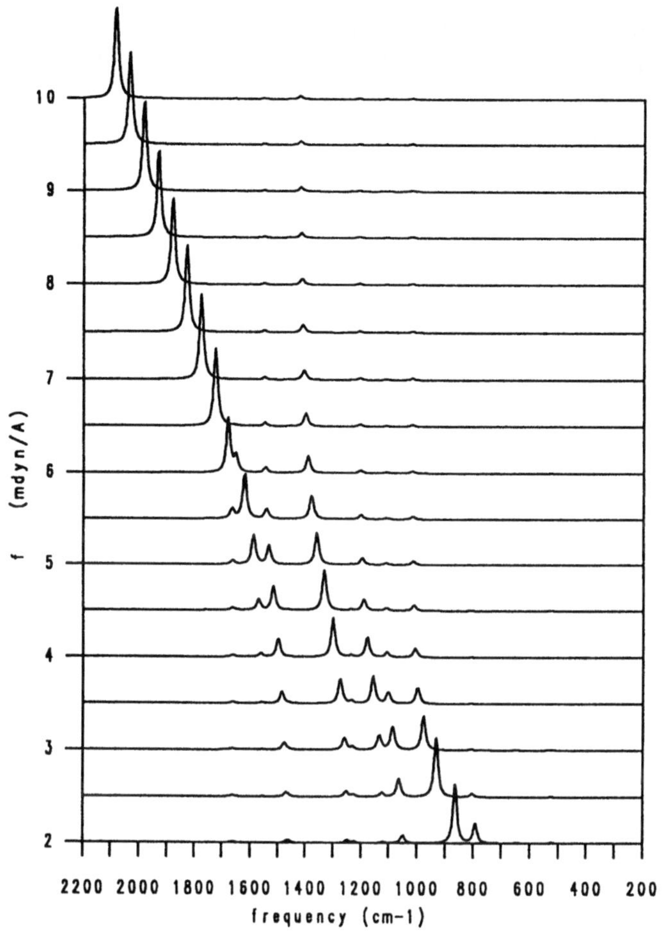

Fig. 4 Dependency of the calculated Raman intensities on the effective force constant f_{R} in alkyl-polypyrrole

5. i) Y. Xie, K. Fan and J. E. Boggs, Mol. Phys., 58 401 (1986); ii) D. W. Scott, J. Mol. Spectrosc., 33, 77 (1971) and references therein.
6. G. Zerbi and S. Sandroni, Spectrochimica Acta, 24A, 511 (1968).
7. M. J. S. Dewar and W. Thiel, J. Am. Chem. Soc., 99, 4899 (1977).
8. i) B. Tian and G. Zerbi, Synthetic Metals, 28, C1 (1989); ii) B. Tian and G. Zerbi, to be published.
9. P. Pulay, G. Fogarasi, F. Pangor, J. E. Boggs and A. Vargha, J. Am. Chem. Soc., 105, 7037 (1983).
10. Y. Furukawa, S. Tazawa, Y. Fuji and I. Harada, Synthetic Metals, 24, 329 (1988).

11. B. Horovitz, Solid State Comm., 41, 792 (1982).
12. A. C. Albrecht, J. Chem. Phys., 34, 1476 (1961); A. C. Albrecht and M. C. Hutley, *ibid*, 55, 4438 (1971).
13. B. Street, Handbook of Conducting Polymers, vol.1, Dekker, New York, 1986, p 256.
14. J. Ruhe, private communication.

Dynamics of Heterocyclic Polymers

E. Faulques[1], *W. Wallnöfer*[2], *and H. Kuzmany*[2]

[1]IPCM, Université de Nantes,
2 rue de la Houssinière, F-44072 Nantes Cedex 03, France
[2]Institut für Festkörperphysik der Universität Wien,
Strudlhofgasse 4, A-1090 Wien, Austria

ABSTRACT. We present a vibrational analysis of polythiophene, polypyrrole and polyisothianaphthene. Based on monomer force constants and MNDO geometries, the A_g frequencies for aromatic and quinoid forms are calculated and compared with experimental Raman lines.

1. INTRODUCTION

In this paper we present an in-plane dynamical calculation of heterocyclic polymers. In a detailed work (1), we applied a method based on complex coordinates to investigate the aromatic (A) structure and the quinoid (Q) structure of polythiophene (PT), polypyrrole (PPy) and polyisothianaphthene (PITN) (fig.1). From the force constants we defined some magnitudes depicting quantitatively the concepts of aromaticity, quinoidicity and alternation in heterocyclic polymers. We present here the calculation of the Raman Ag modes of PT, PPy, PITN obtained by direct transfer of thiophene, pyrrole and benzene force constants (2,3). We describe the model used to perform the calculation of the A and Q forms and we show that the monomers valence force fields are able to account with a good approximation for the experimental Raman data. We discuss briefly the capacities of this simple model to evaluate the most probable ground states of these heterocyclic polymers.

2. MODEL

2.1 Cell parameters and coordinates

For PT, PPy and aromatic PITN, we introduced the bond lengths and the angles calculated by Brédas et al. (4,5) on oligomers of thiophene and pyrrole. For quinoid PITN the values of ref.6 were transfered in our dynamical calculation. The bond lengths are indicated in fig.1 for PT and PITN. Note that for this last polymer the bond lengths of the fused benzene ring are not equivalent in both forms. In PPy the bond lengths are: l(t)= 1.424, l(t')= 1.363, l(u)= 1.385, l(r)= 1.474, N-H= 0.99 A in the A form and: l(t)= 1.38, l(t')= 1.437, l(u)= 1.428, l(r)= 1.357 A in the Q form. All these parameters must not be confused with the internal coordinates (fig.2). The internal coordinates are expressed in terms of small displacements of the atoms with respect to their equilibrium positions in the harmonic approximation. In order to avoid the earlier method of Shimanouchi (7) commonly used in the vibrational analysis of

Springer Series in Solid-State Sciences, Vol. 91
Electronic Properties of Conjugated Polymers III
Editors: H. Kuzmany · M. Mehring · S. Roth

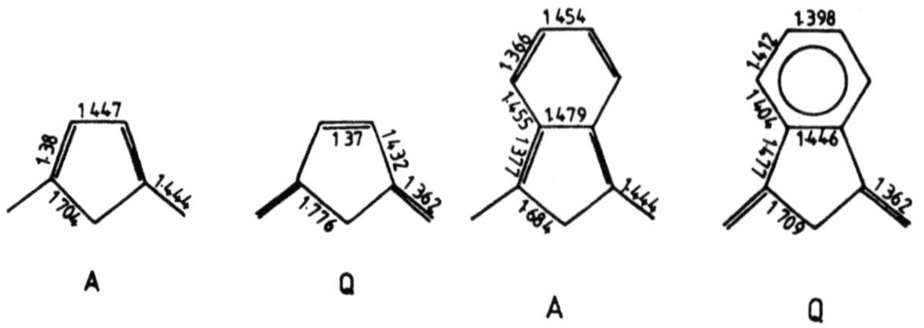

Fig. 1. Aromatic (A) and Quinoid (Q) forms in PT and PITN

polymers we preferred to work with space groups arguments derived from crystal lattice dynamics (8) to take advantage of the zig-zag arrangement of the monomer units along the chain axis. This is valid only in the case of planar chains free of defects. Under these assumptions, the heterocyclic polymers belong to the $L_{21/mcm}$ line group isomorphous to the point group D_{2h} (9). We construct complex coordinates (also called symmetry coordinates for k=0) and express the valence force field in them. To obtain the Fourier's dynamical matrix in complex coordinates we symmetrize the internal coordinates α and the cartesian coordinates X relatively to the glide plane (translation a/2) which is perpendicular to the chain plane. Therefore we get $2^{1/2}I = \alpha \pm \alpha'\exp(-ika/2)$ for the complex internal coordinates and $2^{1/2}C = X \pm X'\exp(-ika/2)$ for the complex cartesian coordinates.

These coordinates are related by I = BC where B is the transformation matrix.

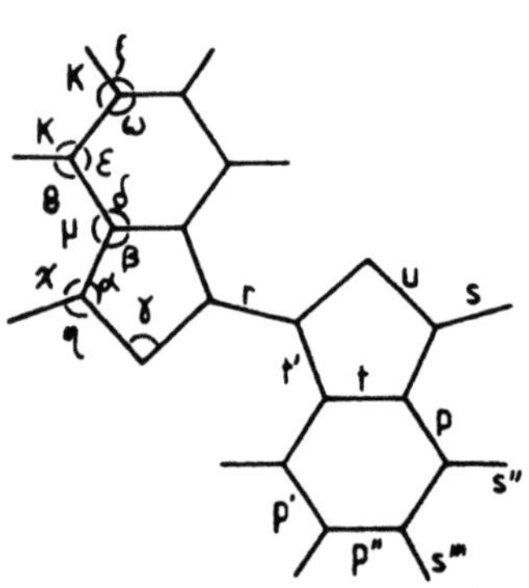

Fig. 2. Internal Coordinates in the Unit Cell of PITN

2.2 Valence force fields

We decided to take for the interring interaction a value close to 4.6 mdyn/A recently determined in the aromatic polyparaphenylene (3). For calculating the Q structures we have mutually exchanged the stretching force constants $F(r^2)$ and $F(t'^2)$ in PT and PITN. In the quinoid PPy we have approximately equalized the stretching force constants $F(r^2)$ and $F(t^2)$ with the $F(t'^2)$ of A structure. The other stretching and bending interactions remain the same for both forms of PT and

PPy. For PITN, in absence of any data concerning the MNDO valence force field in the monomer isothianaphtene we were constrained to scale approximately the force constants with the bond lengths of the fused benzene ring. The off-diagonal interactions are the same in all the cases.

The second-neighbor interactions between two consecutive monomer units are of particular importance for calculating the chain phonons. For example the off-diagonal interaction relative to the internal coordinate t' can be written as:

$$F^2(t't'^*) + 1/2F^2(t't'^{**})\exp(ika/2)$$

where the first term describes the coupling (t't'*) inside the thiophene ring and the second term the coupling (t't'**) between two consecutive rings.If we include the third-neighbour coupling $F^3(rr)$ between two consecutive interring bonds for long range effects then we get the squared matrix element:

$$F(r^2) + 1/2F^3(rr)\cos(ka/2).$$

These k-dependent terms are matrix elements of the force constant matrix expressed in complex internal coordinates.

More details on the valence force fields of unrefined and refined structures and about the symmetrized Fourier's matrices can be found in ref.1. The IR and Raman modes are obtained by setting k=0 in B and F.

3. RESULTS AND DISCUSSION

For $k=\pi/a$ the phonons branches must stick two by two. This is due to the twofold degeneracy at the zone boundary which is a consequence of the non-symmorphic structures of the chains.
For $k\neq0$ if we exclude the acoustic modes we have respectively for PT, PPy and PITN 26, 30 and 50 phonons dispersions.
For k=0 we obtained the Ag calculated frequencies listed in Table 1. They are compared with experimental values found for neutral and doped polymers. The most intense Raman bands were assigned with Ag although some uncertainty remains in this classification because of the lack of polarized experiments on stretched samples. In ref.10 the 1500 cm-1 band of PT was assigned with the Bg symmetry. However, some trials we performed showed that the normal modes are not strongly affected with this kind of exchange. It may be seen that the agreement between experimental and theoretical frequencies is quite good with respect to the approximations introduced in the calculation. In PT and PPy the experimental values are very close to the calculated A forms. So the ground state of these polymers can be depicted with the aromatic geometry as already pointed out by MNDO calculations. The case of PITN is interesting. We found at first sight a good fit between the experimental Raman bands of the undoped (neutral) polymer and the calculated frequencies of the A structure. However a careful examination of the values presented here shows that the calculated frequencies in the A structure are always higher than those found in the Q stucture, except for the 11th mode. For the experimental frequencies the values of the doped polymer are always higher than those of the undoped material except for the 11th mode

Table 1. Raman Ag modes of PITN, PT and PPy: comparison of calculated and experimental frequencies in cm-1

Mode	PITN Cal A	PITN Cal Q	PITN Exp neut	PITN Exp dop	PT Cal A	PT Cal Q	PT Exp neut	PPy Cal A	PPy Cal Q	PPy Exp neut
1	195	184	197	-	246	250	-	326	346	-
2	455	449	446	460	673	678	703	967	916	996
3	640	626	672	672	1003	996	1050	1022	1007	1044
4	949	920	991	1005	1183	1105	1224	1179	1247	1319
5	1030	1030	1065	1065	1289	1447	1464	1426	1438	1526
6	1184	1123	1169	1169	1577	1799	1500	1649	1888	1562
7	1232	1204	1196	1220	3082	3083	-	3107	3106	-
8	1333	1293	1307	1313				3535	3529	-
9	1467	1460	1448	1448						
10	1491	1496	1470	1480						
11	1586	1809	1590	1520						
12	3024	3023	-	-						
13	3077	3075	-	-						

which has a lower frequency in the doped polymer. So we can conclude finally that the Q form is a good candidate to describe the ground state of the neutral PITN. This observation is supported by the MNDO work of Lee and Kertesz who claimed that the undoped polymer exhibits a quinoid geometry (6).

We thank Dr. J. P. Buisson for useful discussions.

REFERENCES

1. E. Faulques, W. Wallnöfer and H. Kuzmany: J. Chem. Phys. in press (June 1989).
2. D. W. Scott: J. Molecular Spectrosc. 31, 451 (1969) and J. Molecular Spectrosc. 37, 77 (1971).
3. J. P. Buisson, S. Krichene and S. Lefrant: Synthetic Metals 21, 229 (1986)
4. J. L. Brédas, B. Thémans, J. G. Fripiat, J. M. André and R. R. Chance: Phys. Rev. B29, 6761 (1984).
5. J. L. Brédas, B. Thémans, J. M. André, A. J. Heeger and F. Wudl: Synthetic Metals 11, 345 (1985).
6. Y. S. Lee and M. Kertesz: J. Chem. Phys. 88, 2609 (1988).
7. T. Shimanouchi, M. Tsuboi and T. Miyasawa: J. Chem. Phys. 35, 693 (1961).
8. M. Born and K. Huang, Dynamical Theory of Crystal Lattices Clarendon Press (1966).
9. I. B. Bozovic and N. Bozovic: J. Phys. A: Math. Gen. 14, 1825 (1981).
10. Y. Furukawa, S.Tazawa, Y. Fujii and I. Harada: Synthetic Metals 24, 329 (1988).

Raman Spectra of Polyphenylene-Vinylene and Similar Heterocyclic Compounds

S. Lefrant[1], *J.Y. Mévellec*[1], *J.P. Buisson*[1], *E. Perrin*[1], *H. Eckhardt*[2], *C.C. Han*[2], *and K.Y. Jen*[2;*]

[1]Laboratoire de Physique Cristalline, I.P.C.M.**,
Université de Nantes, F-44072 Nantes Cedex 03, France
[2] Allied-Signal Inc., Corporate Technology, Morristown, NJ 07960, USA
*Present address: Enichem Americas Inc., Monmouth Junction, NJ 08852, USA
** Unité Mixte de Recherche CNRS–Université de Nantes

Abstract. In this paper, we report on the experimental Raman spectra of a series of polymers composed of aromatic or heteroaromatic rings separated by a vinylene group, namely polyphenylene-vinylene (PPV), polyfurylene-vinylene (PFV) and polythienylene-vinylene (PTV). Most of our studies have been done on PPV for which the main Raman bands are observed at 966, 1174, 1330, 1550, 1586 and 1628 cm^{-1}. No "frequency dispersion" effect is seen as the excitation is changed from the red to the violet range. In PTV, the Raman bands are peaked at 654, 932, 1045, 1287, 1410 and 1586 cm^{-1} and in PFV, we observe the corresponding ones which are slightly shifted. In the latter two cases, a small band shift is detected if the 457.9 nm laser line is used as the excitation wavelength.
We have built a dynamical model in order to assign the IR and Raman modes in all cases. Among the numerous force constants used in the calculations, we have focused our attention on those affected by the different carbon-carbon bonds. The parameters which are determined to give a good fit for the Raman spectra are consistent with a benzenoid structure. Moreover, the modified Raman spectrum of $FeCl_3$-doped PPV is well interpreted if we assume a quinoid character along the chains.

1. Introduction

Poly-(arylenevinylenes) have attracted increased interest in recent years in the field of conducting polymers. Most of the studies have been developed on polyphenylene-vinylene (PPV) prepared from a soluble sulphonium polyelectrolyte precursor (1). Similar methods have been applied to synthetize poly-heteroaromatic compounds (2) such as polythienylene-vinylene (PTV) and polyfurylene-vinylene (PFV). All these polymers become conductive after doping with appropriate agents.

In order to gain information on the structure of the polymer, we have used the Resonance Raman Scattering (RRS) technique which has been shown to be particularly powerful in the case of other compounds such as poly-paraphenylene (PPP) (3). Since the technique is not destructive, the modifications induced in the

Springer Series in Solid-State Sciences, Vol. 91
Electronic Properties of Conjugated Polymers III
Editors: H. Kuzmany · M. Mehring · S. Roth

Raman spectra by doping the polymeric chains can be followed and this point turns out to be of particular importance if one wants to evidence the appearance of the quinoid structure on the chains as predicted by the theory (4).

2 . Experimental Results

2.1. Polyphenylene-vinylene (PPV)

Preliminary results have already been published (5). Let us recall the main features of the Raman spectra. In Table 1, we have collected the Raman frequencies of undoped PPV together with those of $FeCl_3$-doped PPV for an excitation wavelength in the red range λ_l = 676.4 nm.

Table 1. Raman frequencies of PPV (in cm^{-1})

Undoped PPV	1174	-	1304	1330	1550	1586	1628
$FeCl_3$-doped PPV	1174	1120	1300	1330	1530	1566	1595

It must be noticed that for an excitation wavelength in the near U-V , i.e. λ_l=351.1nm, no change is detected in the Raman spectrum of undoped PPV indicating that unlike trans-polyacetylene or other polymers like polythiophene, no "frequency dispersion" is seen in this case. For $FeCl_3$-doped PPV, upon excitation with 351.1 nm, a Raman spectrum similar to that of the undoped polymer is recorded. Such an observation was already made in the case of polyparaphenylene for example (3).

2.2. PTV and PFV

Figure 1 shows the Raman spectra of both polythienylene-vinylene (PTV) and poly-furylene-vinylene (PFV) for the two excitation wavelengths λ_l=457.9 nm and λ_l=676.4 nm. It is seen that, unlike for PPV, some changes are observed in the Raman spectra on going from the red to the violet excitation. The most significant modifications consist in a broadening and in a shift of some Raman lines. This effect is more pronounced in PFV than in PTV, as can be seen in Table 2 where the Raman frequencies are collected for both polymers.

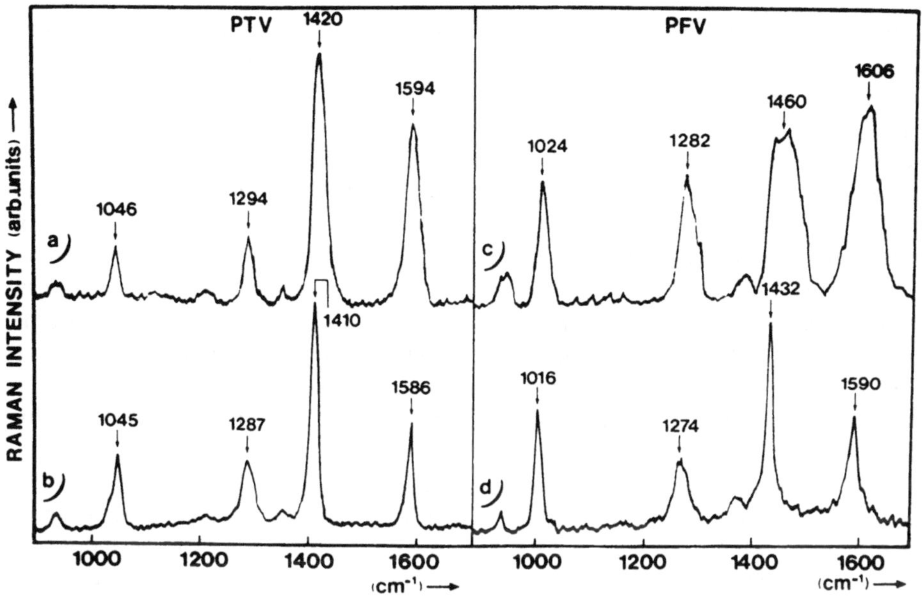

Fig.1 Raman spectra of : a) PTV, λ_l=457.9 nm ; b) PTV, λ_l=676.4 nm ; c) PFV, λ_l=457.9 nm ; d) PFV, λ_l=676.4 nm

Table 2. Raman frequencies of PTV and PFV for two different excitation lines

PTV	λ_l=676.4 nm	654	932	1045	1212	1287	1354	1410	1586
	λ_l=457.9 nm	654	936	1046	1212	1294	1354	1420	1594
PFV	λ_l=676.4 nm	765	952	1016	--	1274	1374	1432	1590
	λ_l=457.9 nm	765	958	1024	--	1282	1384	1460	1606

3. Interpretation and Discussion

A complete description of the dynamical calculations which have been done to assign all IR and Raman modes in PPV will be given in a forthcoming paper (6). They are based on a valence force field model characterized by force constants which can be divided into four different groups : the first one is associated with the unperturbed benzene ring, the second one with the benzene ring perturbed by the para-substitution, the third refers to the coupling between the ring and the vinylene group and finally the fourth to the vinylene group itself. Notice that our theoretical study has been restricted to the in-plane vibrations. For PPV, the point group is C_{2h}, the unit cell is composed of 14 atoms, yielding active vibrations of the following type : 14 A_g and 14 B_u. In order to achieve a better assignment of both IR and Raman modes, we have used a model compound, namely trans-stilbene. In the framework of our dynamical model, a good fit between experimental and calculated frequencies can be obtained and among the numerous force constants needed for the calculations, the values found for the four parameters associated with the carbon-carbon bonds agree well with a benzenoid type of structure along the chain (6.32 mdyn.A^{-1} inside the benzene ring, 4.8 and 7.06 mdyn.A^{-1} for the single and double bonds respectively for the vinylene group).

The modified Raman spectra which are observed after doping can also be accounted for, provided the four force constants already mentioned are subjected to some modifications which are consistent with a quinoid-like structure along the chains.

Concerning PTV and PFV, preliminary calculations based on a similar model have already led to a good interpretation of the main vibrational modes. Starting with the parameters used in polythiophene for example, a preliminary set of force constants has been obtained, but some uncertainties still exist due to the lack of experimental results in model compounds associated with the two polymers. Further studies are in progress to achieve a definitive assignment of the IR and Raman vibrational modes.

In conclusion, the dynamical model used in this study gives significant information on the structure of the polymers considered as individual and infinite chains in the case of undoped and doped compounds. The "frequency dispersion" effect, which is not accounted for in this type of calculations, is clearly related to the π electron delocalization along the polymeric chains.

References

(1) I.Murase, T.Ohnishi, T.Vogushi and M.Hirooka, Polym.Comm., 25,237 (1984)
(2) K.Y.Jen, M.Maxfield, L.Shacklette and R.Elsenbaumer,Chem.Comm.309(1987
(3) J.P.Buisson, S.Krichene and S.Lefrant, Synth.Met., 21, 229 (1987)
(4) J.L.Bredas, B.Themans, J.G.Fripiat, J.M.Andre and R.R.Chance, Phys.Rev B 29, 6761 (1984)
(5) S.Lefrant, E.Perrin, J.P.Buisson, H.Eckhardt and C.C.Han, Synth.Met.29 (1989)
(6) S.Lefrant, E.Perrin, J.P.Buisson, H.Eckhardt and C.C.Han, in preparation

Resonance Raman Spectroscopy of Accumulation Layers in Durham-Route Polyacetylene

R.A. Lawrence, J.H. Burroughes, and R.H. Friend

Cavendish Laboratory, University of Cambridge,
Madingley Road, Cambridge CB3 0HE, UK

Abstract. Semiconductor devices with a polyacetylene active layer can show charge depletion, inversion and accumulation. In field effect transistors variation of the bias voltage gives rise to known changes in charge concentration in the polyacetylene layer. Charge is stored in solitonic self-localised excitations of the polymer chain. Two main features are seen in Raman difference spectra between the accumulation regime and the zero bias case. The A_g modes seen in the normal Raman spectrum of polyacetylene are bleached in accumulation - one defect bleaching the Raman signal for approximately 30 carbon atoms. New Raman-active modes are also observed at 1065 cm^{-1} and 1455 cm^{-1}. These are attributed to the amplitude mode of the soliton. These frequencies are higher than predicted by theory, but may be raised by disorder and pinning.

1. Introduction

In conjugated polymeric semiconductors such as *trans*- polyacetylene, the presence of charge on a chain causes a structural relaxation about the charge to form a localised state. The self-localised excitation in polyacetylene is the soliton, a domain wall separating two regions of the chain with opposite sense of bond alternation.

The use of polyacetylene as the active semiconductor in device structures such as the MISFET provides a new and controlled means for introducing charge on to the polymer chains [1,2]. Semiconductor devices with a polyacetylene active layer can be driven to accumulation or depletion, providing a controlled concentration of charged soliton defect states.

In this paper we describe the use of Raman spectroscopy to study the vibrational modes of a polyacetylene chain with a controlled charge concentration. The polyacetylene was employed as the active semiconductor layer in a metal-insulator semiconductor field effect transistor (MISFET) [1,2]. The polyacetylene was prepared by the Durham route, which allows the sample to be prepared while the polymer is in soluble precursor form. The Raman spectrum of the defect states was studied by a difference measurement between different biasing states of the transistor gate. The results obtained are compared with existing model calculations for the Raman spectrum of polyacetylene with similar levels of charge concentration.

Springer Series in Solid-State Sciences, Vol. 91
Electronic Properties of Conjugated Polymers III
Editors: H. Kuzmany · M. Mehring · S. Roth

2. Experimental System and Results

The devices studied in this experiment are constructed on a Si substrate, with an n-silicon gate layer, an insulator layer of 120 nm silicon dioxide and interdigitated source and drain contacts of poly n-silicon coated with aluminium. The precursor polymer is then spin-coated on to the substrate from 2-butanone solution in an argon-filled glove box. Film thicknesses are controlled by variation of spin speed and solution concentration. Films of 100-200 Å were preferred for this series of experiments. The precursor is then transformed to *trans*-polyacetylene under vacuum at 370 K for 15 hours. Samples were then mounted in a purpose-built evacuable vessel with optical windows. Raman experiments were performed in dynamic vacuum ($p < 10^{-4}$ mbar).

A Dilor Omars 89 double spectrometer, with a multi-diode array detector, was used for these measurements. Spectra are obtained from subtraction of a "background" scan from a "signal" scan. Here the "signal" was the measured scattering from a device with the gate at -30 V with respect to the source and drain, and the "background" was the same scattering signal with the gate set to 0 V. Under these bias conditions the charge carriers will be positive. The difference in the hole concentration between thes two gate voltages can be calculated from the field across the insulating layer [1], $n_s = \varepsilon_r \varepsilon_o V / q d$, which gives a value of 5.3×10^{-12} charges/cm^2. If we compare this to the number of carbon atoms per unit area, found from the density (1.05 g/cm^3), we have a calculated fractional density of 2.21×10^{-4} charges/carbon atom for a 100 Å film. As the thickness of the film is substantially less than the absorption depth (~500 Å), the whole of the film is accessible to the laser excitation.

The resulting difference spectrum gives the signal from the introduction of charges into the polyacetylene. Spectra were taken over periods of up to 16 hours, with normal Raman spectra taken before and after each difference measurement. The spectral region from 900 cm^{-1} to 1700 cm^{-1} was studied at a range of laser wavelengths from 476.5 nm to 584.0 nm.

Durham polyacetylene when prepared as an unoriented film, as in the present work, is poorly crystalline, and the Raman scattering from the A_g modes is at higher frequencies than found in oriented or Shirakawa-prepared samples [3,4].

Two different classes of feature are observed in the difference spectrum of accumulation layers in Durham polyacetylene (Figure 1) .

The main class of feature is negative and arises from the bleaching of the resonance-enhanced Ag modes that dominate the normal Raman spectrum of *trans*-polyacetylene. The two main modes are near 1100 cm^{-1} and 1500 cm^{-1} with the exact frequency a function of the excitation energy. The bleaching of these two main modes dominates the difference spectrum. The dispersion of the bleaching is the same as that of the normal spectrum, but the bleaching fraction (the ratio of the integrated bleaching signal to the normal signal) changes with excitation energy, being 7×10^{-3} at 476 nm, 6.9×10^{-3} at 488 nm, 7.7×10^{-3} at 514 nm and 1.6×10^{-2} at 584 nm.

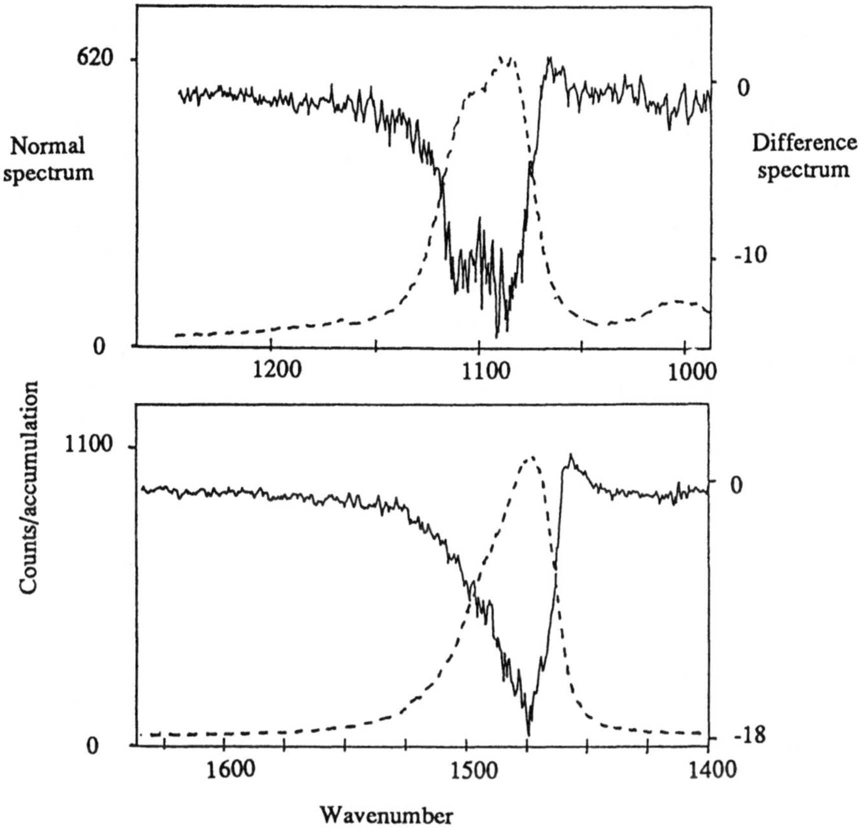

Figure 1 Difference spectra of polyacetylene accumulation layers. The solid line is the difference signal; the dashed line is the normal Raman spectrum. The excitation wavelength is 584 nm

There are also small positive features to the low energy side of each of the bleaching features. These are much less intense, do not disperse with excitation energy, and occur at 1065 cm^{-1} and 1455 cm^{-1}.

3. Discussion

We seek to explain these observations by the use of the multi-mode model of Mele and Hicks [5], an extension of the amplitude mode model of Horovitz [6]. This multi-phonon approach is used to determine the vibrational modes of the *trans*-polyacetylene chain both with and without chain defects. Initially the bare phonon modes of an undimerised chain are considered. Certain of these modes have the correct symmetry to couple to the dimerisation when π electrons are introduced - these are the totally symmetric Raman-active A_g modes. The resulting excitations for the dimerised chain are amplitude modes, consisting of linear combinations of those modes which couple to the dimerisation, renormalised by the electron-phonon coupling.

This approach was originally devised to explain the Raman spectrum of *trans*-polyacetylene, and the infra-red modes introduced on doping or photoexcitation. The Raman dispersion is a resonant effect - it results from variation of electron-phonon coupling with the band gap. The new doping-induced infra-red modes are pinned translational modes of the defect as a whole. Different classes of defects should however have their own vibrational excitations. The vibrational frequencies for a soliton have been calculated in single phonon Hamiltonians, such as SSH and TLM, with some consistency [7,8]. Vibrational frequencies in the multi-mode model can then be calculated using an appropriate phonon propagator: the factor involved in the renormalisation is the ratio of the mode frequency to the bare mode frequency, both in a single phonon model

$$D_0(\omega) = \sum_n \frac{\lambda_n}{\lambda} \frac{\omega_n^{o2}}{\omega^2 - \omega_n^{o2}} = \frac{-1}{1 - \omega_m^2 / \omega_Q^2}$$

where ω_n^o are the frequencies of the totally symmetric modes of the bare polyacetylene chain (with no π electrons present), λ_n/λ are relative electron-phonon coupling constants for each such mode, $\tilde{\lambda}$ is the effective electron-phonon coupling constant, ω_m is the frequency of the amplitude mode in the single phonon model, ω_Q is the frequency of the bare mode, also in the single phonon model.
Such calculations agree that there should be a vibrational mode representing the fluctuation in the amplitude, or width of the soliton; we shall call this the A mode. From the calculations of Terai and others [8], the A mode frequency in the single mode model is calculated as $\omega_m{}^2 = 0.723\ (2\ \lambda\ \omega_Q{}^2\ )$.

The new Raman-active modes at 1065 cm^{-1} and 1455 cm^{-1} are at rather higher frequencies than the values of equal to 973 cm^{-1} and 1414 cm^{-1} calculated with a value of electron-phonon coupling of 0.200 appropriate to unstretched Durham polyacetylene [3,4]. There is clear evidence from the non-zero frequency of the Goldstone mode for the strong effect of pinning in this system - this could readily account for the increased frequency observed for the A mode.

The presence of a defect on the polyacetylene chain is also expected to bleach the normal Raman spectrum. Terai et al calculate [9] that this bleaching will extend over a range of 30 Carbon atoms - that is, a 30% bleaching for a 1% soliton density. This large bleaching effect corresponds to the enhancement in the band to band optical absorption coefficient as calculated by Kivelson et al [10].

The bleaching data can be modelled quantitatively. Excitation in the red gives a larger bleaching fraction than blue excitation. This suggests that the soliton defects prefer to reside on polyacetylene with a smaller band gap - the less disordered chains. Quantitatively, the areal soliton density was calculated simply from the field across the insulating layer to be 2.21 x 10^{-4} charges/carbon atom. Using the calculated enhancement factor of 30, this should give us a bleaching of 0.007 - which is as we observe in the blue, though below that observed in the red.

4. Conclusions

Solitons in polyacetylene are accessible to Raman spectroscopy by means of voltage modulated difference measurements on devices with a polyacetylene active semiconductor layer.

Solitons give rise to a bleaching of the normal Raman signal approximately 30 times larger than the soliton density - the increase in the bleaching fraction with decreasing excitation energy indicates that solitons prefer to reside on the less disordered chains.

This experiment allows the first observation of the soliton A mode in *trans*-polyacetylene. The vibrational frequencies associated with the mode have been raised considerably by disorder and pinning.

We thank British Petroleum plc for financial support for this work.

References

1. J. H. Burroughes, C. A. Jones, and R. H. Friend, *Nature* **335,** 137 (1988)
2. J. H. Burroughes, C. A. Jones, R. A. Lawrence and R. H. Friend, these proceedings
3. R. H. Friend, D. D. C. Bradley, C. M. Pereira, P. D. Townsend, D. C. Bott and K. P. J. Williams, *Synthetic Metals* **13**, 101 (1986)
4. R. H. Friend, H. E. Schaffer, A. J. Heeger and D. C. Bott, *J. Phys. C.* **20**, 6013 (1987)
5. E. J. Mele and J. C. Hicks, *Phys. Rev. B* **32**, 2703 (1985)
6. B. Horovitz, *Solid State Comm.* **41**, 729 (1982)
7. J. C. Hicks and G. A. Blaisdell, *Phys. Rev. B* **32**, 919 (1985)
8. A. Terai, Y. Ono, and Y. Wada, *J. Phys. Soc. Jpn.* **55**, 2889 (1986)
9. A. Terai, Y. Ono, and Y. Wada, *Synthetic Metals* **28**, D353 (1989); A. Terai, Y. Ono, and Y. Wada, *J. Phys. Soc. Jpn.* (to be published)
10. S. Kivelson, Ting-Kuo Lee, Y.R. Lin-Liu, I. Peschel, and Lu Yu, *Phys. Rev. B* **25**, 4173 (1982)

Laser-Impact Induced Isomerisation of a Film-Contained cis-Polyacetylene as Studied by Multichannel Raman Spectroscopy

V. Tabacik[1], *A. Djebaili*[2], *and M.J.M. Abadie*[2]

[1]Institut de Chimie Fine de l'USTL, Unité de Recherche 1097 du CNRS*
[2]Laboratoire d'Etudes des Matériaux Polymères – Matériaux Avancés Organiques, LEMP-MAO, Service 102*.
*Université des Sciences et Techniques du Languedoc (USTL), 2, place Eugène-Bataillon, F-34060 Montpellier Cedex, France

Although the impact of the laser photons (λ = 514.5 nm) on polyacetylene (PA) solid film essentially produces a thermal effect, a photochemical effect is also observed [1]. The laser induces the Raman scattering from the two film-components simultaneously with the conversion of PA_{Cis} macromolecules into PA_{Trans}. The total Raman-Stokes scattering from the selected frequency region, containing only two bands (at 1248 cm^{-1} from PA_{Cis} and at 1110 cm^{-1} from PA_{Trans}), was measured simultaneously [2]

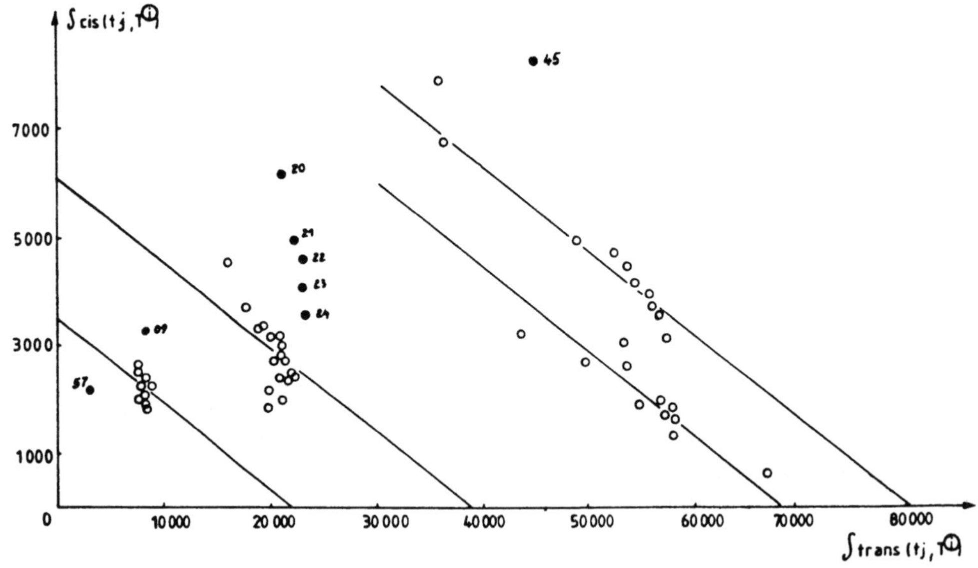

Fig. 1. According to four laser powers (i.e. 30, 120, 200 and 300 mW for i=1;2,3,6;4 and 5 respectively), four clusters of experimental points are obtained. The four multilinear regressionstraight-lines have a common slope -0.157 and N_i/f_{Cis} values of 3481.5, 6079.2, 10727.7 and 12603.2 respectively. The black points are artefacts

Springer Series in Solid-State Sciences, Vol. 91
Electronic Properties of Conjugated Polymers III
Editors: H. Kuzmany · M. Mehring · S. Roth

taking advantage of multichannel technique. After ten repetitions of the cycle: (1 sec laser shot + 1 sec obscurity), the mean value of the collected spectra (j) was registered. Periodically (Δt=20s) about twelve such measurements (each at the time $t_j = j^* \Delta t$; j=1,...,12) were performed at constant laser power (P_i) on the same area of the sample. Six sets of twelve spectra were obtained consecutively from six different and well separated sample areas with P_i (i=1,...,6) being respectively 30, 120, 120, 200, 300 and 120 mW. The sample (circular; diameter ≈ 0.8 cm) was maintained at room temperature except for the measurements at the sixth sample area where liquid-N_2 temperature was used. The laser beam axis was normal (90°) to the

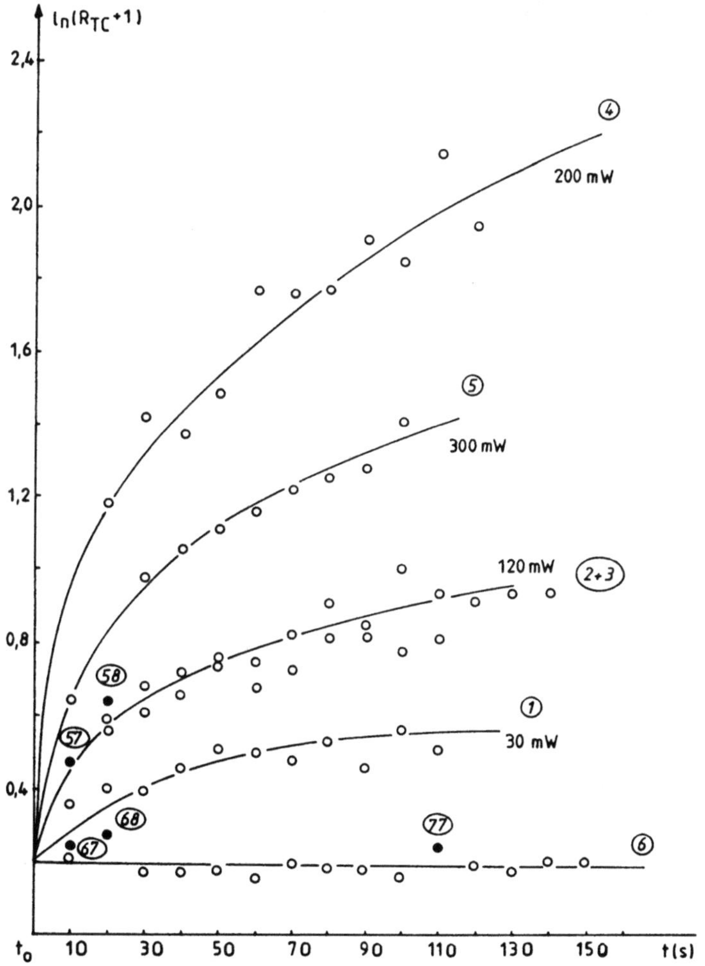

Fig. 2. The six isotherms (i.e. i=1,...,6) in a (n=1;m=p=0)-representation [5]. At 1(N_2)-temperature (i=6;P_6=120 mW) the R_{TC}-ratios remain constant, because k(1(N_2))≈0. The black points are artefacts

PA-film surface and the retro-Raman (180°) scattering was collected. The integrals (J) of Raman intensities over the band, $J_{Cis}(t_j,P_i)$ and $J_{Trans}(t_j,P_i)$, were numerically calculated each one starting from a previously defined straight base-line. Assuming that under different circular laser spots (diameter ≈ 50 μm),

i) the number N_0 of laser-photon-accessible PA molecules is constant;

ii) the number $N_i = N_{iCis} + N_{iTrans}$ ($\leq N_0$) of Raman-scattering-active molecules (for each j) equals to $J_{Cis}(t_j,P_i)^* f_{Cis} + J_{Trans}(t_j,P_i)^* f_{Trans}$,

we obtain six linear relationships (fig.1):

$$J_{Cis}(t_j,P_i) = N_i/f_{Cis} - (f_{Trans}/f_{Cis})^* J_{Trans}(t_j,P_i) = a_i + b^* J_{Trans}(t_j,P_i),$$

each one binding all the points with different t_j and the same P_i. The ratio of each set of integrals multiplied by (-b) is R_{TC}= (%Trans)/(%Cis) i.e. the

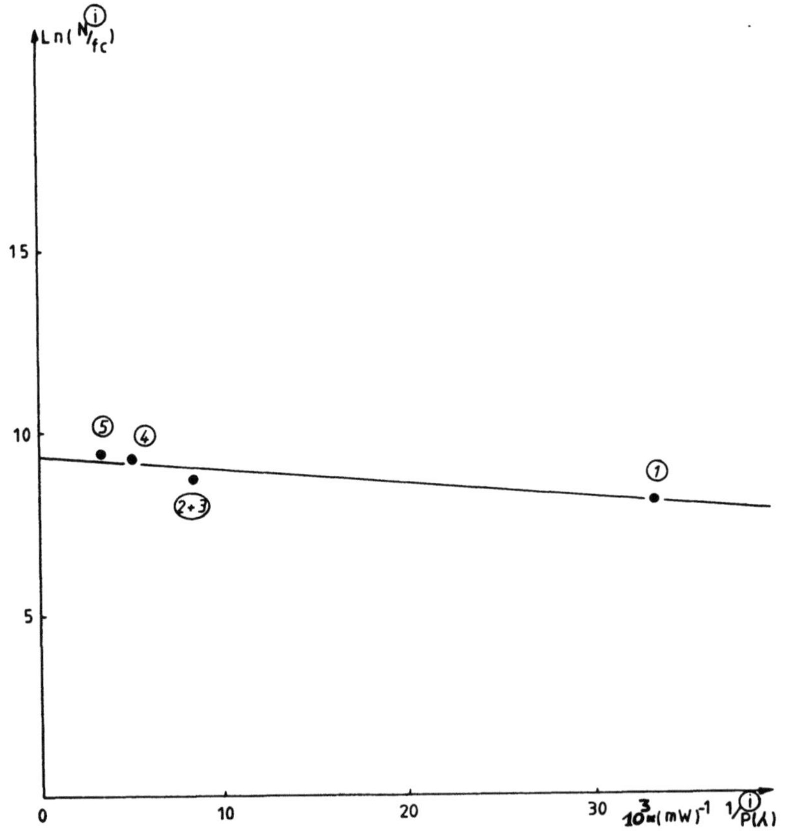

Fig. 3. The natural logarithms of N_i/f_{Cis} correlate linearly with the inverse laser powers: $\ln(N_i/f_{Cis}) = 9.374062 - 38.122/P_i$ (r=-0.914)

isomeric composition of the PA-film as determined by multichannel Raman spectroscopy. The six series of ($R_{TC}(t_j,P_i)$ vs. t_j)- points represent typical isotherms of PA_{Cis} isomerization in a solid PA-film. Thus isotherms in a time-scale of seconds were obtained for the first time. They enlarge upon the isotherms obtained by thermal isomerisation and IR spectroscopy which are in a time scale of minutes [3] and of hours [4]. Our isotherms were first analysed using the theory of an irreversible first-order chemical process. It was shown (fig. 2) that the ($\ln\{R_{TC}(t_j,P_i)+1\}$ vs. t_j)-points do not form straight lines and consequently the isomerization of PA_{Cis} in a solid PA-film is not a first order process. Applying SESTAK & BERGGREN's [5] mathematical model of the three-step chemical process in a solid, we found that the isomerization propagation is 3-dimensional ($n = 2/3$). Moreover, the effects of both the initiation and the anihilation steps on the rate-constant control cannot be neglected ($m \neq 0$ and $p \neq 0$).
The observed Raman scattering seems to obey some activated process since the numbers N_i/f_{Cis} (fig. 3) follow the expression:

$$N_i/f_{Cis} = c^*\exp(-d/P_i), \quad \text{where } c = N_o/f_{Cis} \approx 12^*10^3 \text{ and } d \approx 40 \text{ mW}.$$

In order to determine the unknown temperatures T_i of our isotherms we

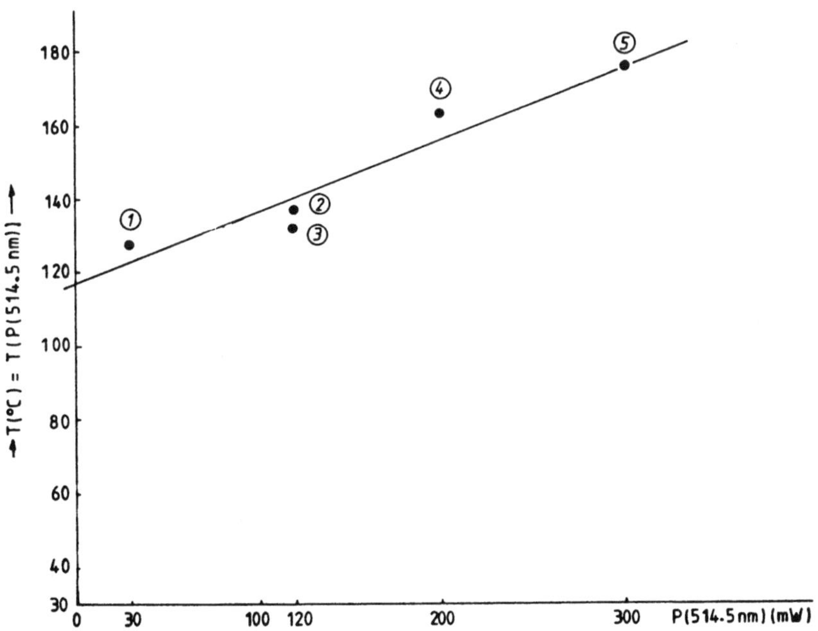

Fig. 4. Temperatures, corresponding to our isotherms (Fig. 2) correlate linearly with the laser powers: $T_i(°C) = 116.8 + 0.1955^*P_i(mW)$ (r=0.955). The interpretation of T(P=0) value is not straightforward

have to estimate first the initial (t=0) rate constants $k_0(P_i)$. If we adjust the polynomials $F_n(t) = \sum B_{nl}*t^l$ with n= 2,3,4 and l= 0,...,n to the experimental $F(t_j,P_i)$-points (fig. 2) then the B_{n0} and the B_{n1} coefficients lead to $k_0(P_i)$ and to the initial isomeric sample composition $(\%Cis)_0$. The temperatures (in K) were determined from the relationship:

$$T(P_i) = E_{al}/R/\ln(A/ k_0(P_i))$$

where A ($7.142*10^6\ s^{-1}$) and E_{al} (17181.3 cal/mole) constants have been deduced from ITO et al [3] isotherms. It was then established (fig. 4) that these temperatures are linearly bound to their corresponding laser powers:

$$T_i(°C) = 116.8 + 0.1956*P_i \quad (\text{with } r= 0.955).$$

Putting $(-d/P_i) = (-E_{ar}/R/T_i)$ and substituting for P_i from $P_i \approx const*T_i$, the "energy of activation of the Raman scattering" (E_{ar}) was found amounting to some few calories/mole.
Proposing a parallelepipedic model (VdW-dimensions: 950 Å*3.5 Å*4.715 Å) of an average-sized PA_{Cis} macromolecule ($M_w \approx (11.00 \pm 1.54)*10^3$) [6] with 1.165 g/cm^3 density, we can obtain for our PA-film (density: 0.4 g/cm^3) an estimate of constants N_0($\approx 2*10^9$ molecules) and then of f_{Cis}(1.4 to $1.7*10^5$) and f_{Trans}(0.2 to 0.3 $*10^5$), which are proportional to the Raman-scattering cross-sections.

References

[1] H. SHIRAKAWA, T. ITO and S. IKEDA, Polymer J., 4 (1973) 460
[2] A. DJEBAILI, Ph.D. Thesis, University of Montpellier (USTL)(1987)
[3] T. ITO, H. SHIRAKAWA and S. IKEDA, J. Polym. Sci., Polym. Chem. Ed., 13 (1975) 1943
[4] H. GIBSON, R.J. WEAGLEY, W.M. PREST Jr., R. MOSHER and S. KAPLAN, J. Physique, 44 (1983) C3-123
[5] J. SESTAK and G. BERGGREN, Thermochim. Acta, 3 (1971) 1
[6] J.C.W. CHIEN, Polyacetylene: Chemistry, Physics and Material Science, Academic Press, New York (1984), pp. 56,58

Specific Problems Concerning In Situ External Reflection Absorption Spectroscopy of Conducting Polymers

H. Neugebauer[1], *N.S. Sariciftci*[2], *H. Kuzmany*[2], *and A. Neckel*[1]

[1]Institut für Physikalische Chemie, Universität Wien, Währinger Str. 42, A-1090 Wien, Austria

[2]Institut für Festkörperphysik, Universität Wien, and Ludwig Boltzmann Institut für Festkörperphysik, Strudlhofgasse 4, A-1090 Wien, Austria

Abstract. Specific spectral phenomena concerning in situ FTIR spectroscopy in the external reflection absorption arrangement for the study of the electrochemical reactions of conducting polymers are discussed in this contribution. The IR beam reaching the detector is a complex function both of the reflection and the absorption properties of the materials. Concentration changes in the thin electrolyte layer during electrochemical reactions can lead to disturbing spectral features.

1. Introduction

In situ FTIR spectroscopy has proved to be an excellent tool for the study of electrochemical redox processes. Particularly external reflection absorption spectroscopy [1], where an electrode is pressed against an IR transparent window causing a thin electrolyte layer between the electrode and the window, is well established in the study of electrochemical reactions, where only small currents are involved [1]. When this method is applied for the study of reactions with higher currents, e.g. redox processes of conducting polymers, some specific problems emerge, which are connected with the complex overlapping of reflection and absorption of the materials and with the thin electrolyte layer between the electrode and the window. In this paper these problems are discussed, showing the in situ FTIR investigations of polyaniline (PANI) [2] as an example.

2. Experimental

The electrochemical polymerization process of aniline and the spectroelectrochemical setup are described elsewhere [3]. The electrolyte solutions for the in situ experiments were 0.5 M $LiClO_4$ in CH_3CN or aqueous 1 M $HClO_4$/0.1 M $NaClO_4$ solution.

Springer Series in Solid-State Sciences, Vol. 91
Electronic Properties of Conjugated Polymers III
Editors: H. Kuzmany · M. Mehring · S. Roth

In situ infrared spectra were recorded during slow cyclovoltammetric sweep experiments. The spectra are ratioed against a reference spectrum recorded just before the considered oxidation process showing only the spectral changes during the process.

The ex situ spectra were obtained in reflection absorption geometry in the spectroelectrochemical cell without electrolyte as well as in transmission geometry using a transparent electrode for the preparation.

3. Results and Discussion

Figure 1 shows the arrangement for in situ external reflection absorption spectroscopy. The intensity of the IR beam reaching the detector is modified by: reflection at the window, absorption of the electrolyte layer, reflection at the polymer surface and absorption of the polymer. Since only the absorption of the polymer is the quantity of interest in this technique, the spectral influences of the other contributions have to be eliminated or considered when analysing the IR spectra.

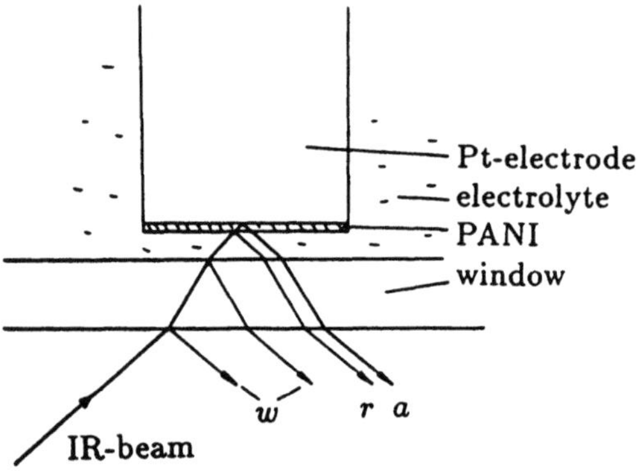

Fig. 1: Schematical arrangement for in situ external reflection absorption spectroscopy. *w* reflection at the window, *r* reflection at the polymer surface, *a* reflection at the mirror electrode containing information on the absorption of the polymer

3.1 Effect of the Window Reflection

The calculation of an absorbance spectrum is based on the ratio of two single beam spectra. The window reflection (*w* in Fig. 1) is a constant quantity added to the part of the IR beam containing the spectral information of the polymer (*r* and *a* in Fig. 1). If *r* and *a* changes between the two single beam spectra (e.g. absorption changes by increasing the conductivity of the polymer) two effects occur (curve a in Fig. 2):

1) The changes of *r* and *a* appear to be smaller.

2) Electrolyte "ghost" bands appear.

The influence of the window reflection can be eliminated by the registration of *w* before the in situ experiment (IR spectrum of the spectroelec-

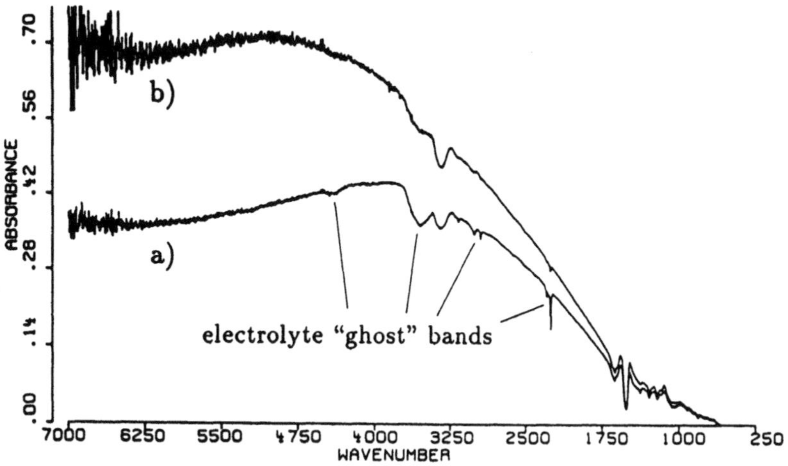

Fig. 2: In situ spectrum of the middle oxidized form of PANI (reference spectrum: reduced form) in acetonitrile/$LiBF_4$ electrolyte. a) uncorrected spectrum, b) corrected spectrum (see text)

trochemical cell without the electrode) and subtraction of *w* from each single beam spectrum before performing the ratio (curve b in Fig. 2) [4].

3.2 Effect of the Polymer Reflection

When PANI is converted into the conducting form, ex situ transmission spectra show strong enhancements of the absorption intensities of vibrational bands [5]. These enhancements are not seen in the reflection absorption geometry [6]. The reason for this effect is the overlapping of the IR beam reflected at the surface of the polymer (*r* in Fig. 1) with the IR beam containing the spectral information of the polymer absorption (*a* in Fig. 1). Model calculations of the absorption of a Lorentz-type oscillator without and with Drude type charge carrier contribution (in terms of a composite medium response with Maxwell-Garnett formalism) [5] confirm the spectral discrepancies between transmission geometry and absorption reflection geometry.

3.3 Effect of Changes in the Electrolyte Composition

Figure 3 shows the spectral changes otained in situ during the second oxidation process of PANI in aqueous 1 M $HClO_4$/0.1 M $NaClO_4$ electrolyte. In this experiment the increase in absorption intensity of ClO_4^--ions at 1100 cm^{-1} is due to the migration of these ions from the surrounding electrolyte solution into the thin electrolyte layer between the electrode and the window by the electrochemical current rather than from an en-

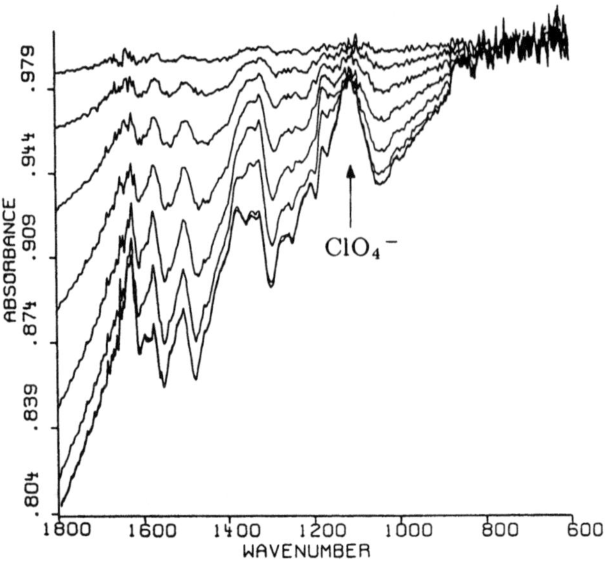

Fig. 3: In situ spectra during the second oxidation process of PANI in aqueous 1 M $HClO_4$/0.1 M $NaClO_4$ electrolyte. Reference spectrum: middle oxidized form

hancement of the anion concentration in the polymer itself [6]. Spectral effects due to changes in the amount of intercalated anions in the polymer are covered by the IR absorption of the migrating ions.

Acknowledgement. This work was supported by the Jubiläumsfonds der Österreichischen Nationalbank (Proj. No. 2505) and by the Fonds zur Förderung der wissenschaftlichen Forschung of Austria (Proj. No. 5735).

References

1. A.Bewick, in "Trends in Interfacial Electrochemistry", (Reidel, Dordrecht, 1986), p.331
2. H.Kuzmany, N.S.Sariciftci, H.Neugebauer and A.Neckel, Phys. Rev. Lett. **60**, 212 (1988)
3. H.Neugebauer, A.Neckel, N.S.Sariciftci and H.Kuzmany, Synthetic Metals **29**, E185 (1989)
4. H.Neugebauer, 39th ISE-Conference, September 4 – 9, 1988, Glasgow, Poster 129
5. N.S.Sariciftci, H.Kuzmany, H.Neugebauer and A.Neckel, J. Chem. Phys., submitted
6. N.S.Sariciftci, M.Bartonek, H.Kuzmany, H.Neugebauer and A.Neckel, Synthetic Metals **29**, E193 (1989)

Evidence of Polaron Defects in Halogen Bridged Mixed Valence Compounds from FIR Reflectivity Measurements

L. Degiorgi[1], *P. Wachter*[1], *M. Haruki*[2], *and S. Kurita*[2]

[1]Laboratorium für Festkörperphysik ETH-Zürich, CH-8093 Zürich, Switzerland

[2]Yokohama National University, Laboratory of Applied Physics, Faculty of Engineering, Hodogaya, Yokohama 240, Japan

Abstract. We report far infrared reflectivity measurements on Pt–halogen chains, discussing in particular the phonon modes along the chain axis. Furthermore we present experimental evidence of the existence of polaron defects, which are of intrinsic nature. A phenomenological lattice vibrational calculation of the phonon spectra supports our conclusions.

1. Introduction

Compounds of the type $[M(en)_2][M(en)_2X_2](ClO_4)_4$, with M=Pt or Pd, X=Cl, Br or I and (en)=ethylenediamine ($C_2H_8N_2$), are members of a large family of quasi–one–dimensional halogen bridged mixed valence insulators. Their skeletal structure consists of linear chains –M(II)–X...M(IV)...X –M(II)– of alternating metal–halogen ions, where the halogen X ions are closer to the M(IV) than to the M(II) ions. Along the chain axis, each M metal ion is coordinated by four nitrogen atoms of the ethylenediamines [1].

The above–mentioned displacement of the halogen ions from the midpoint results in the appearance of a commensurate charge–density–wave (CDW) with a period of twice the M–M separation. Thus these chain compounds are considered a realized Peierls system.

We have recently performed reflectivity measurements on these systems (which hereafter will be abbreviated to MX–chains) between the far infrared (FIR) and the UV energy range, regarding in particular the former. We have investigated the FIR–active phonon modes and a tentative assignment was proposed, with the help of a phenomenological lattice vibrational calculation [2]. In fact, besides the well–known absorption band in the visible energy range, ascribed to the charge transfer (CT) between d_{z^2} orbitals of the M(II) and M(IV) ions, four main modes were detected in the FIR, when measuring with light polarized along the chain axis. Considering our MX–chains for the phenomenological calculations in the linear harmonic approximation, where the lattice interactions were described by spring force constants, we were able to assign one mode to a stretching phonon, a second one to the zone boundary acoustic phonon (i.e. in the Γ point as a result of the Brillouin zone folding), a third one to the IR–active bending mode, since we have considered also the out–of–chain interactions, and finally the fourth one to a mode of external type [2].

It was then possible to find a connection between these lattice dynamical properties and the electronic ones (i.e. related to the CT band), within the so–called Peierls Hubbard (PH) model. Baeriswyl and Bishop proposed a microscopic version of such a PH model, which is really suitable in order to describe these systems theoretically [3]. In fact, within their approach, they focused their attention on the calculation of the ground state and of the configuration of the excited midgap states, due to the presence of defects, such as polarons, bipolarons, kinks or solitons. Kurita et al. have measured photoinduced absorption bands near the midgap energy in Pt–Cl

Springer Series in Solid-State Sciences, Vol. 91
Electronic Properties of Conjugated Polymers III
Editors: H. Kuzmany · M. Mehring · S. Roth

and they identified these structures with the excitations induced by metastable polaron defects [4]. These experimental results, which also look like the pressure induced band near the midgap of Kuroda et al. [5], are in full agreement with the theoretical predictions of Baeriswyl and Bishop [3].

The possible existence of defects, either metastable or intrinsic, then motivated our further interest in these systems. The purpose of this communication is to study the presence of defects along the chain axis from the phonon point of view. In fact it is well known that polarons, solitons or excitons have as their fingerprints in the phonon spectrum the appearence of local or gap modes. We present the "phonon–evidence" for such excited midgap states, discussing once again our experimental results with our model lattice vibrational calculation.

2. Experiment and Results

The reflectivity R(ω) with light polarized parallel and perpendicular to the chain axis of large single crystals of Pt–halogen compounds was measured in the energy range from 12 eV down to 1 meV at 300 K and 6 K. In order to cover the whole energy range, we made use of four spectrometers, and in the FIR a Fourier spectrometer was employed with a He–cooled Ge bolometer as detector. Furthermore the reflectivity along the chain axis was measured at 77 K with a special N–cryostat, which also permits one to photoexcite the samples for a desired time. For the recrystallization procedure we refer the reader to our previous work [2].

Due to the imposed limited space for this contribution and to the similarity between the MX–chains, we will concentrate our efforts on only one Pt–halogen compound, namely Pt–I. We will review the whole Pt–halogen series in an extended and amplified forthcoming publication [6].

Fig. 1 presents the FIR spectrum of Pt–I along the chain axis at 6 K. We can recognize five distinct modes (see arrows), which clearly increase with decreasing temperature, especially the one at 14 meV. We will return in the following discussion to these structures. Here, we finally want to mention that measuring at 77 K we obtained the same spectrum as at 6 K and besides, after a photoexcitation of many hours at the energy of the CT band (i.e. in the visible region), we did not detect any changes. At the present time we claim that there are two reasons for this negative result: first the physical incapacity of light to photoinduce enough bulk defects to increase the intensity of the corresponding phonon modes, and (or) second an intrinsic experimental difficulty due to the reduced power of our source (i.e. approximately a factor of ten less than the source used by Kurita et al. [4]). From this point of view experimental improvements are in progress.

3. Discussion

The experimental results of Fig. 1 confirm and strengthen the first spectra, presented in our previous work [2]. The mode structures at 9.3, 21.7, 31.6 and 36 meV were assigned, as mentioned in the introduction, to the zone boundary acoustic, the stretching, the IR bending and the external–type phonon. The present discussion will, however, underline the peculiar mode at 14 meV, which becomes increasingly sharp with decreasing temperature, and which is tentatively assigned to a gap mode, due to the presence of defects.

In fact, it is well known that it is possible to produce defects along a 1 dimensional chain by an isotopic substitution, a vacancy or a different charge distribution. Of course these defects perturb the phonon modes in different ways and new ones such as local, gap or resonating modes also arise. The difference between the former modes depends on the type of the substitution and the definition of these localized modes depends on the position of the modes relative to the phonon dispersion of the unperturbed chain [7].

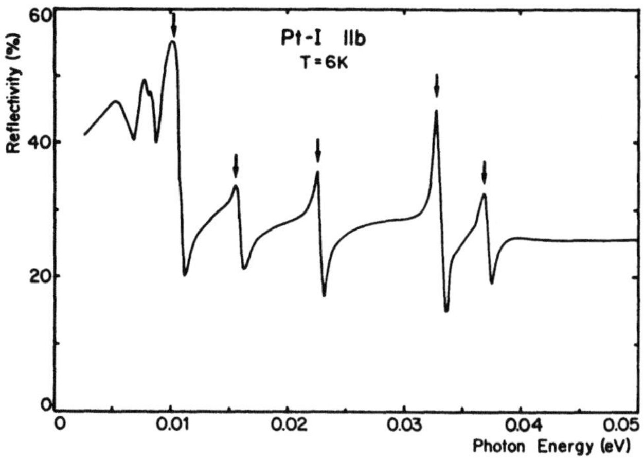

Fig. 1: Reflectivity spectrum of Pt–I in the FIR energy range at 6 K

Following our previous phenomenological approach [2], we develop our lattice vibrational calculation, based on the linear harmonic approximation, with the introduction of the defects along the chain axis. In the unperturbed chain the lattice interactions between the halogen and the Pt ions are described by the spring force constants f_1 and f_2 around the Pt(IV) and the Pt(II) ions, respectively. The defects are then represented by a new charge disproportionation, given by the presence of Pt(III) ions, which are considered in our phenomenological calculation with a new spring force constant distribution around the Pt(III) ions [6,7]. Taking into account a defect of polaronic type and the same spring force constants for the Pt–I as used in Ref. 2, which are halved around the defects, we can calculate the IR–active phonon modes along the chain. The numerical result is presented in Table I, where a comparison with the experimental one is also proposed. The agreement is very good and, from the eigenstates calculation, the new phonon at 14 meV can be definitely assigned to the localized mode around the polaron. We should perhaps mention that, for the sake of simplicity, we do not consider the out–of–chain bending interactions (as in Ref. 2) and consequently we do not calculate the IR–active bending mode. In fact the normal modes along the chain axis essentially depend on $f_1 \pm f_2$ [6].

The choice of polaron is imposed by the work of Kurita et al. [4], who demonstrated the existence of such a kind of defects as metastable states. However, our polarons, which are not photoinduced, seem to be more of intrinsic type. This could be explained by the fact that not exactly equimolar solutions of Pt(IV) and Pt(II) complexes may have been mixed and excess (or less) halogen atoms are then introduced in the process of synthesis. Furthermore even at 300 K we can detect the

Tab. I: Experimental phonon structures compared with the phenomenological fit result.
Spring force constants used in the calculation:
$f_1 = .73$, $f_2 = .61$, $f_3 = .40$, $f_4 = .30$, [mdyn/Å]

	[eV]				
exp	.01[a]	.016[b]	.0225[c]	.032[d]	.036[e]
fit	.012	.015	.021	——	.039

a: Acoustic, b: Gap, c: Stretching, d: External, e: IR–Bending Mode corresponding electronic and phononic excitations of the polarons (i.e. as midgap state and as gap mode), which look like a broad bump and which would confirm the intrinsic nature of these defects.

From the theoretical point of view Baeriswyl and Bishop [3] have also considered the case of polarons along the chain within the PH model, from which it results that the lattice dimerization around the Pt(III) ion is only half. This could explain in a first approximation the numerical choice for the spring force constants around the polaron (i.e. approximately a factor two less than f_1 and f_2). In fact around the defect due to the reduced charge disproportion, there is a weak coupling limit and consequently screening of the lattice interactions arises.

Of course our phenomenological calculation should be discussed in more detail, particularly regarding the number of defects along the chain, their distribution and the type (i.e. polaron, soliton etc.). These aspects, which, however, would not qualitatively influence the above conclusions, and the formal details of the phenomenological calculation will be the subject of a forthcoming publication [6].

4. Conclusions

We have discussed our FIR spectrum of Pt–I within the phenomenological calculation, based on the linear harmonic approximation. Our analysis supports very well the existence of intrinsic defects along the chain, which we consider to be of polaronic type. The experimental evidence consists in the detection of the corresponding localized phonon around these defects.

Acknowledgement

The authors are grateful to Dr. D. Baeriswyl and Dr. A. R. Bishop for stimulating discussions and to E. Jilek, J. Müller and H. P. Staub for technical assistance.

References

[1] R.J.H. Clark, in R.J.H. Clark and R.E. Hester (Eds.), Advances in Infrared and Raman Spectroscopy, Wiley Heyden, (1984) Ch. 3, p. 95 and Refs. therein.
[2] L. Degiorgi, P. Wachter, M. Haruki and S. Kurita, submitted for publication to Phys. Rev. B.
[3] D. Baeriswyl and A. R. Bishop, J. Phys. C2 (1988) 339.
[4] S. Kurita, M. Haruki and K. Miyagawa, J. Phys. Soc. Jap. 57 (1988) 1789.
[5] N. Kuroda, M. Sakai, Y. Nishina, M. Tanaka and S. Kurita, Phys. Rev. Lett. 58 (1987) 2122.
[6] L. Degiorgi, P. Wachter, M. Haruki and S. Kurita, in preparation.
[7] A.S. Barker and A. J. Sievers, Rev. Mod. Phys. 47 (1975) S1.

Part V

Electrochemistry, Polyanilines I

Structural Changes in Conducting Polymers upon Electrochemically Induced Redox Processes

J. Heinze, K. Meerholz, and R. Bilger

Institut für Physikalische Chemie der Universität Freiburg,
Albertstr. 21, D-7800 Freiburg, Fed. Rep. of Germany

Abstract. The cis/trans isomerization of cis-PA upon cathodic reduction or anodic oxidation can be detected voltammetrically. The analysis of the redox data reveals that the trans-form is thermodynamically more favorable in the charged than in the neutral state. Cyclic voltammetry also reflects geometric relaxation processes which occur in conjugated materials such as defined oligomers of the p-phenylene series or polymeric PPy and PTh during charging and discharging. Additional spectroelectrochemical studies unambiguously show that the charging process in the whole accessible potential range is faradaic.

1. Introduction

The bipolaron model suggests that in conjugated materials such as polypyrrole (PPy), polythiophene (PTh) or poly-*p*-phenylene (PP) geometric distortions occur during switching from the neutral to the doped state [1,2]. It further assumes that after discharging the system relaxes into its initial conformation. Up to now, all electrochemical measurements have provided only indirect evidence of structural changes of this type during doping and undoping. The only documented example of an electrochemical induced geometric change is the cis/trans isomerization during the electrochemical oxidation of cis-polyacetylene (PA) for which there are IR, Raman and UV measurements [3-5].

Our studies of the cathodic reduction of PA in liquid ammonia reveal that such conformational changes can be also detected by cyclic voltammetry. Moreover, from our knowledge of conformational changes during electrode reactions of monomeric species [6,7] we were able using spectroelectrochemical data to interpret asymmetric cyclic voltammograms of PPy in terms of geometric relaxation processes which take place during charging and discharging of conductiong polymers. Measurements at oligomers (PP, PTh) of defined chain length reveal that the charging of the

Springer Series in Solid-State Sciences, Vol. 91
Electronic Properties of Conjugated Polymers III
Editors: H. Kuzmany · M. Mehring · S. Roth

conjugated chain involves potential-separated redox steps which produce for large systems plateau-like voltammetric waves.

2. Electrochemical Isomerization of Polyacetylene

The reduction of freshly prepared cis-PA films in liquid ammonia ($T < -40°C$) was studied on the basis of cyclic voltammetric measurements. Fig. 1 shows the first three voltammetric cycles in the potential range between -0.5 V and -1.9 V vs. Ag/AgCl. In the first cathodic scan a sharp wave (I) appears at E_{pc} = -1.78 V. Upon scan reversal after a plateau-like anodic signal a small oxidation wave at E_{pa} = -1.23 V is observed. The second cathodic scan shows a new reduction wave II, E_{pc} = -1.62 V, with the original wave (E_{pc} = -1.78 V) significantly decreased in height. In subsequent sweeps the peak current for the front wave II

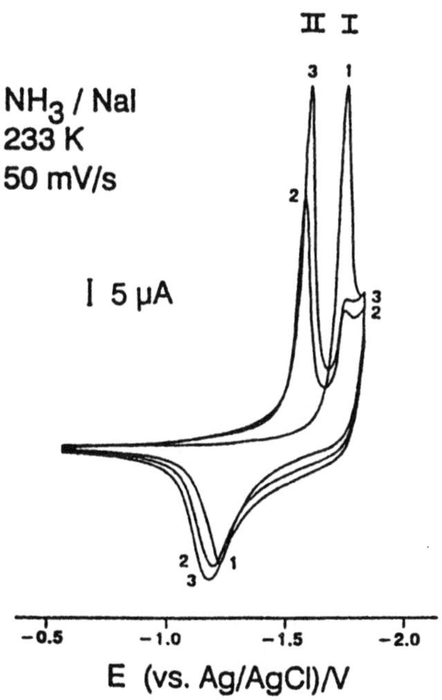

Fig 1: CV for the reduction of *cis*-PA in liquid NH_3/0.1 M NaI, switching potential E_λ= -1.87 V

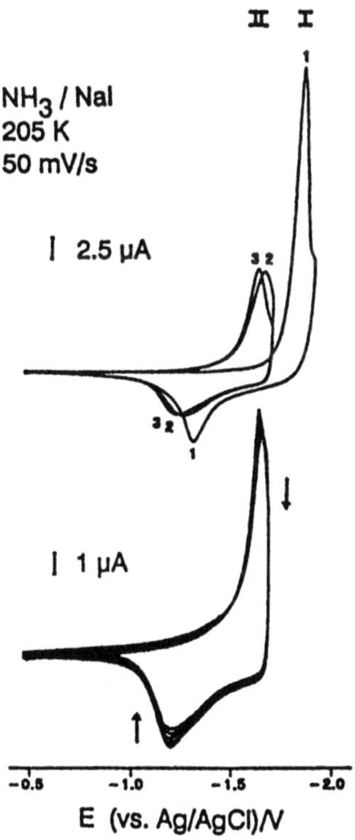

Fig. 2: CV for the reduction of *cis*-PA in liquid NH_3/0.1 M NaI, switching potential for MSCV E_λ = -1.70 V

increases further, while the original wave becomes smaller. The current of the front wave reaches its maximum value when the current of the original wave has totally disappeared. The maximum peak current is dependent on film thickness and sweep rate. During this experiment the red colour of the cis-PA film in the neutral state gradually changes to blue, which is typical for the formation of the trans-component [4].

The increase of the wave at E_{pc} = -1.62 V during multisweep experiments was only observed when the scan was reversed at switching potentials (E_{λ}) of less than -1.80 V which lie negatively to the peak potential of the original wave I. The variation of the sweep rate documents that in the case of a slow scan (<20 mV/s) the maximum of the peak current of wave II is reached within a few cycles, while in the case of a fast scan (>50 mV/s) a high number of cycles is necessary. Upon scan reversal at switching potentials of more than -1.60 V the current of wave II slowly decreases with repeated cycling (Fig. 2). PA films which were first anodically oxidized and subsequently reduced to their n-doped form already show the wave at E_{pc} = -1.62 V during the first cathodic scan. The limiting peak current of this wave will be reached within 3 to 4 cycles.

The voltammetric and optical results obtained during charging/discharging of cis-PA clearly indicate that upon reduction the film isomerizes from the cis- to the trans-form. Obviously, the kinetics of the isomerization strongly depend on the applied scan rate and on the selected switching potential. Using low scan rates (<20 mVs^{-1}), the isomerization is completed after a few charging/discharging cycles. Wave II at E_{pc} = -1.62 V, which is correlated with the charging of the trans-form, then exhibits a limiting peak current, indicating that the isomerization has terminated. The reason for this rapid geometrical transformation is that a slow potentiodynamic scan facilitates the complete charging of the polymer film, which allows enough time for the incorporation of the counterion into the solid layer by diffusion so as to fulfill the condition of electroneutrality.

Of great importance is the proper choice of the switching potential. The isomerization only proceeds when the cis-segments of the PA film can be charged. As the reduction potential of the cis-component lies negatively to that of the trans-component, it is not sufficient to reduce the partially isomerized film in the potential range where the trans-form is electroactive (Fig. 2). To induce further cis/trans isomerization the cis-segments must be charged, which is only achieved when the potential is scanned more than 200 mV negatively to the reduction potential of the trans-form.

This finding is in complete agreement with earlier experiments done by Heeger et al. [4]. They found that PA films could not be isomerized by "doping to a low dopant level". As the doping, which corresponds from a modern point of view [8-10] to a sequence of redox reactions, is a potential dependant process, a low dopant level implies that the electrode potential does not exceed values at which only the trans-form can be charged. On the other hand, the conclusions of Heeger et al. [4] that there is no significant doping of cis-PA and that any given level of doping requires a certain amount of trans-PA must be amended.

The voltammetric experiments unambiguously prove that an isomerization only takes place via the charged cis-segments of PA. The charged (doped) cis-component is thermodynamically unstable and transforms through a fast chemical step to the trans-component (EC mechanism).

$$\text{cis-PA} \underset{}{\overset{E^o_{cis}}{\rightleftharpoons}} \text{cis-PA}^{y-}, \qquad (1)$$

$$\text{cis-PA}^{y-} \xrightarrow{k_f} \text{trans-PA}^{y-}, \qquad (2)$$

$$\text{trans-PA} \underset{}{\overset{E^o_{trans}}{\rightleftharpoons}} \text{trans-PA}^{y-}, \qquad (3)$$

where $E^o_{cis} < E^o_{trans}$.

This mechanistic phenomenon explains why, on the one hand, it is impossible to detect any cis-material after full charging of a cis-film at a potential which lies negatively to the reduction potential (E^o_{cis} = - 1.80 V vs. Ag/AgCl) of cis-PA and, on the other hand, the isomerization stops when charging/discharging cycles are carried out in the range positive to E_{cis}^o. Moreover, it becomes clear that the initiation of the isomerization does not require the presence of any trans-PA

3. Structural Relaxation of PPy

During our studies on conducting polymers we have found that the typical difference of the anodic and cathodic peak potentials of PPy or PTh or even PA in cyclic voltammetry is not due to slow heterogeneous electron transfer processes but is caused by structural rearrangements [11]. We have presented experimental and theoretical arguments which show, in agreement with the predictions of the bipolaron model, that during charging a polymer chain with its twisted units undergoes a conformational interconversion, forming a partially coplanar chain with localized semi-chinoid segments. This produces an energetic stabilization of the charged states. In the reverse voltammetric scan during discharging, the chain segments remain in the coplanar form up to the end of the discharging process. Therefore, the energetically stabilized charged states are reduced at lower redox potentials and, consequently, asymmetrical charging/discharging curves result. In some cases, the polymer may in its neutral state already convert at slow rates from the twisted to the planar form and vice versa. Using slow scan rates, equivalent to a long time scale, this process leads to more or less symmetrical cyclic voltammograms.

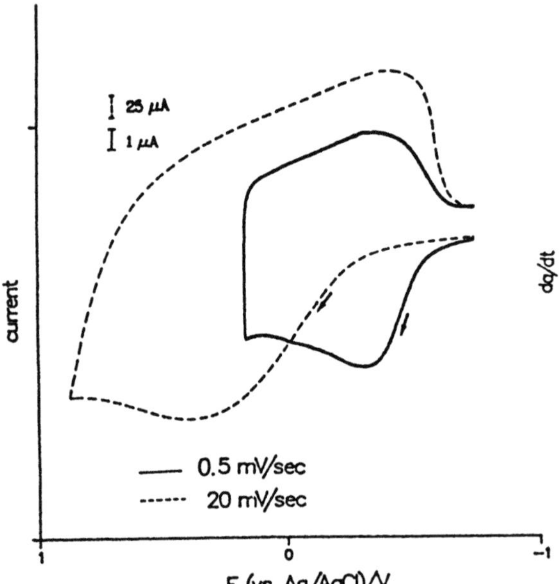

Fig. 3: CV of the oxidation of potentiodynamically prepared PPy films in PC/0.1 M $LiClO_4$

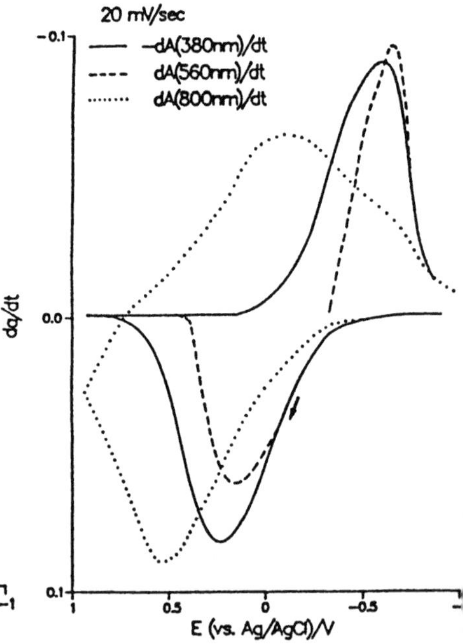

Fig. 4: Voltammetrically measured dA/dt curves during the charging/discharging of PPy in PC/0.1 M $LiClO_4$, A is the optical absorptivity. dA (560 nm)/dt is only partially plotted

Similar effects are observed when potentiodynamically prepared PPy films are studied using scan rates of 0.5 mVs^{-1} and 20 mVs^{-1} (Fig. 3). At a scan rate of 0.5 mVs^{-1} the charging and the discharging waves are nearly symmetrical, whereas with a scan rate of 20 mVs^{-1} the oxidation wave shifts to positive potentials and the reduction wave remains in its position. Consequently, in the latter case an asymmetrical voltammogram results.

Additional information is received when the time derivatives of the absorption bands for the interband (λ = 380 nm), a polaron (λ = 560 nm), and a bipolaron (λ = 800 nm) transition are measured [12]. Fig. 4 shows the results obtained during a cyclic voltammetric experiment with a scan rate of 20 mVs^{-1}. Negative values of dA/dt indicate the decrease, positive values the increase of the absorptivity for the corresponding electronic transition. (It should be noted that for the sake of clarity the dA/dt curve for the polaron state (λ = 560 nm) is only partially plotted.) The maxima in the plot are characteristic for those potentials where the spectral changes of the respective absorption occur most rapidly. Obviously, at positive potentials, when both the interband and the polaron transition are depleted, the bipolaron band at 800 nm continues to increase.

Our voltammetric and spectroelectrochemical data confirm some assumptions of the bipolaron model. The formation of the monoionic state (polaron) is obviously coupled with a structural relaxation process, provided that the time scale is short enough to preclude a significant formation of planar segments in the neutral state. However, the experimental dA/dt curves (Fig.4) also clearly reveal that, in contrast to the prediction of the bipolaron model, an additional stabilization of the bipolaron level cannot be deduced. On the contrary, it must be concluded that the bipolaron state is only as or even less stable than the polaron state. The dA/dt curve for the bipolaron evolution (Fig.4) also shows that further diionic states are generated, despite high oxidation potentials. This rules out the old interpretation of capacitive charging at high oxidation potentials [13].

4. References

1. J. L. Brédas, Mol. Cryst. Liq. Cryst. 118, 49 (1985).
2. J. L. Brédas and G. B. Street, Acc. Chem. Res. 18, 308 (1985).
3. R. J. Nigrey, A. J. Heeger, and A. G. MacDiarmid, Mol. Cryst. Liq. Cryst. 83, 309 (1982).
4. A. Feldblum, A. J. Heeger, T.-C. Chung, and A. G. MacDiarmid, J. Chem. Phys. 77, 5114 (1982).

5. H. Kuzmany, Pure Appl. Chem. 57, 235 (1985).
6. D. H. Evans and K. M. O'Connell, in: "Electroanal. Chemistry", Vol. 14, p. 113, ed. A. J. Bard, M. Dekker, New York 1986.
7. M. Dietrich, J. Heinze, H. Fischer, and F. A. Neugebauer, Angew, Chem. 98, 999 (1986); Angew. Chem. Int. Ed. Engl. 25, 1021 (1986).
8. A. F. Diaz, J. I. Castillo, J. A. Logan, and W.-Y. Lee, J. Electroanal. Chem. 129, 115 (1982).
9. E. M. Genies and A. A. Syed, Synth. Met. 10, 21 (1984).
10. J. Heinze, Top. Curr. Chem. (1989), in press.
11. J. Heinze, M. Störzbach, and J. Mortensen, Ber. Bunsenges. Phys. Chem. 91, 960 (1987).
12 A. O. Patil, A. J. Heeger, and F. Wudl, Chem. Rev. 88, 183 (1988).
13. S. W. Feldberg, J. Am. Chem. Soc. 106, 4671 (1984).

The Electrochemistry of Polypyrrole Derivatives

S. Pugh and D. Bloor

Queen Mary College, University of London, Mile End Road, London E1 4NS, UK

1. Abstract

The mechanism for the movement of ions through the conducting polymer, polypyrrole, is of great importance for those making use of the polymer in electronic or electrochemical devices. The use of polypyrrole in electrochemical biosensors [1] and in supercapacitors[2] relies heavily on the ionic conductance of the polymer film. The mechanism of this conductance could be of primary importance to the efficiency of these devices. The use of large counterions to produce soluble oligomers of polypyrrole shows there is more importance in the correct use of counterion than previously thought.

2. Introduction

In a recent paper Paulse and Pickup[3] maintained that the movement of ions through thicker polymer films was by migration rather than diffusion. The ionic conductivity being due to the migration of the counterion through the polymer under the influence of the electronic field, rather than simple diffusion of the counterion from the polymer through the loss of electrostatic bonding with the polymer.

In this experiment we measure the ionic conductivity of polypyrrole incorporating a large counterion. In this polypyrrole derivative the migration effects will dominate, as diffusion becomes more difficult through entanglement of the counterion with the polymer chain. Thus if the transport mechanism is migration there should be only small differences between the ionic conductivities with small or large counterions. If the mechanism is by diffusion there should be large differences.

The anion systems used were perchlorate as the small anion, the ionic conductivities and diffusion coefficients are well documented for this system and it can be used as a standard[3,4,5,6,7], and dodecylbenzene sulphonate, a large surfactant molecule, sufficiently larger than the perchlorate to highlight any differences or similarities.

3. Theory

The current from the chronoamperometric experiment can be mathematically modelled in two ways, depending on how the ions are transported through the polymer matrix. Model 1 presumes a diffusion process and the variation in current with time from the voltage step is described by the equation:[8]

$$i = nFAD^{1/2}.c/\pi^{1/2}.t^{1/2}$$

in which c is the concentration of the active sites in the polymer, and can be calculated from the number of coulombs passed. If the equation holds a plot of current density against $t^{-1/2}$ should give a straight line which

Springer Series in Solid-State Sciences, Vol. 91
Electronic Properties of Conjugated Polymers III
Editors: H. Kuzmany · M. Mehring · S. Roth

passes through the origin. D, the diffusion constant, can be obtained from the slope.

Model 2 [5] assumes that the conducting polymer behaves like the single pore model for a porous electrode (migration model). The transient current for the chronoamperometric experiment can then be described using the equation

$$i = AE(aC/\pi\tau_f t)^{\frac{1}{2}} .$$

Again a plot of current density against $t^{-\frac{1}{2}}$ gives a straight line through the origin. However this model assumes an infinitely thick electrode and neglects any solution resistance, clearly a situation not found experimentally. Compensating for the solution resistance(R_s) and the film thickness, by measuring specific film resistance(R_f), the chronoamperometric current is given by the equation [3]

$$\ln i = \ln R_f/E + \ln K - m^2/\tau \, t$$

where $\tau = R_f C_f$, and m is the positive roots of the equation m.tan(m)=1/p where p is R_s/R_f.

$K = \sin(m)/(1 + p)\sin(m) + p.m\cos(m)$.

If the film resistance is known the ion resistance, R_f, can be calculated from either the intercept or the slope.

4. Experimental Section

The chronoamperometric experiment involves stepping the voltage between the polymer film deposited on a platinum electrode, and a reference electrode, in this case Ag/Ag^+, and measuring the transient current between the working and auxiliary electrodes. The experiment is carried out in an electrochemical cell. The electrolyte was tetraethylammonium perchlorate, purified by recrystallization from water and vacuum dried, dissolved in acetonitrile, distilled from calcium hydride. The films were grown on a platinum disc, mounted in glass, of area $7.9*10^{-3}$ cm^2. The thickness of each film was estimated by the number of coulombs passed during growth (240mC = 1μm of film[10]). A variety of films between 1.0μm and 5.0μm were tested. The films were grown in either 0.1M tetraethylammonium perchlorate or 0.1M tetraethylammonium dodecylbenzene sulphonate,[9] depending on the film studied, solutions in acetonitrile using 0.1M pyrrole. The chronoamperometric measurements were taken using a Thompson potentiostat. The current was measured using a Farnell electrometer and recorded on a Bryan chart recorder. The overall results for many films are shown in table 1.

5. Discussion

These results are in good agreement with previously published works[10]. The diffusion coefficients are smaller than those works using a diffusion model for the interpretation of results, however many of the results were taken using sub-micron film.

The small size of the diffusion coefficient with model 1 shows that for the thicker films the model is not really appropriate. Although for very thin films diffusion might be the overriding process.

A comparison of the results for the two counterion show a similarity between the two of around an order of magnitude, for the ionic resistance. This would seem to show that the motion of the counterion is due to migration rather than diffusion. The diffusion constant being determined by the mobility of the counterion in the polymer. It is not, therefore, just a simple diffusion process, a point that is observed in cyclic voltammetry experiments using large counterions[11]. The closeness of the two values for the two counterions

Table 1. Ionic Resistances calculated from Chronoamperometric Data

		MODEL 2			MODEL 1
Ion	Thickness of film	Infinite film	Solution comp	Diffusion coeff.	Diffusion coeff.
		(Ω.cm)	(Ω.cm)	(cm^2sec^{-1})	(cm^2sec^{-1}
ClO_4	4.0μm	$3.02*10^4$	$3.9*10^4$	$2.1*10^{-9}$	$3.02*10^{-5}$
	6.0μm	$4.0*10^4$	$4.91*10^4$	$1.33*10^{-9}$	$2.5*10^{-5}$
SOAP	5.0μm	$3.08*10^5$	$2.39*10^5$	$2.12*10^{-10}$	$3.33*10^{-5}$
	2.0μm	$7.58*10^5$	$8.62*10^5$	$8.62*10^{-11}$	$1.09*10^{-5}$
	6.0μm	$9.35*10^4$	$3.0*10^5$	$2.18*10^{-10}$	$7.0*10^{-7}$

SOAP = dodecylbenzene sulphonate

would suggest that the upper limit of size for mobility of ions within the polymer matrix has not yet been reached and that the pore size within the polymer is of sufficient size to allow anions much larger than dodecylbenzene sulphonate to migrate through the polymer. Thus allowing a large number of different and larger counterions to be incorporated in the polymer matrix while still maintaining the electrochemical properties of the polymer.

The ability of the larger counterion to solubilise the oligomer suggests that there is some bonding between the oligomers and the counterion and this limits the free movement of the counterion. This work bears this out but also suggests that there is a similar bonding for the perchlorate anion.

7.References

1 C.Lowe, This meeting
2.D.Naegele, This meeting
3.C.Paulse & P.G.Pickup,J.Phys.Chem,92,24,(1988)
4.B.Herburg & J.P.Pohl,Ber.Bensen.Phys.Chem,92,(1988),1279
5,P.G.Pickup & R.A.Osteryoung,J.Electroanal.Chem. 195,(1985),271
6,J.B.Schlenoff & J.C.Chien.J.Am.Chem.Soc.109,(1987),6269
7,L.Curtin et al,J.Phys.Chem,92,(1988),12
8,E.M.Genies,G.Bidan,A.F.Diaz,J.electroanal.Chem,149,(1983), 101
9,Supplied by Albright and Wilson,Whitehaven,Cumbria
10,C.K.Baker & J.R.Reynold, J.Elecroanal.Chem., 251,(1988),307,and references therein.
11,K.Cheung, D.Bloor, G.C.Stevens to be published

Synthesis and Electrochemical Properties of Substituted Polyanilines

E.M. Genies and J.F. Penneau

Electrochimie Moléculaire, Laboratoires de Chimie,
Département de Recherche Fondamentale, Centre d'Etudes Nucléaires,
85X, F-38041 Grenoble Cedex, France

INTRODUCTION

Polyanilines have been the subject of intensive research [1,2,3], with many workers trying to answer the fundamental questions related to this material, because they are probably the cheapest conducting polymers that can be prepared.

The interest in studying substituted polyanilines is as follows :

- to give a better understanding of the polymerization and redox mechanisms of polyaniline.
- to give information on the structure-property relations in order to increase conductivity and stability.
- to improve the processability of polyanilines.
- to give way to the synthesis of new functionalized polymers.

DISCUSSION ON SUBSTITUTED POLYANILINES

We have exploited the excellent suitability of the $NH_4F,2.3HF$ anhydrous medium for the oxidative polymerization of aniline [4], and have used this to investigate many aromatic amines and related nitrogenated compounds.

The list of the investigated species is as follows :

- the asterisk is an indication that the monomer yields a polymer which presents some potential as an electronic conducting material - electrochemical activity and insolubility in common organic solvents in doped form -

- o*,m*, p-toluidines [5], o and m-ethylaniline*, m and o-propylaniline* [6], o*,m*, p-anisidines [7], azobenzene* [8], hydrazobenzene* [9], m-phenylenediamine*, phenazine [9], o-isopropylaniline*, 2,3, 2,4, 2,5, 3,4, 3,5 and 2,6-dimethylaniline*, 2,4,6-trimethylaniline*, N-methylaniline*, N,N-dimethylaniline, o*,m* [9] and p-phenylenediamine, 2,3 and 4-aminoacetophenone*, 2*,3 and 4*-aminobenzoic acids, o*,m, p-aminophenol, o,m and p-aminothiophenol, 2*,3 and 4-aminobenzonitrile, 2*,3* and 4-trifuoromethylaniline, o,m, p-nitroaniline, 2 and 3-hydroxmethylaniline*, o* and p-anilinosulfonic acid, 2* and 4-aminobiphenyl, 4-aminodiphenylamine*, benzidine, 1-aminonaphthalene, 1,2, 1,5* and 1,8*-diaminonaphthalene, 1 and 2-aminoanthracene, 1 and 6-aminochryzene, 1-aminopyrene, 1 and 2-aminoanthraquinone, 4-amino 1,2,3-benzothiadiazole*, 9,10-diaminophenanthrene, 2,6-diaminoanthraquinone, acriflavine*, 9-aminoacridine, 6-amino 3,4-benzocoumarine, 2, 3 and 4-amino-

Springer Series in Solid-State Sciences, Vol. 91
Electronic Properties of Conjugated Polymers III
Editors: H. Kuzmany · M. Mehring · S. Roth

pyridine, carbazole*, indole*, 2-aminopyrazine, 2-aminopyridazine, benzimidazole, benzotriazole, adenine, 4-amino 1, 2,4-triazole, phenazine, triazine, melanine.

It is of course impossible to develop here all the results obtained with such compounds. However, it is possible to make the following remarks, with regard to the influence of substituting groups on the aniline ring :

- Electronic acceptor groups strongly decrease the oxidative polymerization reaction efficiency.
- Inductive donor groups in ortho- or meta- position make the polymerization reaction easier.
- Mesomeric donor groups favour the polymerization only if they are in the ortho- position.
- Alkyl chains in the ortho- position give a very stable polymer with organophilic properties.
- Two alkyl groups in 2 and 5 positions [10] are probably very suitable for the conductivity and stability of the corresponding polyaniline.

The most important fact that we observe with many aniline derivatives concerns the spectroscopic detection in the visible range of the corresponding phenylnitrenium cation (C_6H_5-N^+-H) as an intermediate species [11]. The presence of this cation suggests the following: First of all it is a very reactive electrophile which gives addition compounds on every aromatic ring . It can be used to incorporate specific groups, such as redox, photosensitive or biologically active species, into polyaniline chains. We had used it to incorporate phenazine rings into polyaniline [9]. Secondly, this nitrenium cation can exist in polyaniline itself at high levels of doping. It can also be considered as nitrogen atoms with two substituent groups and bearing a strong positive charge density. We have attributed a cross-linking reaction to the presence of such an electrophile in the chain, which is responsible for the middle peak in the cyclic voltammetry of polyaniline [9]. A full bidimensional polymer corresponding to a polyaniline with a maximum of cross-linking bonds between chains was obtained by the oxidation of hydrazobenzene which gave a very high concentration of phenylnitrenium cations [9].

A significant contribution to the elucidation of the polymerization mechanism of aniline was also made as a consequence of an electrochemical study of 4-aminodiphenylamine (dimer of aniline) [12]. This study made it possible to quantitatively prepare the tetramer of aniline by the oxidation of the dimer at low potential [12].

In this paper let us take the opportunity to focus more precise results on a substituted aniline, o-propylaniline before concluding on the effect of substituent groups on conducting polymers.

RESULTS ON POLY(O-PROPYLANILINE) - POP -

The potential cycling of a platinum electrode in a 0.1 M 2-propylaniline solution in an anhydrous NH_4F,2.3HF mixture between -0.2 V and 0.6 V (vs. Cu/CuF_2) leads to the regular

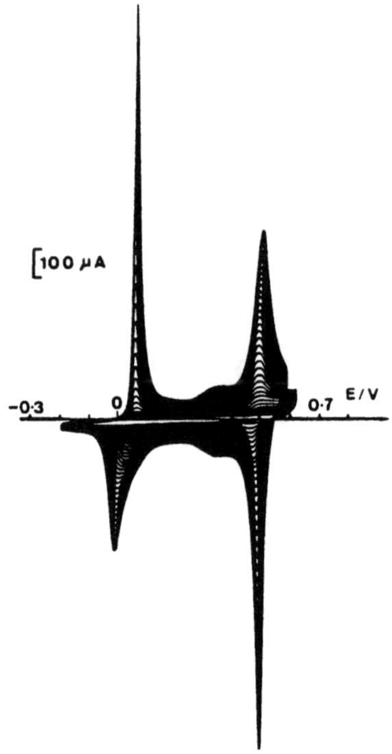

Fig.1: Successive cyclic voltammograms at 100 mV/s, on a platinum electrode ($1cm^2$), in NH_4F, 2.3 HF, 0.1M 2-propylaniline solution. Ref. Cu/CuF_2

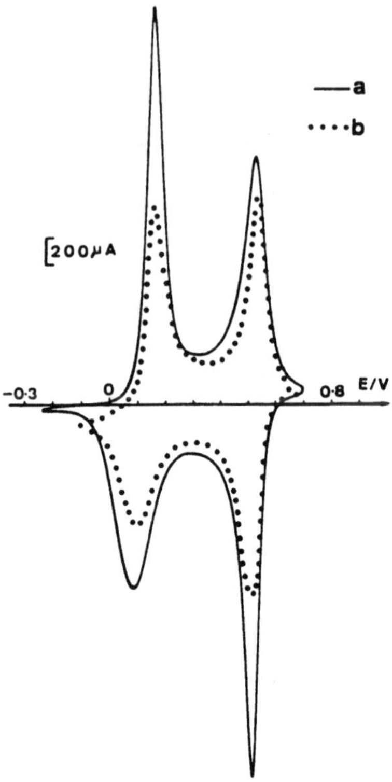

Fig.2: Voltammograms, 100 mV/s, POP film (Q=5.5 mC/cm^2), NH_4F, 2.3HF.
(a) After polymerization.
(b) After dissolution in acetone and redeposition on a clean platinum electrode

growth of an electroactive film of POP (Fig.1). The FTIR and visible spectra were studied as function of the doping level. After transferring to the hydrofluoric medium without monomer, the Pt/POP modified electrode (Fig.2, curve a) exhibits an electrochemical response which is different from that in Fig.1 probably because the conformation of the polymer chains change after washing and transfer the film. We note in Fig.2, curve a, that the potential of first electron transfer in POP is very close to that of polyaniline and the second charge transfer is 0.2 V more cathodic. This is probably due to the difference in pKa between the two polymers [13].

In comparison with the polyaniline, this POP is remarkable from three points of view. First of all it is soluble in the reduced state in organic solvents such as benzene or acetone. Secondly, the films are very stable upon electrochemical cycling of all redox systems even in sulfuric acid solution. Effectively we do not observe any hydrolysis to quinone as is the case for polyaniline itself. Finally the electrochemical behaviour of POP in an organic medium such as acetonitrile displays an interesting feature (see Fig. 4).

Fig.2, curve a, shows the cyclic voltammogram of a POP film in the fluorhydric medium and curve b after dissolution in a drop of acetone and casted on a new naked platinum electrode. Since POP films are not soluble in an organic solvent with a high dielectric constant such as acetonitrile (32) the solubility of POP is not due to low molecular weight but to the hydrophobic properties afforded by the propyl substituent groups. The solubility of reduced POP in DMF was 20g/l (+/- 5). The solubility of conducting polymers has been demonstrated to be an important parameter for the processability and development of such materials [15].

Fig.3 shows the first scan (a) and the 35th scan (b) of a POP film in 5 M sulfuric acid solution. Good POP films can also be prepared in this aqueous medium but not a lower concentration of acid. In contrast to the classical influence of a donor group such as propyl, which usually increases the basic properties, we observed here that the pKas of POP are lower than that of PANi, because of the screening effect of the propyl group on the nitrogen.

Fig.4 shows the voltammogram in 0.5 M $LiClO_4$ acetonitrile solution of a polyaniline (PANi) film (a) and POP film (b) which was prepared in the same way. We observe that the electrochemical behaviour of POP is much better than that of PANi. These results indicate the suitability of POP as electrode for secondary batteries in organic media [14],

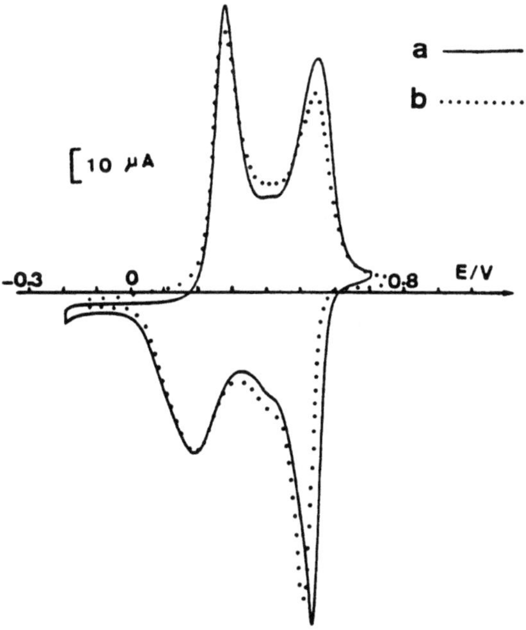

Fig.3: Voltammograms of a POP film (made in NH_4F, 2.3HF, Q=4 mC/cm^2) in H_2SO_4 5 M solution, vs SCE, at a scan rate of 10 mV/s. (a) 1st scan, (b) 35th scan

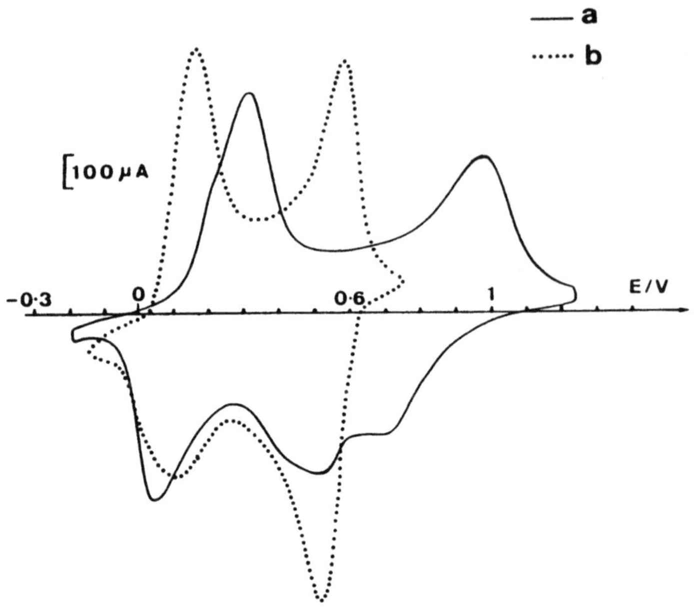

Fig.4: Cyclic voltammogram at 100 mV/s, acetonitrile 0.5 M $LiClO_4$ solution, Ref. Ag/10^{-2} M Ag^+.
(a) platinum electrode modified with PANi film (Q=4.5 mC/cm^2). (b) platinum electrode modified with POP film (Q=5 mC/cm^2)

particulary because the level of doping is the same as that for PANi and it is possible to cycle it to a higher potential.

CONCLUSION

The interesting properties of POP when compared with PANi: solubility, stability and electrochemical behaviour, must be of course attributed to the influence of the propyl chain in the ortho- position. This alkyl group protects the nitrogen versus hydrolysis, and gives better compatibilities of the polymer with organic media than with aqueous media.

Unfortunately the conductivity of POP is 2 to 3 orders of magnitude lower than that of PANi. This is attributed to the role of water on the process of conductivity of PANi [16] which is lower in the case of POP. On the other hand we have observed, using a molecular modelling software [10], that a propyl group does not play any role in favorising the planarity and conjugation between electrons of aromatic rings as does the octyl group in thiophene [15]. However, in the case of aniline two alkyl chains in the 2 and 5 positions seem very promising [10].

References

1- A.G. MacDiarmid, J.C. Chiang, A.F. Richter and A.J. Epstein, Synth. Met. 18 (1987) 285. Y.H. Kim, S.D. Phillips, M.J. Nowak, D. Speigel, C.M. Foster, G. Yu, J.C. Chiang and A.J. Heeger, Synth. Met., 29 (1989) E291.
2- A. Kitani, J. Izumi, J. Yano, Y. Hiromoto and K. Sasaki, Bull. Chem. Soc. Jpn., 57 (1984) 2254.
3- A. Litzelmann, B. Scheerer, J. Fink, K. Meerholz, J. Heinze, N.S. Sariciftci, and H. Kuzmany, Synth. Met., 29 (1989) E313.
4- E.M. Genies, M. Lapkowski and C. Tsintavis, New J. Chem., 12 (1988) 181.
5- E.M. Genies, J.F. Penneau and M. Lapkowski, New J. Chem., 12 (1988) 765.
6- E.M. Genies and J.F. Penneau, J. Electroanal. Chem., in press.
7- J.F. Penneau, M. Lapkowski and E.M. Genies, New J. Chem., in press.
8- E.M. Genies, J.F. Penneau and M. Lapkowski, J. Electroanal. Chem., 260 (1989) 145.
9- E.M. Genies, M. Lapkowski and J.F. Penneau, J. Electroanal. Chem., 249 (1989) 97.
10- E.M. Genies, unpublished results.
11- E.M. Genies and M. Lapkowski, J. Electroanal. Chem., 236 (1988) 189.
12- E.M. Genies, J.F. Penneau, M. Lapkowski and A. Boyle, J. Electroanal. Chem., in press.
13- E.M. Genies and E. Vieil, Synth. Met., 20 (1987) 97.
14- E.M. Genies, P. Hany and Ch. Santier, J. Applied Electrochem., 18 (1988) 751.
15- polyoctthio santa fe
16- M. Nechtschein, C. Santier, J.P. Travers, J. Chroboczek, A. Alix and M. Ripert, Synth. Met. 18 (1987) 311.

Proton Exchange Processes in the Electrochemical Reactions of Polyaniline

M. Lapkowski

Institute of Inorganic Chemistry and Technology,
Silesian Technical University, PL-44-100 Gliwice, Poland

ABSTRACT

Thin layer transmission spectroelectrochemical method has been applied to study the mechanism of proton exchange accompanying the electrode reactions of polyaniline in aqueous and non-aqueous media. The results obtained for aqueous solutions seem to suggest that after polyaniline oxidation to the first polaron state a small insertion of protons from the solution occurs.

In non-aqueous media no proton exchange is observed during the first oxidation step whereas in the potential range corresponding to the second oxidation step the acidity of the electrolyte solution increases. Protons are strongly bound within the polyaniline matrix.

INTRODUCTION

Polyaniline has been known for many years [1], but only in the last decade a significant progress in the studies of its electrochemical properties has been achieved. This revival of the interest in polyaniline is mainly associated with the facility of its preparation, its high stability and high potential for technological applications [2]. The elucidation of the polymerization and redox reaction mechanisms is therefore vital for better understanding of this extremely interesting polymer.

Polyaniline is the only conducting polymer whose properties are not only dependent on its oxidation state but also on its protonation state which in turn are strongly associated with the acidity of the medium being in contact with polyaniline electrode. The polymer is electroactive in solutions of pH < 4 and its conductivity can be varied over 10 orders of magnitude depending on the protonation level [3].

According to many authors [4] the redox reactions are accompanied by proton exchange processes. However no direct experimental evidence for this process in the course of the redox reaction has been published to date. This work is therefore focused on the "*in situ*" observation of proton transfer (local pH of the solution being in contact with polyaniline film) induced by the change of the redox state of polyaniline.

RESULTS AND DISCUSSION

The details of the electrochemical and spectroelectrochemical sets used in the present research can be found elsewhere [5]. Polyaniline film washed in aqueous, sulphuric acid solution was washed with distilled water and then protonated in a solution consisting of sulphuric acid (pH = 1.4) and 1M of sodium

Springer Series in Solid-State Sciences, Vol. 91
Electronic Properties of Conjugated Polymers III
Editors: H. Kuzmany · M. Mehring · S. Roth

sulphate. For the studies in acetonitrile polyaniline was also washed with distilled water, then protonated for 1h in perchloric acid solution (pH = 1.4) containing 1M of lithium perchlorate and finally washed with acetonitrile.

The indicator selected for this research had to fulfill three requirements:

1) it had to change its color in the pH range of polyaniline electroactivity (pH < 4);
2) in addition it had to be soluble in aqueous as well as in acetonitrile;
3) it had to be electroinactive in the potential range of polyaniline oxidation.

Methyl violet (tetra-, penta- and hexa-methyl-o-rosaniline) was the best of readily available indicators fulfilling this requirements since it changes color in the pH range 0.15 - 3.5.

Aqueous solution

In Fig.1 a pseudo-three-dimensional diagram of spectral changes observed in aqueous solution of methyl violet is presented. The potential of the electrode, covered with 8 μm thick polyaniline film, was scanned in the range from -0.5 to +0.9 V. For E = -0.5 V polyaniline is in the totally reduced so-called polyleuco-emeraldine state .

The electrolyte solution (sulphuric acid + sodium sulphate) with uniform concentration of protons in the bulk (pH = 1.4) is violet-yellow (λ_{max} = 600 nm, λ_{max} = 450nm) because for this pH value methyl violet is in the transition state. The change of its color is not sharp and takes place in the pH interval from 0.15 to 3.2.

The spectral features of the solution in contact with polyaniline electrode are strongly dependent on its potential. It may be therefore stated that local pH varies during the potential change due to the variation of the polyaniline protonation state. When the potential of the working electrode is raised to 0 V

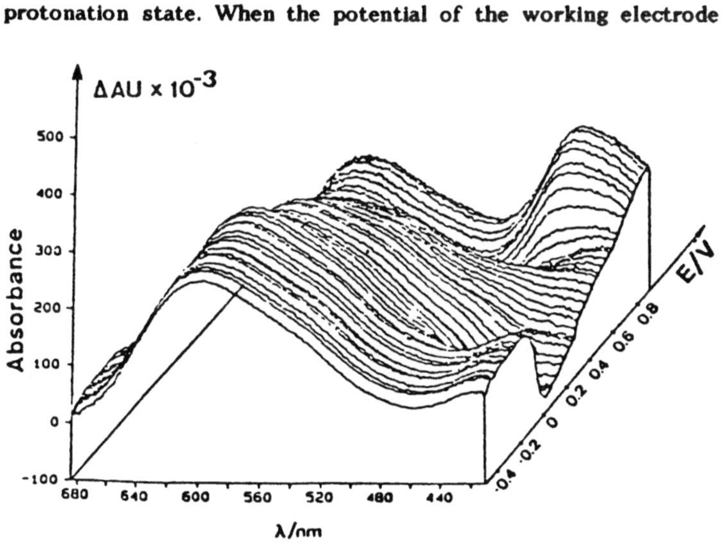

Fig. 1. Pseudo-three-dimensional diagram of absorbance changes in sulphuric acid solution (pH = 1.4) containing 0.02 M of methyl violet and 1 M sodium sulphate. The curves were registered during voltammetric scans of potential of the working electrode covered with 8 μm thick polyaniline film. Scan rate 2 mV/s

the absorption at λ = 450 nm virtually disappears. Further increase of the potential results in a decrease of the absorption due to the = 600 nm peak together with renewed formation of the = 454 nm absorption which is observed at potentials E > 0.65 V.

From this figure it is clear that there exist two potential ranges for which significant changes associated with the indicator occur. They correspond strictly to the two electrode processes recorded for polyaniline film (voltammetric curve is represented as the dotted line).

First electrode process

The following explanation can be proposed:

In totally reduced polyaniline (polyleucoemeraldine), at pH = 0, only 1/4 of amine groups (-NH-) are protonated. Therefore for pH = 1.4 the protonation level should be much lower. In the initial stages of the first electrode process polyaniline is being oxidized to the first polaron (shoulder preceding the first peak in the cyclic voltammogram). At this moment only 1 per 10-12 nitrogen atoms is being oxidized, the rest remaining in the reduced state (-NH-). The oxidation changes significantly pK of polyaniline which becomes more basic. Thus the protonation level of -NH- groups increases. The local pH of the solution being in direct contact with the polyaniline electrode must therefore increase since protons are transferred from the solution to the polymer according to the following scheme:

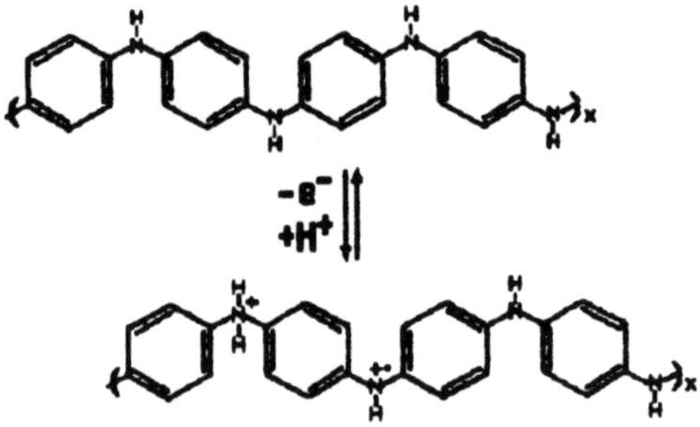

(1)

There are three possible reasons for such behaviour:

a) a change of polyaniline pK_a under the influence of doping,
b) stimulated protonation by double charged SO_4^{2-} anion,
c) doping by HSO_4^- anion.

A rather insignificant increase of the acidity occurring during further oxidation can be rationalized by the diffusion of protons from the bulk of the electrolyte.

It should be noted that the oxidation from the polaron state to the bipolaron state does not change the protonation level. Most probably non-protonated -NH- groups after being oxidized recombine to bipolarons [5].

Second electrode process

During the second electrode process a new clear absorption band at λ = 454 nm grows. This band is responsible for the absorption of the acidic form of methyl violet, it therefore manifests a drop of the local pH of the electrolyte solution to values < 0.15. Polyaniline deprotonation starts at potentials E = +0.35 - +0.40 V corresponding to the additional (third) oxidation peak which is responsible for the crosslinking of linear polymer or for the formation of phenazene structures [6]. A clearer increase of the electrolyte solution acidity occurs immediately prior to the second oxidation potential at E = 0.5 V. The second electrode process is therefore accompanied by polyaniline deprotonation:

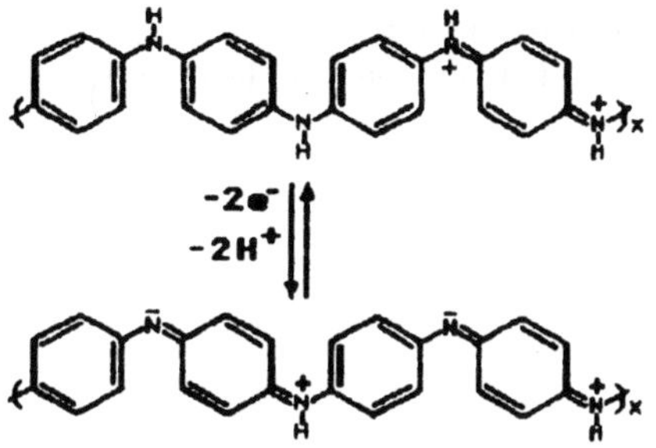

(2)

Till now it was postulated that polyaniline deprotonation occurs in the potential range E = 0 - +0.5 V, between the two electrode processes.

Non-aqueous solutions

From the spectroelectrochemical measurements [5] it is clear that proton exchange between polyaniline and the electrolyte is, in the non-aqueous solutions, associated with much higher oxidation level of the polymer. It means that proton are more strongly bound to the polymer and less readily released to the acetonitrile solution. The deprotonation is not simultaneous with the second oxidation process but follows it [5].

CONCLUSIONS

Spectroelectrochemical measurements carried out in this research enabled us to elucidate the mechanism of redox processes occurring during polyaniline cycling and to verify the models presented in the literature [2,4].

Proton exchange occurring during the two oxidation processes has been determined in both aqueous and non-aqueous solutions. The presented research has also induced us to verify the possibility of the construction of a new type of electrochemichromic sensors based on reversible color changes of a selected acid-base indicator which in turn are induced by proton exchange between the polymer and the electrolyte.

ACKNOWLEDGEMENTS

We would like to thank Prof. E.M. Genies for many helpful discussions. The present work is financially suported by the C.P.B.P. 01.15.

REFERENCES

1. J. Fritzche, J.Prakt.Chem., 20, 453, (1840); 28, 198, (1843).
2. E.M. Genies, M. Lapkowski and C. Tsintavis, New J. Chem., 12, 181, (1988).
3. (a) A.J. Epsten, J.M. Ginder, F. Zuo, R.W. Bigelow, H.-S. Woo, D.B. Tanner, A.F. Richter, W.-S. Huang and A.G. MacDiarmid, Synt.Met., 18, 303, (1987); (b) J.M. Ginder, A.F. Richter, A.G. MacDiarmid and A.J. Epstein, Solid State Commun., 63, 97, (1987).

4. (a) W.-S. Huang, B.D. Humphrey and A.G. MacDiarmid, J.Chem. Soc., Faraday Trans. 1, 82, 2385, (1986); (b) A.G. MacDiarmid, J.C. Chiang, A.F. Richter and A.J. Epstein, Synth.Met., 18, 285, (1987).
5. M. Łapkowski, Bull. Electrochem., in print.
6. E.M. Genies M. Łapkowski and J.-F. Penneau J.Electroanal.Chem., 249, 97, (1988).

Electrochemical Investigations of Electronic Conducting Polymers

J.W. Schultze, M.M. Lohrengel, and A. Thyssen

Institut für Physikalische Chemie und Elektrochemie,
Arbeitsgemeinschaft Elektrochemischer Forschungsinstitutionen
Heinrich-Heine-Universität Düsseldorf,
Universitätsstr. 1, D-4000 Düsseldorf, Fed. Rep. of Germany

1. Introduction

All the technical applications of metallic conducting polymers, like batteries /1/, displays /2/, catalysis /3/ and protection against corrosion /4/ are systems metal/polymer/electrolyte. Electrochemical **in situ** measurements are therefore very useful for investigating these systems. Problems like aging and contamination of the probe are negligible. In this paper we would like to present some electrochemical results concerning the electrochemical polymerization and characterisation of aniline and some methyl derivatives.

2. Experimental

All experiments were carried out with gold as the substrate (working electrode WE) and counter electrode (CE). The electrode potential was measured against a mercury/mercury sulfate reference electrode (RE). All potentials given in this paper refer to the standard hydrogen electrode (SHE). Current densities were measured with a fast-rise potentiostat. The charge was calculated by numeric integration of i over t. The impedance of the electrode was measured by modulating the electrode potential by an AC signal (1 mV, 1 kHz). The changes in mass of polymer were measured simultaneously by an Electrochemical Quartz Micro-Balance EQMB /5/ by changes in frequency. A 5 MHz-quartz was covered by a 300 nm Au film with a total area of 0.4 cm². Fig. 1 shows the experimental set-up.

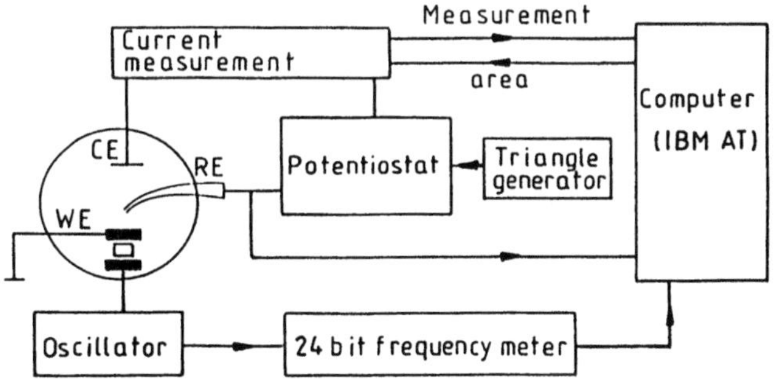

Fig. 1 Experimental set-up

Springer Series in Solid-State Sciences, Vol. 91
Electronic Properties of Conjugated Polymers III
Editors: H. Kuzmany · M. Mehring · S. Roth

The decrease of f corresponds to a positive mass coating, referred to the initial frequency at $t = 0$ or $m_{poly} = 0$. The average thickness $d°$ is calculated from

$$d^{\circ} = -\frac{\Delta f \cdot K}{\rho} \qquad (1)$$

with $\rho = 1$ g · cm^{-3} and the proportional factor $K = 20$ ng · Hz^{-1} · cm^{-2}, determined by Cu deposition in acid solution /6/.

3. Electrochemical polymerization

The oxidation and reoxidation of polyaniline (PANI) between 0 and 1 V is combined with an exchange of counterions and electrolyte. Fig. 2 shows the i/ϵ (a) and $\Delta m/\epsilon$ plots (b) for the redox reaction of polyaniline in pure sulfuric acid, beginning at 1 V with $d\epsilon/dt = 2$ mV · s^{-1}. From XPS measurements /7/ we know that the percentage of HSO_4^- and H_2O molecules in the PANI film is $[HSO_4^-]_{ox} \approx 25\%$ and $[HSO_4^-]_{red} \approx 0\%$. The reduction of PANI is combined with a loss of mass up to 15 μg · cm^{-2}, referred to a m_o of the total polymer film after preparation of about 100 μg · cm^{-2}. The loss of mass during reduction - caused by migration of the counterions - is smaller than expected, obviously compensated by the insertion of water molecules.

In our previous work /8/ we investigated the kinetics of the anodic polymerization of aniline and its methylderivatives. The results show various periods of the reaction : the

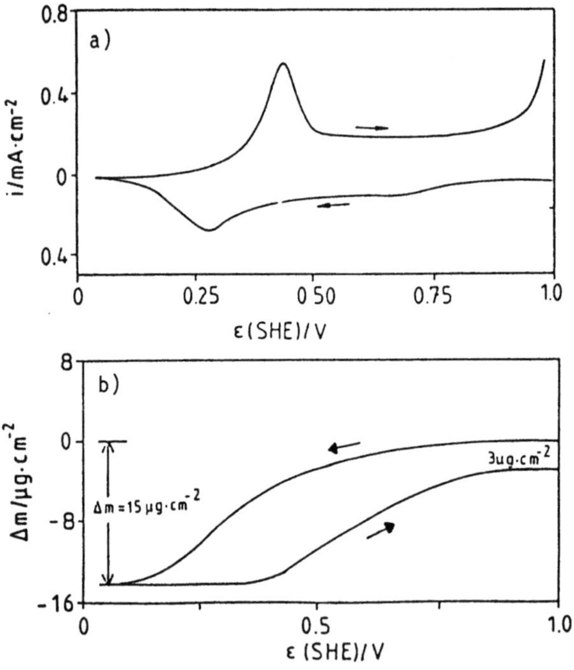

Fig. 2 Reduction and re-oxidation of polyaniline in pure 0.5 M H_2SO_4, $d\epsilon/dt = 2$ mV · s^{-1}, a) i/ϵ and b) $\Delta m/\epsilon$

adsorption and formation of oligomers, the nucleation and growth and finally the steady state of the growing film. Branching reactions at the o-position of the ring are responsible for the autocatalytical growth of the polyaniline hemispheres. The in situ current yield γ

$$\gamma = \frac{z \cdot F \cdot r}{M} \cdot \frac{dm}{dq} \tag{2}$$

for the polymerization of aniline, 2,6-dimethylaniline and p-toluidine, determined with the EQMB measurements, is shown in fig. 3 as a function of time. γ_{PANI} is about 0.7 after 500 s and much greater than the values of the derivatives. After this maximum the current yield is constant at 0.3 during the polymerization.

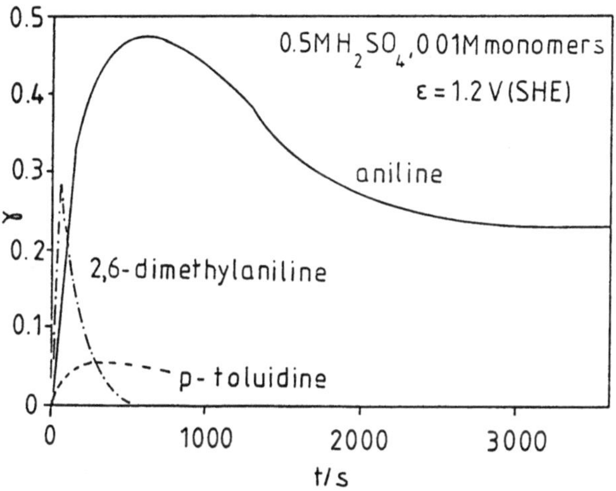

Fig. 3 In situ current yield γ from EQMB measurements as a function of time for the polymerization of aniline, 2,6-dimethylaniline and p-toluidine, 0.01 M monomer, 0.5 M H_2SO_4, 1.2 V

4. Electrochemical characterisation

During our investigations concerning the electronic in situ conductivity of polyaniline /9/, we observed the metallic conductivity of the black-blue oxidized form. The colourless reduced form has a small electronic conductivity, which can be increased by strong oxidants, e.g. Fe^{3+}. The overoxidized form of the polyaniline at potentials higher than 1.3 V is irreversible non-conductive. A conductivity of protons through the PANI film of about $\kappa > 10^{-3}\ \Omega \cdot cm^{-1}$ could be determined by the investigation of the hydrogen evolution /10/ on a bare Au-electrode in comparison with the Au/PANI-system. The same Tafel factors prove that the H_2 evolution takes place at the interface Au/polymer and not at the interface polymer/electrolyte. We tried to determine the porosity of a polymer film in contact with the electrolyte by metal deposition. Fig. 4 shows the lg$i$ / lg$t$ and Δm / lg$t$ plots for the Cu deposition on Au and Au/PANI. We see a strong increase in the mass deposited at the bare Au without any induction period. There is a constant

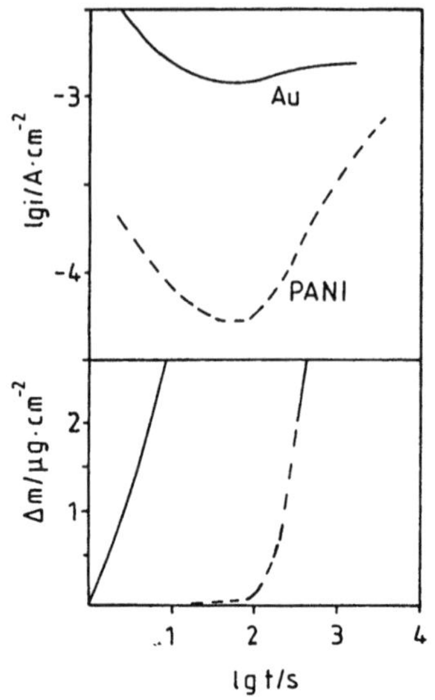

Fig. 4 lgi/lgt and m/lgt plot for the Cu deposition on Au and Au/polyaniline, 0 V, 0.5 M H_2SO_4, 0.01 M Cu^{2+}

current density during the induction period from t ≈ 20 to 100 s. The changes in frequency in the same time period are about 10 Hz, which corresponds to 200 ng · cm^{-2}. We calculate the radius of the pores with

$$r = \left(\frac{\Delta f \cdot K \cdot 10^{-9}}{\rho \cdot N \cdot d \cdot \pi}\right)^{\frac{1}{2}} \quad (3)$$

and ρ = 8.93 g · cm^{-3} as r ≈ 0.7 · 10^{-4} cm. The *m*/lg*t* plot shows the inhibition of the Cu deposition on Au/polyaniline in comparison to that on bare Au. This proves the migration of the Cu^{2+} ions through the polymer film to the electrode.

5. References

1. G. Weddingen in S. Roth, H.J. Mair (Ed.), Elektrisch Leitende Kunststoffe, Carl Hanser Verlag 1986

2. M. Gazard in T.A. Skotheim (Ed.), Handbook of Conducting Polymers, Marcel Dekker Inc. 1988

3. F.T.A. Vork, L.J.J. Janssen, E. Barendrecht Electrochim. Acta 30 (1985) 1433

4. R. Noufi, D. Tench, L.F. Warren J. Electrochem. Soc. 127 (1980) 2310

5. a) S. Bruckenstein, S. Swathirajan
Electrochim. Acta 30 (1985) 851
b) B. Schreck, J. Heitbaum
Dechema Monographie 112 (1988) 49
c) R. Schuhmacher
Dechema Monographie 102 (1986) 107

6. D. Ebling, M.M. Lohrengel, A. Meyer, A. Thyssen
Dechema Monographie 1989, in press

7. R. Kessel, G. Hansen, J.W. Schultze
Ber. Bunsenges. Phys. Chem. 92 (1988) 710

8. a) A. Thyssen, A. Hochfeld, J.W. Schultze
Dechema Monographie 112 (1988) 441
b) A. Hochfeld, R. Kessel, J.W. Schultze, A. Thyssen
Ber. Bunsenges. Phys. Chem. 92 (1988) 1406
c) A. Thyssen, A. Hochfeld, R. Kessel, A. Meyer, J.W. Schultze
Synth. Met. 29 (1989) E 357

9. A. Thyssen, A. Borgerding, J.W. Schultze
Macromol. Chem., Macromol. Symp. 8 (1987) 143

10. B. Pfeiffer, A. Thyssen, J.W. Schultze
J. Electroanal. Chem. 260 (1989) 393

Memory Effect in Conducting Polymers: Case of Polyaniline

C. Odin and M. Nechtschein

DRF/SPh/Dynamique de Spin et Propriétés Electroniques (UPR CNRS 216),
Centre d'Etudes Nucléaires de Grenoble,
85X, F-38041 Grenoble Cedex, France

Abstract. The study of the charging/discharging behavior of conducting polymers shows evidence for very slow relaxation processes and dependence of the response upon the electrochemical history of the sample. Preliminary results on the time behavior of these effects are reported, based on two chronoamperometry methods. Connections between memory and hysteresis effects are discussed.

1. Introduction

Conducting polymers can be cycled between an insulator neutral state and a conductive charged state by electrochemical doping. It has been noticed that the doping behavior depends on the electrochemical history of the sample, showing the existence of slow evolution (relaxation) processes. For instance, the first anodic or cathodic wave always differs in shape and peak position from the stationary wave [1-3]. Such effects have been observed on many different compounds (polythiophenes, polypyrrole, PTCNQ, polyaniline...), therefore we believe them to be rather general features of electroactive polymer modified electrodes. These phenomena have been attributed to the incorporation of ions and solvent molecules into the polymer matrix, as well as to the potential dependent morphological changes upon the doping process (swelling and slow shrinkage, twist...) which may relax slowly [2,4,5]. Nevertheless, a general understanding of these phenomena has not been yet achieved.

We report here preliminary studies of the time behavior of this relaxation process. The experiments have been carried out on polyaniline because of its easy processability and its high stability. The films were synthetized on a planinium wire by three methods (electrochemical and chemical polymerization, and step by step coating [6]), and cycled in HCl 1M in air at room temperature with a PAR273 piloted by a PC/AT computer. Thicknesses were less than 1μm. The following experimental procedure has been used. At first, the film is "prepared" by a double pulse sequence as shown in Fig.1a. The first step (V_1 = 0.5V vs SCE) leads the film to a known oxidized state, whereas the second one (V_2 = -0.2V vs SCE) brings it back to its reduced (neutral) form. The film is then left relaxing at this potential for a waiting time T. Next, the current is recorded by two methods : a) with a potential step b) with a symmetric triangular potential waveform.

Springer Series in Solid-State Sciences, Vol. 91
Electronic Properties of Conjugated Polymers III
Editors: H. Kuzmany · M. Mehring · S. Roth

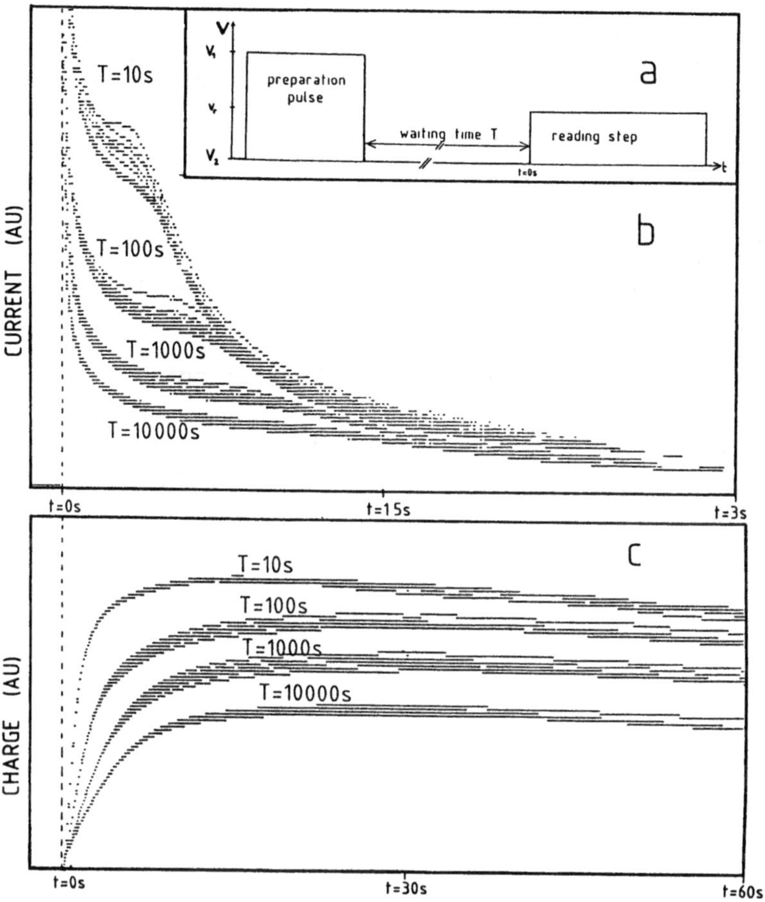

Figure 1 : a) Double Pulse Potential Sequence (V_1 = 0.5V, V_2 = -0.2V ; V_r = 0.1V vs SCE) ; b) Current during the reading step V_r for different waiting times T ; c) Charge integrated from the current (notice the different time scales)

2. Chronoamperometry with a potential step

The current response is followed during a "reading" potential step : V_r = 0.1V vs SCE. The response pointedly depends on T (Fig.1b), in a reproductible way. As a matter of fact, many experiments with the same waiting time sequence lead to superposable curves. The charge was obtained by integrating the current (Fig.1c). It turns out that both the kinetics speed and the "asymptotic" charge decrease as the waiting time increases. A shadow has to be put on the latter point, since leak currents make long time measurements doubtful. Despite this fact, these results seem to be consistent with EPR experiments which avoid this leak current problem [7].

We suggest that the asymptotic charge behavior as a function of T could be connected to the hysteresis effect which is observed for the charge as a function of the potential. Namely, for a given applied potential, say V_r, the charge injected in the polymer depends on whether V_r has been reached from increasing (oxidation), or decreasing (reduction) potentials. In our double pulse experiment, the limiting case $T \longrightarrow 0$ should correspond to the reduction half cycle (since for $T = 0$, V_r is reached directly from the oxidized state). On the other hand, in the case of long T, the system has time to relax in the neutral state, and the charge of the system at V_r should correspond to the oxidation half cycle.

3. Chronoamperometry with a symmetric triangular potential waveform

A symmetric triangular potential waveform is used instead of a step (Fig.2a). It allows us to compare the first wave with the stationary ones. All other experimental parameters are the same as previously.

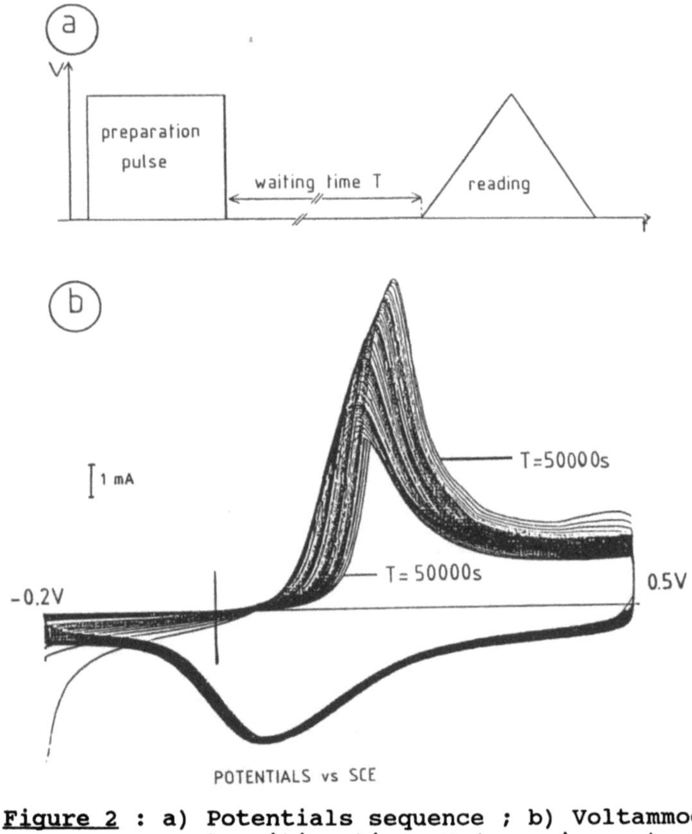

__Figure 2__ : a) Potentials sequence ; b) Voltammograms for different waiting times T (sweeping rate = 100mV/s)

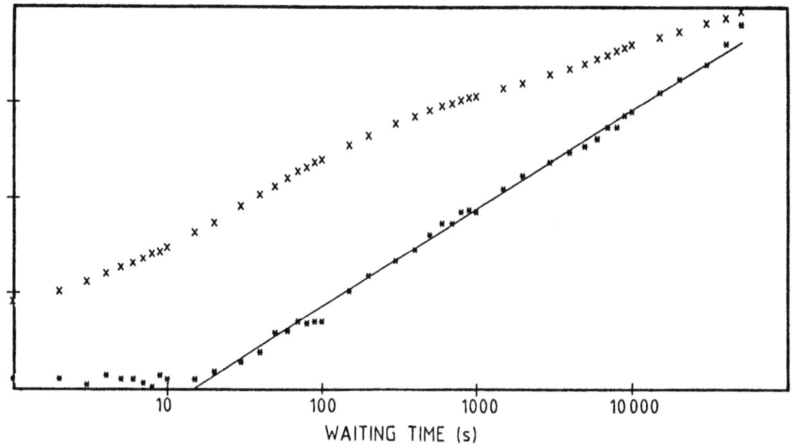

Figure 3 : Potential shift (*) and current amplitude (x) of the oxidation peak as a function of T. Solid line is a fit with a logarithmic law for T ⩾ 10s ; notice that the slopes are arbitrary

The waiting time has been increased between 0 up to $5x10^4$ s. Two features were observed (Fig.2b) :

a) the oxidation wave becomes sharper,

b) the oxidation peak potential shifts to higher values.

One should notice that almost no effect is seen on the reduction wave. It can then be inferred that the memory was lost during the oxidation (charging). We can also conclude that, with longer T, the hysteresis is increased, and, consequently, the reaction becomes more irreversible.

Furthermore, the shift of the maximum of the oxidation peak is well fitted by a logarithmic law (Fig.3) : $E_p \sim \text{Log } T$ for T ⩾ 10s. Therefore the relaxation seems to involve no characteristic time. We notice that a logarithmic law is consistent with the dynamic of the decay of a system having a random distribution of energy barriers.

References

1. G. Tourillon, F. Garnier, J. Electroanal. Chem **161**, 51-58 (1984)
2. D. Orata, D.A. Butry, J. Am. Chem. Soc. **109**, 3574-3581 (1987)
3. J. Heinze, M. Störzbach, J. Mortensen, Ber. Bunsenges Phys. Chem. **91**, 960-967 (1987)
4. J.L. Baudoin, F. Chao, M. Costa, J. Chim. Phys. **86**, n° 1, 181 (1989)
5. J. Heinze, R. Bilger, K. Meerholz, Ber. Bunsenges Phys. Chem. **92**, 1266-1271 (1988)
6. Y. Nicolau, M. Nechtschein, these proceedings
7. B. Villeret, M. Nechtschein, to be published

Part VI

Magnetic Resonance

Spin Dynamics and Conductivity in Conducting Polymers

M. Nechtschein, F. Devreux, K. Mizoguchi**, and J.P. Travers*

DRF/SPh/Dynamique de Spin et Propriétés Electroniques (UPR CNRS 216),
Centre d'Etudes Nucléaires de Grenoble,
85 X, F-38041 Grenoble Cedex, France

Abstract. Spin dynamics studies have been performed in polyaniline and polypyrrole, and the results have been correlated to the transport properties. In polyaniline the data agree with the conducting island picture, but arguments are given that a given island just consists of a single conducting chain. The conductivity is controlled by interchain hoppings. In polypyrrole, both NMR and ac conductivity data can be interpreted in terms of diffusion of the spin-charge carriers on a dynamical percolation network, whose effective dimension increases with temperature.

1. Introduction

Unlike trans-$(CH)_x$, in which the charge carriers are spinless solitons, a clear connection between electrochemical doping and spin concentration has been demonstrated in conducting polymers with non-degenerate ground state [1-3]. Beyong the static viewpoint of the spin-charge relationship, it is interesting to search for a possible connection between the dynamical properties of spins and charges. Namely, can spin dynamics and transport properties be correlated ? We here present the main points of two studies achieved in polypyrrole (PPy) and polyaniline (PANI).

2. Magnetic Resonance and Spin Dynamics

Magnetic Resonance techniques enable one to probe the spin motion and, if the spin carriers are also charge carriers (e.g. polarons), the transport properties can be analyzed at the scale of the polymer chains. Of particular interest is the case of low-dimensional systems [4]. A unique advantage of the method is the possibility to determine the anisotropy of the motion, even for powder or amorphous samples.

The spin dynamics is described by the motion spectrum

$$f(\omega) = \int_{-\infty}^{+\infty} P(t)\ e^{i\omega t}\ dt \qquad (1)$$

* Present address : Laboratoire PMC, Ecole Polytechnique, 91128 Palaiseau Cedex, France

** Permanent address : Dept. of Physics, Tokyo Metropolitan Univ., Setagaya-Ku, Tokyo 158, Japan

Springer Series in Solid-State Sciences, Vol. 91
Electronic Properties of Conjugated Polymers III
Editors: H. Kuzmany · M. Mehring · S. Roth

where P(t) is the probability that a given spin comes back to its starting point after time t.

Assuming that the spin carrier has equal probability to be at time t at any accessible point of a volume V(t), one has

$$P(t) \propto V(t)^{-1} \propto \langle r^2(t) \rangle^{-d/2} \quad (2)$$

where r(t) is the distance covered and d the space dimension. For a standard random walk, one has $\langle r^2(t)\rangle = Dt$, where D is the diffusion coefficient. This yields $P(t) \propto (Dt)^{-d/2}$. Then, for one-dimensional systems $f(\omega)$ exhibits a divergence as $(\omega D)^{-1/2}$ for $\omega \rightarrow 0$. However, for real pseudo one-dimensional systems this divergence is truncated in particular by inter-chain couplings, and one has

$$\begin{aligned} f(\omega) &\sim (\omega D_{/\!/})^{-1/2} \quad \text{for} \quad D_{/\!/} > \omega > D_{\perp} \\ f(\omega) &= \text{cte} \quad \text{for} \quad \omega < D_{\perp} \end{aligned} \quad (3)$$

where $D_{/\!/}$, and $D_{\perp}$ are the intra-, and inter-chain diffusion coefficients, respectively.

For high spin concentration the motion spectrum $f(\omega)$ governs both the electronic and nuclear spin relaxation processes. We give two expressions with explicit dependence on $f(\omega)$. First, the nuclear spin-lattice relaxation time T_{1N} is given by

$$T_{1N}^{-1} = \frac{c}{4}\left[\frac{3}{5}\, d^2\, f(\omega_N) + \left(a^2 + \frac{7}{5}\, d^2\right) f(\omega_e)\right] \quad (4)$$

where c is the spin concentration per unit cell ; a and d are the scalar and hyperfine couplings ; ω_e and ω_N are the electronic and nuclear Larmor frequencies, respectively. Second, the electronic spin-spin relaxation time can be expressed as

$$T_{2e}^{-1} = c\,\frac{R^2}{10}\,[3f(0) + 5f(\omega_e) + 2f(2\omega_e)] \quad (5)$$

where R^2 is the average sum of the spin-spin dipolar interactions. From expressions (4) and (5) it appears that measuring T_{1N} and T_{2e} as a function of frequency, provides two independent methods for determining $f(\omega)$.

3. The Case of Polyaniline

Polyaniline is in fact a family of compounds rather than a simple polymer. For instance the conductivity of the emeraldine form of PANI changes by ten orders of magnitude upon protonation [5-6], which shows that every given protonation level yields to a definite compound with specific properties. Experimental evidences have been given that, instead of being homogeneous, the protonation process gives rise to phase segregation into fully protonated, conducting, and high-spin-concentration domains (conducting islands) embedded in an unprotonated, insulating sea [7-9].

Proton T_1 and ESR linewidth ($\Delta H \sim T_{2e}^{-1}$) have been measured as a function of frequency for samples of different doping levels, y. The data can be unambiguously interpreted in terms of pseudo one-dimensional spin diffusion with the two parameters, $D_{/\!/}$ and $D_{\perp}$ [10,11]. These two parameters are given in Fig.1 as a function of the protonation level. It appears that $D_{/\!/} \sim 10^{12}-10^{13}$ rad sec^{-1} is almost not dependent on y. In other words, whatever the protonation level, and the total spin concentration, the spins have basically the same on-chain diffusion rate. This statement is in agreement with the conducting island picture : the on-chain diffusion appears to be an intrinsic property of the conducting islands. However, the behavior of the transverse diffusion rate is quite different : $D_{\perp}$ exhibits a sudden drop of 2-3 orders of magnitude for y ~ 0.2-0.3, a value which appears as a percolation threshold : y_p.

To explain that inter-cluster interactions can dramatically influence the inter-chain couplings, we have proposed that the protonated-conducting domains are composed of single polymer chains [10,11]. The protonation process in PANI would give rise to fully protonated chains embedded in an unprotonated matrix.

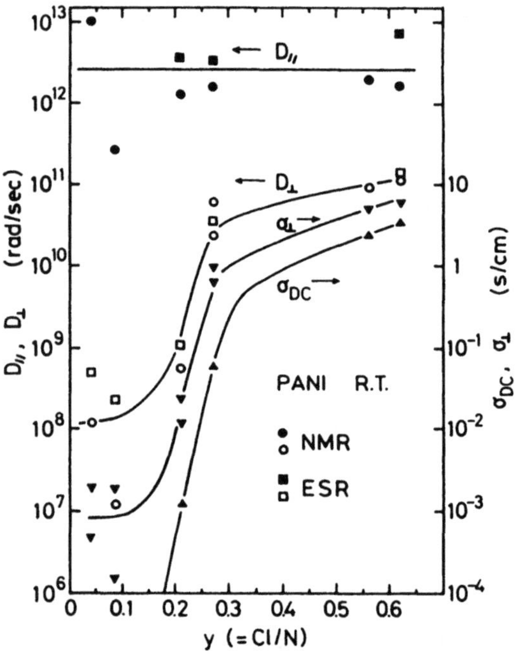

Figure 1 - Room temperature spin diffusion rates and conductivity in PANI as a function of the protonation level. Room temperature spin diffusion rates, as obtained from NMR (•, ○) and from ESR (■, □), and conductivity in PANI as a function of the protonation level. σ_{DC} is the measured conductivity and $\sigma_{\perp}$ has been calculated from $D_{\perp}$, using $\sigma_{\perp} = ne^2 D_{\perp}/kT$, with n, the carrier concentration, proportional to y

Let us now consider the transport properties. In Fig.1 the conductivity calculated from $D_\perp$ is compared to the measured value at different protonation levels. The remarkable agreement which is obtained (no adjustable parameter has been used), gives evidence that the macroscopic conductivity is governed by inter-chain hoppings.

4. The case of Polypyrrole

The proton spin-lattice relaxation time has been measured in PPy-ClO_4 as a function of frequency for different temperatures between 50 and 300K [12]. For T $\leqslant$ 150K the data can be fitted with a one-dimensional diffusion law : $T_{1N}^{-1} \propto (\omega D_{/\!/})^{-1/2}$. However, for T > 150K a departure from such a behavior is observed. The data can also not be fitted with the use of Eq.3 by introducing an inter-chain diffusion coefficient $D_\perp$.

Interpretation of the data in terms of anomalous diffusion has been proposed [15]. Namely, generalization of Eq.2 to effective dimension $\tilde{d}$ yields :

$$f(\omega) = A\,\omega^{\tilde{d}/2} + B \quad , \quad \tilde{d} < 2$$
$$f(\omega) = A \ln (B/\omega) \quad , \quad \tilde{d} = 2 \; . \qquad (6)$$

The data of Fig.2 can be fitted with the use of Eq.(6), and with $\tilde{d}$ = 1 for T $\leqslant$ 150K ; and $\tilde{d}$ = 1.2, 1.5, and 2 for T = 200, 250, and 293K, respectively. Thus, it appears that at low temperature (T $\leqslant$ 150K) the spins undergo a one-dimensional motion,

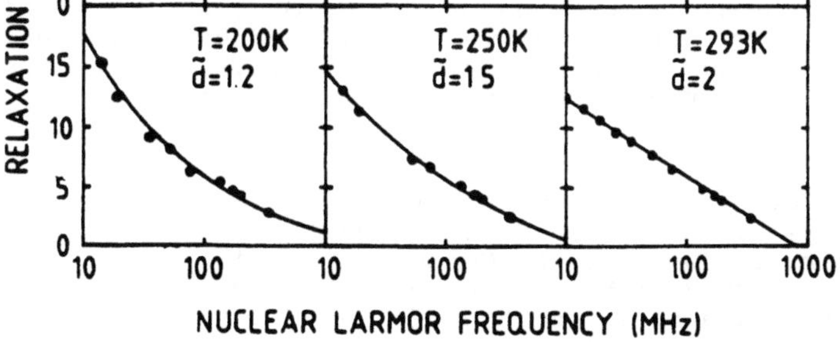

Figure 2 - Proton spin-lattice relaxation rate in PPy, ClO_4 as a function of frequency at different temperatures. Full lines are fits with Eq.(6) using the displayed value of $\tilde{d}$

which likely reflects the confinement of the polarons on the polymer chains. At higher temperature, they are allowed to execute interchain hoppings giving rise to an increasing effective dimensionality.

This dimension crossover reflects spin dynamics properties basically different from the behavior observed in PANI and other anisotropic conductors. In the latter systems, the crossover occurs at a definite frequency, which corresponds to a well-defined interchain hopping rate. In the case of PPy, no definite interchain hopping rate can be defined. Instead, the data can be accounted for by a continuum of the interchain rates. Such a continuum could be likely originated in the disordered nature of the material. As T increases, more and more transverse paths are allowed and the effective dimension $\tilde{d}$ increases.

Considering the transport properties, as in other disordered systems the frequency dependence of the conductivity in PPy has been shown to obey a power law : $\sigma \propto \omega^s$. The exponent s can be related to the effective dimension [12] : $s = 1 - \tilde{d}/\bar{d}$, where $\bar{d}$ is the standard space dimension (usually $\bar{d} = 3$). Thus, it should be possible to determine $\tilde{d}$ from measurements of the conductivity as a function of frequency. From data available in literature [13] satisfactory agreement with the NMR result is obtained [12].

5. Conclusion

Evidence hase been presented for a strong correlation between the motion of spin carriers and that of charge carriers in two conducting polymers. Two different behaviors have been observed. In PANI, the motion can be described as one-dimensional diffusion along conducting chains isolated from each others in an insulating matrix. Interchain hoppings take place at a rate determined by the protonation level and they govern the macroscopic conductivity. In PPy, the concept of effective dimension has been introduced, similarly to the spectral dimension for the dynamics of fractal systems. The motion of the spin-charge carriers can be described as diffusion on a temperature and frequency-dependent percolation network. We think that the behavior of PPy, being controlled by the disordered nature of the material, is more representative of the conducting polymer class. The case of PANI is probably more peculiar, due to the specific role of the protonation process in this compound.

References

1. F. Genoud, M. Guglielmi, M. Nechtschein, E. Geniès and M. Salmon, Phys. Rev. Lett. **55**, 118 (1985)
2. M. Nechtschein, F. Devreux, F. Genoud, E. Vieil, J.M. Pernaut and E. Geniès, Synth. Met. **15**, 59 (1986)
3. F. Devreux, F. Genoud, M. Nechtschein, and B. Villeret, Springer Series in Solid State Sc. **76**, 270 (1987)
4. M. Nechtschein, F. Devreux, R.L. Greene, T.C. Clarke, and G.B. Street, Phys. Rev. Lett. **44**, 356 (1980)
5. J.P. Travers, J. Chroboczek, F. Devreux, F. Genoud, M. Nechtschein, A. Syed, E. Geniès, and C. Tsintavis, Mol. Cryst. **121**, 195 (1985)

6. W.S. Huang, B.D. Humphrey, and A.G. MacDiarmid, J. Chem. Soc. Faraday Trans. **82**, 2385 (1986)
7. J.M. Ginder, A.F. Richter, A.G. MacDiarmid, and A.J. Epstein, Sol. St. Comm. **63**, 97 (1987)
8. F. Zuo, M. Angelopoulos, A.G. MacDiarmid, and A.J. Epstein, Phys. Rev. **B 36**, 3475 (1987)
9. M. Helmle, M. Mehring, M. Nechtschein and A. Pron ; this conference
10. K. Mizoguchi, M. Nechtschein, J.P. Travers, and C. Menardo, Synth. Met. **29**, E417 (1989)
11. M. Nechtschein, F. Genoud, C. Menardo, K. Mizoguchi, J.P. Travers, and B. Villeret, Synth. Met. **29**, E211 (1989)
12. F. Devreux and H. Lecavelier, Phys. Rev. Lett. **59**, 2585 (1987)
13. F. Legros and A. Fourier, communication at the meeting on Electroactive Polymers, Kerallic, France, 1984 (unpublished) ;
M. el Kadiri, J.P. Parneix, and A. Chapoton, Ann. Phys. (Paris) **11**, 89 (1986)

Pulsed ENDOR Spectroscopy of Stretch-Oriented Feast-Type Polyacetylene

H. Käss[1], *A. Grupp*[1], *P. Höfer*[1], *P.K. Kahol*[1;§], *M. Mehring*[1], *R. Weizenhöfer*[2], *and G. Wegner*[2]

[1]2. Physikalisches Institut, Universität Stuttgart, D-7000 Stuttgart 80, Fed. Rep. of Germany

[2]Max-Planck-Institut für Polymerforschung, D-6500 Mainz, Fed. Rep. of Germany

Abstract: We have investigated pulsed ENDOR spectra of stretch-oriented Feast-type polyacetylene up to a stretching ratio l/l_0 = 14.6. Highly oriented chains appear already at low stretching ratios.

1. Introduction

Polyacetylene is the simplest conjugated polymer and still attracts considerable interest. In particular the conjugational defect, termed "neutral soliton" by Su, Schrieffer and Heeger [1] has been investigated in the past extensively. The Electron Nuclear Double Resonance (ENDOR) technique has played an important role in unravelling the hyperfine structure of this defect [2-9]. The ENDOR technique allows to determine the hyperfine splitting constants of all the nuclei connected to the radical defect. We have advocated in the past strongly different pulsed-ENDOR techniques which allow to determine the distribution function of the hyperfine coupling and therefore the "shape" of the soliton [5-9].

In the case discussed here we have investigated the proton ENDOR spectra of stretch-oriented polyacetylene. We have used the same soliton shape and the same hyperfine parameters in all spectral simulations and have varied only the degree of alignment. From TRIPLE resonance experiments it is known that the soliton exhibits alternating positive and negative spin-density[9]. We have used the same spin-density distribution function as in refs. [6-9], where the full width at half maximum is 2N and the total width is 2J. The ratio ug of negative to positive spin-density was determined independently by pulsed-TRIPLE resonance experiments. We found in all samples a value of ug = 0.42 [7-9].

The spin-density was connected with the hyperfine interaction by using a McConnell type relation for the hyperfine tensor with the characteristic values (A_{11} = -32.2 MHz, A_{22} = -64.4 MHz, A_{33} = -96.6 MHz) [5-9].

---

§ Current address: Physics Department, Wichita State University, Kansas USA.

Springer Series in Solid-State Sciences, Vol. 91
Electronic Properties of Conjugated Polymers III
Editors: H. Kuzmany · M. Mehring · S. Roth

2.ENDOR spectra of Feast-$(CH)_x$.

The ENDOR spectra of the usually unoriented polyacetylene films are broad and show only little structure. The precursor route developed in Durham by Feast and co-workers can be used to make stretch-oriented polyacetylene films with high stretching ratios[10].

Davies-ENDOR spectra were obtained for increasing stretching ratios l/l_0 up to $l/l_0 = 14.6$ as shown in Fig.1(left). The spectra show a slight asymmetry with respect to the free proton frequency (15 MHz) and a "spectral hole" in the center.These features are well understood and have been discussed elsewhere [5-9]. Here we focus on the development of spectral shoulders with increasing stretching ratio. These come about by the increasing orientation of the chains and the anisotropic hyperfine interaction. Since the hyperfine anisotropy is known and the spin-density distribution function was determined independently

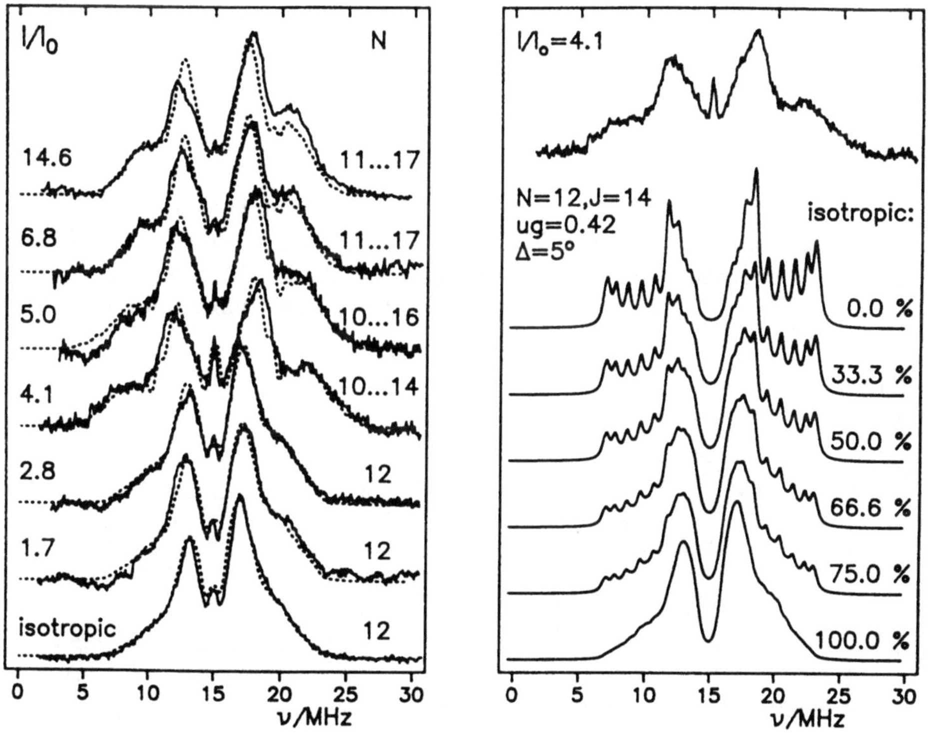

Fig.1 Left:Davies ENDOR spectra of stretch-oriented polyacetylene for different stretching ratios up to $l/l_0 = 14.6$. Calculated spectra (see text) are shown as dotted lines. Right: Calculated spectra for highly oriented chains and a certain amount of isotropic contribution, given in %. Comparison with typical experimental spectra(top) is discussed in the text.

[7,9], model calculations can be made which fit the experimentally observed spectra.

The chain alignment is described by a Gaussian distribution function $P(\Theta) = A \exp(-\Theta^2/\Delta^2)$ where Δ determines the degree of alignment and A is the appropriate normalization constant. In Fig.1(left) we compare the calculated (dotted lines) ENDOR spectra with the experimental spectra for different parameter sets. Appreciable agreement between experimental and calculated spectra can be achieved. All essential features of the experimental spectra show up also in the calculation. The small peak in the middle of the spectra is attributed to dipolar interactions with neighbouring chains. This has been taken into account in a quantitative way.

For comparison we have also included in Fig.1(right) calculated spectra for a highly oriented chain ($\Delta = 5^0$) and added to this different amounts of the isotropic spectrum. The variation of the spectra show that similar features appear, however, in contradiction to most of the experimental spectra. The "hanging shoulders" and the "rounding-off" of the central peaks seen in the calculation cannot be observed in the experimental spectra. We therefore believe that amorphous parts in the sample play a minor role. We estimate their contribution to be less than 30%.

3.Results

The stretch-oriented Feast-type polyacetylene films all show the same spin-density distribution with the following characteristics: modified SSH-type distribution, alternating spin-density with ug = 0.42,half width at half maximum N = 10...17 (maximum range),total extension 2J = 22...44 (maximum range).The

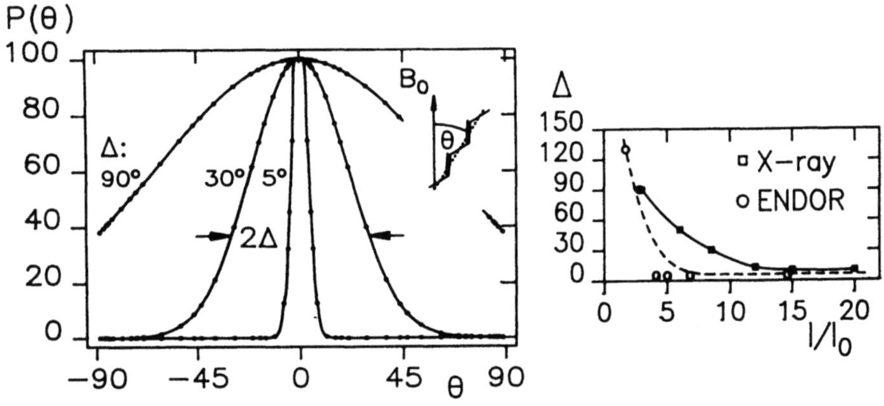

Fig.2 Gaussian distribution function of chain orientation with respect to the stretching direction for different degrees of alignment. Experimentally determined degrees of alignment are shown for different stretching ratios l/l_0 as determined from ENDOR spectra as well as from X-ray analysis [11].

degree of chain orientation was determined from fitting the experimental spectra with calculated spectra, based on a Gaussian distribution function of alignment angles as shown in Fig.2. The characteristic alignment angle Δ decreases rapidly with the stretching ratio l/l_0 as is also evident from Fig.2. A less pronounced onset of alignment is observed for the x-ray data (Fig.2). However, these data were taken for different samples which were also stretched under slightly different conditions. In both cases a large degree of alignment ($\Delta = 5^0$) could be achieved for stretching ratios in excess of $l/l_0 = 10$.

Acknowledgments: We gratefully acknowledge financial support by the Stiftung Volkswagenwerk and the Deutsche Forschungsgemeinschaft (SFB 329).

References

1 W.P.Su, J.R.Schrieffer and A.J.Heeger, Phys.Rev.Lett. 42 (1979) 1698.
2 a)H.Thomann, L.R.Dalton, Y.Tomkiewicz, N.S.Shiren and T.C.Clarke, Phys.Rev.Lett. 50 (1983) 533. b)H.Thomann, L.R.Dalton, M.Grabowski and T.C.Clarke, Phys.Rev. B32 (1985) 3141.
3 a)S.Kuroda and H.Shirakawa, Solid State Commun. 43 (1982) 591,b)S.Kuroda, H.Bando and H.Shirakawa, Solid State Commun. 52 (1984) 893, c)S.Kuroda and H.Shirakwa, Phys.Rev. B35 (1987) 9380.
4 P.K.Kahol and M.Mehring, J.Phys.C: Sol.Stat.Phys. 19 (1986)1045.
5 M.Mehring, P.Höfer and A.Grupp, Ber. Bunsenges. Phys.Chem. 91 (1987)1132.
6 A.Grupp, P.Höfer, H.Käss, M.Mehring, R.Weizenhöfer and G.Wegner, in Electronic Properties of Conjugated Polymers, Editors M.Kuzmany, M.Mehring, S.Roth, Solid-State Sciences 76, Springer Verlag 1987
7 H.Käss, P.Höfer, A.Grupp, P.K.Kahol, R.Weizenhöfer, G.Wegner and M.Mehring, Europhys. Lett. 4 (1987) 947.
8 M.Mehring, P.Höfer, H.Käss and A.Grupp, Europhys.Lett. 6 (1988) 463.
9 M.Mehring, A.Grupp, P.Höfer and H.Käss, Synthetic Metals 28 (1988) D399.
10 a) J.H.Edwards and W.J.Feast, Polymer 21 (1980) 595.
b)G.Leising, Polymer Bulletin 11 (1984) 401.
11 R.Weizenhöfer, PhD Thesis Mainz 1987

Cluster Versus Random Doping in Polyaniline: A NMR Approach Based on the ^{19}F–^{19}F Dipole–Dipole Coupling

M. Helmle[1], *C. Mayer*[1], *M. Mehring*[1], *M. Nechtschein*[2], *and A. Pron*[2]

[1]2. Physikalisches Institut, Universität Stuttgart, D-7000 Stuttgart 80, Fed. Rep. of Germany
[2]Centre d'Etudes Nucléaires de Grenoble, 85 X, F-38041 Grenoble Cedex, France

Abstract A long standing problem is the question whether doping in conducting polymers occurs in a random fashion or in clusters (metallic islands). We have investigated this question in polyaniline (PANI) by doping with counterions (FC_6H_4-SO_3^-) which contain ^{19}F nuclear spins.

1. Doping Process in Polyaniline

One way to obtain fairly highly conducting polyaniline is to protonate the insulating emeraldine base in aqueous HCl. The corresponding emeraldine salt shows a conductivity in the range of several S/cm, which is an increase by 10 orders of magnitude with respect to the conductivity of the emeraldine base. By following the conductivity and EPR intensity as a function of doping (protonation) Epstein et al. [1] proposed a phase segregation of PANI into metallic islands within the undoped bulk material. This view of doped PANI is supported by a combined EPR/NMR investigation of Mizoguchi et al.[2], who concluded from analysis of spin dynamics that these metallic islands are formed along single PANI chains. However, in contradiction to this interpretation are the electron energy loss data of Litzelmann et al. [3]. The present NMR investigation is focused on the following topics:

- evidence of cluster formation in PANI
- structural information on these clusters

2. NMR Approach to the Problem

The ^{19}F NMR experiment on protonated PANI utilizes the fact that for reasons of Coulomb interaction the counterion of the dopant has to stay close to the protonated nitrogen. Therefore one can study the doping process by looking at the counterion. In order to distinguish the counterion from the PANI chain in a NMR experiment the counterion has to carry a spin species not present on the PANI chain. A convenient choice for this purpose is ^{19}F. The dipolar interaction among the ^{19}F spins on the counterion is therefore directly correlated to the

Springer Series in Solid-State Sciences, Vol. 91
Electronic Properties of Conjugated Polymers III
Editors: H. Kuzmany · M. Mehring · S. Roth

doping behaviour at the chain. By studying the dipolar interaction as a function of doping level all the desired information is achievable.

3. Theoretical Description of the Dipolar Interaction in a Dilute Multiple Spin System

Measurable quantities of the dipolar interaction can be calculated by the so-called moment expansion [4]. The n-th moment is here defined as $M_n = \int(\omega-\omega_0)^n \cdot f(\omega)d\omega$ where $f(\omega)$ is the NMR spectrum. To a first approximation it is sufficient to take into account only the 2nd moment M_2. The 2nd moment M_2 per spin for a dipolar coupled spin system in a powdered sample with random distribution is given by

$$M_2 = \frac{4}{15}\left(\frac{\mu_0}{4\pi}\right)^2 \gamma^2\hbar^2 I(I+1) \cdot \sum_j r_{jk}^{-6}.$$

Here it is assumed that differences in Larmor frequency due to chemical shift anisotropy quench the so called flip-flop contribution of the dipolar interaction. Therefore the second moment is reduced by a factor $\frac{4}{9}$.

For a dilute spin system one has to take into account the probability of the occupation of a site k: $M_2 \sim \sum p_k \cdot r_{jk}^{-6}$.

a) Random Doping Process

For a random doping process the probability of site k being occupied is equal to the doping concentration f. Therefore the second moment is given by $M_2 = \text{const} \cdot f \cdot \sum r_{jk}^{-6} = f \cdot M_2^0$, where M_2^0 is the second moment of the completely doped lattice.

b) Formation of infinite clusters along single chains

In this case the probability of site k being occupied is

$$p_k = p_{site} \cdot p_{chain} = \begin{cases} 1 \cdot p_{chain} & \text{if k is located on a doped chain} \\ 0 & \text{elsewhere} \end{cases}$$

Here p_{site} represents the probability that a certain site on a chain is doped whereas p_{chain} gives the probability of this chain being doped. The sum can be divided into two contributions depending on whether sites j,k are located on the same chain or not:

$$M_2^{cluster} = \text{const} \cdot \sum_{j,k \in (\text{same chain})} r_{jk}^{-6} + \text{const} \cdot \sum_{j,k \in (\text{diff chains})} r_{jk}^{-6}.$$

The first contribution is the second moment of one cluster. If j is located on a doped chain then the probability p_{chain} for k is equal to one. Therefore this expression is independent of doping concentration. In the second contribution

P_{chain} for k is proportional to the dopant concentration. Taking into account the limiting case of a complete doped system consisting of clusters one obtains $M_2^{cluster} = M_2^{chain} + f \cdot (M_2^0 - M_2^{chain})$.
Clearly the main difference between the cluster and random doping model appears at very low doping concentrations: for random doping then M2 approaches zero whereas in the case of cluster behaviour one obtains a finite value M2 corresponding to the second moment of one cluster.

4. Experimental Section

Sample Preparation

The samples were protonated in an aqueous solution of 4-fluoro-benzene-sulfonic acid at different pH values. The protonation level was checked by elemental analysis. During the experiments the samples were kept under dry nitrogen.
So far only the dipolar interaction among the fluorine spins has been mentioned. In reality there are a couple of additional interactions that mask the dipolar interaction. The solution to this problem is the use of a special pulse sequence to eliminate the following interactions:

- ^{19}F chemical shift anisotropy
- dipolar coupling between fluorine and nitrogen spins
- dipolar coupling between fluorine and proton spins

One can solve the problem by performing a spin echo double resonance experiment(Fig.1). The first two interactions are eliminated by a spin echo sequence, the dipolar couplings to the protons are eliminated by a pulsed proton decoupling.

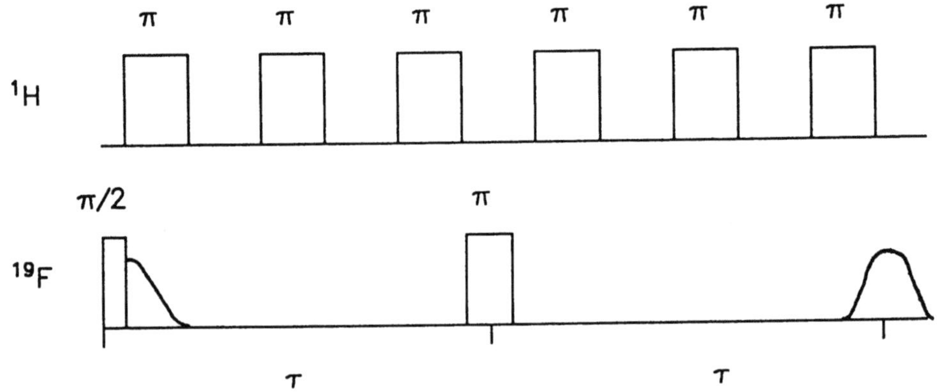

Fig.1: Schematic representation of the double resonance spin echo experiment

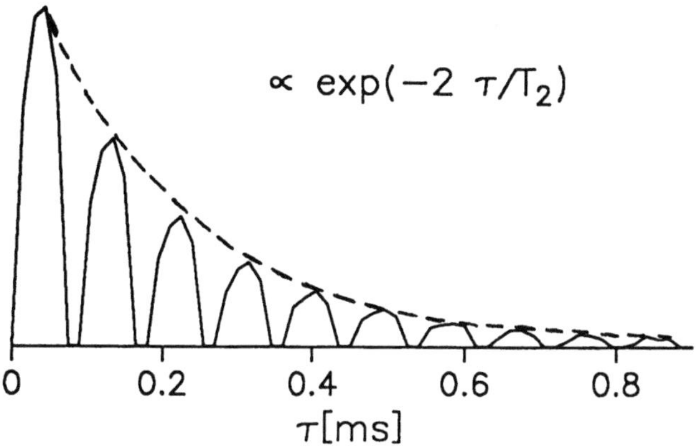

Fig.2: Spin echo signals for different τ's: the envelope is governed by the relaxation rate $1/T_2$,which reflects the strength of the dipolar coupling among the ^{19}F spins on the dopant molecules

The experimental parameter which reflects the dipolar coupling is the relaxation rate $1/T_2$ of the spin echo signal for different times τ. The decay rate is indicated in Fig.2 as the envelope of the different spin echo signals.

To compare the theoretical behaviour of M_2 versus dopant concentration with the measured tranverse relaxation rate one has to link M_2 to $1/T_2$. This is by no means straightforward because the exact connection includes a complete lineshape calculation. To a first approximation one can use the following relation: $1/T_2 = \text{const} \cdot \sqrt{M_2}$.

5. Results and Discussion

Fig.3 shows the experimental data recorded with the double resonance spin echo sequence at ambient temperature. The solid lines represent the behaviour for clustering and random doping respectively. Thereby the two values M_2^0 and M_2^{chain} are fitted to the data. Clearly our data show a finite value M_2^{chain} supporting thereby 1-d clustering of the dopant.

To extract the average distance between the dopant molecules one needs to know the prefactor in the equation above. Taking this factor equal to one we can obtain an approximate value for the average distance because of the steep dependence of $1/T_2$ on the average distance r_{jk}. For example a change of the constant by a factor of two will affect the average distance r_{jk} by only 20%. We obtain from our data in this manner an average distance of about 3.2 Å. This is a considerably smaller value than approximately 9 Å for the polaron lattice model proposed by Epstein et al. [1].

It should be emphasized that due to the steep dependence of $1/T_2$ on the distance between the dopant molecules we are not able to fix the size of the one

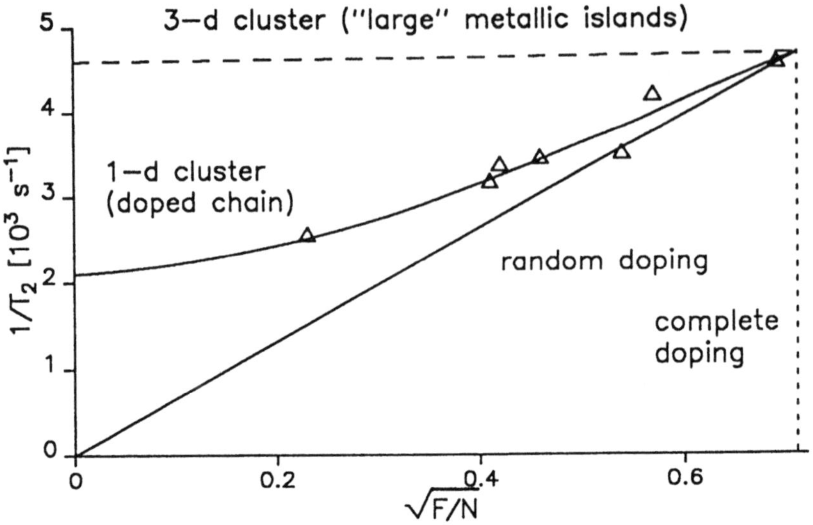

Fig.3: $1/T_2$ versus dopant concentration: The dopant concentration is given as the ratio of ^{19}F content per nitrogen. Therefore F/N = 0.5 corresponds to a completely doped lattice. Triangles show experimental data (RT), solid lines represent the behaviour for 1-d cluster model and random doping. For 3-d clusters (metallic islands [1]) the data should follow the dashed line.

dimensional chain cluster. The direct neighbouring molecules on each side contribute together already 99% of the M_2 value of an infinite cluster. So the clusters found in our investigation might be quite small.

Acknowledgement: Financial support by the Deutsche Forschungsgemeinschaft (DFG: SFB 329) is gratefully acknowledged.

6. References

1. A.J.Epstein,J.M.Ginder,F.Zuo,R.W.Bigelow,H.-S.Woo,D.B.Tanner, A.F.Richter,W.S.Huang,A.G.MacDiarmid, Synthetic Metals 18,303(1987)
2. K.Mizoguchi,M.Nechtschein,J.-P.Travers,C.Menardo, to be published
3. A.Litzelmann,B.Scheerer,J.Fink,K.Meerholz,J.Heinze,N.S.Sariciftci, H.Kuzmany, Synthetic Metals 29,E313(1987)
4. A.Abragam, Principles of Nuclear Magnetism, Oxford University Press 1961

Intercalation of Alkali Ions in Polyacetylene: NMR Studies

P. Bernier, K. Zniber, F. Rachdi, H. Bleier, and C. Fite***

GDPC-USTL, Place E. Bataillon, F-34060 Montpellier Cedex, France
*Permanent address:
Max-Planck-Institut für Festkörperforschung,
Heisenbergstr. 1, D-7000 Stuttgart, Fed. Rep. of Germany
**Present address:
IPOS-University of California, Santa Barbara, CA 93106, USA

Abstract. The characteristics of the alkali ions Li^+, Na^+, Cs^+ during their intercalation in polyacetylene have been studied using NMR of the dopant species. In the cases of Li^+ and Na^+, the use of electrochemistry has allowed us to study the reversibility of the observed behaviour, while in the case of Cs^+, high temperature vapour phase doping and solution chemical doping have been used.

The ^7Li and ^{23}Na resonance positions, which are close to the purely ionic values, do not depend on the dopant concentration. On the other hand the ^{133}Cs spectra present a resonance -200 ppm from the reference at any concentration, with an additional line (at +12100 ppm at room temperature) for the highest concentration (≈16 mol. %), which therefore suggests a high density of delocalized electrons on the dopant site.

1 Introduction

As a consequence of the intercalation of alkali dopants in the pristine $(CH)_x$ structure, rearrangements of the polymer chains and ordered structures appear during the doping process, while important changes in electronic properties are also observed [1,2,3]. At high dopant concentration (≈0.16) the alkali atoms are assumed to be arranged in channels parallel to the chain direction. The distance between channels, as well as the distance between alkali dopants along a channel, depends on the nature of the dopant [4,5]. The situation is expected to be simple in the case of heavy alkali dopants (K, Rb, Cs), which can be intercalated in any case in polyacetylene without solvent molecules. The case of Cs doping is particularly clear, having a tetragonal unit cell with a lateral spacing of the dopant channels of 6.48Å, while the distance between Cs atoms in the channel is around 4Å and thus slightly incommensurate with the polyacetylene host lattice [5]. For this compound the formula is close to the ideal expression $Cs(C_3H_3)_2$.

One key point concerning the electronic properties of these systems is to understand the role played by the dopant itself. ESR experiments have clearly shown that conduction electrons are sensitive to the nature of the dopant

Springer Series in Solid-State Sciences, Vol. 91
Electronic Properties of Conjugated Polymers III
Editors: H. Kuzmany · M. Mehring · S. Roth

species, the linewidth increasing as the fourth power of the dopant atomic number [6]. While such a result suggests that there exists a significant conduction electron density on the dopant site, preliminary experiments on the ^{7}Li NMR in these systems are not very conclusive about the possible existence (or not) of a resonance shift caused by the delocalized electrons [7]. In the present paper we have reexamined this problem by investigating the NMR of ^{7}Li, ^{23}Na, ^{133}Cs nuclei intercalated in polyacetylene, in order to determine whether or not a Knight (metallic) shift [8] occurs in these systems.

2. Experimental

In all cases we have used films of standard Shirakawa $(CH)_x$ (85% cis, ≈100μm thick). Specially designed electrochemical cells have been used for Li and Na doping, permitting continuous observation during doping of the alkali metal NMR [6,7]. In the case of Cs, for which electrochemical doping has not been done before, we used standard solutions of Cs-benzophenone and Cs-naphthalide in THF, yielding a doped system with concentrations $Y \approx 0.04$ and $Y \approx 0.16$ respectively [9].

^{7}Li, ^{23}Na, and ^{133}Cs NMR spectra were recorded with a Bruker CXP 200 spectrometer. The chemical shift references were aqueous solutions of LiCl, NaOH, and $CsNO_3$ for Li, Na, and Cs respectively. Shifts are positive when the resonance is displaced towards low magnetic field versus the reference. Variable temperature was obtained using a flow of cold nitrogen through the NMR probe, covering the range 200 to 360 K.

3. Results and Discussion

The advantage of electrochemical doping is to allow a continuous recording of the properties of the sample which act as an electrode during doping as well as during undoping. The reversibility and reproducibility of the observed properties can then be studied.

In figure 1 the concentration Y dependence of the measured shift on the ^{7}Li resonance is shown. While the system varies between semiconducting and highly metallic (dc conductivity ≈ 200 Ω^{-1} cm^{-1}) states, no significant variation of the shift is observed. Furthermore its absolute value depends on the choice of the reference. Taking the Li^{+} reference to be an aqueous solution of LiCl we found

$$\delta = -0.7 \pm 1 \text{ ppm}.$$

This value is inside the range of values measured for the various Li^{+} solutions which extend over 5-6 ppm around 0 ppm. On the other hand, its absolute value is very small compared with the shift measured in metallic Li (≈260ppm) [8]. Therefore we may conclude that there is no significant Knight shift of the resonance in this case.

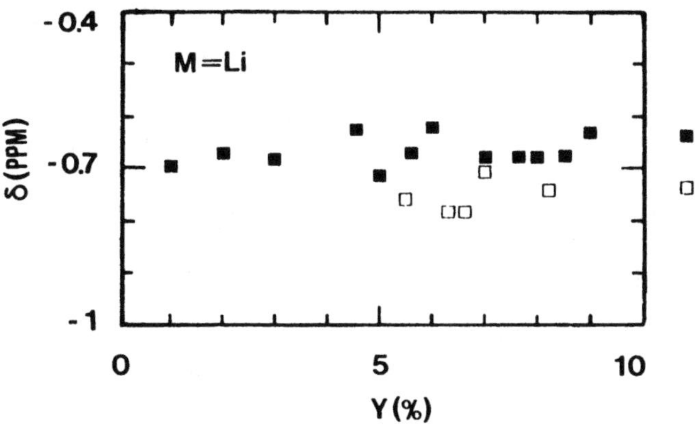

Figure 1. Variation of the resonance shift versus doping level for Li doped $(CH)_x$: ■ doping, □ undoping

In figure 2 the concentration dependence of the ^{23}Na resonance shift is shown. As above for ^{7}Li, no significant variation of the shift is observed both during doping up to Y=0.12 and during a complete undoping. It is worth noting that the reversibility is good, as also observed (partially) in the case of Li. The absolute value of the shift also depends on the choice of the reference, and taking the standard Na+ resonance to be an aqueous solution of NaOH we measure

$$\delta = -9.5 \pm 1.5 \text{ ppm.}$$

In this case also the measured shift is inside the range of values usually obtained for Na^+ solutions, which extend over 15-20 ppm around 0 ppm. Furthermore it is very small (in absolute value) compared with the metallic

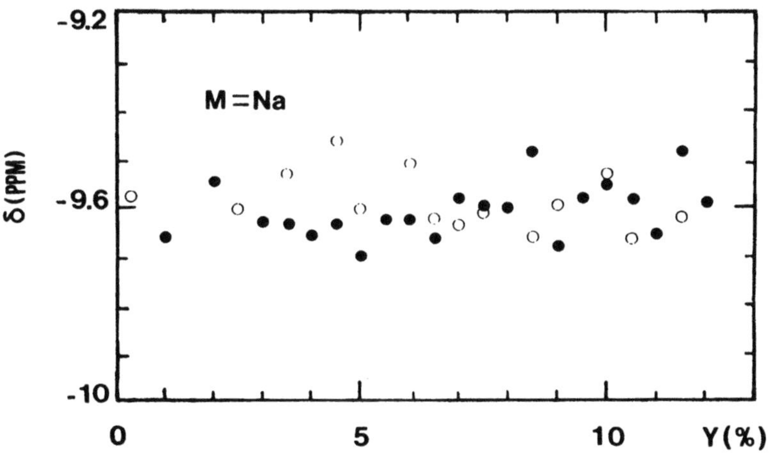

Figure 2. Variation of the resonance shift versus doping level for Na doped $(CH)_x$: ● doping, ○ undoping

shift of Na (≈1120ppm) [8]. Then there is an uncertainty from the present results whether if the measured shift is a chemical or a Knight one.

As stated above, only discrete values of Y are accessible in the case of Cs doping. For low dopant concentration (≈0.04) we observe a resonance at -260 ppm from the standard reference (aqueous solution of $CsNO_3$) but the accuracy is rather poor (±100ppm). We notice that this value is just outside the range of values usually measured for Cs^+ in solution which extend over ≈200ppm around 0ppm. Neverthless the absolute value is still small and has the opposite sign to the Knight shift of metallic Cs (+14900ppm at room temperature) [8].

For the highest Cs concentration (≈0.16) a new resonance appears in addition to the preceding one which is strongly shifted (≈12000ppm at room temperature) from the reference. The relative intensity of both resonances is difficult to quantify, due to the weak signal-to-noise ratio and the large spectral distance between them. Nevertheless it is reasonable to suggest these intensities to be of the same order of magnitude.

To understand the origin of such a strongly shifted resonance, we can first suggest the existence of clusters of metallic Cs which would be formed during the chemical doping process. We also notice that the temperature dependence of the shift (figure 3) is considerably larger than in the pure metal. Furthermore while the melting of metallic Cs appears as a small decrease of the Knight shift K around T=305 K, no singularity is observed in the same temperature range for the Cs intercalated polyacetylene (figure 3).

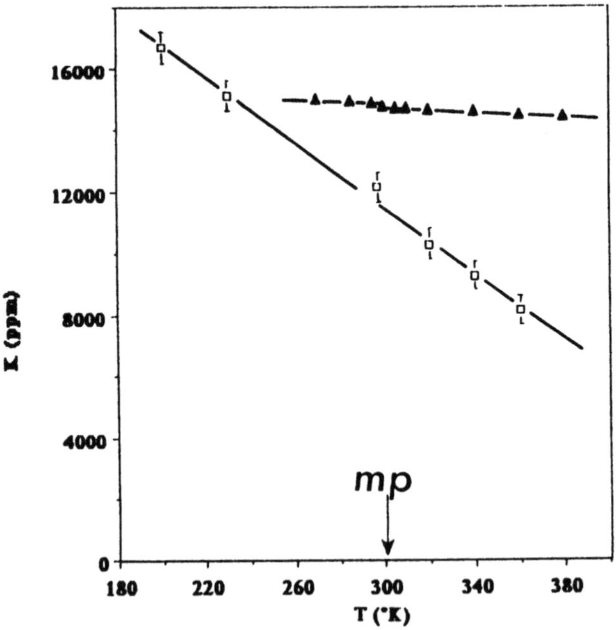

Figure 3 Variation of the Knight shift versus temperature for metallic Cs (upper curve) and for Cs doped $(CH)_x$ (lower curve)

We then suggest that the observed behaviour comes from the existence, at the Fermi level of the doped system, of a significant s-electron density of states. We can point out some arguments which favour this interpretation:
(i) This strongly shifted line appears only for high Cs concentration, a situation which favours the overlap between orbitals.
(ii) We notice that for the highest Cs concentration the distances between dopant atoms are 4 and 6.4 Å [4,5], while the smallest distance between Cs atoms in the metal is around 5.2 Å. This situation can be favourable for the formation of an s-type band close to the Fermi level of the system.
(iii) A similar, but smaller, effect has been observed on the first stage of Cs intercalated graphite, and interpreted in terms of the existence of an s-type band at the Fermi level of the conducting system [10].

References

1. L.W. Shacklette and J. E. Toth, Phys. Rev. B, 32 (1985) 5892.
2. M. Winokur, Y. B. Moon, A. J. Heeger, J. Baker, D. C. Bott, Phys. Rev. Lett., 58 (1987) 2329.
3. N. S. Murthy, L. W. Shacklette, R. H. Baughman, J.Chem. Phys., to be published.
4. R. H. Baughman, N. S. Murthy, G. G. Miller, J. Chem. Phys. 79 (1983) 515.
5. O. Leitner, H. Kahlert, G. Leising, J. Fink, H. Fritzsche, Synth. Met., 28 (1989) 225.
6. F. Rachdi, P. Bernier, Synth. Met., 17 (1987) 401. P.Bernier, F. Rachdi, A. Elkhodary, C. Fite, Synth. Met., 24 (1988).
7. A. K. Whittaker, C. Fite, K. Zniber, P. Bernier, Electronic Properties of Conjugated Polymers, Springer Verlag (1987) p 170.
8. L. E. Drain, Met. Rev., 119 (1967) 195.
9. C. Mathis and B. François, J. Phys. , C3 (1983) 21.
10. P. Lauginie, Synth. Met., 23 (1988) 311

Radical Anion Salts of Dicyanochinondiimine (DCNQI) with Metallic Counterions: New Low-Dimensional Conducting Systems

J.U. von Schütz[1], *M. Bair*[1], *U. Langohr*[1], *H.-P. Werner*[1], *H.C. Wolf*[1], *P. Erk*[2], *and S. Hünig*[2]

[1]3. Physikalisches Institut, Universität Stuttgart
Pfaffenwaldring 57, D-7000 Stuttgart 80, Fed. Rep. of Germany
[2]Institut für Organische Chemie, Universität Würzburg,
Am Hubland, D-8700 Würzburg, Fed. Rep. of Germany

Abstract. Radical anion salts of DCNQI with metallic counterions were examined in view of their prospective applications as wires in microscopic, molecular dimensions. A feature therefore is high, but strongly anisotropic conductivity. Evaluating the proton spin relaxation, which is proportional to $\omega^{-1/2}$, it is concluded that in the salts $(DCNQI)_2X$, X = Li, Na, K, Rb, Ag, a one-dimensional electron migration is well established. In contrast to that, the Cu salts have an effective interstack motion.

1. Introduction

The salts of DCNQI with metallic counterions have raised considerable interest since the first publications [1]. Reasons for this are the extremely high conductivity of some salts (up to 500 000 S cm^{-1}) [1,2], strange pressure effects on σ (pressure leads to a metal-to-insulator transition [3,4], the high susceptibility ($\chi(T)$ is not correlated with $\sigma(T)$ [5]) and the ESR lines, which are narrow in the case of 1-d motion [6] and very broad [7] if effective interstack scattering mixes Bloch and Zeeman states via the spin-orbit contribution [8] as in the case of Cu complexes.

One key to these considerations is given by the crystal structure, which is represented in Fig.1 (right): stacks of DCNQI molecules encircle that of the metal ions.

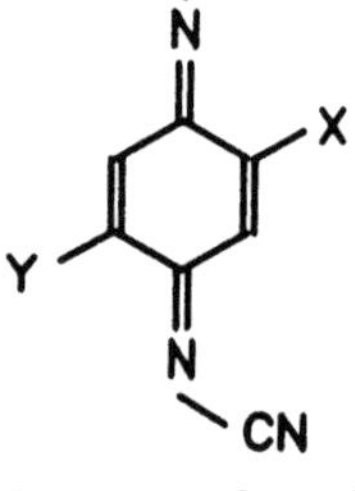

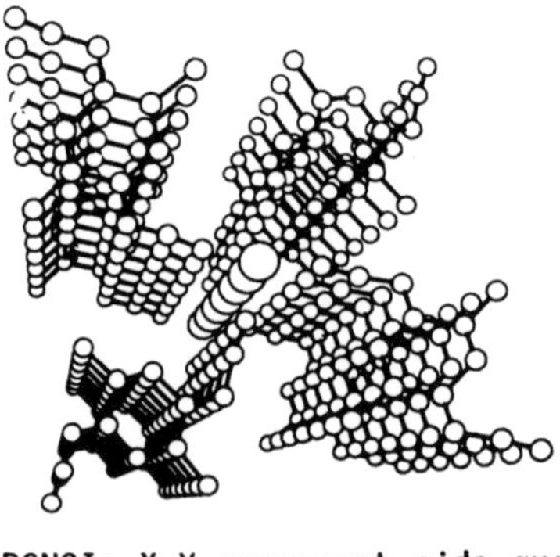

Fig.1: Left: skeleton of DCNQI; X,Y represent side groups such as CH_3, Cl, Br or I. Right: crystal structure of the salts

Springer Series in Solid-State Sciences, Vol. 91
Electronic Properties of Conjugated Polymers III
Editors: H. Kuzmany · M. Mehring · S. Roth

Assuming a closed shell configuration for the cations, the conductivity is established in the DCNQI stacks solely. Pushing these further apart from each other via increasing cation size should result in a larger anisotropy of the conductivity, as confirmed by the angle dependence of the ESR [9].

2. Anisotropy

The determination of the anisotropy of σ would be - of course - the direct measurement of the conductivity tensor by means of contacts parallel or perpendicular to the stacking direction. Because of the smallness of the crystals, which are only some microns thick and till now too thin for the dimensions of our contacts in the orthogonal direction, we were limited to contactless methods such as Dysonian lineshape analysis (Fig.2) and the angle dependence of ESR lines in the case of completely penetrating microwave fields [9].

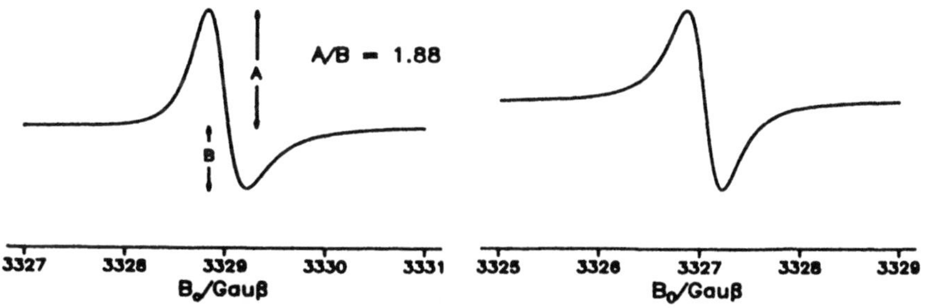

Fig.2: ESR linewidth of $(DMe\text{-}DCNQI)_2Li$ at 300 K and 9.6 GHz. Left side: conducting crystal axis in the B_0-plane. Right side: conducting crystal axis perpendicular to the B_0-plane

The asymmetrical lineshape is due to a limited penetration depth of the B_1-field, resulting in the superposition of absorptive and dispersive components of χ.

Further insights - also in a contactless manner - can be obtained by NMR experiments, in which we determine the shifts and the relaxation rates of the nuclei such as protons, ^{13}C (MAS [10]) and the nuclei of the individual metallic counterions. These counterions as well as the protons of the DCNQI should "feel" in some sense the presence of the conducting electrons via the electronic paramagnetic properties.

3. Nuclear relaxation

In general, the relaxation rates of the nuclei are determined by several contributions, being individually dominant in different temperature ranges: For example, in systems undergoing a metal-to-insulator transition, the proton relaxation is composed of (coming from low temperatures) a) the interaction with localized paramagnetic states, b) the reorientation of methyl groups, which are commonly used as substituents X,Y on DCNQI (see Fig.1) and c) the interaction with delocalized spins in the conducting region. (The interaction between non-equivalent nuclei, in our case between the hydrogens of the DCNQI and the metallic counterions, is not effective due to the great distance and the low g-factors of the latter.)

For the proton relaxation rates we have

$$T_1^{-1} = \underset{\text{localized param.}}{a \cdot n_1 \cdot J(\omega_1)} + \underset{CH_3\text{-groups}}{b \cdot J(\omega_H)} + \underset{\text{mobile electrons}}{c \cdot n_e \cdot J(\omega_e)}. \qquad (1)$$

a, b, c are coupling constants containing distances and g-factors, n_i are densities and the power spectra $J_i(\omega)$ depend on the correlation times of the interactions.

In the following we concentrate on the high temperature term c, which is due to the conducting electrons. The corresponding power spectrum is given by the dimensionality of the electron migration and the electron scattering times within the chain $\tau_\parallel$ and perpendicular to it, $\tau_\perp$. Thes rates contain the anisotropy via $A = \tau_\parallel^{-1}/\tau_\perp^{-1}$. In an elegant manner, these considerations are summarized in the work of Soda et al. [11] for preferential 1-d motion:

$$(T_{1H} \cdot T)^{-1} = C \cdot \frac{\tau_\perp}{\tau_\parallel} \left\{ \frac{1 + \sqrt{1 + \omega_e \tau_\perp}}{2\,(1 + \omega_e^2\, \tau_\perp^2)} \right\}^{1/2} \qquad (2)$$

which means for $\omega_e \cdot \tau_\perp > 1$: $T_{1H}^{-1} \sim \omega_e^{-1/2}$.

4. Experimental results and discussion

The proton spin relaxation rates of $(DCNQI)_2Li$ and $(DCNQI)_2Cu$ are compared in Fig.3. The Li salt is representative of most non-copper salts i.e. Na, K, Rb and Ag.

On $(DCNQI)_2Li$ the three contributions mentioned above (eqn.1) are clearly seen. The CH_3-group peak at about 70 K is most pronounced, the interaction with the electrons (above 100 K) is frequency dependent and Korringa-like, that means $T_1 \cdot T \cdot X^2$ = const. The peak in T_{1H}^{-1} at about 30 K is due to localized paramagnetic centers, the dynamics of which is governed by the interaction with "free" spins of temperature dependent contribution [12].

Concerning the results for the copper salts (Fig.3 right), the relaxation rates due to paramagnetic species are orders of magnitude lower and - even up to $\omega_H/2\pi$ = 300 MHz - not frequency dependent. The lack of any dispersion is one key to our statements on the dimensionality of the conductivity in this salt.

Evaluating the correlation and scattering times in eqn.2 a rough estimation of $\tau_\parallel$ is possible via the Drude approximation of the conductivity:

$$\sigma_\parallel = \frac{n \cdot e^2 \cdot \tau_\parallel}{m^*} \qquad (3)$$

(n: density of the electrons, m^*: effective mass).

With common conductivity values of $\sigma = 100\ S\ cm^{-1}$ we obtain $\tau_\parallel < 10^{-14}$ sec and consequently $\omega_e \cdot \tau_\parallel << 1$ even at the highest magnetic fields available. Therefore just $\tau_\perp$, which is usually much larger than $\tau_\parallel$, can cause a dispersion. In the case of $(DCNQI)_2Cu$, there is no dispersion at all, that means $\omega_e \cdot \tau_\perp < 1$ even at B_0 = 5.4 Tesla ($\omega_e/2\pi$ = 300 MHz) and consequently $\tau_\perp < 8 \cdot 10^{-13}$ sec.

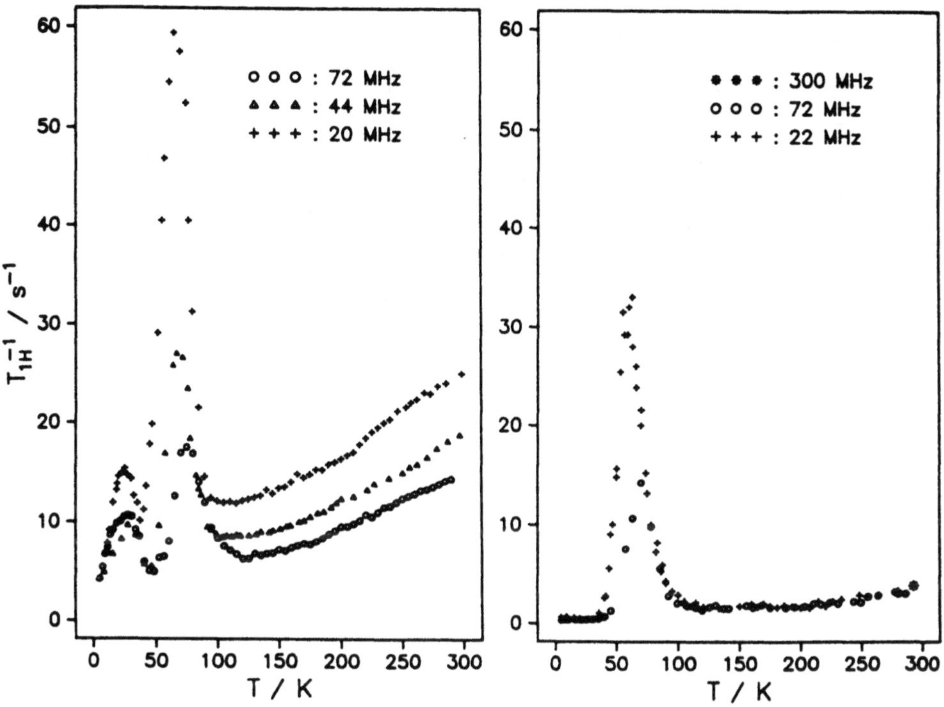

Fig.3: T_{1H}^{-1} as a function of T and $\omega_H/2\pi$: $(DCNQI)_2Li$ left, $(DCNQI)_2Cu$ right

For $(DCNQI)_2Li$ $T_{1H}^{-1} \sim \omega^{-1/2}$ even at the lowest frequency measured. That means that $\omega_e\tau_\perp > 1$ at 0.5 Tesla ($\omega_H/2\pi$ = 22 MHz) and $\tau_\perp > 10^{-11}$ sec. If only these values enter in the conductivity, the anisotropy $(\tau_\parallel/\tau_\perp)^{-1}$ is larger than 10^3 in the case of $(DCNQI)_2Li$ and less than 10 for the copper complex, confirming the assumed higher dimensionality of the latter [5,9].

References

[1] A. Aumüller, P. Erk, G. Klebe, S. Hünig, J.U. von Schütz and H.-P. Werner; Angew. Chemie 98 (1986) 159; Angew. Chemie Int. Ed. Engl. 25 (1986) 740.

[2] A. Kobayashi, R. Kato, H. Kobayashi, T. Mori and H. Inokuchi; Solid State Comm. 64 (1987) 45.

[3] S. Tomic, D. Jerome, A. Aumüller, P. Erk, S. Hünig and J.U. von Schütz; Europhys. Lett. 5 (1988) 553.

[4] S. Tomic, D. Jerome, A. Aumüller, P. Erk, S. Hünig and J.U. von Schütz; J. Phys. C 21 (1988) 203.

[5] H.-P. Werner, J.U. von Schütz, H.C. Wolf, R. Kremer, M. Gehrke, A. Aumüller, P. Erk and S. Hünig; Solid State Comm. 65 (1988) 809.

[6] J.U. von Schütz, M. Bair, H.-P. Werner, H.C. Wolf, A. Aumüller, P. Erk and S. Hünig in: Organic and Inorganic Low-Dimensional Crystalline Materials, ed. by P. Delhaes and M. Drillon, NATO ASI Series, Series B; Physics Vol. 168 (1988) 297.

[7] J.U. von Schütz, H.-P. Werner, H.C. Wolf, A. Aumüller, P. Erk and S. Hünig in: Magn. Reson. Relat. Phenom. Proc. Congress Ampere 23 (1986) 158; ed. by B. Maraviglia, F. DeLuca and R. Campanella.
[8] R.J. Elliott; Phys. Rev. 96 (1954) 266.
[9] J.U. von Schütz, M. Bair, H.J. Gross, U. Langohr, H.-P. Werner, H.C. Wolf, D. Schmeißer, K. Graf, W. Göpel, P. Erk, H. Meixner and S. Hünig; Synth. Metals 27 (1988) 249.
[10] D. Köngeter, F. Hentsch, H. Seidel, M. Mehring, J.U. von Schütz, P. Erk and S. Hünig; Solid State Comm. 65 (1988) 453.
[11] G. Soda, D. Jerome, M. Weger, J. Alizon, J. Gallice, H. Robert, J.M. Fabre and L. Giral; J. Physique 38 (1977) 931.
[12] M. Bair; Dissertation, University of Stuttgart, in preparation.

Part VII

Nonlinear Optics

The Role of Solitons in the Nonlinear Optical Properties of Polyenes and Polyacetylene Chains

J.L. Brédas

Service de Chimie des Matériaux Nouveaux,
Département des Matériaux et Procédés,
Université de Mons, B-7000 Mons, Belgium

Abstract We discuss the importance of the role of solitons in short polyenes and long polyacetylene chains. Emphasis is put on a comparison of: (i) the nonlinear optical properties of these compounds and (ii) the geometry relaxations that occur in the first Bu excited state.

Since over a decade ago, conjugated polymers have provided fascinating challenges [1]. Chemists and materials scientists have faced the task of preparing polymers with increasingly better solubility properties and environmental stability. Physical chemists and physicists have uncovered a field of choice to advance our understanding of the organic solid state, in particular the consequences of strong electron-lattice coupling. The way to best characterize conjugated polymers is to stress the intimate interplay between the electronic structure and the geometric structure, which constitutes the basic ingredient for the exciting physics in these compounds.

Any electronic process (we may think of charge transfer upon doping to lead to highly electrically conducting systems, or charge photogeneration upon laser illumination) results in very fast geometry relaxations (in a subpicosecond time scale). In turn, these lattice relaxations induce modifications in the electronic structure. In particular, new localized electronic states appear in the gap. The consequence is that the same elementary excitations which have been discovered to play a major role in the conduction properties of doped conjugated polymers, can also be of importance in the nonlinear optical properties of these systems.

Here, we focus our attention to trans-polyacetylene and its oligomers. Extension to non-degenerate ground-state systems can, however, be made; e.g. the role of polarons in polyaniline [2] or bipolarons in polythiophene [3] has already been pointed out from photoinduced absorption experiments.

In the case of trans-polyacetylene, a number of experimental works [4,5] have allowed to grasp an understanding for the mechanism of electron-hole relaxation after photogeneration. Following the early theoretical study of Su and Schrieffer [6], it has been shown that the photogenerated hole and electron are efficiently separated through lattice relaxations. This process leads to the formation of a positively charged soliton and a negatively charged soliton, respectively. The time scale for the relaxation is on the order of 10^{-13}s; such photogenerated solitons can last for several picoseconds. Important aspects of this resonant regime behavior are to be stressed:

Springer Series in Solid-State Sciences, Vol. 91
Electronic Properties of Conjugated Polymers III
Editors: H. Kuzmany · M. Mehring · S. Roth

(i) The relaxation process induces the presence of localized solitonic electronic states around midgap. Therefore, new subgap optical absorptions appear at the expense of the original band-to-band transition. There occur major shifts in oscillator strengths, indicating that trans-polyacetylene inherently possesses a high nonlinear optical response. (The same is valid for the other conjugated polymers).
(ii) The relaxation process provides for an efficient charge separation mechanism.

Such a charge separation has been advocated to be a key for understanding the nonlinear optical properties of short polyene molecules in the parametric (i.e., non-resonant) regime. Garito and co-workers [7] have presented a number of elegant theoretical works on short polyenes where they calculated the third-order microscopic polarizability using configuration interaction techniques under rigid lattice approximations. Their results indicate that the major contributions, e.g. to the longitudinal third-order polarizability in octatetraene, come from virtual excitations leading first from the ground state to the ^1Bu (one-photon allowed) excited state and then, from there to a high lying Ag excited state. The important role of this latter state originates from its significant charge redistribution with respect to the 1Bu state, which results in a very strong transition dipole moment (with charge flowing towards the ends of the molecule) [7].

We thus come to the conclusion that there are striking similarities between the two types of systems discussed above: the infinite polyacetylene chain in the resonant regime and short polyenes in the parametric regime. In both cases, the origin of important nonlinear responses can be traced back to the presence of a mechanism of charge redistribution insuring efficient charge separation.

However, there remain areas of apparent discrepancies. A rigid lattice approximation seems to be sufficient to describe the properties of small polyenes whereas lattice relaxation is an essential ingredient in the nonlinearities of polyacetylene in the resonant regime. A next step is therefore to consider what could be the situation for trans-polyacetylene in the parametric regime.

The key to the understanding in that case may well be the presence of quantum fluctuations along the polyacetylene chains. These quantum fluctuations allow the lattice in one degenerate ground state to tunnel its way to the other degenerate ground state. In a more common terminology, we might say that the vibrations that exist along the chains (in particular, the carbon-carbon stretching modes) lead to some elongation of the double bonds and some shortening of the single bonds. This very process can be modeled as a contribution of the second degenerate ground state to the first degenerate ground state [8]. These quantum fluctuations have been theoretically estimated to lead to about a 15% decrease in bond dimerization with respect to a purely rigid lattice situation [9]. Note that similar phenomena can also arise in nondegenerate ground-state systems [10].

The fluctuations can also be pictured as arising from the appearance of virtual soliton-antisoliton pairs [8]; however, the more important the fluctuation, the smaller its probability. At any given time, a photon interacting with a polyacetylene chain therefore probes the presence of more or less important fluctuations. These fluctuations constitute a source of enhancement in hyperpolarizability [11]. Indeed, the interaction of the photon with virtual soliton-antisoliton pairs again provides for an efficient charge separation mechanism at energies significantly lower than would be the case were it for a rigid lattice. This effect can also be easily understood from another standpoint. Since solitons can be characterized as islands of reduced bond dimerization along a trans-polyacetylene chain, they naturally provide for larger hyperpolarizabilities, as has been shown by De Melo and Silbey [12].

An extension of the work of Garito and co-workers to evaluate precisely the consequences, on small polyenes, of removing the rigid lattice approximation would thus prove most appropriate.

Another aspect that illustrates the fact that solitons can play an important role even in short polyene chains, is the geometry relaxation that takes place in the first Bu excited state. As mentioned above, upon photoexcitation of trans-polyacetylene, the electron-hole pair rapidly dissociates into a pair of charged solitons. In short polyenes, experimental evidence also points to the presence of strong geometry relaxations in the first Bu excited states of molecules containing up to six double bonds [13]. In spite of this, a vast majority of the calculations on polyene excited states are performed assuming frozen ground-state geometries [14-16]. Excited-state geometries are then simply inferred from bond length-bond order relationships. Such calculations indicate that as chain length increases, the geometric relaxation in the first Bu excited state attenuates to become very small in intermediate-sized polyene chains [14-16]. These results imply negligible relaxation in long chains, in contrast to what is observed in polyacetylene [4].

We have investigated [17] the relaxation process in the first Bu excited state of polyene chains ranging in size from 10 to 58 carbon atoms, in order to try to understand the evolution on going from short to long chains. We make use of a Hückel Hamiltonian with bond-length-dependent transfer integrals and with sigma-bond compressibility [18,19]. We find that subsequent to excitation to the Bu state, relaxation to form a pair of soliton-like reversals in the bond-alternation pattern minimizes the energy for polyenes of all lengths; the width of the bond-alternation reversal simply rescales to fit into the polyene chain. In other words, even in the shortest chains, the first Bu excited state is always found to relax optimally in such a way as to produce a pair of soliton defects. In order to accommodate the soliton pair, the width of the optimized soliton defects tends to shrink as chain length decreases.

In more recent calculations, we have combined an ab initio approach (to calculate the ground state energies) with a Pariser-Parr-Pople approach including configuration interaction (to calculate the excited state energies) [20]. We also find at this level that the two-soliton geometry is more favorable in medium-sized polyenes.

The overall picture that we obtain is that there exists a smooth evolution of the relaxation of the first Bu excited-state geometry in going from short polyenes to long chains, which is consistent with the strong geometry relaxations that have been experimentally observed for short polyenes such as decapentaene as well as polyacetylene. In all cases, the relaxation is such as to produce a pair of soliton topological defects on the chains. We note that a relaxation tending to produce a two-soliton geometry is not unlike the ionic valence-bond picture used to describe the first Bu excited state [21].

In our opinion, a thorough understanding of the nonlinear optical properties of polyacetylene and other conjugated polymers currently constitutes one of the major tasks facing the scientists interested in these conjugated systems. From the theoretical standpoint, the complete unraveling of the excited state properties requires the dedication of important efforts. Indeed, one-electron pictures have proven useful to rationalize those basic ground-state properties which are governed by electron-lattice interactions; however, they may turn out to be far from sufficient [21] when trying to discriminate between various excited states which can at least partly originate from typical many-body elementary excitations such as excitons or polar-excitons (these are known to play a major role e.g. in the nonlinear optical properties of polydiacetylenes [22] and polyanilines [2]). Impetus for such work is strong since this field is at the forefront of basic research as well as applied research with enormous potentialities.

REFERENCES

[1] «Handbook of Conducting Polymers», ed. by T.A. Skotheim (Marcel Dekker, New York, 1986).

[2] M.G. Roe, J.M. Ginder, P.E. Wigen, A.J. Epstein, M. Angelopoulos, and A.G. MacDiarmid, Phys. Rev. Lett. 60, 2789 (1988).

[3] Z. Vardeny, E. Ehrenfreund, O. Brafman, M. Nowak, H. Schaffer, A.J. Heeger, and F. Wudl, Phys. Rev. Lett. 56, 671 (1986).

[4] A.J. Heeger, S. Kivelson, J.R. Schrieffer, and W.P. Su, Rev. Mod. Phys. 60, 781 (1988).

[5] S. Roth and H. Bleier, Adv.Phys. 36, 385 (1987).

[6] W.P. Su and J.R. Schrieffer, Proc. Natl. Acad. Sci. USA 77, 5626 (1980).

[7] J.R. Heflin, K.Y. Wong, O.Zamani-Khamiri, and A.F. Garito, Phys. Rev. B 38, 1573 (1988); J. Opt. Soc. Am. B4, 136 (1987).

[8] J. Yu, H. Matsuoka, and W.P. Su, Phys. Rev. B 37, 10367 (1988).

[9] J.E. Hirsch and E. Fradkin, Phys. Rev. B 27, 1680 & 4302 (1983).

[10] J. Yu, H. Matsuoka, and W.P. Su, Phys. Rev. B, to be published.

[11] M. Sinclair, D. Moses, D. McBranch, A.J. Heeger, J. Yu, and W.P. Su, Synth. Met. 28, D 655 (1989).

[12] C.P. De Melo and R. Silbey, Chem. Phys. Lett. 140, 537 (1987); J. Chem. Phys. 88, 2558 & 2567 (1988).

[13] M.F. Granville, B.E. Kohler and J.B. Snow, J. Chem. Phys. 75, 3765 (1981).

[14] K. Schulten, I. Ohmine and M. Karplus, J. Chem. Phys. 64, 4422 (1976).

[15] S. Ramasesha and Z. Soos, Synth. Metals 9, 282 (1984).

[16] J.L. Brédas, M. Dory and J.M. André, J. Chem. Phys. 83, 5242 (1985).

[17] J.L. Brédas and A.J. Heeger, Chem. Phys. Lett. 154, 56 (1989).

[18] W.P. Su, J.R. Schrieffer and A.J. Heeger, Phys. Rev. Lett. 42, 1698 (1979); Phys. Rev. B 22, 2099 (1980).

[19] J.L. Brédas, R.R. Chance and R. Silbey, Phys. Rev. B 26, 5843 (1982).
[20] J.M. Toussaint, A.J. Heeger, and J.L. Brédas, to be published.
[21] B.S. Hudson, B.E. Kohler and K. Schulten, in «Excited States», vol. 6, ed. by E.C. Lim (Academic Press, New York, 1982) p. 1.
[22] B.I. Greene, J. Orenstein, R.R. Millard, and L.R. Williams, Phys. Rev. Lett. 58, 2750 (1987).

Electronic Structure and Nonlinear Optical Properties of Polythiophene, Polythieno[3,4-C]thiophene and Polyisothianaphthene Oligomers and Their Vinylene Derivatives

M. Dory, B. Thémans, J.L. Brédas, J. Delhalle, and J.M. André*

Laboratoire de Chimie Théorique Appliquée,
Facultés Universitaires Notre-Dame de la Paix,
rue de Bruxelles 61, B-5000 Namur, Belgium

Abstract. The electronic structure of oligomers of thiophene, isothianaphthene, thieno[3,4-c]thiophene and related vinylene compounds are calculated by the semi-empirical MNDO method. Based on the MNDO-optimized geometries structures, the electrical polarizabilities are evaluated at the STO-3G *ab initio* level by the Finite Field method.

1. Introduction

Large non-resonant third-order nonlinear optical susceptibilities, $\chi^{(3)}$, so far observed in polymers are related to highly π-conjugated structures. For instance, polyacetylene exhibits a $\chi^{(3)}$ value of ~10^{-9} e.s.u., i.e. the largest value observed to date for a polymer. Other systems, e.g. polyaromatics, are intrinsically attractive for their nonlinear optical properties. Recent α and γ measurements have been reported for thiophene oligomers of increasing length up to the hexamer. The maximum γ value reached in this case is 10^{-10} e.s.u. [1]. Polythiophene vinylene, which can be cast into conducting films with very good mechanical properties and suitable environmental stability, has a measured third harmonic generation, $\chi^{(3)}$, of 3.2 10^{-11} e.s.u. [2].

In this contribution, we compute the geometrical structure of oligomers of thiophene (T_3), thieno[3,4-c]thiophene (TT_3), isothianaphthene (IT_3), thiophene vinylene (TV_3), thieno[3,4-c]thiophene vinylene (TTV_2), and isothianaphthene vinylene (ITV_2), where the subscript n in the abbreviations indicates the maximum number of repeated monomers in the structure. The most important structural features are discussed in relation to the calculated electric polarizability.

2. Methodology

The molecular geometries are determined with the semi-empirical MNDO (Modified Neglect of Diatomic Overlap) SCF (Self-Consistent Field) MO (Molecular Orbital) technique, as developed by Dewar and coworkers [3,4]. This method is known to provide good estimates of the geometrical parameters of polycyclic hydrocarbons and heterocyclic compounds. The optimized geometries are then used as input for polarizability calculations. The electric polarizabilities of the systems are evaluated at the Hartree-Fock *ab initio* minimal basis set (STO-3G) level. The validity of this approach has been argued previously [5]. All the

* presently at : Service de Chimie des Matériaux Nouveaux, Université de Mons, avenue Maistriau 21, B-7000 Mons (Belgium).

Springer Series in Solid-State Sciences, Vol. 91
Electronic Properties of Conjugated Polymers III
Editors: H. Kuzmany · M. Mehring · S. Roth

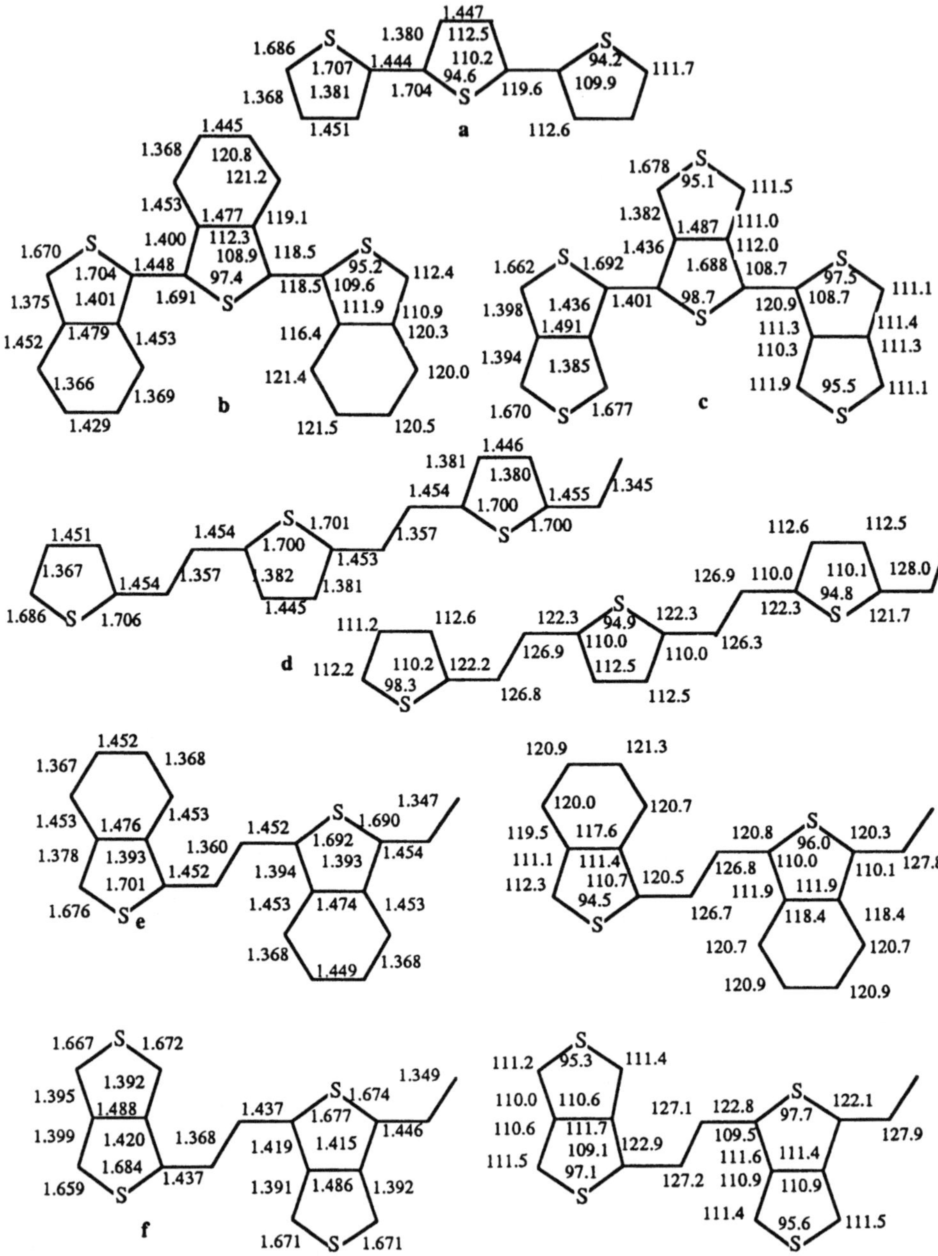

Figure 1. MNDO optimized geometries of trimers of T3 (a), IT3 (b), TT3 (c), TV3 (d), and dimers of ITV2 (e), and TTV2 (f). Bond lengths are in Å, bond angles in degrees

polarizabilities calculations are performed with a modified version of the Gaussian 82 series of programs [6]. The thresholds are set to 10^{-6} atomic unit (a.u.) for the cutoff of two-electron integrals and 10^{-5} a.u. for the requested convergence on the density matrix. The Finite-Field approach of Cohen and Roothaan [7] is chosen to calculate the polarizability; in this paper, all the polarizabilities are expressed in a.u. (1 a.u. of polarizability being equal to 1.6488 x 10^{-41} $C^2\,m^2\,J^{-1}$).

3. Electronic and geometric structures

MNDO results on thiophene (monomer, dimer and trimer), isothianaphthene (monomer) and thieno[3,4-c]thiophene (monomer and dimer) were reported previously [8]. In this work, we have computed the optimized parameters for IT_3, TT_3, TV_3, ITV_2, and TTV_2. Because of computer limitations, the trimers ITV_3 and TTV_3 could not be considered. The results are displayed in Fig.1. For çomparaison, the T_3 geometry obtained by Brédas et al. [8] is included. A more complete analysis of all oligomers, n equal 1 to 3, will be reported elsewhere [10]. All systems were kept coplanar during the optimization. An analysis of Fig.1 shows that in contrast to T_3 and IT_3, the TT_3 structure adopts a quinoid-like structure. Indeed, the inter-ring C-C bond length is calculated to be equal to 1.401 Å while in the case of the other systems, this bond length is larger, ~ 1.445 Å. This geometrical feature is also reproduced at the *ab initio* STO-3G level [9]. The presence of the vinylene linkages between the aromatic units does not induce significant modifications in the geometries of TV_3 and ITV_2. However,

Table 1. Polarizabilities (α_{ii}; i = x,y,z), mean polarizability (α), and volume (V) for thiophene (T_1,T_2,T_3), thiophene vinylene (TV_1,TV_2,TV_3), isothianaphthene (IT_1,IT_2,IT_3), isothianaphthene vinylene (ITV_1,ITV_2), thieno[3,4-c]thiophene (TT_1,TT_2,TT_3) and thieno[3,4-c]thiophene vinylene (TTV_1,TTV_2) oligomers. Experimental values are from Ref 1

Molecule	α_{xx}	α_{yy}	α_{zz}	α	V(Å^3)	$\alpha_{exp.}$
T_1	35.25	5.38	37.92	26.18	74.49	66.13
T_2	104.32	10.26	74.95	63.18	135.35	168.71
T_3	199.93	15.13	109.71	108.26	194.58	303.67
TV_1	65.53	7.47	60.41	44.47	101.42	
TV_2	194.18	14.46	125.26	111.30	189.60	
TV_3	360.69	21.43	186.61	189.58	280.97	
IT_1	68.15	9.15	92.19	56.50	111.50	
IT_2	206.87	17.67	210.89	145.14	217.43	
IT_3	421.77	26.12	301.79	249.89	324.88	
ITV_1	101.11	11.18	123.62	78.64	138.24	
ITV_2	318.37	21.75	279.49	206.54	265.36	
TT_1	82.69	8.84	79.90	57.14	105.84	
TT_2	341.36	16.87	200.98	186.40	203.83	
TT_3	1182.85	24.65	263.37	490.29	290.00	
TTV_1	130.12	10.84	110.08	83.67	132.50	
TTV_2	577.28	20.95	264.12	287.45	256.09	

the double and simple bond lengths of the TTV_2 vinylene linkage are respectively longer and shorter than in the case of the other two structures. The TTV_2 structure is reminiscent of a quinoid-like structure but this tendency is less important than the one observed in TT_3.

4. Molecular electrical polarizabilities

The calculated *ab initio* component polarizabilities (α_{ii}) and the molecular volume (V) (estimated from van der Waals radii) of thiophene, isothianaphtene, thieno[3,4-c]thiophene, and related vinylene compounds are presented in Table 1. Experimental values for thiophene oligomers are included. A comparison of the evolution of the polarizability between the three sets of compounds with respect to the chain length indicates that the largest increase in the longitudinal component α_{xx} is found in the thieno[3,4-c]thiophene (the other components increasing linearly). Going from the TT_2 dimer to the TT_3 trimer, there is an increase of 280 % and 200 % for α_{xx} and α, respectively. This strong effect in TT_3 can be related to the aforementioned quinoid-like structure which leads to a better delocalization than in the T_3 and IT_3 structures. The incorporation of a vinylene linkage between the aromatic units increases the mean polarizability by a factor of 1.7 and 1.4 for thiophene and isothianaphthene, respectively. Thieno[3,4-c]thiophene vinylene also displays the same enhancement as in TT_2. (It would have been interesting to be able to compute TTV_3 to confirm the similarity of trends found in TT_2 and TT_3.) Since the overall effect in terms of nonlinear optical efficiency is dependent on the number and nature of the atoms, we have reported in Table 1 the average polarizability normalized to the calculated volume (in Å^3) of the systems. In Fig. 2, the mean polarizabilities per volume (α/V) are presented for all the systems. There is a fast saturation occurringfor the normalized polarizabilities of thiophene, isothianaphthene and their related vinylene structures while those of TT_3 and TTV_2 are still increasing with the number of unit. If we consider the significant $\chi^{(3)}$ values (~ 10^{-10} e.s.u.) already observed for the thiophene oligomers, it can be anticipated from our results that the TT and TTV oligomers should be even more effective.

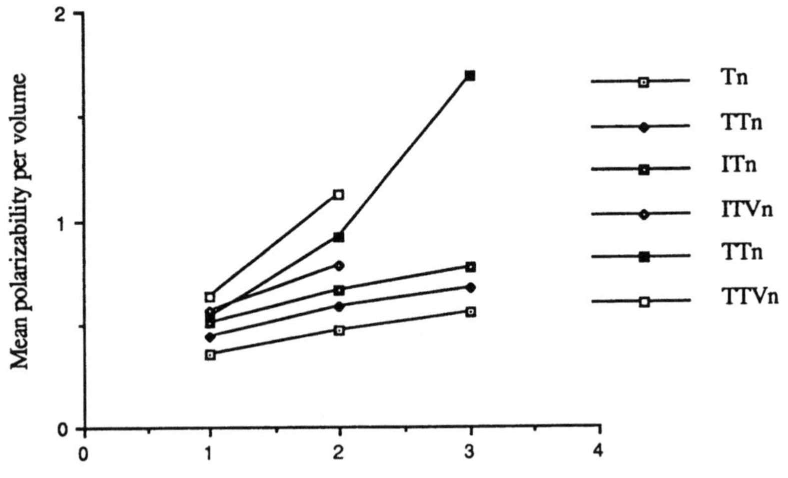

Figure 2. Mean Polarizability per volume versus the number of units (n)

5. Conclusions

To summarize, we can stress that :

(i) the thieno[3,4-c]thiophene trimer appears to present a quinoid-like structure which leads to large polarizability values;

(ii) the vinylene derivatives turn out to be better candidates for optoelectronics than the purely aromatic structures;

(iii) the mean polarizabilities per volume rapidly saturate except for the thieno[3,4-c]thiophene derivatives which still display a strong enhancement for three repeating units.

6. Acknowledgements

The authors are most grateful to J.G. Fripiat whose collaboration in the course of this work has been invaluable. This work has been supported in Namur under the ESPRIT-EEC contract N°2284 on "Optoelectronics with Active Organic Molecules". The authors thank the Belgian National Fund for Scientific Research (FNRS), IBM Belgium, and the Facultés Universitaires Notre-Dame de la Paix for the use of the Namur Scientific Computing Facility (SCF).

7. References

[1] M.T. Zhao, B. P. Singh, and P. N. Prasad, J. Chem. Phys. **89**, 5535 (1988).

[2] T. Kaino, K-i Kubodera, H. Kobayashi, T. Kurihara, S. Saito, T. Tsutsui, S. Tokito, and H. Murata, Appl. Phys. Lett. **53**, 2002 (1988).

[3] M.J.S. Dewar and W. Thiel, J. Am. Chem. Soc. **99**, 4899 and 4907 (1977).

[4] M.J.S. Dewar and M.L. McKee, J. Comput. Chem. **4**, 84, (1983).

[5] (a) V.P. Bodart, J. Delhalle, J.M. André, and J. Zyss, in Polydiacetylenes : Synthesis, Structure and Electronic Properties, edited by D. Bloor and R.R. Chance (Martinus Nijhoff Publishers, Dordrecht, 1984), p. 125; (b) V.P. Bodart J. Delhalle, J.M. André, and J. Zyss, Canad. J. Chem. **63**, 1631 (1985).

[6] J.S. Binkley, M. Frisch, K. Raghavachari, D.J. DeFrees, H.B. Schlegel, R.A. Whiteside, E.Fluder, R. Seeger, and J.A. Pople, **Gaussian 82,** Carnegie-Mellon University (1983).

[7] H.D. Cohen and C.J. Roothaan, J. Chem. Phys. **43**, 534 (1965).

[8] J.L. Brédas, B. Thémans, J.M. André, A.J. Heeger, and F. Wudl, Synt. Metals **11**, 343 (1985).

[9] J.L. Brédas, M. Dory, B. Thémans, J. Delhalle, and J.M. André, SPIE Proceedings **971**, 42 (1988).

[10] M. Dory, B. Thémans, J.L. Brédas, J. Delhalle, and J.M. André, to be published.

Nonlinear Optical Properties of Thin Films of Polymers with a One-Dimensional Conjugated π-Electron System

C. Bubeck, A. Kaltbeitzel, D. Neher, J.D. Stenger-Smith, G. Wegner, and A. Wolf

Max-Planck-Institut für Polymerforschung,
Postfach 31 48, D-6500 Mainz, Fed. Rep. of Germany

Abstract. Third harmonic generation and degenerate four wave mixing experiments have been performed with poly (p-phenylene vinylene) (PPV), poly (p-naphthalene vinylene), and three derivatives of poly (phenyl acetylene). Subpicosecond response times are observed. PPV shows a fluorescence that is induced via a two photon process.

1. Introduction

Polymers with extended π-electron delocalization are not only interesting with respect to electrical applications, but may be also very useful in the field of nonlinear optics. The large hyperpolarizability might be used for new optoelectronic or pure optical devices. Therefore a search for new materials with high nonlinear optical susceptibilities $\chi^{(2)}$ and $\chi^{(3)}$ is under way [1-7]. This contribution is restricted to $\chi^{(3)}$ effects that are visible via third harmonic generation (THG) or a nonlinear refractive index

$$n = n_o + n_2 I \qquad (1)$$

where n_o is the linear refractive index, I is the laser power density at the sample and n_2 is proportional to $\chi^{(3)}$.

Potential applications of $\chi^{(3)}$ phenomena are mostly based on the nonlinear refractive index. Examples are discussed in the fields of all-optical signal processing, waveguides and phase conjugation [8-10]. Therefore materials are needed that have good nonlinear optical properties and can be processed to thin films of homogenous thickness. Recently it was shown that the conjugated polymer poly (p-phenylene vinylene) (PPV) is a promising material [11-14]. We present some further studies of PPV, poly (p-naphthalene vinylene) (PNV) [15] and poly (phenyl acetylene) (PPA) derivatives with various substituents at the phenyl ring. These polymers or their precursors are soluble and can be processed to thin films by casting from solution or spincoating.

2. Survey on Theory and Resonance Phenomena

The macroscopic susceptibility $\chi^{(3)}$ is related to the microscopic hyperpolarizability γ of the molecular unit via

$$\chi^{(3)}(-\omega;\omega_1,\omega_2,\omega_3) = N\, f(\omega)\, f(\omega_1)\, f(\omega_2)\, f(\omega_3)\, \gamma(-\omega;\omega_1,\omega_2,\omega_3). \qquad (2)$$

Springer Series in Solid-State Sciences, Vol. 91
Electronic Properties of Conjugated Polymers III
Editors: H. Kuzmany · M. Mehring · S. Roth

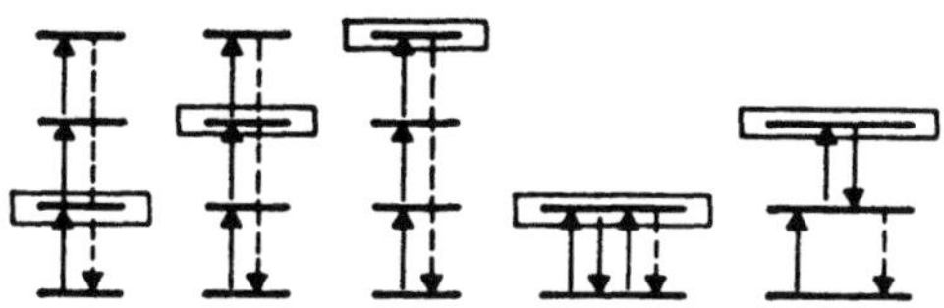

Fig. 1: Energy level diagrams showing resonances with real electronic excitation states (in frames). The nonlinear optical processes are third harmonic generation (THG) and degenerate four wave mixing (DFWM)

The incident optical fields with frequencies ω_1, ω_2, ω_3 create a new optical wave with frequency ω that leaves the sample. N is the number of polarizable units per unit volume and $f(\omega)$ are the Lorentz-Lorenz local field factors for the different frequencies. Following the diagramatic techniques of Yee and Gustafson [16], γ is given by

$$\gamma(-\omega;\omega_1,\omega_2,\omega_3) = P\sum_{A,B,C}\left(\frac{1}{4\hbar^3}\right)\frac{\langle G|\mu|A\rangle\langle A|\mu|B\rangle\langle B|\mu|C\rangle\langle C|\mu|G\rangle}{(\omega_1-\Omega_{AG}+i\Gamma_{AG})(\omega_1+\omega_2-\Omega_{BG}+i\Gamma_{BG})(\omega-\Omega_{CG}+i\Gamma_{CG})} \quad . \quad (3)$$

G, A, B, C are molecular ground and excitation states, μ: dipole moment operator, Γ: relaxation parameter and P indicates that a summation over all permutations must be performed [17]. γ is resonantly enhanced, if optical frequencies coincide with the frequencies of molecular transitions Ω. This situation of multiple resonances is visualized in Fig. 1. Consequences of such resonances are an enhanced and complex $\chi^{(3)}$ value and the response time of the nonlinear process becomes a function of the intrinsic relaxation processes of the excited states of the molecular units. In the absence of resonances $\chi^{(3)}$ is real and the response time is ultrafast (shorter than 10^{-13}sec).

Treating the conjugated polymers as one-dimensional systems with a π electron delocalization length L_d, Flytzanis et al have derived a scaling law for the nonresonant γ being proportional to $(L_d)^6$ [18,19]. If L_d is inversely proportional to the optical gap, γ should scale with $(E_o)^{-6}$.

3. Experimental

The experimental setup of the third harmonic generation was described recently [13]. Infrared light pulses (1064nm, 0.4mJ and approximately 20ps) are generated by an active/passive modelocked Nd:YAG laser. They are focused on the sample mounted on a rotation stage in an evacuated chamber. The degenerate four wave mixing (DFWM) experiments were performed using a folded BOXCARS configuration [20]. The output of a synchronously pumped and cavity dumped dye laser system was split into three beams with variable delay and focused on the sample(laser wavelength 650nm, intensity in the order of 1-2GW/cm^2, pulse duration 1ps).

Thin films of PPV were obtained by spin coating or casting from aqueous solution of the tetramethylene sulfonium chloride precursor polymer [21,22], followed by thermal treatment in vacuum (24h at 220°C). PNV films were prepared in a similar way [15,22]. The PPA derivatives shown in Tab. 1 were synthesized

Tab. 1: List of conjugated polymers and a survey of their linear and nonlinear optical properties. THG was measured with films obtained by spin coating (thickness on the order of 50 nm) and λ(laser) = 1064 nm. The DFWM results were obtained with cast films (thickness in the order of 1-2 μm) and λ(laser) = 650 nm. For the absolute values of the DFWM data only the order of magnitude can be given

Polymer			λ_{max} [nm] spin cast	$\chi^{(3)}$ [esu] THG	$\chi^{(3)}$ [esu] DFWM	τ [ps]
PPV	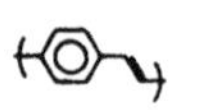		460	$1.2 \mp 0.6 \; 10^{-10}$	$2 \; 10^{-11}$	< 1
PNV	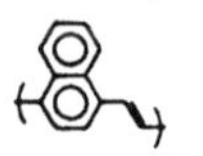		440		$5 \; 10^{-12}$	4
PPA-1		R:$Si(CH_3)_3$	525	$1.9 \mp 0.4 \; 10^{-11}$	$2 \; 10^{-11}$	< 1
PPA-2	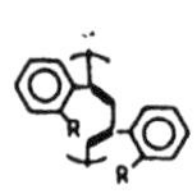	R: CH_2CH_3	450	$8.4 \mp 1.7 \; 10^{-12}$	$7 \; 10^{-12}$	< 1
PPA-3		R: CH_3	450	$1.5 \mp 0.4 \; 10^{-11}$	$3 \; 10^{-12}$	< 1

following the method of Masuda [23] with the tungsten catalyst. Thin films of PPA were prepared by casting from solution. All film thicknesses were measured with a step profiler.

4. Results

The optical absorption spectra of the thin films are shown in Fig. 2 and 3. The wavelengths of the absorption maxima of the PPA films depend on the preparation method . The maxima of the thicker films, cast without spinning are shifted to shorter wavelength by approximately 5-20 nm as compared to the thinner films. All samples show some residual absorption below the bandgap that is not due to scattering.

The THG experiments are evaluated following the procedure of Kajzar and Messier [24,25] and as described recently [13] taking into account the film thickness and refractive indices at ω and 3ω. The results are displayed in Tab.1.

For comparison the order of magnitude of the $\chi^{(3)}$ values determined from DFWM are also shown. Their larger experimental error comes mainly from uncertainties of the thickness of the cast films and the CS_2 solvent used as reference. The response time τ is determined from the decay of the transient grating of the nonlinear refractive index. With the exception of PNV the response times are shorter than the pulse width of our laser system.

During the DFWM experiments with PPV films, a fluorescence of the sample was excited by laser wavelengths between 580 nm and 760 nm. The fluorescence has a quadratic dependence on the laser intensity in the range of 1 GW/cm^2. The spectrum of the

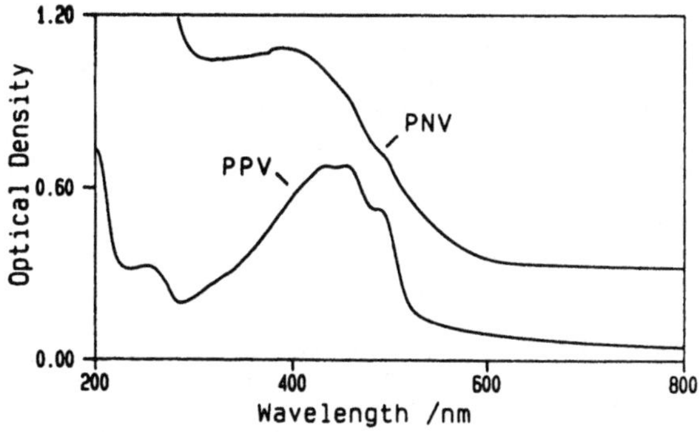

Fig. 2: Absorption spectra of a PPV spin cast film and a PNV film cast from solution The PNV spectrum is shifted by OD = +0.3

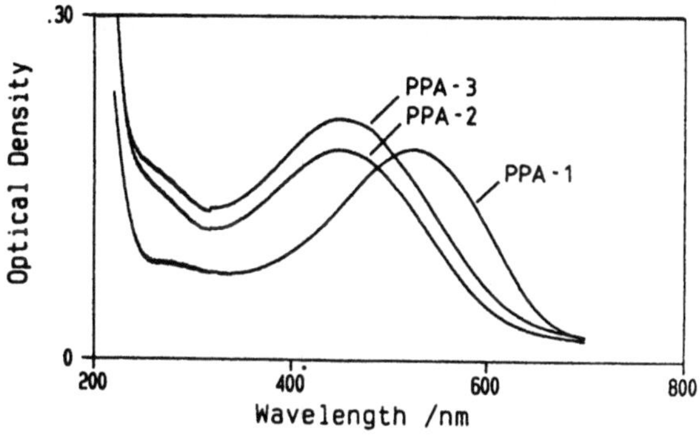

Fig. 3: Absorption spectra of PPA films prepared by spin coating

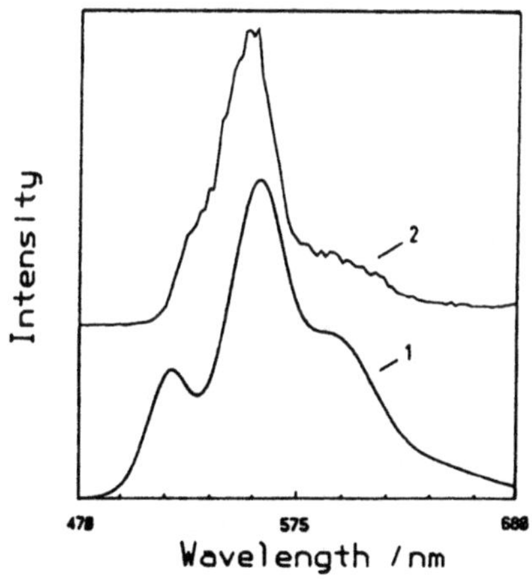

Fig. 4: Fluorescence spectra of PPV films. Spectrum 1 is measured with a fluorescence spectrometer using an excitation wavelength of 440 nm. Spectrum 2 is obtained with pulsed laser excitation at 700 nm and approximately 1 GW/cm^2 at the sample position

PPV fluorescence is shown in Fig. 4 together with the spectrum excited by UV light that has been observed earlier [26].

5. Discussion

We have presented a series of conjugated polymers that can be processed to form thin films by spin coating. The magnitude of $\chi^{(3)}$ observed for PPV, PNV and the PPA derivatives is in the range of 10^{-12} to 10^{-10} esu typical for conjugated polymers [4-7,27,28]. The values presented in Tab. 1 may contain resonance contributions. The third harmonic frequency falls into the range of the polymer absorption bands. However, wavelength dependent investigations would be necessary to determine the nonresonant $\chi^{(3)}$ values. Therefore a comparison with the scaling laws described above is not yet possible.

With the exception of PNV we observe a subpicosecond response time. This might indicate an absence of resonance contributions to DFWM at 650 nm for PPV and the PPA derivatives. However the possibility to excite a fluorescence of PPV with excitation energies well below the energy gap and the quadratic dependence on the laser intensity indicates a two photon process in the excitation pathway of the fluorescence. Therefore a two photon resonance contribution to $\chi^{(3)}(-\omega;\omega,\omega,-\omega)$ cannot be excluded.

Acknowledgements

We thank V. Percec for the supply of PPA-1 and H.J. Menges for considerable technical support. Financial support by the BMFT under the project 03M4008E9 is gratefully acknowledged.

References

1. D.J. Williams, Angew. Chem. **96**, 637 (1984)
2. D.J. Williams (ed.), Nonlinear Optical Properties of Organic and Polymeric Materials, ACS Symp. Ser. **233**, (1983)
3. C. Flytzanis, J.L. Oudar, Nonlinear Optics: Materials and Devices, Springer (1986)
4. D.S. Chemla, J. Zyss (eds.), Nonlinear Optical Properties of Organic Molecules and Crystals, Academic Press (1987)
5. A.J. Heeger, J. Orenstein, D.R. Ulrich (eds.), Nonlinear Optical Properties of Polymers, Materials Research Soc. (1988)
6. P.N. Prasad, D.R. Ulrich (eds.), Nonlinear Optical and Electroactive Polymers, Plenum Press, (1988)
7. J. Messier, F. Kajzar, P.N. Prasad, D.R. Ulrich (eds.), Nonlinear Optical Effects in Organic Polymers, Kluver Acad. Publ., (1989) in press
8. D.R. Ulrich, Mol. Cryst. Liq. Cryst. **160**, 1 (1988)
9. J. Zyss, J. Molec. Electr. **1**, 25 (1985)
10. R.A. Fisher (ed.), Optical Phase Conjugation, Acad. Press (1983)
11. T. Kaino, K. Kobodera, S. Tomaru, T. Kurihara, S. Saito, T. Tsutsui, S. Tokito, Electron. Lett. 23, 1095 (1987)
12. B.P. Singh, P.N. Prasad, Polymer, submitted
13. C. Bubeck, A. Kaltbeitzel, R.W. Lenz, D. Neher, J.D. Stenger-Smith, G. Wegner, in Ref. 7
14. D.D.C. Bradley, Y. Mori, Jap. Journ. Appl. Phys. **28**, 174 (1989)
15. J.D. Stenger-Smith, PhD thesis, Univ. of Mass., Amherst (1988)
16. T.K. Yee, T.K. Gustafson, Phys. Rev. **A 18**, 1597 (1978)
17. J. Zyss, D.S. Chemla, in Ref.4, Vol. 1, p.66

18. G.P. Agrawal, C. Cojan, C. Flytzanis, Phys. Rev.**B17**, 776 (1978)
19. C. Flytzanis in Ref.4, Vol. 2, p. 121
20. G.M. Carter, J. Opt. Soc. Am. **B4**, 1018 (1987)
21. R.W. Lenz, C.C. Han, J.D. Stenger-Smith, F.E. Karasz, J. Polymer Sci. **A26**, 3241 (1988)
22. J.D. Stenger-Smith, R.W. Lenz, G. Wegner, Polymer, submitted
23. T. Masuda, K. Hasegawa, T. Higashimura, Macromolecules **7**, 728 (1974)
24. F. Kajzar, J. Messier, Phys. Rev. **A32**, 2352 (1985)
25. P.A. Chollet, F. Kajzar, J. Messier, Solid State Commun. **61**, 11 (1985)
26. D.D.C. Bradley, R.H. Friend, K.S. Wong, W. Hayes, H. Lindenberger, S. Roth in "Electronic Properties of Conjugated Polymers" H. Kuzmany, M. Mehring, S. Roth (eds.), Springer (1987)
27. C. Sauteret, J.-P. Hermann, R. Frey, F. Pradere, J. Ducuing, R.H. Baughman, R.R. Chance, Phys. Rev. Lett. **36**, 956 (1976)
28. F. Kajzar, J. Messier, Polymer Journ. **19**, 275 (1987)

Influence of the Film Preparation on the Nonlinear Optical Properties of Phthalocyanine Films

A. Kaltbeitzel[1], *D. Neher*[1], *C. Bubeck*[1], *T. Sauer*[1], *G. Wegner*[2], *and W. Caseri*[1]

[1]Max-Planck-Institut für Polymerforschung,
Postfach 31 48, D-6500 Mainz, Fed. Rep. of Germany
[2]Institut für Polymere, ETH Zürich,
Universitätsstr. 6, CH-8092 Zürich, Switzerland

Abstract: The nonlinear optical susceptibility $\chi^{(3)}(-\omega;\omega,-\omega,\omega)$ of different phthalocyanine films was measured by degenerate four wave mixing (DFWM). The relaxation time T_1 varies between several picoseconds and nanoseconds depending on the film preparation.

1. Introduction

Materials with a large value of the third order nonlinear optical susceptibility $\chi^{(3)}(-\omega;\omega,-\omega,\omega)$ can be used for optical signal processing. For this application a short relaxation time and good material properties are also essential. The theoretical calculation shows that $\chi^{(3)}$ is proportional to the fourth power of the transition dipol moment [1,2]. Therefore a strong absorption gives rise to large values of $\chi^{(3)}$. Phthalocyanines in solution have a sharp absorption around 700 nm in combination with a high absorption coefficient. The molecules are thermally stable up to 400°C. These facts have raised interest in the investigation of the nonlinear optical properties of thin films of phthaloycyanines [3,4,5]. The literature shows that the absorption spectra of these films differ from the solution spectra and depend strongly on the preparation method [3,6,7]. Therefore we have investigated the influence of the film preparation on the optical properties with respect to the absorption spectra, the value of $\chi^{(3)}$, the relaxation time and the material properties.

2. Experimental Setup

The nonlinear susceptibility $\chi^{(3)}(-\omega;\omega,-\omega,\omega)$ was measured by degenerate four wave mixing in a folded-BOXCARS configuration [8]. With this kind of experiments the amount of $\chi^{(3)}$ as well as the relaxation time T_1 of the excited state can be determined. A pulsed laser beam is devided in three parts and focused onto the sample. Two beams create a transient grating of the refractive index inside the sample. The amplitude of the grating is proportional to the square of $\chi^{(3)}$. The third beam is diffracted from this grating. By a variable time delay between the creation of the grating and the diffraction process the decay of the grating can be probed. Thus the relaxation time τ_r

Springer Series in Solid-State Sciences, Vol. 91
Electronic Properties of Conjugated Polymers III
Editors: H. Kuzmany · M. Mehring · S. Roth

is calculated from a measurement of the diffraction efficiency versus time delay of the pulses. In the case of an electronic nonlinear susceptibility the relaxation of the grating τ_r is given by $\tau_r = 2 \cdot T_1$.
The light source consists of a cw mode-locked Nd-Yag laser that is used to synchrounously pump a cavity-dumped dye laser.The resulting pulses are 0.5 - 1 picosecond wide with a peak power of ~ 30 kW and a repetition rate between 0 and 76 MHz. Use of three different dyes leads to a tuning range between 560 nm and 760 nm.

3. Sample preparation

Thin films of the Phthalocyanine derivates depicted in Tab. 1 were prepared by different methods:

Tab. 1: Chemical structure and preparation method of the investigated phthalocyanine films

Chemical Structure		System	R^3
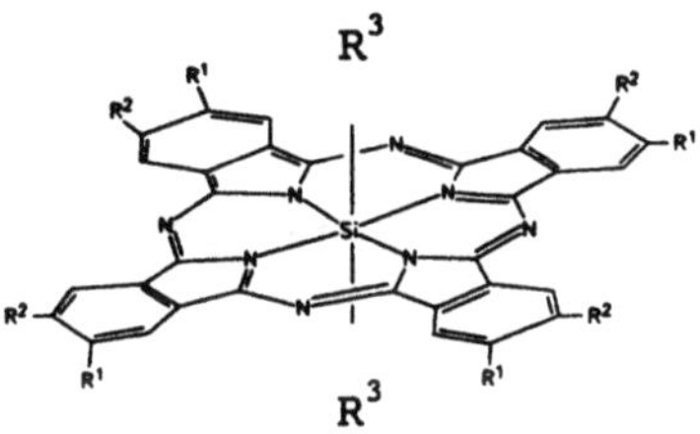	1	LB-Film of the polymer molecules 50 layers	see text
	2	LB-Film of the monomer, 80 layers	Cl
	3	Polystyrene doped by monomer	Cl
R^1= O-CH_3 R^2= O-C_8H_{17} in all permutations	4	Styrene copolymer in bulk	O-C_4H_7

Thin films of the first two materials were prepared by Langmuir-Blodgett-technique [9,10]. System 2 consisted of 80 layers of the molecule described in Tab. 1. System 1 consisted of 50 layers of a polymer with a phthalocyanine moiety as depicted in Tab. 1 connected by a flexible spacer with the chemical structure [10]:

R^3 = -O-(C H_2)$_4$- Si(Me)$_2$ 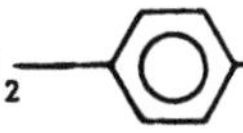 Si(Me)$_2$ - (C H_2)$_4$-O-

In system 4 the phthalocyanine moiety was attached as sidegroup in a copolymer with a styrene. The viscous solution of the copolymer was cast between glass slides. Evaporation of the residual monomer gave a solid film of several micron thickness. System 3 was prepared in a similar way by polymerizing styrene containing the phthalocyanine 1.

4. Experimental Results

Fig. 1 and 2 show the absorption spectra of these samples and the corresponding decay of the transient grating as a function of the delay time. From a semilogaritmical plot the relaxation times were calculated. System 2 shows a monoexponential decay with a relaxation rate of approximately 7 ps. The nonlinear signal of the other samples decays multiexponentially. The value of $\chi^{(3)}$ is frequency dependent. For the LB-films of the systems 1 and 2 the maximum value lies in the range between 10^{-9} and 10^{-10} esu. This is an order of magnitude estimate because a tight focusing in the sample leads to large error bars in the comparison with a CS_2 reference cell. A direct comparison of the different LB-Films shows that the values of $\chi^{(3)}$ are the same within a factor of two. Because of the dilution of phthalocyanine molecules the polystyrene films have a lower $\chi^{(3)}$ value that depends on the concentration.

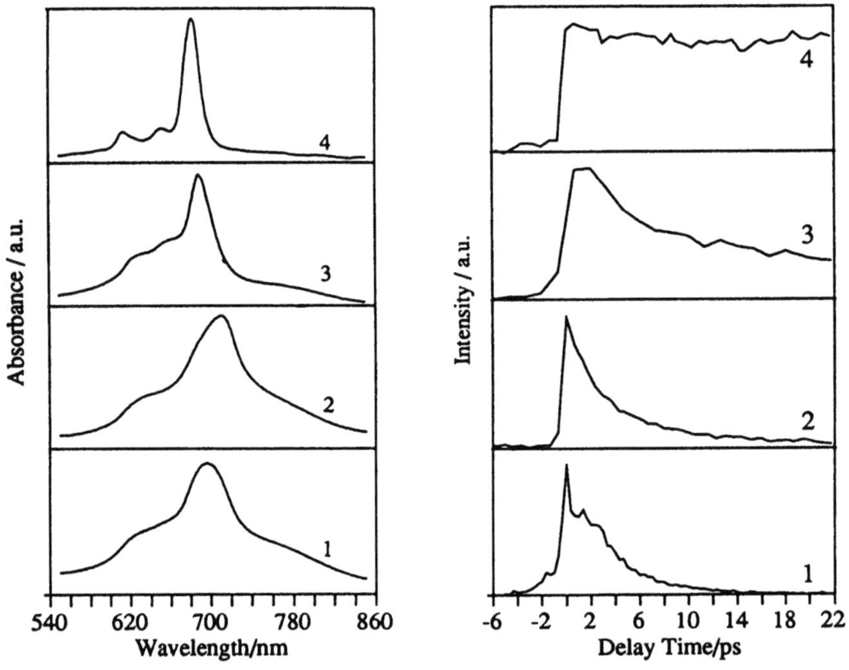

1 Polymer LB-Film 2 Monomer LB-Film 3 Doped Polystyrene
4 Copolymer

Fig. 1: Absorption spectra of the systems 1 - 4

Fig. 2: Diffraction intensity as a function of the time delay

5. Discussion

The comparison of the absorption spectra with the sample preparation shows that an inhomogeneous broadening occurs in densely packed phthalocyanine films. This fact is due to a dipol-dipol interaction [11] between adjacent phthalocyanine rings [10]. The same process is probably responsible for a drastic shortening of the relaxation time of these films. In isolated phthalocyanine molecules fluorescence leads to a decay of the excited state. This gives rise to a large relaxation time on a nanosecond time scale. In densely packed films an energy transfer between the rings is possible. If the absorption wavelength of the molecules is different due to inhomogeneous broadening, this process leads to a fast recovery of the ground state population at the excitation wavelength. Therefore the energy transfer leads to a fast decay of the nonlinear signal in a DFWM experiment.

Besides the systems 1 and 2 other LB-Films of phthalocyanine molecules have been investigated. A further broadening of the spectrum is observed in the LB-film of a metal-free phthalocyanine. This broadening is due to a line splitting that is induced by the lower symmetry of the molecule. The nonlinear susceptibility $\chi^{(3)}$ is again within a factor of two of the value measured for the LB-films of the systems 1 and 2. The nonlinear signal decays monoexponentially with a relaxation time of 6 ps.

Slight changes in the chemical structure of the flexible spacer lead only to minor changes in the optical properties of the phthalocyanine LB-films [10]. Obviously, the flexible spacer does not reduce the electronic coupling between the phthalocyanine rings. Replacement of the flexible spacer by oxygen leads to a reduction of the distance between the phthalocyanine moieties. This results in a further broadening of the spectrum [12,7,10]. The rod-like geometry of the Polyphthalocyaninatosiloxanes induces an alignment of adjacent molecules. The orientation of the rods varies within the film but there is a preferred orientation due to the film preparation. Because of the large absorption anisotropy of the molecules fluctuation of the orientation gives rise to large light scattering in the samples. The intensity of the scattered light exceeds the nonlinear signal so that the nonlinear optical properties of this sample could not be investigated by DFWM.

All samples show thermal degradation at high repetition rates. The maximum repetition rate that can be used without damage depends on the excitation wavelength. For LB-films the corresponding value lies in the range of 50kHz. Polystyrene films doped by phthalocyanine molecules have a lower stability that leads to a degradation above a repetition rate of several KHz. This is due to a melting of the polystyrene matrix at higher rates.

6. Conclusion

The optical properties of phthalocyanine films depend strongly on an electronic coupling between the molecules which can be controlled by the molecular architecture. This interaction leads to a spectral broadening and to a reduction of the relaxation time. Degradation of the films at elevated repetiton rates is a problem.

7. Acknowledgement

This work was supported by the BMFT under project Nr.03M4008E9

8. References

1. T.K.Yee and T.K.Gustafson, Phys.Rev.**A18**, 1597 (1978)
2. G.P.Agrawal, IEEE J.Quantum Electron. 17, 2335 (1981)
3. Z.Z.Ho, C.Y.Ju and W.M.Hetherington III, J.Appl.Phys. **62**, 716 (1987)
4. Z.Z.Ho and N.Peyghambarian, Chem.Phys.Lett.**148**,107 (1988)
5. Garito, private communications
6. J.P.Lloyd, C.Pearson and M.C.Petty, Thin Solid Films **160**, 431 (1988)
7. E. Orthmann and G. Wegner, Angew.Chem.Int.Ed.Engl. **25**, 1105 (1986)
8. G.M.Carter, J.Opt.Soc.Am. **B4**, 1018 (1987)
9. H.Kuhn, D.Möbius and H.Bücher in "Techniques of Chemistry" (A.Weissberger, Ed.) Vol.1 Part IIIB Interscience, New York (1972) 577
10. T. Sauer, PhD Thesis, Mainz (1989)
11. Kasha, M. in "Spectroscopy of the Excited State", B.DiBartolo (Ed.) Plenum Press (1976) 337
12. C.Bubeck, A.Kaltbeitzel, R.W. Lenz, D.Neher, J.D.Stenger-Smith and G. Wegner, in "Nonlinear optical effects in organic Polymers" J. Messier, F. Kajzar, P.N. Prasad and D.R. Ulrich (Eds.) Kluver Acad. Publ.(1989) in press

Third Order Nonlinear Optical Properties of Precursor Route Poly(p-phenylene vinylene)

D.D.C. Bradley and Y. Mori*

Toshiba R&D Center, 1, Komukai Toshiba-cho, Saiwai-ku, Kawasaki 210, Japan
*and Cavendish Laboratory, Madingley Road, Cambridge CB3, 0HE, UK

Abstract. This paper concerns investigations of the third order nonlinear optical response of precursor route poly(p-phenylene vinylene) [PPV]. Previous results for photoexcitation measurements provide information on the response time and the nature of the nonlinearity in the single-photon resonant excitation regime. Transmission third harmonic generation [THG] measurements yield complimentary information on the non-resonant regime.

1. Introduction

Poly(arylene vinylene) polymers of general formula [-Ar-CH=CH-]$_n$ (where Ar is an aromatic or heteroaromatic moiety such as phenylene, napthalene or thiophene cf. figure 1) are one class of conjugated polymers currently under investigation as suitable materials for third order nonlinear optics applications. In addition to satisfying the apparently general requirement for a significant third order optical susceptibility in polymers, namely the possession of a highly delocalised, and hence readily polarisable, electronic structure, the poly(arylene vinylene) polymers have the advantage of being prepared via a synthetic route that allows considerable flexibility for sample preparation [1]. This so-called "precursor-route" involves the synthesis of a solution processible, non-conjugated precursor polymer which can be readily converted to the conjugated material by an appropriate thermal treatment. Fabrication into the desired sample form (e.g. fully dense homogeneous thin films) is carried out prior to the thermal conversion step, thus circumventing the difficulties associated with the rigidity of the conjugated structure. Moreover, suitable modification of the fabrication and conversion conditions allows control over the state of order and sample morphology and enables preparation of highly oriented films [1]. The latter have already been used to study the anisotropic nature of physical processes in these quasi-one-dimensional materials [2] and they offer the possibility for polarisation sensitive operation in device structures. Furthermore, these materials are very robust, having both good environmental and thermal stability and good mechanical properties. This combination of many of the features desirable for practical usage makes the poly(arylene vinylene)s of great interest as test materials for the assessment of the viability of conjugated polymers in nonlinear optical device applications.

Springer Series in Solid-State Sciences, Vol. 91
Electronic Properties of Conjugated Polymers III
Editors: H. Kuzmany · M. Mehring · S. Roth

Structure	Name	Energy Gap
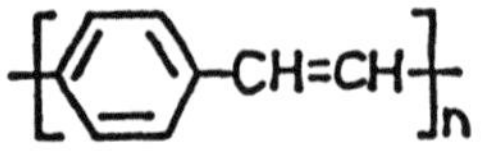	Poly(p-phenylene vinylene) [PPV]	2.5 eV
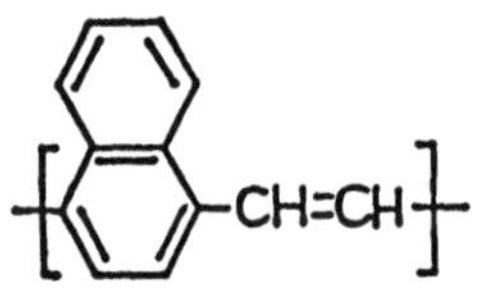	Poly(1,4-naphthalene vinylene) [PNV]	2.3 eV
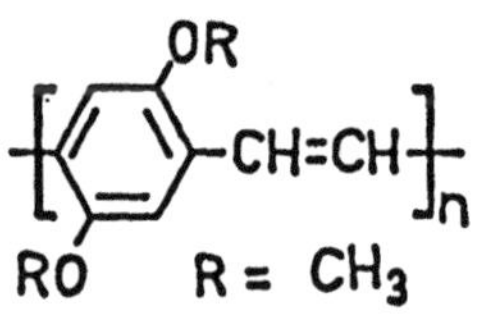	Poly(2,5-dimethoxy-p-phenylene vinylene) [PMPV]	2.1 eV
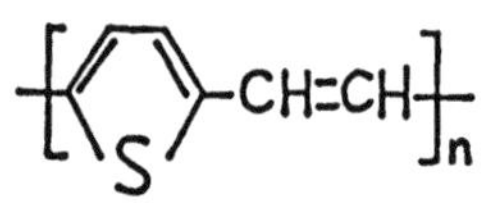	Poly(2,5-thienylene vinylene) [PTV]	1.8 eV

Figure1: Chemical structures, names and energy gaps of several poly(arylene vinylene) polymers

2. Resonant Nonlinear Response

In the single-photon resonant excitation regime the nonlinearity originates from the generation of real excitations of the π-electron system. The nature of these photoexcited states and their generation and recombination processes have been studied via photoinduced absorption, photoconductivity and photoluminescence measurements [2,3,4]. For PPV and, in principle, for all non-degenerate ground state polymers both excitonic and polaronic states can arise. The branching ratio for polaronic states is, however, always low and probably of order a few percent at most, even when polarising the excitation light perpendicular to the chain direction. Exciton formation is the dominant photoexcitation process and the contribution associated with exciton transitions is thus expected to be foremost in determining the resonant nonlinearity. Phase space filling by excitons leading to a saturation of absorption has been advanced as a mechanism to explain the behaviour of polydiacetylene p-toluene sulphonate [5] and a similar mechanism might be expected to apply for other materials including PPV. Assuming that this is the case, we can make an estimate of the recovery time for the resonant nonlinearity based upon measurements of the luminescence decay. The room temperature luminescence in PPV falls off to e^{-1} of its initial value after a time of approximately 60 ps [2] indicating that the excitonic nonlinearity should have a rapid recovery time.

We also expect that there will be a response associated with the generation of polaronic states which, although perhaps less significant in the vicinity of the absorption edge than that arising from excitons, will give rise to an additional nonlinearity deep within the energy gap. The self-localisation of charge during the generation of a polaronic state produces two energy levels that lie within the gap. These are formed by removal of states from the conduction and valence band edges, one state from each for each polaron or bipolaron that is generated. Creation of polaronic states is thus accompanied by the appearance of new optical transitions with large oscillator strength at energies below the absorption edge and by a corresponding bleaching of the interband transition. As such, the photoexcitation of such states is an intrinsically nonlinear response.

Unlike the case of phase space filling by excitons in polydiacetylene [5], the bleaching of the interband transition due to the photogeneration of polaronic states occurs without the necessity of reaching a certain excited state saturation density. In fact, once saturation is reached for polaronic states, the bleaching of absorption ceases. Hence, we might expect the importance of the polaronic contribution to the resonant nonlinearity to be greatest when the excitation intensity is relatively low. An estimate of the recovery time for the polaronic nonlinearity in PPV is obtained from the decay time for the fast component in the photoconductive response. This has been shown to be of order 100 ps or less [3], once again indicating a rapid recovery for resonant nonlinearities in PPV.

3. Non-resonant Nonlinear Response

THG measurements, using a standard transmission geometry, were performed in order to study the single-photon non-resonant response of PPV. Excitation was by 1.064 μm pulses (12ns) from a Q-switched Nd:YAG laser operated at 10 Hz with maximum energy 200 mJ per pulse. The third harmonic generated in the sample was separated from the fundamental using an IR-cut filter, monochromated (355nm) and then detected by a photomultiplier tube whose output was averaged in a digital oscilloscope. The samples used in this study were typically of order 40 to 100 nm thickness and were prepared via the sulphonium polyelectrolyte route as previously described [1]. Such thin films are necessary to avoid excessive absorption of the third harmonic (355 nm) which falls within the π-π^* absorption band of PPV (cf. figure 2). That the observed signal originates from harmonic generation in the film was confirmed by its cubic dependence upon the intensity of the fundamental and by its sharp spectral profile centred at 355 nm. The magnitude of $\chi^{(3)}$ $(-3\omega; \omega, \omega, \omega)$ was determined by comparison of the THG signal measured for the polymer sample on a fused silica substrate with that measured under identical experimental conditions for the substrate alone. Taking $\chi^{(3)}$ for the substrate alone to be 3.1×10^{-14} esu [6] and taking $n(\omega) = 2.1$ and $n(3\omega) = 2.6 + i\,0.22$ for the polymer refractive indices at the fundamental and harmonic frequencies [7] leads, via the analysis of Kajzar and Messier [8], to a value of $\chi^{(3)}$ $(-3\omega; \omega, \omega, \omega) = 1.2 \times 10^{-10}$ esu. This value was obtained

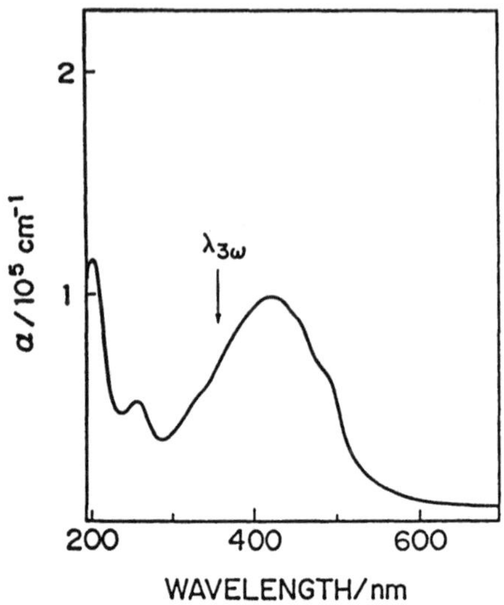

Figure 2: Optical absorption spectrum for fully (i.e. 300 °C) converted PPV. The arrow indicates the third harmonic wavelength (355 nm) for excitation at 1.064 μm

for an unoriented film and is thus an orientational average of the different tensorial components of $\chi^{(3)}$. For a quasi-one-dimensional π-electron system the $\chi^{(3)}$ tensor should be dominated by a single component, namely, $\chi^{(3)}{}_{z,z,z,z}$ which has all indices parallel to the polymer chain axis (z direction). In order to estimate its value we need only multiply the orientationally averaged one by an appropriate numerical factor. With the assumption that the polymer chains lie parallel to the substrate surface but are randomly oriented within the film plane, we estimate $\chi^{(3)}{}_{z,z,z,z} = 3.2 \times 10^{-10}$ esu. This is a large value for a single-photon non-resonant nonlinearity but we note that a very similar value is reported by Bubeck et al. [9].

An important insight into the origin of the large $\chi^{(3)}$ at 1.064 μm can be gained by comparison with the THG results of Kaino et al. [10]. The $\chi^{(3)}$ values that they derive for excitation at wavelengths in the range 1.85 μm to 2.15 μm are significantly smaller than reported here, clearly showing that $\chi^{(3)}$ disperses strongly as the excitation wavelength is increased. Their values, which fall in the range 7.8×10^{-12} esu to 5×10^{-12} esu, are in fact in better agreement with the non-resonant values obtained for other polymers. The dispersion is readily understood if allowance is made for multiple-photon resonance with the π-π* transitions. As already noted above and as illustrated in figure 2, with excitation at 1.064 μm, the harmonic wavelength overlaps the absorption band and hence the resultant $\chi^{(3)}$ value will be enhanced by three-photon resonance. From a closer look at figure 2 we would expect that the maximum three-photon resonantly enhanced $\chi^{(3)}$ should occur for excitation somewhere in the range 1.2 to 1.5 μm (depending in detail on the way in which contributions from different conjugation length segments combine). We can also clearly see why the values reported by Kaino et al. [10] are much smaller. For excitation at wavelengths longer than 1.8 μm the third harmonic falls below the onset of the absorption edge and hence there should no longer be any three-photon resonance effect.

The presence of such three-photon resonances offers the possibility of being able to optimise the single-photon non-resonant nonlinearity at a desired operation wavelength by chemically tuning the energy gap. Moreover, unlike the enhancement achieved through single-photon resonant excitation, we do not expect the three-photon resonance to produce a large increase either in the absorption at the fundamental wavelength or in the recovery time for the nonlinear response. Tunability of the energy gap can be achieved for the poly(arylene vinylene) polymers through, for example, an alteration of the Ar moiety (cf. figure 1) or in a more subtle fashion through preparation of copolymers containing units of the different poly(arylene vinylene) homopolymers.

In order to examine the dependence of the nonlinearity upon conjugation length we have measured THG at various stages in the conversion of a single film from saturated precursor to fully conjugated polymer. The optical absorption spectra as recorded at each stage are shown in figure 3. The spectra contain a number of features that have been assigned in detail elsewhere [1, 11]. The region of interest to the present study is that at wavelengths longer than 300 nm, where the absorption arises from the presence of conjugated sequences generated during the thermal conversion. This long wavelength absorption comprises a superposition of contributions from different conjugation length components as may most readily be seen from the resolved structure at low degrees of conversion. Increasing conversion causes an overall increase in the absorption strength (indicating the formation of a larger

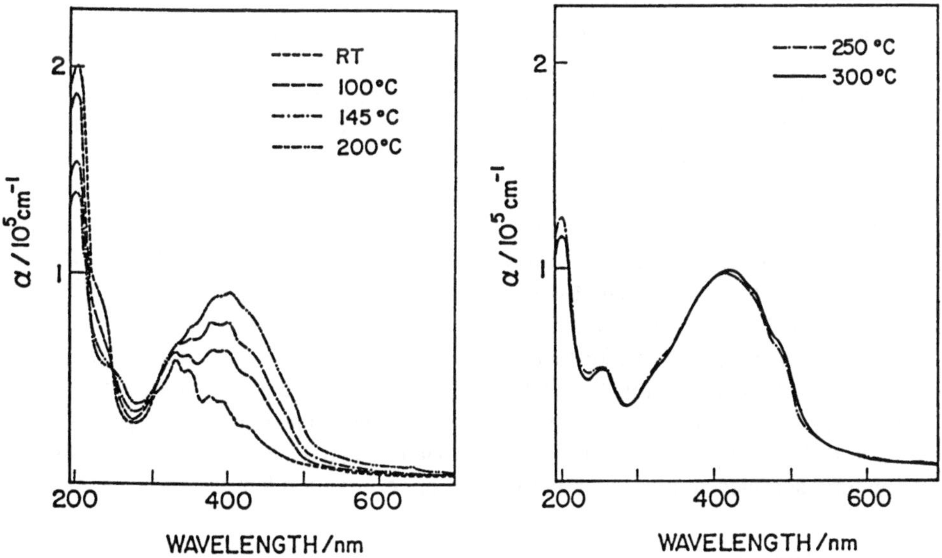

Figure 3: Optical absorption spectra recorded at successive stages in the thermal conversion of a single 45 nm thick film from saturated precursor to fully converted conjugated polymer. With the exception of storage at room temperature (RT), the film was heated for 3 hours at each temperature

Table 1: $\chi^{(3)}$ values derived for a 45 nm thick film at various stages in the thermal conversion (heating for three hours at each subsequent temperature) from precursor to fully converted conjugated polymer. Note that the room temperature (RT) stored film has already undergone partial conversion as seen from the absorption in figure 3 at wavelengths greater than 300 nm

Conversion Conditions	$\chi^{(3)}$ / 10^{-11} esu
RT stored	0.8
100 °C	1.6
145 °C	2.2
200 °C	3.1
250 °C	6.3
300 °C	7.5

number of phenylene vinylene repeat units) and a red shift in its weight (indicating a shift in the conjugation length distribution towards progressively longer conjugated sequences). The corresponding values of $\chi^{(3)}$ determined after each stage of conversion are listed in table 1. These values are uncorrected for harmonic absorption (i.e. derived assuming no absorption at the harmonic wavelength) but since the absorption coefficient at 355 nm changes little (cf. figure 3), the trend that they display should be indicative of what would be seen with more accurately determined values (for which a knowledge of the refractive indices at each stage of conversion is required). Consideration of the effect of three-photon resonances suggests that the $\chi^{(3)}$ values at higher degrees of conversion are underestimated relative to those at low conversion. This situation arises because the contributions from progressively longer conjugated sequences are expected to be increasingly less resonantly enhanced (due to the red shift of their absorption away from the the third harmonic wavelength). Hence, a more representative comparison e.g. for measurements with excitation at wavelengths of 2μm or longer, might be expected to show a greater increase in $\chi^{(3)}$ during conversion. Nevertheless, the observation of a $\chi^{(3)}$ that increases with increasing conversion is entirely consistent with an assignment of the origin for the nonlinearity to the conjugated π-electron system.

Ideally, we would also like to be able to determine the functional form of the dependence on conjugation length for comparison with theoretical calculations. In order to undertake such a determination it would seem to be necessary to prepare a set of well-defined oligomer samples since an extensive analysis is not really feasible for polymeric samples containing a distribution of conjugation lengths.

4. Conclusion

In this paper we have reported our initial studies of the nonlinear response in PPV which seem encouraging for the possible future use of PPV in optical devices. Further work is needed in order to accurately determine the magnitude of the single-photon resonant response and to clarify the relative importance of the contributions arising from both excitonic and polaronic states (picosecond time resolved transient absorption and bleaching measurements should prove especially useful in this regard). Similarly, more work is needed in order to fully understand the role of multiple-photon resonances in determining the $\chi^{(3)}$ dispersion in the single-photon non-resonant excitation regime. Finally, detailed measurements should also be made of the polarisation dependence and of the conjugation length dependence for well defined oligomeric samples so as to improve the understanding of the origins of the single-photon non-resonant response.

5. Acknowledgements

DDCB thanks Toshiba Corporation for the provision of a Toshiba Fellowship that enabled his stay in Japan and Corpus Christi College and Unilever plc for support through the Unilever Research Fellowship in Chemical Physics.

References

[1] D.D.C. Bradley : J.Phys. D: Appl.Phys. 20, 1389 (1987) and references therein
[2] R.H. Friend, D.D.C. Bradley and P.D. Townsend : J.Phys. D: Appl.Phys. 20, 1367 (1987) and references therein
[3] D.D.C. Bradley, Y.Q. Shen, H. Bleier and S. Roth : J.Phys.C: Solid State Phys. 21, L515 (1988)
[4] N. F. Colaneri, D.D.C. Bradley, R.H. Friend and C.W. Spangler : these proceedings
[5] B.I. Greene, J. Orenstein, R.R. Millard and L.R. Williams : Phys.Rev.Lett 58, 2750 (1987)
[6] F. Kajzar, J. Messier and C. Rosilio : J. Appl. Phys. 60, 3040 (1986)
[7] Derived from EELS data. J. Fink : private communication
[8] F. Kajzar and J. Messier : Thin Solid Films 132, 11 (1985)
[9] C. Bubeck, A. Kaltbeitzel, D. Neher : these proceedings [10] T. Kaino, K-I. Kubodera, S. Tomaru, T. Kurihara, S. Saito, T. Tsutsui and S. Tokito : Electronics Lett. 23, 1095 (1987)
[11] J. Obrzut and F. E. Karasz : J. Chem. Phys. 87, 2349 (1987)

Development and Application of Optically Nonlinear Polymers

G.R. Möhlmann

Akzo Research Laboratories Arnhem, Corporate Research Department,
P.O. Box 9300, 6800 AB Arnhem, The Netherlands

Abstract Optically nonlinear materials change their refractive index if placed in external (electrical, optical, etc.) fields, and can be used to manipulate light thus permitting optical processing, etc. Amorphous side chain polymers have been developed, showing attractive optically nonlinear properties. The value for the electro-optic coefficient r_{33}, achieved in polymers made by Akzo, equals 28 pm/V; for comparison, in single crystalline lithium niobate the r_{33}-value is equal to 32 pm/V. Device applications of optically nonlinear polymers will be discussed. Attention in this article is focused on electro-optic (second order nonlinear) effects.

1. Introduction

Materials in external electric fields possess a separation of the centres of gravity of their positive and negative charges. For small electric field strengths (E), this charge separation, also called the induced electric polarization (P_i), varies linearly with E; for large E values, P_i may depend nonlinearly on E [1].

Materials with nonlinear P_i's show optically nonlinear effects. Such effects may lead to variations of the refractive index (n), providing a mechanism to manipulate light (phase change). Devices, including optically nonlinear materials and possessing the desired optically nonlinear functions (phase modulation, intensity modulation, spatial optical switching, etc.) enable optical processing functions. Applications are optical telecommunications, optical storage, image recognition, optical computing, etc.

To produce integrated devices for optical processing, optically nonlinear materials with easy processing properties and large nonlinear coefficients are required. It will be shown that modified side chain polymers are attractive candidates for optically nonlinear materials. Optically nonlinear polymers have been produced and simple devices, showing electro-optic effects, have been made from such polymers.

Second and third order optically nonlinear effects are known to take place in organic substances; the second order nonlinear coefficients being the largest.

Springer Series in Solid-State Sciences, Vol. 91
Electronic Properties of Conjugated Polymers III
Editors: H. Kuzmany · M. Mehring · S. Roth

2. Nonlinear Polarizations

The dependence P_i on E can be expressed as follows:

$$P_i = X^{(1)}.E + X^{(2)}.E.E + X^{(3)}.E.E.E + \ldots \qquad (1)$$

in which $X^{(1)}$ and $X^{(2)}$, $X^{(3)}$ are the linear and nonlinear electric susceptibilities, respectively.

Writing equation 1 slightly differently, one obtains:

$$P_i = E\,[\,X^{(1)} + X^{(2)}.E + X^{(3)}.E.E + \ldots\,]\,. \qquad (2)$$

In the optical domain, $X^{(1)}$ is connected to n via $X^{(1)} = n^2 - 1$. Since the terms between the square brackets of equation 2, have the same dimension, materials with $X^{(2)}$, $X^{(3)}$, etc. unequal to zero show a change in n in electric fields.

The n of $X^{(2)}$-materials depends on E; this is called the (linear) electro-optic or Pockels-effect. Besides, such materials also show other nonlinear effects like: frequency doubling, frequency mixing and optical rectification.

$X^{(3)}$-materials show the quadratic electro-optic effect; here n can also be changed light, since the intensity of light depends quadratically on the optical electric field amplitude.

3. Optically nonlinear Materials

Well-known nonlinear materials are the inorganic single crystals and semi-conductors like KDP, lithium niobate, gallium arsenide, etc.

To achieve large and fast nonlinear effects, charges should undergo significant displacement and have a small mass. In inorganic single crystals, heavy and localised ions are to be displaced. Better materials are those with conjugated π-electron systems, in which the small electronic masses can be widely displaced. Organic materials include molecules with conjugated π-electron systems like polyenes, polyphenylenes, etc.

4. Organic nonlinear materials

Organic molecules like ureum, 1,4-nitro-aniline, 2-methyl-4-nitro- aniline and 4-dimethylamino-4'-nitrostilbene (DANS), etc. show considerable optically nonlinear effects [2].

Nonlinear effects increase with the length of the conjugated system. Huijts [3] showed that $X^{(2)}$-effects depends cubicly on the conjugated chain length. Third order effects ($X^{(3)}$), are known to scale with the fifth power [4].

For organic naterials, the nonlinear properties can also be described on the molecular level. Here, P_i is to be replaced by the induced dipole moment (μ_i). The relation between μ_i and E is:

$$\mu_i = \alpha.E + \beta.E.E + \Gamma.E.E.E + \ldots \qquad (3)$$

in which α is the polarizability of the molecule, and β, Γ, etc. the molecular hyperpolarizabilities.

Every material possess an α and Γ, and consequently a $X^{(1)}$ and $X^{(3)}$, unequal to zero. However, for materials to possess β (or $X^{(2)}$) unequal to zero, the system is not allowed to exhibit centrosymmetry.

In the case of molecules, centrosymmetry can be removed by asymmetric substitution. On the macroscopic level it is more difficult to control the noncentrosymmetry. In single crystalline material, the noncentrosymmetry must be spontaneously present, in polymeric materials, the noncentrosymmetry can be induced in a later stage.

Organic nonlinear materials are available in different forms as briefly discussed below.

Single crystals Single crystals must be grown with sufficient optical quality and size. However, the material is difficult to process (realisation of waveguides, etc.) and to integrate in systems. Application will predominantly be in the field of frequency doubling, for which the least processing of the material is required.

Langmuir Blodgett films Here, (time consuming) successive deposition of many monolayers (while maintaining noncentrosymmetry) is required to provide waveguiding and nonlinear properties.

Polymeric Solid Solutions Optically nonlinear molecules are mixed with a polymer host. Heating the sample to above the glass transition temperature (Tg) in a strong electric field alignes the permanent dipoles of the active molecules yielding noncentrosymmetry. Disadvantages are the low maximum concentration of molecules, segregation of the dye molecules into (centrosymmetric) crystallites, and the tendency for orientational relaxation after poling.

Main Chain Polymers To prevent segregation and relaxation, the optically active molecules may be incorporated in the polymer back bone. However, the viscosity of such materials is so high that inducing polar order by electric field is very difficult.

Side Chain Polymers Here, the active molecules are attached as side chains to the polymer back bone. Large amounts of molecules can be incorporated, combined with a reduced tendency for orientational relaxation. Electric field poling can be performed above Tg. The processibility of this type of molecules is very attractive.

5. Side Chain Polymers

Side chain polymers are believed to offer the best solution to achieve materials for nonlinear applications. Attractive properties are:

- large concentration of optically nonlinear groups,
- fast nonlinear response to applied modulating fields,
- "easy" processing via spin coating, dipping, etc.,
- possibility for thin film formation, permitting optical waveguiding,
- compatibility with glass, metals, plastics, Si, III-V materials,
- low temperature processing, allowing integration with IC's, etc.,
- low dielectric constants, permitting use of fast electronics,
- molecular tailorability permits choice of Tg, hardness, colour, etc.

Side chain polymers with Tg's in the range 50-160°C have been produced, withstanding electrical poling fields of about 150 V/μm.

6. Applications in Nonlinear Devices

In order to achieve polymeric nonlinear devices, multilayer structures including nonlinear polymers must be produced. The exact configuration of such structures depends on the desired effect and whether the light is guided or unguided (free space).

A typical example of a multilayer structure is shown in Figure 1. The required processing conditions resemble those of microelectronics. On the substrate, the lower electrode for poling and switching is evaporated or sputtered. Then, an optical (polymeric) buffer layer is deposited by spin coating. This layer prevents undesired interaction (absorption, scattering) of the guided light with the electrode.

The next layer consists of nonlinear polymer; this layer, with the highest refractive index, serves as the wave guide. All the light intensity is not necessarily confined to this region. An upper buffer and top electrode completes the multilayer. The layer thicknesses are of the order of microns.

The structure in Figure 1 resembles a slab waveguide. Monomode channeled wave guides have been produced in the optically active layer exhibiting optical transmission losses smaller than 1 dB/cm @ 1.3 μm. Heating the complete structure above Tg in an electric field of 138 V/μm results in r_{33} = 28 pm/V [5]. Larger nonlinear constants are expected if poled at higher field strengths.

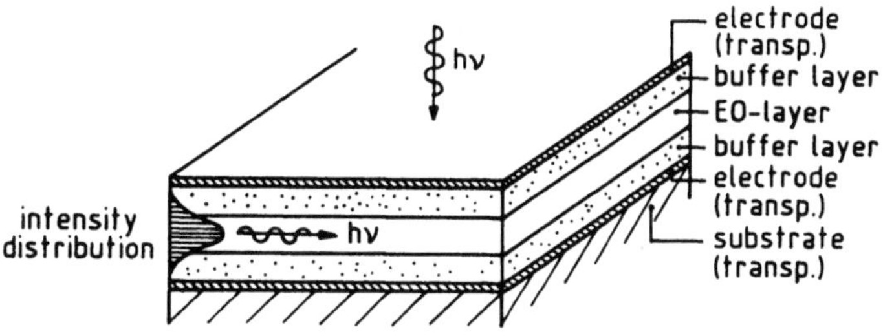

Fig. 1 Schematic representation of a slab waveguiding multilayer

By shaping the lower and/or upper electrodes, matched to the generated wave guides, electro-optic functions/devices can be made.

Concluding, one may say that optically nonlinear side chain polymers have been produced, providing electro-optic coefficients of the same order as that of lithium niobate. Low loss channel wave guides have been made [5].

References

[1] D.S. Chemla and J. Zyss, eds., "Nonlinear Optical Properties of Organic Molecules and Crystals", Vol. 1, Academic Press, Inc., Orlando (1987).

[2] D.J. Williams, ed., "Molecular Optical Properties of Organic and Polymeric Materials", ACS Symposium Series 233, Am. Chem. Soc., Washington (1983).

[3] R.A. Huijts and G.L.J. Hesselink, Chem. Phys. Letters, Accepted for publication.

[4] A.F. Garito, J.R. Heflin, K.Y. Wong and O. Zamani-Khamiri, Proceedings SPIE Conference on Nonlinear Optical Properties of Organic Materials, Vol. 971, 2 (1988).

[5] G.R. Möhlmann, W.H.G. Horsthuis, M.B.J. Diemeer, F.M.M. Suyten, E.S. Trommel, A. Mc.Donach and N. Mc.Fadyen, Proceedings of the Topical Meeting on Nonlinear Guided Wave Phenomena: Physics and Applications, Houston, February 1989, Optical Society of America, Washington, D.C. 1989.

Nonlinear Spectroscopy on Gap States of trans-Polyacetylene

K. Fiedler [1] *and P. Knoll* [2]

[1] Institut für Experimentalphysik, Universität Graz, Universitätsplatz 5, A-8010 Graz, Austria
[2] Alexander von Humboldt Foundation Fellow, Max-Planck-Institut für Festkörperforschung, Heisenbergstr 1, D-7000 Stuttgart 80, Fed. Rep. of Germany

1. Introduction

Nonlinear optical processes used in spectroscopy can provide considerable insight into the electronic structure of conjugated polymers. Such nonlinear optical investigations are in some way complementary to the usual linear optical experiments, and in the case of trans-polyacetylene one can expect to obtain more detailed information about the electronic structure, e.g. to what extent the one-electron approximation will hold. An example of the failure of one-electron molecular orbital theory are finite polyenes [1]. The simple theory predicts a 1^1B_u symmetry state as the lowest-lying excited state, but experiments showed that there is a 2^1A_g state below the 1^1B_u, i.e. the e-e- correlation has to be considered. It is still an outstanding problem whether this behaviour can be extrapolated to infinite and three-dimensionally coupled polyenes (polyacetylene) and also exactly where this 2^1A_g state is located [2]. In this work we want to describe a four-wave mixing experiment, which should offer the possibility of scanning the whole gap in order to investigate gap states by their symmetry.

2. Suggested nonlinear optical process

Because of the inversion symmetry of the idealized trans-polyacetylene chain, single-photon experiments are only allowed for transitions between odd and even states. Within the dipole approximation odd-odd and even-even transitions are strictly forbidden and cannot be seen in single-photon processes. An even order process (as e.g. two-photon absorption) is required in order to detect transitions between two levels of the same symmetry type. So far, photo-induced absorption (PA) in trans-polyacetylene has been interpreted as a single-photon process on a polymer chain excited by a pump laser. In some way PA can also be understood as a second-order process where the one-photon absorption becomes stronger if the two-photon sum or difference process comes in resonance. This mechanism corresponds to the common two-photon absorption method [3]. It is a sophisticated problem to distinguish between these two mechanisms, but the independence of PA peaks from pump laser energy well above the optical gap favours the idea of transitions between the ground state and a photoinduced state of different symmetry. In agreement with these considerations, the 0.45 eV peak of PA has been assigned [4] to a charged soliton pair which changes the symmetry of the undisturbed chain. There are two

Springer Series in Solid-State Sciences, Vol. 91
Electronic Properties of Conjugated Polymers III
Editors: H. Kuzmany · M. Mehring · S. Roth

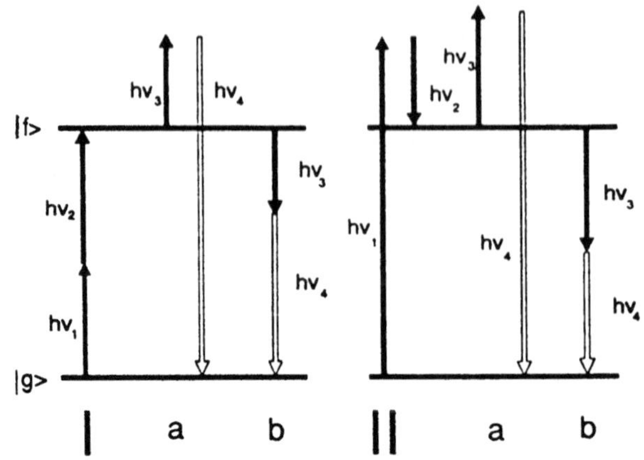

Fig. 1: Four examples of two-photon resonant third-order nonlinear optical processes

ways to occupy higher energy states by a two-photon process, which are indicated by I belonging to the sum-frequency absorption and II belonging to the difference-frequency absorption [5] in Fig.1. The successful occupation of this level can be probed by a second two-photon process, which is either an anti-Stokes Raman process (case (a) in Fig.1) or a difference process (case (b) in Fig.1). $h\nu_4$ denotes the observed radiation. Some types of combinations are of special interest. Case I(a) reduces to third-harmonic generation if the incident photons $h\nu_1$, $h\nu_2$, and $h\nu_3$ become equal. The process II(a), which in its general form is a two-photon difference absorption with up-conversion, becomes the usual CARS process if $h\nu_1$ and $h\nu_3$ are chosen to be equal. Process I(a) has been applied to trans-polyacetylene [6] and resonances have been assigned to a two-photon and a three-photon absorption at the band edge, but at the position of the latter a two-photon resonance at 2/3 of the gap could also be situated. In order to obtain additional information and to clarify the symmetry of the gap states we suggest the use of a process of type II(a), which corresponds to electronic CARS in the case of resonance with a state of the same symmetry type as the ground state. The third-order nonlinear susceptibility is given in its general form in ref.7, but for the special process only the resonant part of $\chi^{(3)}$ is of interest and can be written as given in ref.8. A group-theoretical treatment of trans-polyacetylene (point group C_{2h}) shows that a matrix element $\langle f|er_k|n\rangle$ can only be different from zero if $|f\rangle$ and $|n\rangle$ have opposite parity. Therefore the whole expression of $\chi^{(3)}$ containing terms like $\langle f|er_i|n\rangle\langle n|er_j|g\rangle$ vanishes if $|f\rangle$ and $|g\rangle$ are not of the same symmetry type. If the ground state $|g\rangle$ is assumed to belong to the totally symmetric irreducible representation A_g, such a process probes only states $|f\rangle$ of type A_g where the transition works over intermediate states $|n\rangle$ of B_u symmetry. For these considerations, the symmetry of the ground state is assumed to be that of the atomic arrangement. Within this picture the symmetry of photogenerated states in polyacetylene can be determined by taking the symmetry of the distorted chain. Neutral soliton pairs and bipolarons do not change the symmetry of trans-polyacetylene and belong to A_g irreducible representations. A transition from the ground state of the undisturbed trans-polyacetylene chain to the lowest energy state (ground state) of a chain with one of the mentioned defects should, therefore, be detected by our suggested four-wave experiment. In contrast, single solitons, a positive-negative soliton pair and polarons distort the chain to C_{2v} symmetry which belongs to a B_u irreducible representation in a C_{2h} point group and such states should be observed in PA measurements. Beside the photogeneration of A_g and B_u states also the occurrence of an A_g state due to electronic correlations is expected. The following results can be expected from our suggested four-wave mixing experiment:

a) The experiment leads to the same results as the usual optical absorption. This means that there are no A_g states within the gap. This result is not expected since it would only make sense if all laser energies are well below the energy gap so that no photoinduced states are generated, and no electronic correlation effects occur.

b) The experimental results are different to the usual optical absorption but qualitatively the same as in PA. The same photoinduced states are seen, but the breakdown of the symmetry selection rules suggests that these states are produced in regions of high disorder.

c) If the results are different to PA, the observed resonances occur from states in surroundings with a certain inversion symmetry. The A_g symmetry of these states is proved.

3. The problem of phase-matching in trans-polyacetylene

The advantage of the suggested four-wave mixing experiment is the use of lasers and detectors working in the visible light region. The lasers can easily be tuned over a considerable range with a few suitable dyes in order to cover the whole energy range of the electronic gap. Unfortunately, all laser frequencies are in the region of strong absorption, which enhances the problem of the stability of the polymer in the same way as the problem of phase-matching. In materials with high absorption the only way to observe nonlinear optical signals is a reflection geometry. We studied the problem of phase-matching in such highly absorbing materials in reflection geometry for the simple case of third-harmonic generation, which was quantitatively investigated on polyacetylene by Krausz et al. [9]. A mathematical expression for the intensity of the third harmonic at the point z is

$$I \approx {\chi^{(3)}}^2 e^{-2\beta_{3\omega}z} \frac{1+e^{-2(3\beta_\omega-\beta_{3\omega})d}-2e^{-(3\beta_\omega-\beta_{3\omega})d}\cos((3k_\omega-k_{3\omega})d)}{(3k_\omega-k_{3\omega})^2+(3\beta_\omega-\beta_{3\omega})^2} \quad . \qquad (1)$$

It is assumed that the incoming and outgoing radiations are plane waves with complex wave vectors $k_\omega+i\beta_\omega$ and $k_{3\omega}+i\beta_{3\omega}$, respectively. In Equ.1 only the z-components of the complex wave vectors enter, which are perpendicular to the surface of a film of thickness d. Transmission and reflection geometry are simulated by setting the z-components of the wave vectors of incoming and outgoing waves with the same or opposite signs. Additionally, the place of observation has to be changed from the back side (z=d) for transmission geometry to the front side (z=0) for reflection geometry. In the limit of non-absorbing materials ($\beta_\omega=\beta_{3\omega}=0$) Equ.1 reduces to the well-known formula given by Shen [10]. Taking the values for $k=2\pi/n$ and β at 1064nm and 354.7nm [11] the intensity in a 45° reflection geometry turns out to be a factor 1.5 larger than in collinear transmission. As, in contrast to ref.9, the interference of the reflected fundamental wave is excluded, this enhancement is due to the better phase-matched condition in non-collinear reflection geometry.

4. Experimental results

In this paper experimental results are presented in order to prove the theoretical investigations of section 3. On a thin film of polyacetylene (d=250 nm) the third harmonic of a 0.5 $MWcm^{-2}$ Nd-YAG (1064 nm) pulse (8 nsec) was measured in transmission and reflection geometry. As Fig.2 shows, the intensity obtained in collinear transmission is about a factor 2 less than in 45° reflection geometry. This is in quite good agreement with the theoretical prediction given in section 3.

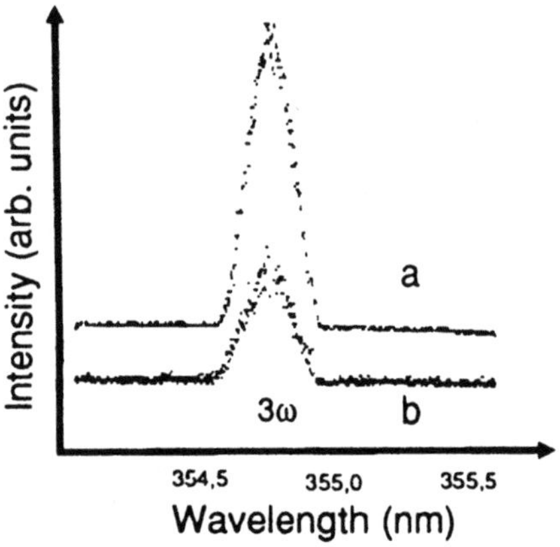

Fig. 2: Intensity of the TH signal in 45° reflection (a) and collinear transmission (b) geometry

Acknowledgements

We would like to thank Dr. G. Leising for providing us with samples and Prof. J. Brandmüller, Prof. F.R. Aussenegg and Dr. A. Leitner for valuable discussions.

References

1. B.E. Kohler, Ch. Spangler, C. Westerfield, J. Chem. Phys.89, 5422(1988)
2. B.E. Kohler, J. Chem. Phys.88, 2788(1988)
3. H. Mahr, in H. Rabin and C.L. Tang, eds., Quantum Electronics, Academic Press, New York, 1975
4. P.D. Townsend, R.H. Friend, J. Phys. C20, 4221(1987)
5. D.C. Hanna, M.A. Yuratich, D. Cotter, Nonlinear Optics of Free Atoms and Molecules, Springer-Verlag, Berlin, 1979
6. W.-S. Fann, S. Benson, J.M.J. Madey, S. Etemad, G.L. Baker, F. Kajzar, Phys. Rev. Lett.62, 1492(1989)
7. N. Bloembergen, H. Lotem, R.T. Lynch(Jr.), Indian J. Pure Appl. Phys.16, 151(1978)
8. pg. 243 of ref.10
9. F. Krausz, E. Wintner, G. Leising, Phys. Rev. B39, 3701(1989)
10. Y.R. Shen, The Principles of Nonlinear Optics, Wiley, 1984
11. G. Leising, Synth. Met. 28,D215(1989)

Part VIII

Molecular Electronics

Concepts of Molecular Information Storage

M. Mehring

2. Physikalisches Institut, Universität Stuttgart,
D-7000 Stuttgart, Fed. Rep. of Germany

Abstract. I will discuss some basic concepts of molecular information storage with specific emphasis on bistable electron transfer arrangements (**BETA**). In particular I propose a **polaron storage supermolecule.**

1. Introduction

In recent years a tremendeous increase in storage capacity (4 MBit)of semiconductor random access memory (RAM) devices has been achieved.Moreover, it seems to be feasible to increase the information density still further by standard semiconductor technology. Nevertheless there is a limit on physical grounds, when the coherence length of the electrons exceeds the spacings of the intra- and intercell structure. Novel techniques like "quantum wells" etc. may overcome these obstacles and may be utilized for information storage in the future.

There might be, however, an alternative route to higher information density and even processing power by working on the molecular level. There has been a number of speculations along these lines recently [1-4]. In this contribution I want to draw attention to these ideas and discuss some principles concerning "molecular information storage" and introduce some concepts which might lead to molecular storage devices. The ideas presented here were discussed before on a number of different occasions [4].

The key feature of molecules which makes them favourable for information storage is the strong localization of electrons on the molecular skeleton. This allows, at least in principle, to use a single molecule as information carrier. It is obvious that a tremendous storage capacity of e.g. 10^{12} Bit/mm^2 could be achieved in this case for a two-dimensional cell. Even if a cluster of e.g. 1000 molecules is used as the elementary storage cell, carrying one Bit of information, the very large storage capacity of 10^9 Bits/mm^2 is attainable. For molecular information storage therefore the two features, namely

- **localization** and
- **storage capacity**

are intimately connected. In the following I want to discuss a few bacic aspects of molecular information storage (section 2) and will emphasize electron transfer

Springer Series in Solid-State Sciences, Vol. 91
Electronic Properties of Conjugated Polymers III
Editors: H. Kuzmany · M. Mehring · S. Roth

molecular switches in section 3. There the concept of the "polaron storage molecule" will be presented. In section 4 a brief outlook on "neuro-molecular networks" will be given.

2.Basic concepts of molecular information storage

Molecules offer a vast variety of features which can be utilized to store one bit of information. Fig.1 lists a few of those features which might be used for such a purpose. They represent a selected number of elementary molecular processes. The spin orientation (Fig.1(a)) is in fact used today in any magnetic storage device. There, however, a large number of spins is necessary to store one bit of information. It might be possible, however, to switch the spin state of a single molecule (or a cluster of e.g. 1000 molecules) and use this for information storage.

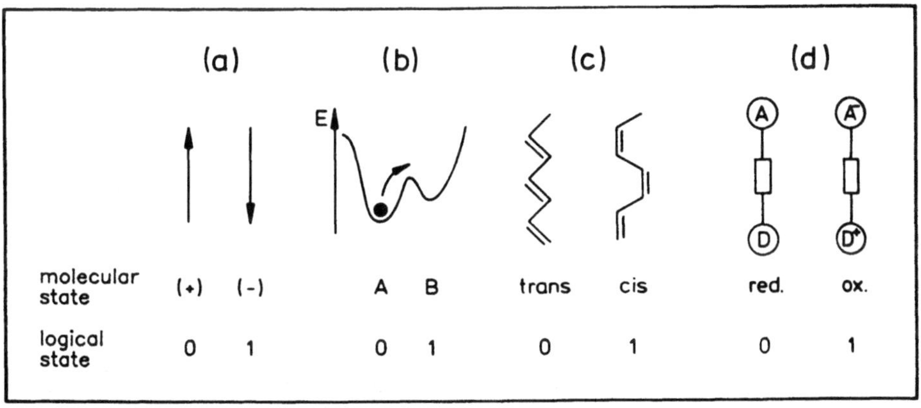

Fig.1:Possible bistable molecular mechanisms for information storage (see text)

A similar two-state process is the switching of a hydrogen bond (Fig.1(b)).It can be represented as a fictitious spin 1/2. Switching could be achieved with electric fields, since an electric dipole moment is connected with each asymmetric hydrogen bond. Also cis-trans isomer states (Fig.1(c)) show two-state behaviour which may be detected by their different optical absorptions. Finally, donor-acceptor supermolecules (Fig.1 (d)) seem to promise interesting switching capabilities which will be discussed in detail in the next section. Electron transfer storage allows in addition to store more than one bit of information on a single molecule when multi-electron transfer is used [5].

All these elementary molecular functions are well known and could possibly be used in a molecular storage device. However, the key problem which still has to be solved is how to address a cluster of molecules or even a single molecule. One possibility is optical excitation and readout. In order to address only a selective number of molecules "frequency selection" [2] and/or "sub-wavelength transmission" by molecular waveguides could be applied. Direct electrical addressing might be feasible by "nano-wires" which can be structured by modern micro-lithography. Finally the principle of the "scanning tunneling microscope" might be applied to address single molecules. Parallel addressing by a large number of concurrently operating "tips" may be technically feasible in the future too.

The basic molecular storage cell may consist of a bistable or multistable molecular array which may be addressed directly by light or electrical fields and receives and transmits information from and to neighbouring cells by "molecular wires". These molecular chains could be made of different conjugated polymers which can carry charged excitations such as solitons, polarons and bipolarons. Intercell information exchange could be perfomed by transmitting such excitations.

A number of supermolecules has been proposed and investigated,which try to mimic the charge transfer in phothosynthetic systems[6-11]. However, no bistable charge separation was achieved.

3. Electron transfer storage

The transfer of a single electron to an individual molecule is the most important elemental step in molecular information storage. It highlights the strong localization of electrons in molecules which can lead to the high density information storage, mentioned in the introduction. It is a well-known fact that electron transfer to a molecule can be achieved either by light or directly from an electrode. In Fig.2 I want to point to a specific example where single and double electron transfer from an electrode to a zwitter-viologen molecule is shown [12]. The viologen molecules accept a maximum of two electrons in consecutive steps when the electrode potential increases in the negative direction. When making this potential more positive again the two electrons are released as is evident from the cyclic voltammogram shown in Fig.2 (top). For the zwitter-viologen there is a difference when increasing the potential in the negative direction. There two electrons are transferred to the molecule in one step when the electrode becomes negative enough. With increasing (positive direction) electrode potential the two electrons are released consecutively as is evident from Fig.2. The reason for his unsymmetrical behaviour is the conformational change of the zwitter-viologen accompanying the two-electron transfer.

This special feature of the zwitter-viologen could be utilized for an elemental information storage as is outlined in Fig.2(bottom). Let us suppose that the molecule is initially in the zero-electron state and a constant negative bias voltage of e.g. -0.5 V is applied. A negative SET pulse would switch the molecule into the one-electron state via the two-electron state. This state can be monitored e.g. by the change in optical absorption and the presence of an electron spin as seen in an ESR experiment[12]. A positive RESET pulse extracts

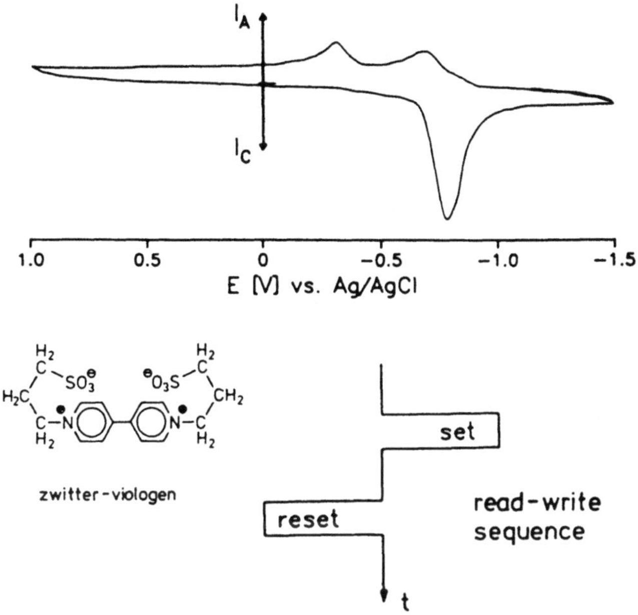

Fig.2: Cyclic voltammogram of a zwitter-viologen (lower trace). SET and RESET operations are indicated (according to [12])

the electron from the molecule and puts it back into the zero-electron state. There are a number of molecules which show similar features and even multi-electron transfer which could possibly be utilized for molecular information storage[5].

Another approach to molecular information storage is the synthesis of "supermolecules" consisting of a donor-bridge-acceptor(D-B-A) unit. The donor may be excited by light resulting in a donor excited state which permits electron transfer via the bridge-molecule to the acceptor. The resulting state D^+-B-A^- may be visualized as a logical "1" state. However, this state has a higher energy than the ground state D-B-A (logical "0"). Due to tunneling processes it is very likely that the D^+-B-A^- state will decay very rapidly into the ground state and therefore loose its information[6-11]. Of course, one may argue that even modern semiconductor RAMs have to be refreshed every few milliseconds and a corresponding "refresh cycle" could be provided also in molecular storage. One should be aware, however, that the timescales for forward and backward electron transfer are not so much different in these "supermolecules" to warrant a stable operation in a molecular storage device. We are therefore seeking a **bistable electron transfer arrangement(BETA)** of donor and acceptor molecules.

As an example of such a **BETA** molecule I propose a "polaron storage molecule" consisting of a donor molecule, a polaron supporting bridge molecule and an acceptor molecule, all with special "redox states" in order to allow a bistable operation as is outlined in Figs.3 and 4. It is important that the HOMO state of the donor molecule is higher in energy than the valence band (VB) of the bridge molecule. Excitation of the donor results in electron tunneling into the conduction band (CB) of the bridge. Due to electron-phonon coupling and electron-electron correlation a polaron P^- is formed in the bridge and can travel to the acceptor where it tunnels to the LUMO of the acceptor molecule provided this has lower energy than the polaron. Under this condition the polaron is

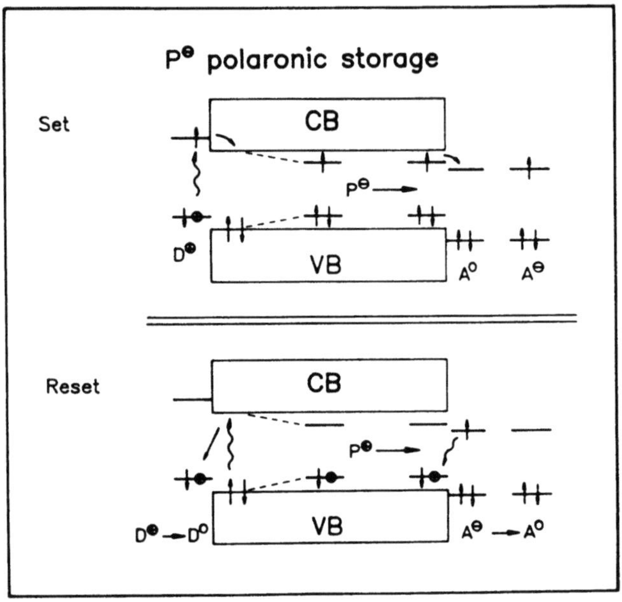

Fig.3: Quantum states for a polaron storage molecule as defined in the text. A special arrangement of donor and acceptor energy levels with respect to the bridge molecule is required for bistable operation

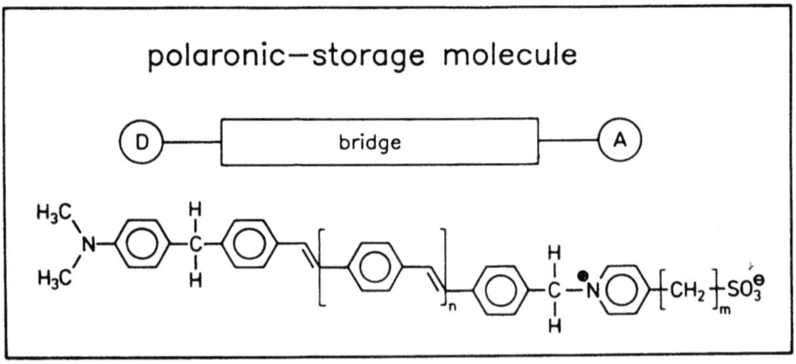

Fig.4: Proposed polaron storage molecule (see text)

trapped at the acceptor molecule and cannot tunnel back to the donor, a necessary requirement for a bistable operation. This process corresponds to a SET operation.

The RESET operation is initiated by interband excitation of the bridge molecule. This immediately leads to the tunneling of the donor hole D^+ to the bridge, forming a polaron P^+ which can travel to the acceptor molecule A^- and discharge the acceptor molecule by recombination. An alternative scenario is possible by exciting the acceptor molecule which then emits a positive polaron to the donor. A possible realization for such a "polaron storage molecule" is presented in Fig.4.

The donor consists of a dimethylanilin molecule and the acceptor is a pyridine cation derivate including counterion. The bridge is made of phenylen-vinylen units which can support polaron transport. An alternative bridge would consist of a polyphenylene chain. It is important to decouple the donor and acceptor molecules from the bridge by "saturated" molecular segments, e.g. by CH_2-groups in the simplest case. Such a storage molecule is currently being synthesized by Müllen and co-workers[13]. A related concept of polaron storage,not within a supermolecule but between different molecular chains was proposed by Zakhidov[14].

4. Neuromolecular networks

Cerebral information storage and processing has stimulated scientists to mimic these mechanisms in artificial devices. Neuronal networks and neuronal computing are current topics of widespread interest among engineers and scientists[14]. See also the lectures of Goser and Mahler at this meeting. Here we want to concentrate on the question if molecular networks could function as

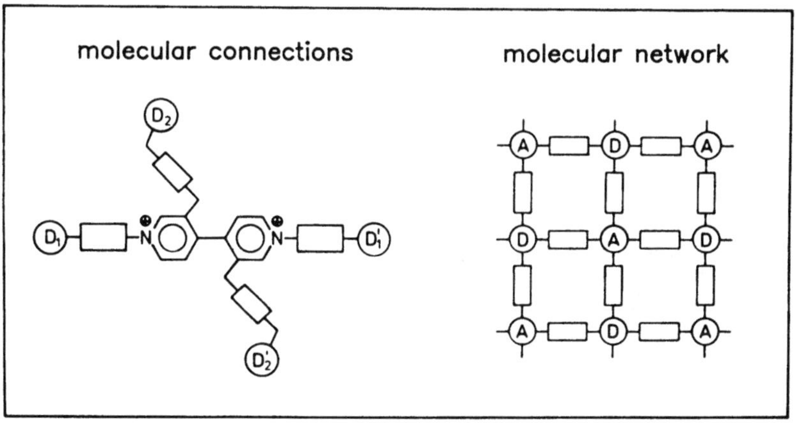

Fig.5: Molecular connectivities (left) and molecular network (right)

synergetic units, performing complex functions on a molecular level. Space does not permit to go into detail and only a few remarks should be sufficient.
Fig.5 hints at the possibility to interconnect molecular active units. There is usually a number of possibilities to connect bridge molecules to an active molecule examplified for a viologen molecule in Fig.5. Electron transfer can be initiated from different donor molecules connected to "high priority sites" (large orbital density of the LUMO) or "low priority sites" (low orbital density) of the acceptor molecule. The acceptor molecule may be switched to a passive state by two electrons from e.g. donors D_2 and D'_2. The electron transfer from D_1 and D'_1 is then inhibited and cannot occur. It could ,however, occur if the acceptor molecule was not "conditioned" by D_2,D'_2. Such a "conditional electron transfer" could be advantageously utilized in a complex network as is also shown in Fig.5.
Some time ago I had proposed a molecular shiftregister based on a chain of multiple viologen molecules separated by spacers and with sidegroups which tailor the energy scale of each molcule in order to obtain descending reduction energy along the chain[4]. Electron transfer could be initiated either from an electrode or a donor molecule. By applying light pulses at the appropriate viologen transitions the reduced state of a particular viologen could be shifted to the right. Similar mechanisms have been proposed by Hopfield and co-workers [16].

In the simplest case such a neuromolecular network would function as a cellular automaton, but more intelligent functions seem to be feasible. It is obvious that rigorous basic research is required in order to prove the usefulness of these speculations.

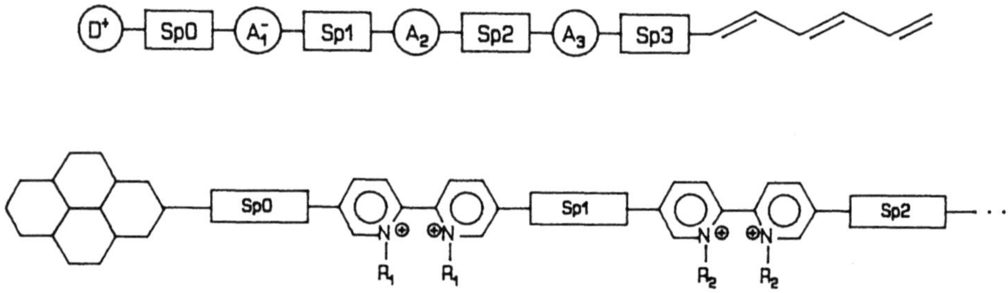

Fig.6 Molecular shiftregister (FIFO)

Acknowledgments: I would like to thank all the members of the Sonderforschungsbereich:"Physikalische und chemische Grundlagen der Molekularelektronik" SFB 369,Stuttgart for discussions. I am indebted to K.Müllen in particular for interesting conversation. The Deutsche Forschungsgemeinschaft (DFG) has supported this project financially.

References

1 F.L.Carter, Molecular Electronic Devices, Proceedings of the First and Second International Workshop on Molecular Electronic Devices, Washington (1981, 1983) Marcel Dekker Inc., New York

2 D.Haarer, Angew.Makromol.Chem. 100/109, 267 (1982)

3 H.Sixl, Phys.Bl. 40, 35 (1984)

4 (a)M.Mehring und H.Sixl, in "Parallel Processing: Logic, Organization and Technology" in: Lecture Notes in Computer Science, Editoren: J.D.Becker und I.Eisele, Springer Verlag 1987,(b)M.Mehring,WEH Seminar über Physikalische und chemische Aspekte der Molekularelektronik, Bad Honnef, Germany Sept.1987,(c)M.Mehring, Proceedings of the ITG meeting Darmstadt (1988).

5 (a) K.U.Klabunde, W.Huber, J.Lex und K.Müllen, Chem.Ber. 120, 1491 (1987),(b)J.Heinze, J.Mortensen, K.Müllen und R.Schenk, J.Chem.Soc.Chem.Commun.Comm, 1720, (1987)

6 (a) G.Dirks, A.L.Moore, T.A.Moore und D.Gust, Photochem. Photobiol. 32,277(1980),(b) A.L.Moore, G.Dirks, D.Gust und T.A.Moore, ibid. 32, 691 (1980),(c) D.Gust und T.A.Moore, J.Photochem. 29, 173 (1985)

7 J.R.Miller, L.T.Calcaterra und G.L.Closs J.Am.Chem.Soc. 106,3047 (1984)

8 M.R.Waselewski, M.P.Niemczyk, W.A.Svec und E.B.Pewitt, J.Am.Chem.Soc. 107, 1080 (1985)

9 H.Heitele, M.E.Michel-Beyerle, J.Am.Chem.Soc. 107, 8286 (1985)

10 (a)J.M.Warman, M.P. de Haas, M.N.Paddon-Row, E.Cotsaris,N.S.Hush, H.Oevering and J.W.Verhoeven, Nature 320, 615 (1986),(b) J.M.Warman, M.P. de Haas, H.Oevering, J.W.Verhoeven,M.N.Paddon-Row, A.M.Oliver and N.S.Hush, Chem.Phys.Lett.128, 95 (1986)

11 F.Effenberger, H.Schlosser, P.Bäuerle, St. Maier, H.Port und H.C.Wolf, Angew.Chemie 100, 274 (1988)

12 (a) K.Beschorner, diploma thesis Stuttgart 1988,(b) K.Beschorner, W.Zweygart, A.Grupp and M.Mehring (to be published)

13 K.Müllen (private communication)

14 A.A.Zakhidov, Synth.Metals 27,A51 (1988).

15 "Neural Networks for Computing", AIP Conference Proceedings 151, Editor J.S.Denker, American Institut of Physics, New York 1986

16 J.J.Hopfield,J.N.Onuchic,D.N.Beratan, Science 241, 817 (1988).

Electronic Interferences Controlled by the Hydroquinone/Quinone Molecule Embedded in a $(CH)_x$ Chain

P. Sautet [1] *and C. Joachim* [2]

[1]Laboratoire de Chimie Théorique, Ecole Normale Supérieure de Lyon, F-69364 Lyon Cedex 07, France,
and Institut de Recherche sur la Catalyse,
2 avenue A. Einstein, F-69626 Villeurbanne Cedex, France
[2]Laboratoire de Chimie des Métaux de Transition, Université Pierre et Marie Curie, 4 place Jussieu, F-75252 Paris Cedex 05, France

Abstract : The hydroquinone/quinone couple is studied as a model active spacer embedded in a $(CH)_x$ chain. The electronic transmission through the spacer is calculated. The modification of the interference pattern introduced by the spacer on going from the hydroquinone to the quinone is analysed. An opposite behavior for para and meta type connexions of the spacer is explained.

1. Introduction

A spacer molecule embedded in a conducting chain modifies the electronic properties of this chain [1]. Beyond the breaking of the translation invariance, the importance of these effects has become a central question in polymer science [2] and for the long term in molecular electronics [3]. From a theoretical viewpoint, calculations on a large number of spacers have already been done [4]. But none of them directly assess how the transmission of an electron along the conducting chain is perturbed by the embedded molecular group. Truly, it is difficult to understand completely the perturbation of the electronic properties as a function of the chemical nature of the spacer only by tracking shifts, vacancies or disorder in the electronic density of states of the chain.

Many active spacers have been proposed in the literature with a view to control the electron propagation along the conducting chain [3,5]. Following the path, this paper presents the hydroquinone/quinone couple as a model of an active controlling unit embedded in a $(CH)_x$ chain. Together with salicylideneaniline [6], this is another example of an attempt to control, by chemical means, the interference pattern introduced in the electronic transmission by the ring geometry of the benzene molecule [7].

The redox equilibrium between the hydroquinone and the quinone can be controlled electrochemically or by photoactivation. In the case of quinone the perturbation of the π system of the benzene ring is strong, while it is much weaker for the hydroquinone. One could expect then that these molecules would yield different electron transmission probabilities when embedded in a $(CH)_x$ chain.

In order to analyse the modification of the interference pattern when going from the hydroquinone to the quinone (and reverse), the electronic transmission through

Springer Series in Solid-State Sciences, Vol. 91
Electronic Properties of Conjugated Polymers III
Editors: H. Kuzmany · M. Mehring · S. Roth

the two forms is calculated with the electron scattering quantum chemistry (ESQC) procedure [7] for a para and a meta type connexion of the two $(CH)_x$ chains on the carbon ring. In these two forms, the variation of this electronic transmission t(E) as a function of the energy E of an incoming electron is explained on the basis of a qualitative molecular orbital analysis.

t(E) is calculated in ESQC by considering the scattering between this incoming electron on the chain and the spacer as an elastic scattering process. This is a serious limitation as far as the propagation of solitons is concerned but not in the case of a purely electronic through-spacer tunneling process.

2. A qualitative analysis of quantum interferences due to a spacer

When embedded in a conducting chain, each molecular orbital of a spacer contributes individually to t(E). However, the superposition of these contributions is not always additive : destructive/constructive interferences between different electron paths through the spacer may occur [6-8]. In order to determine in each case the destructive or constructive nature of the interferences, a qualitative rule has been established [8]. This rule is based on an analytical study of a model system made of a two-level spacer embedded in a one-band chain [9].

In order to apply this rule, each molecular level of the spacer must be characterized by two criteria :

a) the phase relationship in this orbital between the two atomic sites connected to the polymer chains (in-phase or out-of-phase)

b) the position of the level with respect to the electron energy E (above or below).

When two levels are considered together, interferences are destructive if 1 and only 1 criterion coïncides and constructive otherwise. Then if the two levels have the same phase relationship (e. g. both in-phase), the interference is destructive for an energy E between these levels, since in that case criterion b is different, and constructive outside (figure 1a). If the phase relationships are different the interference pattern is reversed, constructive between and destructive outside (figure 1b).

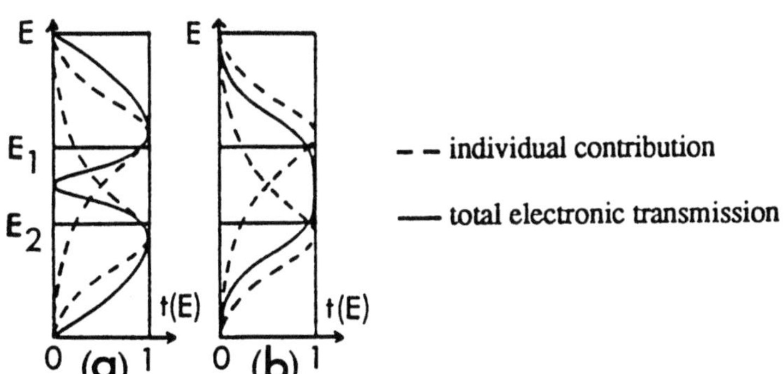

Figure 1 : Destructive/constructive interferences for a spacer with two levels located at energies E_1 and E_2 and with : a) the same phase relationship on the connected sites ; b) an opposite phase relationship

This rule is now applied to the analysis of the interferences between the molecular orbital contributions in the case of the hydroquinone/quinone embedded in a $(CH)_x$ chain.

3. Hydroquinone and quinone embedded in a $(CH)_x$ chain

The π molecular orbitals of hydroquinone and quinone are shown in figure 2, with respect to the energy bands of polyacetylene (PA). Two oxygen p centered orbitals appear at the bottom followed by six "benzene-like" orbitals, destabilized by antibonding interaction with the oxygens. Notice that orbitals ϕ_2 and ϕ_4 do not mix with oxygen p orbitals for symmetry reasons.

The perturbation of the benzene orbitals is weak in the hydroquinone form. Two effects are important when going to the quinone structure. The carbon-oxygen distance is reduced, leading to a stronger interaction, and the ring geometry is modified. In the case of ϕ_3 both effects add and a strong shift of this orbital results. On the other hand, these effects cancel in the case of ϕ_5 and the associated energy level is not significantly shifted. Then, the passage from the hydroquinone to the quinone results in an important modification of the electronic structure of the spacer in an energy range associated with the top part of the π band of PA.

If the spacer molecule is inserted in a $(CH)_x$ chain, we can first notice that the electronic transmission t(E), shown in figures 3 and 4 for para and meta connexions, respectively, presents important variations between hydroquinone and quinone only in this top part of the π band. However, the behaviors of the para and meta

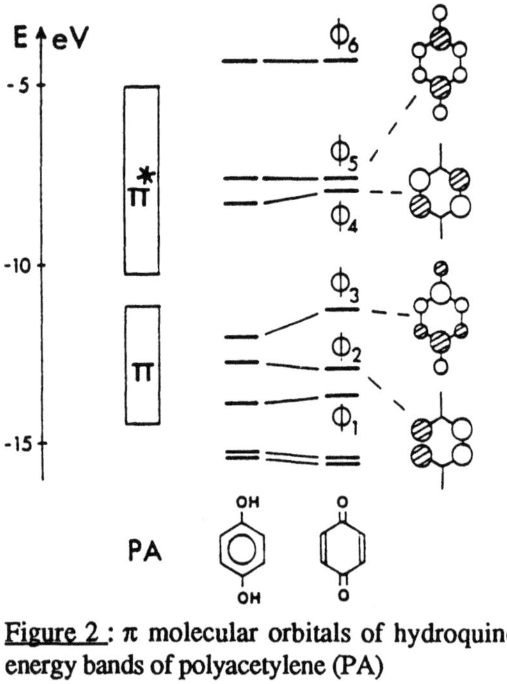

Figure 2 : π molecular orbitals of hydroquinone and quinone with respect to the energy bands of polyacetylene (PA)

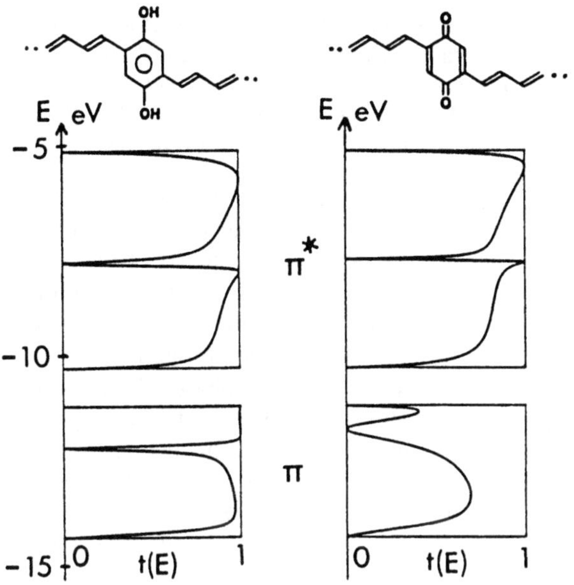

Figure 3 : Electronic transmission t(E) for a para connexion of hydroquinone and quinone embedded in a $(CH)_X$ chain

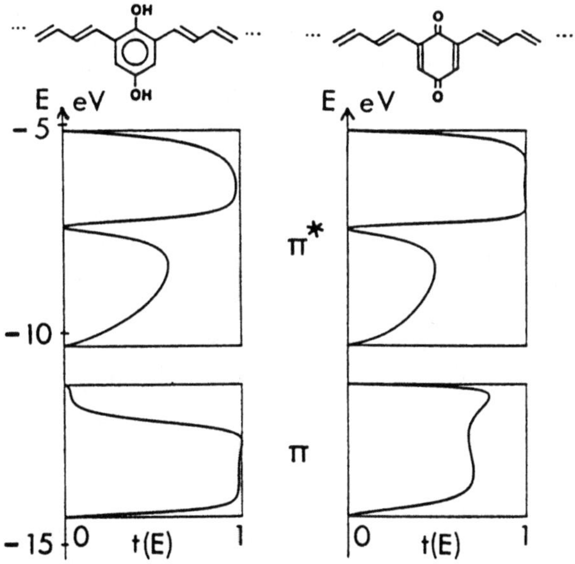

Figure 4 : Electronic transmission t(E) for a meta connexion of hydroquinone and quinone embedded in a $(CH)_X$ chain

connexions appear to be opposite in this energy range. The para connexion (figure 3) is associated with a very high electronic transmission in the top part of the π band for the hydroquinone structure, while this transmission is strongly depleted when one switches to the quinone. On the other hand, the meta connexion (figure 4) shows a transmission hole for the hydroquinone and a transmission maximum for the quinone in the considered energy range.

We shall now analyse in detail the shape of the t(E) curve in the top part of the π band, on the basis of our qualitative interference rule, in order to understand this opposite behavior. In the considered energy range, orbitals ϕ_2 and ϕ_3 of the spacer contribute to the through-spacer electron tunneling. The main contribution comes from ϕ_2 because of its bigger coefficients on the sites connected to the $(CH)_x$ chains. However, strong interferences will result from tunneling through ϕ_3 .

In the case of a para connexion, both orbitals show an out-of-phase relation on the sites connected to $(CH)_x$. The phase relationship criterion is equal and one can conclude from the rule that the interference should be destructive for energies between ϕ_2 and ϕ_3 levels and constructive outside (figure 5a). For the hydroquinone stucture (figure 3), a narrow interference hole appears between ϕ_2 and ϕ_3 (near ϕ_3), while a constructive effect is clearly present at the top part of the band with a high transmission "plateau".

The level ϕ_3 is shifted up in the case of quinone. The associated interference hole appears now at a higher energy. On the other hand, the constructive energy interval has almost disappeared. Therefore a strong decrease of t(E) is apparent with respect to the hydroquinone form.

The behavior is opposite for a meta connexion. The orbital ϕ_3 is now in phase while ϕ_2 is still out of phase. The phase relationship criterion is different and the interference pattern is reversed (figure 5b). The transmission maximum associated with the constructive interference between ϕ_2 and ϕ_3 is then shifted up in the quinone structure, while the destructive region in the top part of the band disappears (figure 4). The quinone form is then associated with a higher transmission probability in this range.

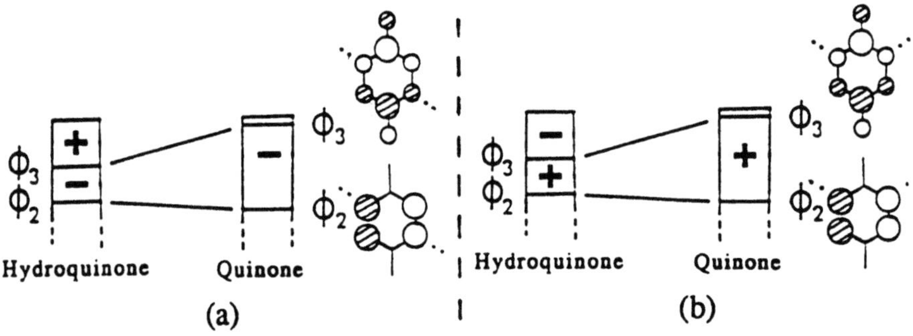

Figure 5 : Destructive (-) or constructive (+) nature of the interference between ϕ_2 and ϕ_3 molecular orbitals of the spacer for an electron energy in the top part of the π band of PA : a) para connexion of the $(CH)_x$ chains on the spacer ; b) meta connexion

The opposite behavior of the quinone and hydroquinone structure for a para or a meta connexion is then clearly understood as a change in the interferences between the tunneling contributions of orbitals ϕ_2 and ϕ_3 when the connexion sites are modified.

4. Conclusion

When a spacer molecule is embedded in a polymer chain, each molecular orbital of the spacer can be considered as an individual path for the through-spacer electron transmission. Quantum interference effects are present between these tunneling channels.

In the case of the quinone/hydroquinone couple as a model active molecule embedded in a $(CH)_x$ chain, distinct interference patterns have been found in the top part of the π band for para and meta type connexions. The shift of this interference pattern, associated with the change of the electronic structure of the spacer between the hydroquinone and the quinone form, results in large variations of the electronic transmission. Therefore, shifting interferences within an active spacer may be a useful way to control the electron propagation along a conducting chain.

References

1. A. Aviram, J. Am. Chem. Soc. 110, 5687 (1988).
   S. Roth, Z. Phys. B 64, 25 (1986).
2. F. Zuo, A.J. Epstein, X.Q. Yang, D.B. Tanner, G. Arbuckle and A.G. MacDiarmid, Synth. Met 17, 433 (1987).
   J.C.W. Chien, G.N. Babu, J. Chem. Phys 82, 441 (1985).
3. F.L. Carter, Physica D 10, 175 (1984).
   C. Joachim, J.P. Launay, J. Mol. Elec., submitted (1989).
4. R. Ruckh, E. Sigmund, H. Sixl and S. Roth, Synth. Met. 21, 305 (1987).
   J.T. Lopez Navarrete and G. Zerbi, Solid State Comm. 64, 1183 (1987).
   J.L. Bredas, J.M. Toussaint, G. Hennico, J. Delhalle, J.M. André, A.J. Epstein, A.G. MacDiarmid, Proceedings of the "IWEPP", H. Kuzmany et al. edt., Springer Series in Solid State Sciences 76, 48 (1987).
5. J.P. Launay, S. Woitellier, M. Sowinska, M. Tourrel and C. Joachim, Proceedings of the 3rd Molecular Electronic Symposium 1986 (North Holland, 1989).
   H. Sixl and D. Higelin, Proceedings of the 2nd Molecular Electronic Symposium, 17 (M. Deckker, 1987).
6. P. Sautet and C. Joachim, Chem. Phys., submitted (1989).
7. P. Sautet and C. Joachim, Chem. Phys. Lett. 153, 511 (1988).
8. P. Sautet, O. Eisenstein and E. Canadell, Chemistry of Materials, in press (1989).
9. P. Sautet, French PhD Thesis, Université de Paris Sud (January 1989).

Photodynamics of Reversible Charge Separation: A Molecular Model for a Cellular Automaton

G. Mahler

Institut für Theoretische Physik, Universität Stuttgart,
Pfaffenwaldring 57, D-7000 Stuttgart 80, Fed. Rep. of Germany

Abstract. Complex molecular structures like donor-acceptor modified polyenes may support complex dynamical processes beyond the realm of conventional relaxation phenomena. In particular, open systems with electron states of various localization types could show a repertoire of different metastable states resulting from charge separation. A quantum-mechanical model is presented for which such a subspace of states can be controlled by a pulsed laser field. Structural requirements and limitations are discussed.

1. Introduction

There are two different approaches to molecular electronics: The more "conservative" way of using unconventional material for a conventional device (or at least a device functioning on a hydrodynamic level), and the up to now "speculative" way of aiming at a device performance on a truely molecular level [1,2]. We will focus on the latter.

Any strictly molecular device faces severe problems related to the control of molecular structure and stability, to the interface between the necessarily macroscopic environment and the microscopic system, and to the quantum fluctuations which might be expected to limit any deterministic single-particle dynamics. The first problem will sensitively depend on the future prospects of material engineering on molecular scales, and is presently hard to assess. The last two problems are actually interrelated, and will be discussed in terms of a model describing not only the microscopic system itself but also its interaction with the environment. It will be shown that the light-field, which on a molecular scale is selective in frequency space rather than in real space can, neverthe-

Springer Series in Solid-State Sciences, Vol. 91
Electronic Properties of Conjugated Polymers III
Editors: H. Kuzmany · M. Mehring · S. Roth

less, be used for control, and that quantum fluctuations can be minimized in a highly hierarchical structure.

2. Hamiltonian Model

Our system is described by [3,4]

$$H = H_S + H_{SP} + H_{SB},$$

where H_{SP} refers to preparation (here the light field), H_{SM} to measurement, and H_{SB} to bath coupling (here the photon vacuum). The molecular system is supposed to decompose into interacting subsystems,

$$H_S = \Sigma H_\nu + \Sigma H_{\nu\nu'}.$$

We start by introducing a minimal model for such a subsystem, H_ν.

3. The Subsystems: DA-modified Polyenes

Let us consider a finite polyacethylen chain, the ground state degeneracy of which has been lifted by two different end groups, a donor D, and an acceptor A [5]. In a first step we reduce the finite chain to a single carbon atom, and the end groups to single hetero-atoms. This simple 3-atom molecule with 4 non-interacting π-electrons, to be sure, is too short to show dimerization: A Hueckel-approach with the diagonal matrixelements α_A=-1eV,α_C=0, α_D=-1.25eV (and double positive charge), the transfer integrals,

$$\beta(r) = \beta_0 \exp[(R_0 - r)/a],$$

(β_0=-1.1eV, R_0=1.4Å, a = 0.31Å), and a nearest neighbor nucleon-nucleon interaction via σ-electrons [6],

$$V(r) = (r - R_b + a)\ \beta(r)\ 2/c,$$

where R_b=1.5Å, c=0.15Å, implies a unique ground state configuration. Bistability, however, can artificially be imposed by constraining the total length of our model molecule to L = 3.8Å, as shown in Fig.1a. This is a

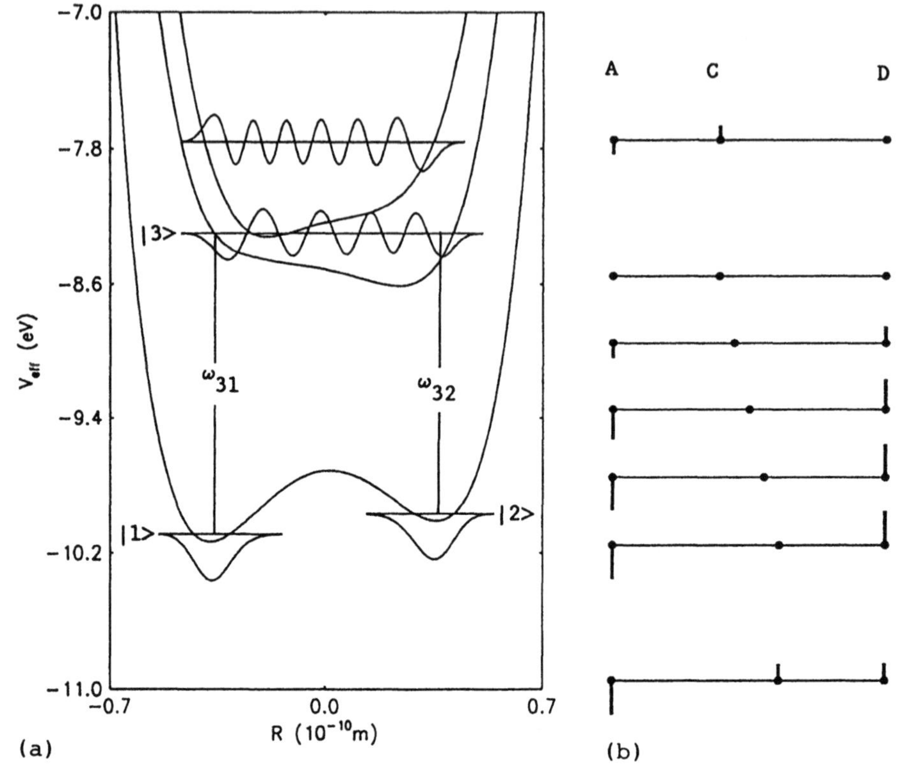

Fig.1a. Minimal model for DA-modified polyene. Shown is $V_{eff}(R)$ for the 3 lowest electronic states and the wave function of the mobile C-atom in state |1>,|2>,|3>. Fig.1b. Correlation between the position R (horizontal line) of the C-atom and the total charge distribution (vertical line). Top: ground state, bottom: metastable state, in between: transient state

configuration coordinate diagram for the three lowest electron states, as it follows from our concrete Hamiltonian. Included are specific wave-functions of the mobile C-atom. Fig.1b shows the correlation between the position of this C-atom and the total charge distribution. The metastable state (bottom of Fig.1b) corresponds to the charge-transfer-state D^+A^-. The two split ground states |1>,|2> together with the first excited state |3> constitute a formal three-level system in electron-nucleon space. Due to the localization behavior of these states, the transition rates $W_{31} = W_{32} = W/4$ between either ground state and the transient state is large compared to the transition rate W_{21} between the ground states ("time-scale-splitting").

The dissipative dynamics of such a three-level system (2 attractor states, 1 transient state) driven by an external light field (detuning δ_{3i}, Rabi-frequency Ω_{3i}) has been discussed in detail in Ref.[3]. The equation of motion for the respective density-matrix is

$$d\rho/dt = -i\ [H_S + H_{SP}, \rho]/h + Z_{incoh}.$$

Here the first part describes the coherent motion, the last part the incoherent motion due to the coupling to the photon bath. Even if the direct decay of the meta-stable state |2> into |1> is safely neglected, the switching-process occurs by principle only with a probability less than 1: The resulting error probability depends on the detuning δ_{32} of the back-reaction [3]. The attractor state reached will be maintained after the laser-field is turned off. A selective switching can be performed also in the reverse direction.

The present three-level dynamics can be applied to a more extended chain, for which N > 1 C-atoms are allowed to move. As will be discussed elsewhere [7], the switching can then be characterized as a transient solitary excitation in an open, externally driven system. This is in contrast to the "closed system"- soliton dynamics proposed, e.g., by Carter [2].

4. Interacting Subsystems

The two ground states differ not only by energy but also by their respective charge distribution, which may thus be exploited for the long-range interaction between different subsystems. In the present case the leading contribution to the Coulomb interaction is the dipole-dipole-term. This term leads to a renormalization of the transition frequencies within system A (states |i>) as a function of the state |i'> of the neighboring system B:

$$\Delta\omega_{31} = \omega_{31}(B=|1\rangle) - \omega_{31}(B=|2\rangle) \equiv 2C.$$

The dynamics of the system A-B in its 9-dimensional state-space is then described by a 2-particle density matrix, <ii'|ρ|jj'>. In Fig.2 we show the resulting conditional

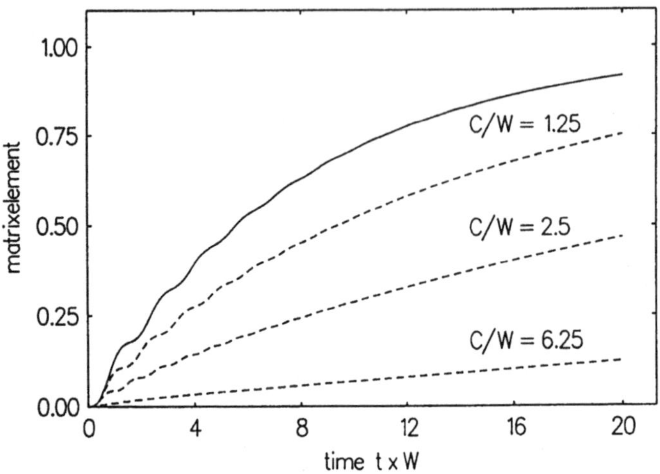

Fig.2. Conditional switching of subsystem A interacting with B, described by a 2-particle density matrix. Solid line: $<21|\rho|21>$, broken lines: $<22|\rho|22>$ for Coulomb renormalizations C/w as specified

switching dynamics: If the renormalization C is large enough (realistic values are around C/W=1000), system A is switched only if B is in state $|1>$ (or, with a different frequency, if B is in $|2>$). The corresponding properties hold for switching of B. The two interacting subsystems are thus controlled by 8 non-degenerate frequencies, by which transitions between the 4 attractors $|ii'>;i,i'=1,2$, can be selected. Again, as the different light frequencies act also non-resonantly, the switching has necessarily a finite error probability [4].

5. Quasi-Molecular Cellular Automaton

From the interacting 2-subsystem model, one may wish to generalize to a N-subsystem model: This could be accomplished by N physically different subsystems, so that each has its unique (conditional) transition frequencies, depending in general, on the states of all neighbors. Obviously, as only a finite frequency range can be used, the number N is then limited by the minimal detuning range required to suppress unwanted transitions (see above). Furthermore, the controlled switching would require a complete knowledge of the initial state out of 2^N possible states resulting in a non-reducible $2^N x 2^N$-transition

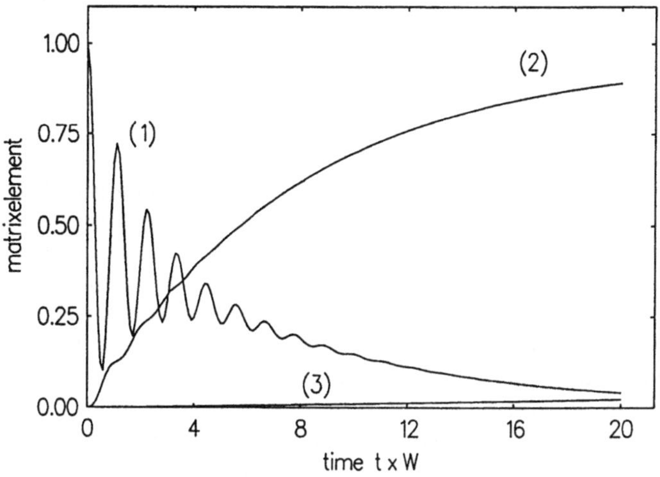

Fig.3. Switching of two identical interacting subsystems: (1): <11|ρ|11>, (2): coherent state <21+12|ρ|21+12>, (3): <22|ρ|22>

matrix. As this matrix represents all the "machine" could do, running it would not produce any new information.

Alternatively, one could think of making all N subsystems physically equal (homogeneous chain). In this case, however, the dynamics would be obscured by large quantum fluctuations, as shown in Fig.3 for two subsystems: The Coulomb-renormalization, which we need for conditional switching, would in this case bring a homogeneous initial state into a coherent (or mixed) state, as the accidental switching of one subsystem drives the second one out of resonance.

To avoid this behavior we must start with an alternating AB-chain, where the repeat-unit consists of two different subsystems. In this case, subsystems A can be switched on condition of B and vice versa. As shown in Ref.[4], the influence of the farther apart neighbors can be averaged out. The resulting recursive dynamics is typical for a linear cellular automaton architecture [8]. Reading-in of the initial state might be carried out in terms of shift operations, reading-out in the same way, followed by a quantum-mechanical measurement (cf.[9]).

6. Conclusions

We have shown that - if structural control is possible down to molecular levels - electronics mapped onto individual localized electrons might be feasible. The architecture based on a highly hierarchical but homogeneous system is that of a quasi-molecular cellular automaton. The interface between the microscopic system and the macroscopic environment is formed by the light-matter interaction, which serves for control and measurement. Control means specification of the local transition rule. Quantum fluctuations can be minimized, though not avoided. The proposed clocked dynamics requires dissipation. Possible materials for realizations include semiconductor quantum dots [3,4], donor-acceptor-modified polyenes [5], and donor-acceptor complex-crystals [10].

This work has been supported by the Deutsche Forschungsgemeinschaft (Sonderforschungsbereich 329).

References

1. L.T.Calcaterra, G.L.Closs, J.R.Miller: J.Am.Chem.Soc. 105,670 (1983)
2. F.L.Carter (ed.): Molecular Electronics, (Marcel Dekker, New York and Basel, 1982)
3. K.Obermayer, W.G.Teich, G.Mahler: Phys.Rev.B37, 8096 (1988)
4. W.G.Teich, K.Obermayer, G.Mahler: Phys.Rev.B37, 8111 (1988)
5. S.Roth, G.Mahler, in: Proceedings of the ICSM'88,Santa Fe', USA, to be published in Synthetic Metals
6. F.Chien: Chem.Phys.125,269 (1988);
   T.Sano: J.Chem.Phys.79,2060(1983)
7. H.Koerner, G.Mahler: in preparation
8. S.Wolfram: Physica D10,1 (1984)
9. W.G.Teich, G.Mahler: Phys.Rev.Lett.62,1 (1989)
10. R.M.Hochstrasser, J.W.Michaluk: J.Molec.Spectry.42,197 (1972)

Progress Towards the Creation of a Three-Dimensional Molecular Memory

P.E. Burrows, K.J. Donovan, and E.G. Wilson

Department of Physics, Queen Mary College, Mile End Road, London E1 4NS, UK

Abstract. Experiments to investigate the motion of photogenerated electrons between Langmuir Blodgett (LB) layers of conjugated molecules are described. Results indicate that for some molecules, carriers can hop between the layers with a hop time per layer of 50ns.

1. Introduction

A three-dimensional molecular memory has been designed [1,2] which seeks to store information in the form of charge on the layers of an LB multilayer. A crucial first step towards fabricating the above device is to understand and control the motion of electrons perpendicular to the LB multilayer. To this end, LB multilayers of molecules of promising dimensions were sandwiched between evaporated metal electrodes. Carriers created uniformly in the layers with a pulse of laser light were induced to hop across the layers by an electric field applied between the electrodes, and the motion detected as a transient photocurrent.

For an LB molecule to be a good candidate for use in the above experiment the inter-layer hop time must be short compared to the time taken to hunt out film defects via in-plane motion. The time taken to hop a distance d between layers is given [3] by τ, where

$$1/\tau = \nu \exp(-2kd) \quad \text{where } k^2 = 2mA/\hbar^2. \qquad \text{....(1)}$$

and ν is an attempt rate, A is the electron affinity of the molecule and m is the electron mass.

Molecular parameters for three molecules considered have been summarised previously [4]. We have succesfully observed electron motion in two molecules; protoporphyrin IX dimethyl ester (PPDME) and tetramethoxytetraoctoxyphthalocyaninato-polysiloxane (Pc). In both cases the macrocycles are asserted to be perpendicular to the plane of the layers [5,6] and electron transport between macrocycles is far easier in the plane than between layers. The multilayers are, therefore, quasi two dimensional. The calculated hop time per layer, τ, is 15ns and 3.8ns respectively [4]

2. Experimental Technique

Figure 1 illustrates the current sample design. The active layers are spaced from the electrodes by layers of dielectric. This protects the active layers during top electrode deposition. It is hoped that carriers created in the active layers would cross the layers and stop at the dielectric boundary, inducing a transient current. Early experiments on porphyrin films used a single shot storage oscilloscope to display the photocurrent or its integral [7]. Carriers were created with a 10ns nitrogen laser pulse.

In the recent design of Fig. 6 the electrodes form a 50Ω stripline [8] terminated at each end with 50Ω resistors. The signal voltage is collected by a boxcar acquisition system of gate width 25ps. Carriers are now created by 20ps pulses of 3.49eV photons from a frequency tripled mode locked YAG laser. The overall resolution of this system is 35ps.

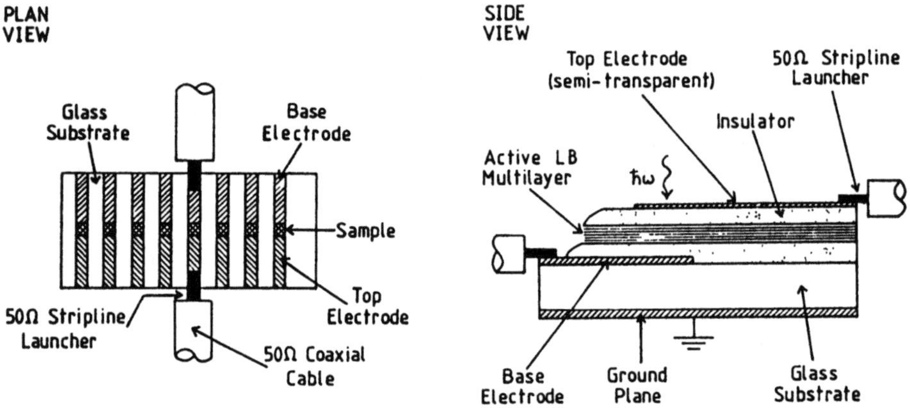

Figure 1. Photoactive LB device

3. Results

Early results [7] with PPDME and SiO dielectric showing the photocurrent are presented in Fig. 2. If carriers are instantaneously created throughout the active layers, then the time evolution of the current due to carriers moving to the dielectric is

$$I(t) = I_0(1-t/T) \qquad \ldots.(2)$$

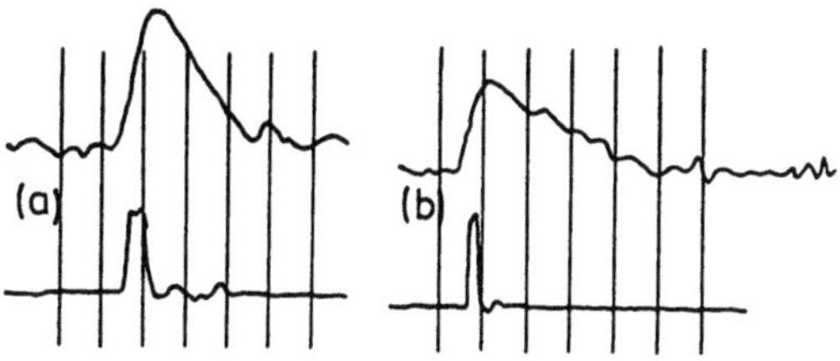

Figure 2. Current response (top) to a 10ns, 590nm laser pulse (bottom) of porphyrin layers sandwiched between 150nm of evaporated SiO. (a) 2 layers, 20ns/div, (b) 4 layers, 40ns/div.

where T is the transit time. If there are N active layers in the device we expect $T=(N-1)\tau$, where τ is the inter layer hop time. If the results in Fig.2 are asserted to have the shape of Eq.(2) τ for the PPDME films is 50ns. Experiments conducted with up to 8 layers gave transit times which went up with N.

The sensitivity, S, defined as the number of electrons circulated in the external system per photon absorbed was also measured. S depends on the product of the quantum efficiency for creating carriers, η, and the probability that they avoid geminate recombination, ϕ, multiplied by a dielectric factor F(N). Assuming electrons and holes cross all the active layers, the N dependence of S is

$$S = (\eta\phi)F(N) = (\eta\phi)(Na)/[Na + (d_1+ d_2)\varepsilon_i/\varepsilon] \quad ....(3)$$

where the active layers have a dielectric constant ε and are spaced from the electrodes by top and bottom dielectric of thickness d_1 and d_2 respectively and dielectric constant ε_i.

Fig. 3 shows that for the PPDME layers the sensitivity goes up with layer number according to Eq.(3). This is further evidence that carriers cross all the PPDME layers.

Photocurrents across layers of the Pc were too fast to be resolved with the storage oscilloscope and necessitated the design of the picosecond experiment. By integrating Eq.(2), the charge induced at the sample electrodes is

$$Q(t) = Q_0(t-t^2/2T) . \quad ....(4)$$

Examples of time resolved Q(t) are shown in Fig. 4. Although these were initially interpreted as indicating τ to be 15ps per layer [4], this is now thought to be misleading as the bulk of later results indicate that Q(t) reaches half height in 70-150ps with no clear dependence on N. This suggests that carriers are not crossing all the Pc layers.

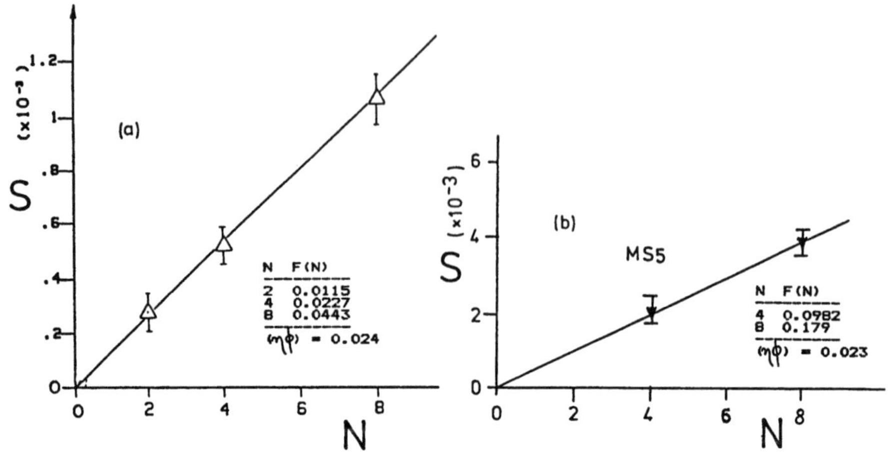

Figure 3. Dependence of sensitivity, S, on number of active layers, N, for porphyrin films between (a) 150nm evaporated SiO and (b) 7 and 8 layers of stearic acid. Both at 55MV/m

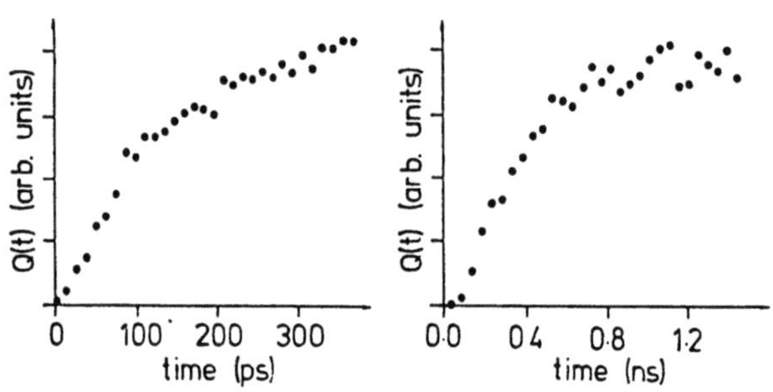

Figure 4. Signal voltage against time for 16 (left) and 48 (right) layers of phthalocyanine polymer sandwiched between 7 and 9 layers of 22-tricosenoic acid

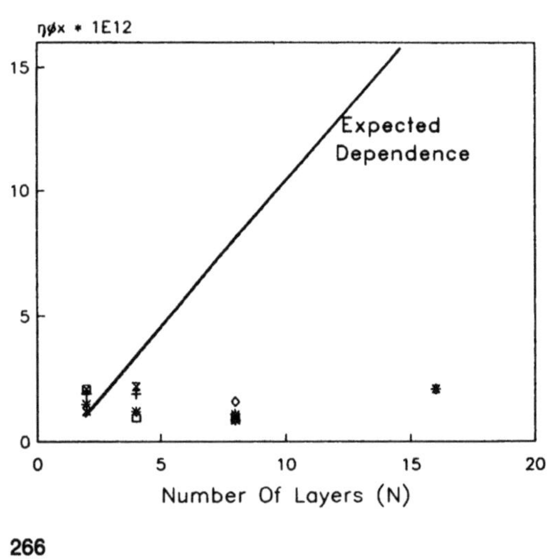

Figure 5 Variation of $\eta\phi x$ with N for phthalocyanine films

For the Pc films, the product $\eta\phi x$ as a function of N, where x is the mean distance moved by the carriers, is presented in Fig. 5. As $\eta\phi$ is constant at a given electric field, $\eta\phi x$ should yield the dependence of x on N, and is expected to increase linearly with N if the carriers are crossing all the Pc layers. Figure 5 suggests that x is constant for all N.

4. Conclusions

In the case of the PPDME films electrons can hop across up to 8 LB layers with a hop time per layer of 50ns. There is no dependence of transit time on the field as the potential drop between adjacent layers is greater than kT/e, inhibiting back jumps against the field.

Results obtained thus far indicate that carriers in Pc films exhibit more complex behaviour. We currently speculate that either carriers trap in one layer, in end states of the siloxane backbone or trap states associated with the aliphatic spacer.

References

[1] Wilson, E.G., Elec Lett 19,237(1983)
[2] Wilson, E.G., Mol Crys Liqu Crys 121,271(1985)
[3] Mott, N.F. & Davis, E.A. Electronic Processes In Non-crystalline Materials p39-41, Clarendon P (1971)
[4] Ahmed, F.R., Burrows, P.E., Donovan, K.J. & Wilson, E.G., Synth Met, 27,B593-B600(1988)
[5] Jones, R., Tredgold, R.H., & Hodge, P., Thin Solid Films 99,25(1983)
[6] Orthmann, E., & Wegner, G., Ang Chem 98(12),114(1986)
[7] Severn, J.K., Sudiwala, R.V., & Wilson, E.G., Thin Solid Films 160,171-175(1988)
[8] Auston, D.H., Lavallard, P., Sol, N. & Kaplan, D., App Phys Lett 35(1),66(1980)

Microelectronics of Artificial Neural Networks

K. Goser

Beuelemente der Elektrotechnik, University of Dortmund,
Postfach 50 05 00, D-4600 Dortmund 50, Fed. Rep. of Germany

Abstract. In the first part we describe the functions and the characteristics of artificial neural networks, which have their roots in biology, physics, and computer science. The second part describes the implementation of neural networks in silicon. Besides connectivity and learning, one of the most severe problems is the functional integration of the connection element (synapse), which has to store the adaptive weights in an analog way. Finally, we survey the potential of the VLSI technique, and new technologies, such as molecular electronics.

1. Introduction

The computational characteristics of artificial neural networks show the interesting features of learning, association, self-organization, optimization, and fault tolerance /1,2,3/. These networks might be described in terms of equivalent electrical circuits. Based on these circuit descriptions silicon integrated circuits have been made, and results have been obtained that remarkably resemble those of the simplest neural networks. This is a fascinating field of research and illustrates the possibilities of improved understanding of neural networks by applying concepts from apparently disparate fields, namely, biology, electronics, physics, and computer science. And, of course, a long-term goal is to configure a computer or a microelectronic component to operate with the advantageous characteristics of neural networks /4/.

2. The roots of artificial neural networks

In the fifties the neurologist Hebb and others developed simple models of a neuron: They characterised it by the inhibitory and excitatory inputs, the synapses, and a threshold stage (Fig.1). We have to emphasize, however, that this equivalent circuit is a very rough approximation only. If we simply draw this circuitry in a symmetrical way, we come up with a matrix, each column of which represents one neuron. The crossovers with the connection elements are the synapses and the end of the columns represent the threshold stages (Fig.2a). A special feature is the feedback from the outputs to the inputs. By this the neurons are interconnected one to

Springer Series in Solid-State Sciences, Vol. 91
Electronic Properties of Conjugated Polymers III
Editors: H. Kuzmany · M. Mehring · S. Roth

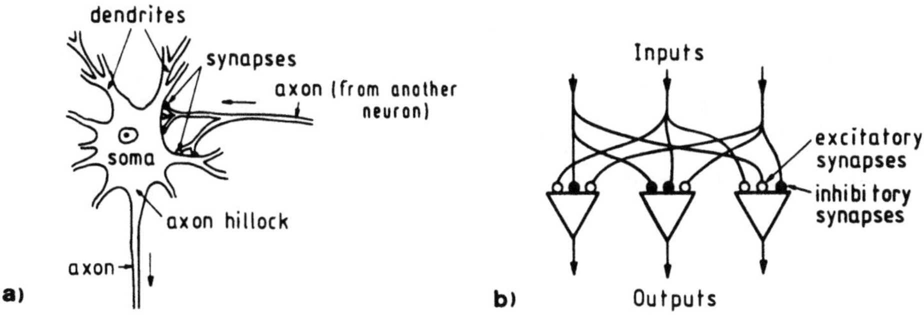

Fig. 1: A neuron (a) can be modelled by a threshold stage (b) the inputs of which can be either excitatory or inhibitory

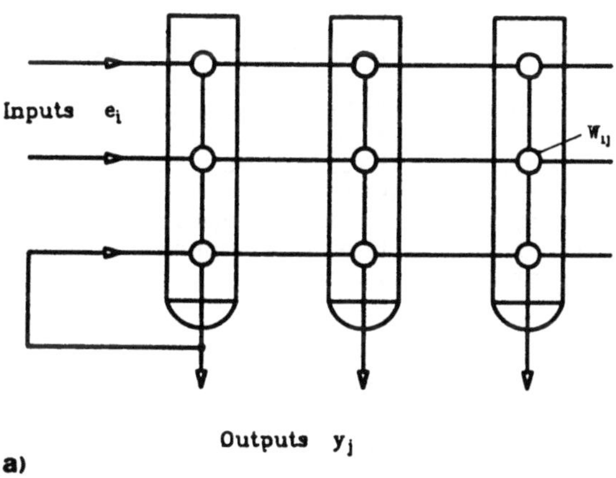

Weighted Sum of the Inputs:

$$S_j = \sum_{i=1}^{n+m} e_i \cdot W_{ij}$$

Transfer Function:

$$y_j = g_j(S_j)$$

Adaptation Function:

$$\Delta W_{ij} = f(W_{ij}, e_i, y_j)$$

b)

Fig.2: The threshold circuitry in Fig.1b can be transformed into a matrix (a). One of the algorithms of a neural network is described in (b)

another in a simple way, and the information stored in the network, too.

Microelectronics uses the matrix as a fundamental structure of a neural network. This architecture distributes the data processing amongst the data storage so we get a low data movement and a highly parallel processing. The input signals are weighted by the strength of the connection elements: The product of an input signal and an adaptive weight represents the output signal (Fig.2b). The sum of these output signals along the column is the input for the transfer function represented as a linear, sigmoid, or threshold function. This function yields the output of the neural network. The network should respond fast to input information; this operation is the recall phase. During a longer time scale the adaption or learning phase takes place. This operation slowly alters the adaptive weights of the connection elements in accordance to an adaptation rule. The limitations of technology, however, restrict the adaptation rules to neighbouring neurons.

The matrix in Fig.2a can be generalized to a network as shown in Fig.3a in which a circle represents a neuron. Each neuron receives signals from outside and neighbouring neurons, and sends signals to other neurons and to outside. The problem is similar to that of thermodynamics but more complex /5/. The physicists try to solve problems such as how many patterns can be stored in a fully interconnected network with N neurons, or under which conditions such a network with feedback is stable.

These questions about such networks concern computer science, too /6/. From its point of view a neural network is some kind of connection machine whose operation can be described in information space (Fig.3b). The different patterns stored in the network are represented as quantities of vectors assembled in a hyperspace. The proper classification of a pattern needs a line (area) separating one pattern from the others. Such an area can be adjusted by learning of a neural network, that means, by altering the adaptive weights of a matrix. A general separation in the hyperspace requires layered networks in which several matrices, at least three, are connected in series.

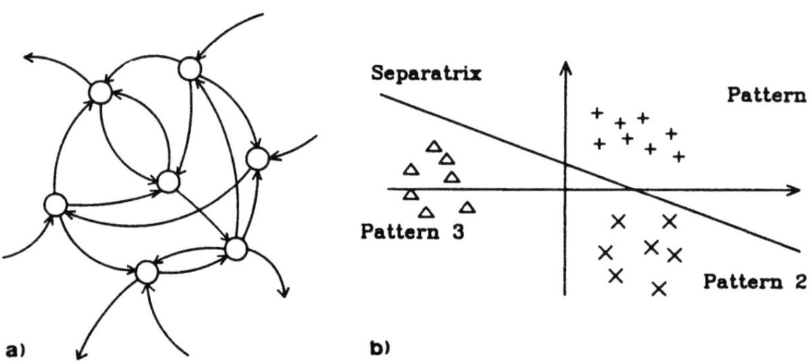

Fig.3: A neural network can be generalized in two ways: (a) a fully interconnected network is a problem of extended thermodynamics, and (b) the patterns shown in an information space can be separated by a neural net, as shown, in computer science

3. Implementation into silicon

If artificial neural networks are to offer a serious alternative to classical information processing we have to implement these systems in hardware using the advantage of real parallel processing. There are three problems in transferring the networks to silicon: The complex connectivity between the neurons, the storing of adaptive weights in an analog technique, and the adaptation of the learning algorithm to small scale networks /9/.

The simplest way to integrate is the use of only digital adaptive weights, which are set to only 0,1 or - 1. One of several digital concepts is the modified Palm matrix /7/. We have integrated 96 rows and 16 columns on a chip 4.7 mm x 7 mm in the 3-μm-CMOS technology (Fig.4). Since we have chosen a modular architecture, we can realize quite a large network with several hundreds of neurons if we interconnect some tens of such chips on a printed circuit board or as a wafer scale integrated system /8/. The performance of such a modular digital network is surprisingly high and sufficient for many applications.

The connection element represents a key subcircuit which should be functionally integrated in one device. It should carry out the multiplication of the input data by the adaptive weights, the storage of the adaptive weight, the selection of the cell and the adaptation of the weights, altogether. The so-called floating-gate MOS transistor offers a good solution for such a functional integration. Based on an EEPROM cell we developed a device with dual drain, dual control gate, and one floating gate (Fig. 5 a) /10/. The electrical charge on the floating gate represents the

Fig.4: Microphotograph of a digital neural network with a (96 x 16)-matrix in 3 μm CMOS

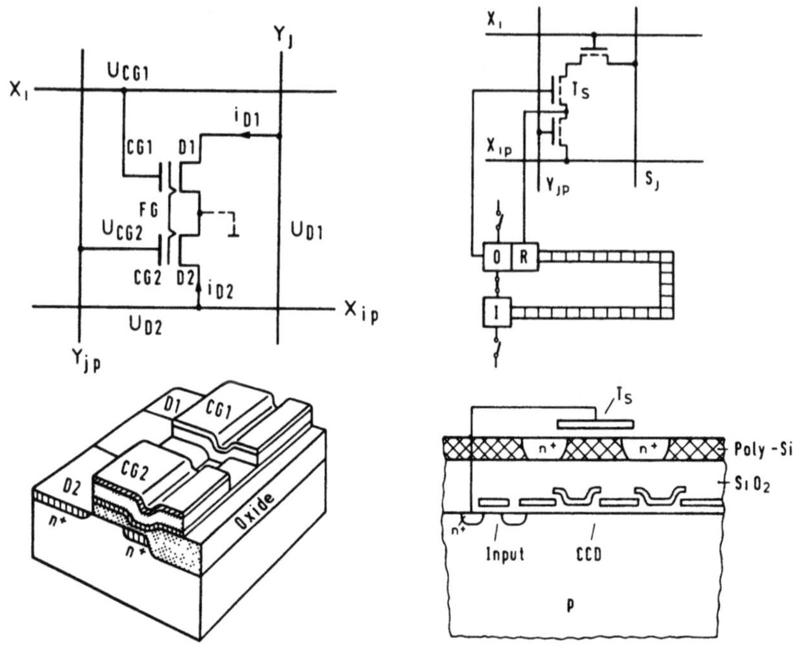

Fig.5: Integrated cells for storing adaptive weights: (a) A functional integrated cell with a floating-gate transistor, and (b) a cell in SOI technology with a CCD loop

adaptive weight, since it determines the threshold voltage of the MOS transistor. The tunneling of electrons to and from the floating gate alters the weight. By this the information is stored in a non-volatile way for years. The dual MOS transistor operates for the selection and the multiplication. A severe drawback of this simple device is the degradation of the insulator in the tunnel element after more than one million adaptation cycles. The device characteristics of today, however, suffice for learning the network one time.

An alternative to this device is much more complex. A three-dimensional integration in SOI technology (silicon on insulator) offers a solution for the interconnections and the integration of the storing circuit as a loop of charge coupled devices (CCD) or shift registers in the substrate (Fig.5b) /10/. Such a connection element is working well but it is area consuming, technologically complex and difficult to test.

4. Outlook

Today we can integrate up to several hundreds of neurons on a chip (Fig.4). The sub-μm-MOS technologies offer several thousands of neurons. At the end of the JESSI program (Joint European Sub-micron Silicon Initiative) with CMOS transistors of 0.3 μm channel length we can expect chips with about 10,000 neurons in analog technique /11/. Even with such highly sophisticated techniques the number of neurons on a chip is quite small in comparison to that of biological systems.

Therefore it will be a challenge to develop integration techniques beyond the scope of JESSI by introducing cooled MOS circuits with HT superconductors, optoelectronics with waveguides for interconnections in the network, and molecular electronics for realizing the adaptive storage element. Molecular electronics offers the vision of self-building and self-organizing in three dimensions, and therefore the outlook of huge neural networks, the performance of which may be similar to their biological counterparts. From this goal, however, we are still far away.

Acknowledgments

The author would like to thank his coworkers, U.Rückert, I.Kreuzer, K.M.Marks, and V.Tryba. Our work in this field is supported by the Deutsche Forschungsgesellschaft, and by the European Community in the projects BRAIN and ESPRIT.

References

/1/ Hopfield, J.: Proc. National Academy of Sciences, 79, 2554 (1982)
/2/ Kohonen, T.: Self-Organization and Associative Memory (Springer-Verlag, Berlin, 1988)
/3/ Shriver, B. D.: Computer, 21, 8 (1988)
/4/ Goser, K.: Mikroelektronik 3, in press (1989)
/5/ Reggia, J.A. and Sutton, G.G.: Proc.of the IEEE, 76, 680 (1988)
/6/ Kemke, C.: Informatik-Spektrum, 11, 143 (1988)
/7/ Palm, G.: Neural Assemblies (Springer-Verlag, Berlin, 1982)
/8/ Denker, J.S.: Neural Networks for Computing (American Institute of Physics, 1986)
/9/ Kreuzer, I.; Goser, K.: Proc. of EUSIPCO-88, Grenoble, Sept 5-8, 1988
/10/ Rückert, U. and Goser, K.: in Delgado-Frias et al., VLSI for Artificial Intelligence (Kluwer, Boston, 1989)
/11/ Mead, C.: Analog VLSI and Neural Systems (Addison-Wesley, Reading, MA, 1988)

From Protein Engineering to Bioelectronics

C. Nicolini, M. Benassai, E. Carrara, C. Gavotti, S. Lucia, and P. Pumilia

Institute of Biophysics – School of Medicine,
University of Genova and CIREF,
Association for Industrial Strategic Research, Agrate, Italy

The engineering of single biomolecules with desired electronic properties, like the proteins described here, appears to be the key critical step (likely a conditio sine qua non) towards bioelectronics.

This represents the messianic promise of the molecular kingdom, where it would become possible to build devices and architectures starting from single biomolecules, that is to say by engineering upwards rather than downwards as is done nowadays.

Introduction

Two major features of modern biology attract considerable attention, namely the coded self-assembly of biopolymers leading to complex structures in the human body and the brain architecture appearing capable of performing computation, which is substantially a process of making a sequence of interacting decisions like those needed to reach for a glass of water or to recall a memory from an associated fact, to a level unmatchable by physical digital computers.

Are these two facts to suggest that (a) a biochip could be assembled in the same way from individual biomolecular electronic components and (b) physical computational devices could be constructed to emulate biological hardware, at least in terms of the essence of neurobiology, like the interaction of neurons with thousands of other neurons, their potentially analogue and non-linear response, and their dynamics and constantly changing behaviour.

However, our current knowledge of the molecular basis of neurobiology is still lacking practically and the extensive exploitation of parallel processing architecture (to make compatible the rapid information processing of the brain with the slow propagation velocity of impulses along the axons and the dendrites) constitutes nothing more than a a poor precedent. Similarly it appears highly questionable at present that biopolymers like proteins or nucleic acids could be the primary building block of electronic circuits, at least as the active element. In general, polypeptides are rather unstable and possess large band gaps typical of electronic insulators; only by means of suitable manipulation in vitro might it be possible to induce optimal semiconducting properties and stability.

The alternative approaches to use as molecular switching chemical and structural modifications in the biopolymers – like those relating to changes in geometry, bonding group or position of localized electrons and protons, or the chemical codification like the coding of genetic information within the DNA – appear presently still inadequate

Springer Series in Solid-State Sciences, Vol. 91
Electronic Properties of Conjugated Polymers III
Editors: H. Kuzmany · M. Mehring · S. Roth

because these chemical processes are relatively slow. However, it should be possible to specify an amino-acid sequence that will yield a protein with the three-dimensional structure needed for the described function (as redox or receptor) or properties (as electron transfer), whether it be for use as a passive matrix or as an active element, we would likely be able to obtain biomaterials with unique semiconducting properties and self-bioassembly thereby making the first step toward the development of a biochip.
Many routes could be taken towards the biochip - which should not be confused with biosensors or chemical sensors - including the one to go strictly and directly for the development of only bioelectronic devices from existing biopolymers. In other words to follow the old obvious route of the so-called molecular electronics starting from existing conductive polymers, like polyacetylane and polyanilines, which were originally synthetized in Italy (by the Italian Nobel Prize winner Natta) and later made conductive in Japan by proper chemical doping. In our estimate however the risk of ending up with this route with little more than a fishing expedition is considerable. Thereby, in addition to further developing adequate tools for a kind of first generation bioelectronics (L&B protein monolayers, tunneling and atomic force microscopy in water, biosensors), we aim initially to develop advanced technologies within two well-focused distinct areas which are complementary to the construction of biochip and biocomputer, and which are at the same time realistic. The first aims to engineer proteins from first principles and the second to build silicon hardware based on a growing understanding of neural networks.

1. Why Engineer Proteins?

Genetic engineering techniques have been developed to the point where we can synthesize DNA fragments that code for precise amino acid sequences which uniquely determine the structure and properties of the resulting proteins and subsequently their spontaneous autoassembly in a monolayer and/or multilayers. From X-ray diffraction (on crystals) and 2DFT NMR (in solution) we can determine the 3D structure at atomic resolution for that protein and then modify the gene (by designing and synthesizing a new base sequence or by site-specific mutagenesis) to alter the protein structure in a highly predictable fashion. Through this revolution presently at its infancy - referred to as protein engineering - we are going to induce desired alterations in active sites, in binding sites for regulatory affector molecules and in other domains of the molecule itself, including the one controlling its electronic properties. At our present stage we have already proved that we are able to genetically alter and functionally enhance different types of protein, such as protease, DNA polymerase, immunotoxins, antibodies, insecticides and various microbiol enzymes that function as immobilized catalysts. The proof of the principle is undoubtedly being obtained; this makes realistic our final goal: to search for the unique relation between primary structure -- tertiary structure --- function and desired properties for any given protein.

In order to establish the ground rules for engineering biopolymers of the desired optical, mechanical and electronic properties, we utilize cytochrome b5 and thioredoxin (Figure 1) as model systems.

Figure 1

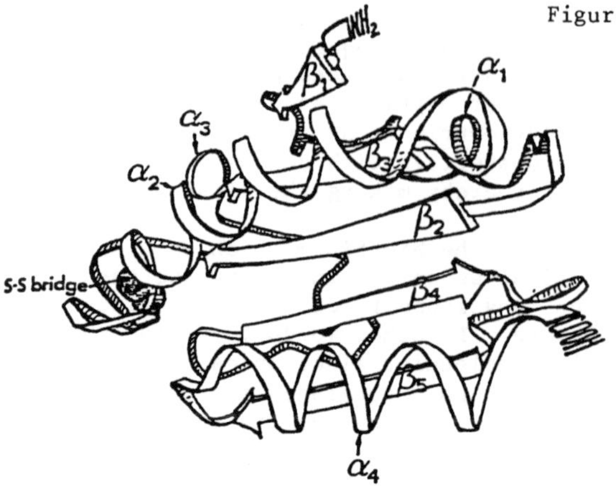

The former, cytohrome b5, a membrane protein that causes fatty acid desaturation and for which two domain structures (a heme-binding and a protein-binding) are known, constitutes one of the optimal examples for studying the electron-transfer process and thereby for making new semiconducting crystal arrays, as suggested by the extensive site-specific mutagenesis being successfully carried out to specifically change surface charges and protein-protein association.

Thioredoxin is a ubiquitous, small (MW 12000 Daltons), acidic and heat-stable protein (1-4) that has the ability to reduce reversibly a large variety of protein disulfides and may be involved in the catalysis of the formation of correct disulfides during protein folding. It exists either in reduced form (thioredoxin-$(SH)_2$) or in oxidized form (thioredoxin-S_2).

Thioredoxin-S_2 from E. coli is the best-known thioredoxin, but this protein has been studied in cyanobacteria, green algae, plant chloroplasts and in many mammalian cells. The active center of thioredoxin from E. coli is made up of the amino acid sequence Cys-Gly-Pro-Cys. The same active center has been found in thioredoxin of other organisms. All thioredoxins studied show many homologies in the amino acid sequence and in the three-dimensional structure. The latter has been solved to 2.8 A resolution by X-ray crystallographic techniques for thioredoxin from E. coli and recently work by 2DFT NMR has been carried out, also in solution (5). It appears then to constitute an excellent model system to ask new fundamental questions about the structure-function relationship and the engineering of biopolymers with new electronic properties.

Toward this end, we have isolated thioredoxin from extreme halophilic (up to 4 M NaCl) and thermophilic (up to 90°C) archeabacteria, to determine the possible changes in both primary and three-dimensional tertiary structure and their correlation with the above thioredoxin functions under prohibitive extreme conditions. Further information is being sought by site-directed mutagenesis in the gene sequence of thioredoxin from archeabacteria.

2. Protein Structural Determination at Angstrom Level

An integrated software system sharing different approaches for protein structure prediction, called IPATA (Integrated Protein Automation Three-dimensional Analysis, is presently being developed by integrating all theoretical (i.e., energy minimization at angstrom level, statistical considerations for assigning each aminoacid to the appropriate secondary structures) and experimental (mostly 2DFT NMR in solution and X-ray on crystals whenever possible, in addition to solution calorimetry and circular dichroism) studies and by considering also the solvent interaction effect on the protein folding.

It consists of:

- MACROST, a modified GOR method (7) to predict a 3D supersecondary structure. It first assigns each aminoacid to the appropriate conformational region (beta, helix, coil, turn) by utilizing the conformational "propensity" of each aminoacid (obtained from a statistical analysis of proteins whose structure is known, taking also into consideration its forward gradient from N-terminus towards C-terminus) and best fitting it to the observed circular dichroism spectra yielding the number of helices and their length in residues. Subsequently all the residue 3D coordinates are obtained by the known geometric values of distances between turns and between residues for each of four possible periodical structural regions. On the basis of the above statistical considerations and of heuristic rules, the procedure computes a tentative supersecondary structure (8). Evaluation of thermodynamics properties, like helix-coil transition, equilibrium constants and enthalpy variation at the melting point, are also taken into consideration.
- DRIVER and BARMAN, two expert systems written to handle a modified version of the static (REFINE) and the dynamic (SHAKE) parts of CHARMM (9); they compute theoretically, at atomic level, macromolecular 3D structure and dynamics from the energy minimization and residue sequences before and after solvent interaction, running on both a 3090/00 IBM supercomputer and on a VAX 708 system. The static and dynamic behavior of thioredoxin proteins in solution is being acquired, with the above programs coded in the Fortran 77 language.
- POLYPEPTO, to automatically derive three-dimensional molecular structures from NMR data; a task of strategic relevance in protein engineering and molecular biology. This process can be schematized as consisting of three steps: 2DFT NMR data acquisition, spin systems assignment and 3-D structure determination at atomic resolution.

First we are developing, in collaboration with the Stanford Magnetic Resonance Laboratories, a new integrated "portable" version of a computer-based expert system called PROTEAN for different and complementary algorithms, with the first release being planned in the middle of 1989 (10). Second, we have completed a PC-based expert system for computer-aided cross-peak assignment (called PEPTO) developed entirely "in house" (11). In the latter case we have achieved the aim of obtaining the full automatization of the peak-assignment and NOE constraints determination from 2DFT NMR data on proteins. Indeed, by pick-picking directly from raw FID data, the validation of PEPTO has been successfully carried out on a large set of proteins, being implemented on Turboprolog of the Borland operating under PC IBM; namely on endorphine (31 aminoacids), BPTI (58 aminoacids) and recently BUSI (57 aminoacids) and glucagon (29 aminoacids), all with well-known assignments. The version of PEPTO is still constantly changing with partial re-elaboration of existing rules and the new implementation of others, leading to optimal automation of this step, which is still extremely time-consuming and critical for a correct 3D protein structural determination at angstrom level in solution.

3. Towards Bioelectronics

Advances on the preparation techniques and on the fundamental properties of L&B films allow organic materials to be transferred one after another from a water surface into solid substrates to form monolayers. Recent advances in designing supermolecular aggregates by manipulating biomolecules and monolayer packages have proven the feasibility utilizing L&B in devices, justifying the attempt to apply protein films of this type within present silicon technology for the production of bioelectronic devices (i.e. biosensors). As they stand, however, these films appear inadequate for a CHEMFET due to their sizeable leakage and alternative approaches are being pursued in parallel, likewise the development of remote fiber optic immuno-sensors and the engineering of protein (i.e. monoclonal antibodies) monolayers.

Towards bioelectronics we are also presently, in addition to the engineering of cytochrome b5 based 2-D and 3-D biomaterials of desired semiconducting properties, investigating solitons and their propagation in model compounds such as acetanilide (ACN) and synthetic polypeptides, and building new apparatus for biopolymer characterization in water, at atomic resolution.

3.1 Soliton Propagation

According to Davydov's theory on self-trapping for excitations in molecular crystals, the main question is to determine the value of coupling parameters, that is to know how a soliton (a kind of deformation along a soft periodic lattice) may be induced by high frequency excitations on the same structure and then propagate along it. We have therefore started a program to examine the structure and dynamics of three-dimensional molecular crystals forming hydrogen bonds inside them (e.g. in successive order: acetanilide, synthetic polypeptide chains wholly arranged in alpha-helix conformation, properly engineered mutants of thioredoxin).

Our work is subdivided into two parts: I) theoretical, using molecular mechanics and dynamics; we aim to determine equilibrium structures and their dynamical behaviour as they carry a high frequency vibrational energy quantum (12); II) experimental, using i.r. absorption and laser-Raman spectroscopy in the 1500-1700 cm^{-1} region, 300 K to 4 K, to detect self-localized C=O stretch excitation peaks (13), NMR spectroscopy to detect H-bond proton movements as C=O excitation is activated and phosphorescence emission to investigate protein dynamics and amino acid residues' environment.

3.2 Development of Atomic Resolution Microscopy in Water

Tunneling microscopy and atomic force microscopy capable of characterizing materials (semiconductors, layered materials such as LB mono- and multilayers, proteins and other kinds of biological matter) appear useful for the development of bioelectronics. To recognize potential functionalities of these objects, one needs indeed to investigate their structure at atomic level.

Scanning Tunneling and Atomic Force Microscopy (STM, AFM) have recently been introduced as devices of ultimate resolution on conducting, semiconducting and insulating surfaces (15). In cooperation with the CNR Biophysics Institute at Pisa we are presently working on a novel possible lever displacement transduction based on evanescent wave coupling contrary to the traditional AFM coupling based on electronic tunneling (Figure 2, top). The evanescent wave (EW) is the optical analogue of the quantum mechanical tunneling effect. Total internal reflection in glass at the air interface may be "frustrated" by the nearness of a third medium (another glass surface). "Virtual" surface electromagnetic waves, generated at the total reflecting interface, become "real" in the presence of a third medium, thereby tunneling through air into this third medium. By using EW-coupled optical fibers, an AFM may be made allowing operation also in liquid environments (Figure 2, bottom).

Finally, starting from the gravitational confinement of water we are utilizing the dynamic program DOP (16) to simulate the behaviour of a solution of lipids in water, up to the formation of L&B monolayers starting from the gravitational confinement of water. On the experimental side by inserting biopolymers, such as thioredoxin, in LB layers we are investigating their overall and local changes in electronic and geometric (by atomic resolution microscopy) properties.

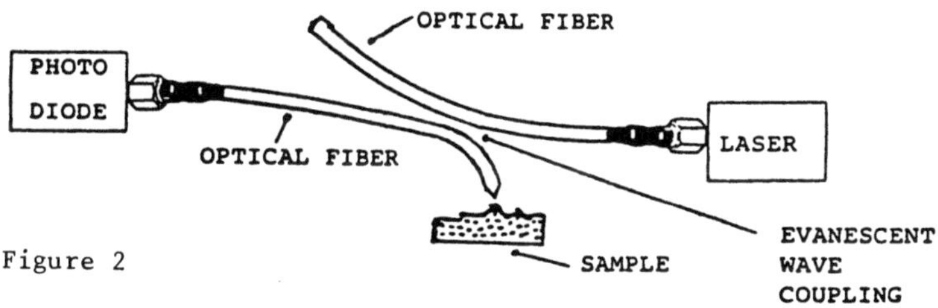

Figure 2

References

1) C. Nicolini, in "Towards the Biochip" , ed. C. Nicolini, World Publishing Co., Singapore-New York, 1989, 1-25
2) Arne Holmgren, Ann. Rev. Biochem. , 54: 237-71, 1985
3) Gleason F.K., Whittaker M., Holmgren A., J. Biol. Chem. , 260:17:9567-73, 1985
4) Johnson T.C. et al., J. of Bacteriology , 2406-08, May 1988
5) Dyson, H.J., A. Holmgren, G. Glippert, D.A. Case and P.E. Wright, Proc. XIII International Conference on Magnetic Resonance in Biological Systems, P15-39, Madison, Wisconsin, 1988
6) M.J.E. Sternberg, Nature , 271: 15-20, 1978
7) B. Robson and Garnier, Introduction to proteins and protein engineering , Ed. Elsevier, 1986
8) G. Rauch, P. Catasti, I.M. Pepe, M. Nizzari, I. Panfoli and C. Nicolini, Bioscience Reports , 1989 (in press)
9) B.B. Brooks; R.E. Bruccolieri, B.D. Olafson, D.J. States, S. Swaminathan, M. Karplus, J. Comp. Chem. 4 (2), 187, 1983
10) Carrara E., "PROTEAN", Technical Report 12/15/88, Stanford Magnetic Resonance Laboratory, California, USA
11) Catasti P., Carrara E. and Nicolini C., Journal of Computational Chemistry, in press

12) Pumilia P., Abbate S., Baldini G., Ferro D.R., Tubino R., "Nonlinear vibrational excitations in model compound for biopolymers", VIII Convegno Nazionale di Scienza delle Macromolecole, Milano
13) Careri G., Buontempo U., Galluzzi F. , Gratton Scott A.C., Shyamsunder E., Phys. Rev. B, 30, 4689, 1984
14) Gabellieri E., Strambini G.B. and P.Gualtieri, Biophys.Chem., 30: 61-67, 1988
15) Hansma P.K., Tersoff J., J. Appl. Phys. 61 (2), 1987
16) Jorgensen W., Chandrasekhar J., Madura J., Impey R., Klein M., J. Chem. Phys , 79, (2), 1983

Part IX

Polyanilines II

Polyaniline: An Old Polymer with New Chemistry, New Physics, and Potential New Technology

A.J. Epstein[1] *and A.G. MacDiarmid*[2]

[1]Department of Physics and Department of Chemistry, The Ohio State University, Columbus, OH43210-1106, USA
[2]Department of Chemistry, University of Pennsylvania, Philadelphia, PA19104-6323, USA

Abstract. Polyaniline is a family of polymers whose initial studies began nearly a century ago. Recent progress in the chemistry and processing of the polymer has now enabled detailed critical physical studies. Among recent advances are the ability to form soluble, high molecular weight polymers, pernigraniline (PN) and (oriented) emeraldine free-standing films and emeraldine base (EB) and salt (ES) fibers. X-ray diffraction studies show that emeraldine base can be prepared with a degree of crystallinity up to fifty percent. Doping of this material occurs initially in the amorphous phase followed by doping into the crystalline regions. These materials are identified as Branch II of the emeraldine family. Emeraldine prepared directly as salt from solution is also up to ~50 percent crystalline (denoted Branch I) though its structure differs from Branch II materials. Dedoping of emeraldine salt I yields amorphous EB-I. Protonation of EB-I yields continuous growth of crystalline ES-I. Corresponding static and dynamic magnetic susceptibility experiments show that the protonated crystalline regions have a Pauli susceptibility corresponding to a metallic state while the protonated amorphous regions are nonmagnetic (spinless defects only). Together these results demonstrate that disordering of the polaron lattice of emeraldine salt does not result in a Fermi glass, but rather a spinless state. Photoinduced absorption studies of emeraldine base and leucoemeraldine base (LEB) show the presence of long-lived, photoexcited entities, including excitons (in EB) and polarons (in EB and LEB). Analysis of the photoinduced IR active vibrations in comparison with the photoinduced electronic transitions shows the polarons created are massive. It is noted that a bond-order or Holstein polaron approach is more appropriate as the energy gaps in EB and LE are not Peierls gaps. Time dynamic studies imply that the longevity of the photoexcited defects is associated with the roles of ring flipping and ring conformation in the polymer system. Extended charge transport studies of oriented and nonoriented films demonstrate the textured, granular metal nature of emeraldine salt. Though the highest dc conductivity measured is approaching 100S/cm at room temperature, elimination of the barriers should lead to substantially higher conductivities. Potential new technologies based on the polyanilines, including nonlinear optics, optical information storage, and controlled microwave absorption are noted.

1. Introduction

The aniline system of polymers was first explored in the late 1800's. Green and Woodhead carried out extensive studies, reported in 1910, that demonstrated its polymeric nature as well as its solubility in H_2SO_4 [1]. Considerable controversy existed between this group and that of Willstätter [2] concerning the actual polymeric nature of the material. In the 1960's Jozefowicz and coworkers picked

Springer Series in Solid-State Sciences, Vol. 91
Electronic Properties of Conjugated Polymers III
Editors: H. Kuzmany · M. Mehring · S. Roth

up studies again, exploring protonic acid doping and the possible utility of polyaniline in aqueous batteries [3]. These electrochemical studies were continued by Genies [4], MacDiarmid [5] and Wrighton [6] in the 1980's. The first attempts at significant physical studies were fraught by problems of dealing with ill-defined compositions of the "polyanilines". The advent of better controlled materials commencing in the mid-1980's enabled the uncovering of unusual new physics and chemistry as well as potential technology in these systems.

Early studies by Nechtschein [7] gave an inkling about the unusual phenomena to be found in this system, while Salaneck et al. [8] and Yang and coworkers [9] and later Choi and Mele [10] clearly show the importance of considering the control of both the protonation level of the polymer (pH) and the electron concentration of the polymer. Figure 1(i) presents the idealized structures for emeraldine base. The idealized structure for leucoemeraldine base is polyphenylene amine or $[(NH\text{-}(C_6H_4)NH\text{-}(C_6H_4)]_x$ while that of pernigraniline is polyphenylene imine or $[N=(C_6H_4)=N\text{-}(C_6H_4)]$. Extensive quantum chemical and band structure calculations show the origin of the band gap of EB and LE to be due to the electronic structure of the C_6 rings [11a, 12]. The large energy gap of ~3.8eV is thus not analogous to the smaller ~1.8eV band gap that arises in polyacetylene due primarily to electron-phonon interaction. In contrast, the energy gap in pernigraniline involves electron-phonon contribution [12, 13]. Early studies of oligomers provided support for the picture of electronic structure in the systems [14, 15] as well as foretold the sensitivity of the electronic structure to ring conformation [16]. The detailed magnetic susceptibility experiments of Ginder et al. [17] demonstrated a reorganization of the electronic structure of EB to form a metallic polaron lattice or polysemiquinone structure upon protonation [11b, 17], Fig. 1. Phase segregation into metallic and nonmetallic regions was proposed [17, 18]. Optical studies confirmed this reorganization, with the 2eV absorption attributed to a charge transfer exciton from benzenoid to quinoid (or equivalently from the benzenoid based HOMO band to the quinoid based LUMO band) disappearing and two new absorptions appearing at 1.5eV and 2.9eV upon protonation of EB to form ES [18].

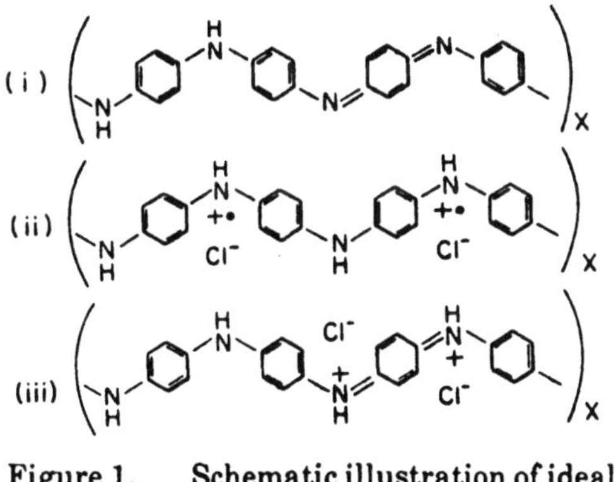

Figure 1. Schematic illustration of ideal repeat unit for (i) emeraldine base, (ii) polaron form of emeraldine salt, (iii) bipolaron form of emeraldine salt

II. New Chemistry, New Physics, Potential New Technology

A key aspect of the new advances in the understanding of polyaniline is the ability to control the chemical composition, oxidation state, derivatization, and physical form (oriented films and fibers) of the polymer so that critical studies can be performed. Oriented polyaniline films with substantial crystallinity are obtained by simultaneous heat treatment and stressing of emeraldine base films which contain a plasticizer such as N-methylpyrrolidinone (NMP) [19]. Samples are observed to elongate by up to 400 percent when held at elevated temperatures (T>110°C) [19] above the glass transition temperature [19, 20]. The resulting films have an anisotropic, x-ray, and optical response, with a misorientation of only a few degrees [19]. IR interference data on thin films indicate an anisotropy in the index of refraction of ~20 percent for parallel and perpendicular polarizations while x-ray studies show a directional enhancement of the Debye-Scherrer rings. An order of magnitude increase of conductivity of the oriented salt is observed (see below). The reflectance at 2eV of stretched, oriented emeraldine base film shows a $\cos^2\theta$ behavior with the maximum reflectance (~0.28) for a light polarized parallel to the chain axis and minimum reflectance (~0.10) for light polarized perpendicular to the chain axis, in accord with expectations for an intrachain charge transfer exciton as the origins of the 2eV absorption for emeraldine base. The oriented emeraldine base films may be subsequently doped to the desired protonation level [19]. Oriented polyaniline may also be obtained by evaporating (in the presence of a heat lamp) a solution of emeraldine base in NMP on polyethylene, polyacetylene or other substrates while the substrate is being mechanically stretched [21].

Exhaustive extraction of the emeraldine base form of polyaniline, first by THF and then by NMP, results in analytically pure emeraldine base. The NMP extract has a maximum molecular weight (g.p.c.) of ~325,000 [22]. Fibers of emeraldine base can be formed by drawing the emeraldine base materials in a water/NMP solution, while fibers of the emeraldine salt can be formed by drawing the material in aqueous HCl. Scanning electromicrographs show both forms of fibers to possess clean, smooth, even surfaces, while x-ray diffraction studies of the thermally stretch-oriented base fibers show the presence of oriented crystallinity. Conductivity in the range of 30-80S/cm is obtained for the emeraldine·HCl fibers [22]. These emeraldine salt fibers differ from those prepared from solutions of polyaniline in concentrated sulfuric acid [23] in that they can be made either by doping the EB fibers or directly by coagulation of a solution of EB in NMP in aqueous HCl [22].

X-ray diffraction studies of the emeraldine form of polyaniline distinguish two classes [24, 25]. Class I consists of materials prepared in the conducting, doped (salt, ES) form. They have formed substantially crystalline salts (ES-I). Dedoping leads to an amorphous insulating base, EB-I. Redoping of EB-I leads again to ES-I. Class II includes materials prepared from EB-I by exhaustive extraction first with THF then with NMP. This material is up to fifty percent crystalline. For partially crystalline EB-II, initial doping occurs in the amorphous region while at higher doping the crystalline components were doped to a new ES-II structure. Figure 2 is a projection along with a chain axis and side view of a single

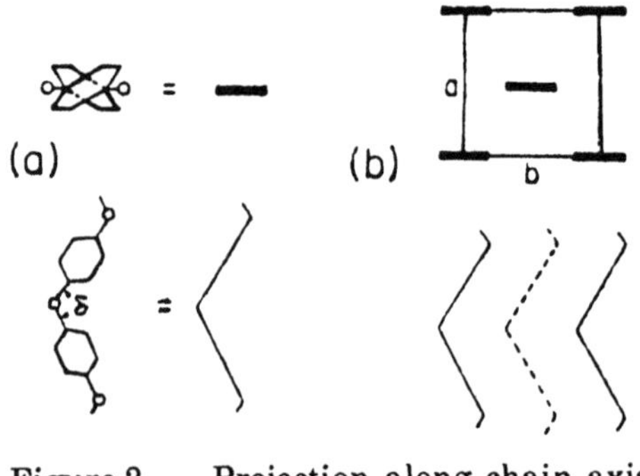

Figure 2. Projection along chain axis and side view of (a) polyaniline chain, and (b) EB-II

polyaniline chain and the proposed EB-II structure. EB-II is orthorhombic as is also ES-II, while ES-I is monoclinic. The coherence lengths in the ordered regions are as large as ~50Å perpendicular to the chain and ~150Å in the chain direction for oriented EB, i.e., coherence lengths are similar to those of polyacetylene [26].

There is a striking difference in behavior for χ^P vs. x for doping into EB-I and EB-II, Fig. 3. The nearly linear increase in χ^P with x=[Cl]/CN [17] for ES-I correlates well with the increased growth of structure ES-I with protonation. On the other hand, the χ^P for doping into EB-II with $x < 0.25$ is essentially zero, with a dramatic increase in χ^P as x exceeds 0.25 [24, 27]. This is in accord with the results of diffraction [24, 25] which show doping only in the amorphous phase of EB-II for $x<0.25$ and a corresponding change in the crystalline structure for $x>0.25$. Thus, while χ^P expected for a polaron lattice [17] is observed when there is either crystalline ES-I or ES-II phase present, protonation into the amorphous regions of EB-II forming ES-II' leads to spinless defects. The charge state introduced by protonation in the amorphous part of EB-II remains spinless until essentially maximum protonation achieved in the amorphous region, and then, protonation begins to proceed in crystalline regions with the onset of χ^P. The likely charge defects within the protonated amorphous region are doubly charged bipolarons, of Fig. 1(iii), as well as protonated amine units ($-NH_2^+-$). The absence of a high spin state in the amorphous region shows the inability of the polyaniline polymer to form a Fermi glass as the polaron metal becomes disordered as earlier proposed [14], instead spinless charge defects form. It is likely that the amorphous regions of doped EB-I, ES-I' also are comprised of similar spinless charged defects in accord with occasional anomalously low χ^P for doped EB-I, Fig. 3.

Dc conductivity studies of the oriented ES-II show a substantial increase in conductivity as compared with unoriented ES-I. For a protonated sample held in vacuum $\sigma_\parallel(295K) \sim 8 S/cm$ with $\sigma_\parallel/\sigma_\perp \sim 2.5$. The temperature dependence of $\sigma_\parallel/\sigma_\perp$ is the same, with $\sigma \propto \exp[-(T_0/T)^{\frac{1}{2}}]$ in accord with charging energy in limited tunneling between metallic islands [28]. Analysis of data for oriented samples yields $T_0 = 7000K$, the same value as for ES-I, implying that the barriers which dominate the charging energy limited tunneling are the same for both unoriented ES-I and oriented ES-II. Earlier analysis of temperature and electric field behavior of the conductivity of ES-I lead to an estimate of a metallic particle size of ~200Å [28] in remarkable correspondence with the coherence length of the crystalline regions in the protonated ES-I and ES-II [24]. Application of

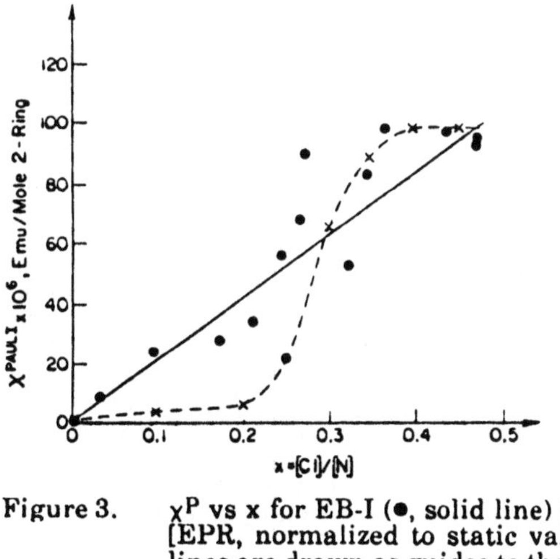

Figure 3. χ^P vs x for EB-I (●, solid line) [static χ] and EB-II (×, dashed line) [EPR, normalized to static value for $x=0.4$]. The solid and dashed lines are drawn as guides to the eye

contactless microwave conductivity and dielectric measurements and infrared spectroscopy lead to the conclusion that the intrinsic conductivity within the metallic islands is substantially higher than the measured microscopic conductivity [29], while audio frequency conductivity studies suggest hopping of charge among positively charged polaron and bipolaron or neutral defect states in the lightly protonated emeraldine polymer [30].

The ability to derivatize polyaniline at ring or nitrogen positions enables an improvement in processibility (for example, ethoxy derivatized polyaniline is water soluble [31]) and the testing of the role of interchain interaction. For example, polyorthotoluidine (replacement of an H by a CH_3 for each ring) leads to an expansion of the crystal structure as would be expected for greater steric interaction. Detailed magnetic, transport, structural, and optical studies [32] of the HCl salt of polyorthotoluidine supports that with the increasing chain separation there is a dramatic increase in the localization of electronic charge in the polymer so that the model of textured metallic islands [29a] developed for the emeraldine salt is no longer applicable and that stronger localization is more appropriate.

Time resolved spectroscopies provide unique insight into the dynamics of excitations of polyanilines and their dynamics. Picosecond scale studies of photoinduced bleaching of the exciton absorption of emeraldine base reveals a temperature-independent dispersive decay $\sim t^{-0.11}$, which differs markedly from the dynamics of earlier studied polyacetylene and amorphous silicon semiconductors [33, 34]. This time dynamics contrasts with the weakly temperature-dependent millisecond decays varying as $t^{-\beta}$, with $\beta \sim 0.5$ [34, 35]. Studies of the photoinduced infrared absorption on the second to hour time scale reveal the presence of very long-lived charged polarons at 80K [36]. The dispersive behavior in recombination reflects the stochastic nature of the spatial and temporal factors controlling recombination. Comparison of the temperature variation of the time dynamics with those obtained by NMR studies of the temperature dependence of ring flipping processes [37] supports the importance of ring rotation processes in the dynamics of polarons formed in polyaniline.

Photoinduced infrared active vibrational modes are substantially weaker than the photoinduced electronic transitions at 1.4eV and above [36], in contrast to the behavior of polyacetylene [38, 39] and polythiophene [40]. Application of the amplitude mode formalism leads to an estimate of a polaron mass M_{pol} of $\sim 56 m_e$ much larger than that of polyacetylene [38] or polythiophene [40]. In view of the large extrinsic gap in EB the use of the bond order polaron formalism of Rice and Phillpott [41] or the Holstein polaron [42] picture are more appropriate. Both these formalisms lead to massive polarons as well [36]. The massive nature of the polarons is in agreement with the very low mobility for photoexcited polarons [43, 44]. These large masses obtained suggest that the origin of the polaron distortions of the lattice are not carbon-carbon bond stretches but localized distortions of the ground state ring torsion angles. The importance of ring rotational defects in polyaniline and other ring-containing polymers such as polyphenylene sulfide and polyphenylene oxide support that a new class of physics is to be found in the study of these systems [12].

The ability to solution process polyaniline and its derivatives into nonoriented and oriented free-standing films, and fibers, as well as the ability to make sublimed films [45] of polyaniline open up the opportunity for technological application of this more stable polymer. In addition to application as a polymeric battery electrode [46] the very large loss tangent of emeraldine salt suggests its use in electromagnetic radiation absorption applications [29a]. Moisture increases both the microwave conductivity and dielectric constant such that the loss tangent of a fully protonated emeraldine polymer under dynamic vacuum increases by a factor of ~ 3.5 upon exposure to moisture [47]. The values of the loss tangent for emeraldine salt compare favorably with those of commercially available carbon-filled insulating polymers [29a].

Other potential uses take advantage of the facile transformation of the electronic structure of polyaniline upon absorption of a photon. The presence of an unusually high nonlinear optical susceptibility $\chi^{(3)}$ has already been reported for both base and salt of emeraldine [48, 49], although polyaniline's use within wave guides is likely to be limited due to optical absorption. The very long lifetimes of some of the photoinduced defect states opens the opportunity for application of polyanilines and related ring-containing polymers in areas of optical information storage [36, 50].

III. Summary

The polyaniline family of polymers differs substantially from earlier studied conducting polymer systems. The air stability and ability to vapor, solution, and solid state process the material enable both critical study and potential application. The physics of polyaniline may be tuned from that of filled energy bands to one where the electron-phonon interaction of the Peierls ground state may be important as a function of polymer structure. The polymer shows a broad range of new and novel transport phenomena as well as optical, nonlinear optical, and time dynamic effects. It has been suggested that a new type of polaron involving ring torsion angles underlies many of the phenomena in these materials. The importance of crystallinity in sustaining the metallic state has been uncovered. The combination of unusual properties and processibility and stability increases the attractiveness of the polyanilines for technological application.

IV. Acknowledgment

The authors thank their research groups for their critical contributions to the studies summarized here as well as acknowledge important collaborations with J.P. Pouget (structural studies) and Y. Wei (molecular weight). The research was supported in part by the U.S. Defense Advanced Research Projects Agency through a contract monitored by the U.S. Office of Naval Research.

V. References

1. A.G. Green, A.E. Woodhead: J. Chem. Soc. Trans. **97**, 2388 (1910); **101**, 1117 (1912)
2. R. Willstätter, C. Cramer: Chem. Ber. **43**, 2976 (1910); **44**, 2162 (1911); R. Willstätter, S. Dorogi: Chem. Ber. **44**, 2166 (1911)
3. R. DeSurville, M. Jozefowicz, L.T. Yu, J. Perichon, R. Buvet: Electrochem. Acta, **13**, 1451 (1968)
4. E.M. Genies, A.A. Syed, C.C. Tsintavis: Mol. Cryst. Liq. Cryst. **121**, 181 (1985)
5. A.G. MacDiarmid, J.-C. Chiang, M. Halpern, W.-S. Huang, J.R. Krawczyk, R.J. Mammone, S.L. Mu, N.L.D. Somasiri, W. Wu: Polym. Prepr. **25** 248 (1984); A.G. MacDiarmid, J.-C. Chiang, M. Halpern, W.-S. Huang, S.-L. Mu, N.L.D. Somasiri, W. Wu, S.I. Yaniger: Mol. Cryst. Liq. Cryst. **121**, 173 (1985); J.C. Chiang, A.G. MacDiarmid: Synth. Met. **13**, 193 (1986)
6. E.W. Paul, A.J. Ricco, M.S. Wrighton: J.Phys. Chem. **89**, 1441 (1985)
7. J.P. Travers, J. Chroboczek, F. Devreux, F. Genoud, M. Nechtschein, A. Syed, E. M. Genies, C. Tsintavis: Mol. Cryst. Liq. Cryst. **121**, 195 (1985)
8. W.R. Salaneck, I. Lundström, W.S. Huang, A.G. MacDiarmid: Synth. Met. **13**, 291 (1986)
9. P.M. McManus, S.C. Yang, R.J. Cushman: J. Chem. Soc. 1556 (1985)
10. H.-Y. Choi, E.J. Mele: Phys. Rev. Lett. **59**, 2188 (1987)

11. (a) D.S. Bourdeaux, R.R. Chance, J.F. Wolf, L.W. Shacklette, J.L. Brédas, B. Thémans, J.M. André, R. Silbey: J. Chem. Phys. **85**, 4584 (1986); (b) S. Stafstrom, J.L. Bredas, A.J. Epstein, W.S. Huang, A.G. MacDiarmid: Phys. Rev. Lett. **59**, 1464 (1987)
12. J.M. Ginder, A.J. Epstein, A.G. MacDiarmid: to be published
13. M.C. Dos Santos, J.L. Bredas: Synth. Met. **29**, E321 (1989)
14. F. Wudl, R.O. Angus, Jr., F.L. Lu, P.M. Allemand, D.J. Vachon, M. Nowak, Z.X. Liu, A.J. Heeger: J. Am. Chem. Soc. **109**, 3677 (1987)
15. R.H. Baughman, J.F. Wolf, H. Eckhardt, L. Shacklette: Synth. Met. **25**, 121 (1988)
16. C.B. Duke, A. Paton, E.M. Conwell, W.R. Salaneck, I. Lundström: J. Chem. Phys. **86**, 3414 (1987)
17. J.M. Ginder, A.F. Richter, A.G. MacDiarmid, A.J. Epstein: Solid State Commun. **63**, 97 (1987); Bull. Am. Phys. Soc. **31**, 582 (1986); B.D. Humphrey, J.C. Chiang, W.S. Huang, A.F. Richter, N.L.D. Somasiri, A.G. MacDiarmid, X.Q. Yang, A.J. Epstein, R.W. Bigelow: Bull. Am. Phys. Soc. **30**, 605 (1985)
18. A.J. Epstein, J.M. Ginder, F. Zuo, R.W. Bigelow, H.S. Woo, D.B. Tanner, A.F. Richter, W.S. Huang, A.G. MacDiarmid: Synth. Met. **18**, 303 (1987)
19. K.R. Cromack, M.E. Jozefowicz, J.M. Ginder, R.P. Mcall, A.J. Epstein, E. Scherr, A.G. MacDiarmid: Bull. Am. Phys. Soc. **34**, 583 (1989), and to be published
20. Y. Wei, G-W.Jang, K. Hsueh, E.M. Scherr, A.G. MacDiarmid, A.J. Epstein: to be published
21. N. Theophilou, A.G. MacDiarmid, D. Djurado, J.E. Fischer, and A.J. Epstein: Solid State Sciences, these proceedings
22. X. Tang, E. Scherr, A.G. MacDiarmid, A.J. Epstein: Bull. Am. Phys. Soc. **34**, 583 (1989), and to be published
23. A. Andreatta, Y. Cao, J.C. Chiang, A.J. Heeger, P. Smith: Synth. Met. **26**, 383 (1988)
24. M.E. Jozefowicz, R. Laversanne, H.H.S. Javadi, A.J. Epstein, J.P. Pouget, X. Tang, A.G. MacDiarmid: Phys. Rev. B **39**, 12,958 (1989)
25. J.P. Pouget, M.E. Jozefowicz, A.J. Epstein, X. Tang, A.G. MacDiarmid, to be published
26. P. Robin, J.P. Pouget, R. Comes, H.W. Gibson, A.J. Epstein: Phys. Rev. B **27**, 3938 (1983)
27. H.H. S. Javadi, R. Laversanne, M. Jozefowicz, P. Vaca, A.G. MacDiarmid, A.J. Epstein: to be published
28. F. Zuo, M. Angelopoulos, A.G. MacDiarmid, A.J. Epstein: Phys. Rev. B **36**, 3475 (1987)
29. (a) H.H.S. Javadi, K.R. Cromack, A.G. MacDiarmid, A.J. Epstein: Phys. Rev. B **39**, 3579 (1989); (b) H. Kuzmany, N.S. Sariciftci, H. Neugebauer, A. Neckel, Phys. Rev. Lett. **60**, 212 (1988); (c) J.M. Leng, R.P. McCall, J.M. Ginder, A.J. Epstein, A.G. MacDiarmid, to be published
30. F. Zuo, M. Angelopoulos, A.G. MacDiarmid, A.J. Epstein: Phys. Rev. B **39**, 3570 (1989)
31. S.K. Manohar, A.G. MacDiarmid, A.J. Epstein: Bull. Am. Phys. Soc. **34**, 583 (1989), and to be published
32. Z. Wang, K.R. Cromack, H.H.S. Javadi, J. Yue, J.M. Ginder, A.J. Epstein, A. Ray, S.K. Manohar, A.G. MacDiarmid: Bull. Am. Phys. Soc. **34**, 824 (1989), and to be published
33. M.G. Roe, J.M. Ginder, T.L. Gustafson, M. Angelopoulos, A.G. MacDiarmid, A.J. Epstein: Phys. Rev. B **40** (15 Aug. 1989)
34. M.G. Roe, J.M. Ginder, R.P. McCall, K.R. Cromack, A.J. Epstein, T.L. Gustafson, M. Angelopoulos, A.G. MacDiarmid: Synth. Met. **29**, E425 (1989)
35. M.G. Roe, J.M. Ginder, P.E. Wigen, A.J. Epstein, M. Angelopoulos, A.G. MacDiarmid: Phys. Rev. Lett. **60**, 2789 (1988)

36. R.P. McCall, J.M. Ginder, M.G. Roe, G.E. Asturias, E.M. Scherr, A.G. MacDiarmid, A.J. Epstein: Phys. Rev. B **39**, 10,174 (1989)
37. S. Kaplan, E.M. Conwell, A.F. Richter, A. G. MacDiarmid: J. Am. Chem. Soc. **110**, 7647 (1988)
38. G.B. Blanchet, C.R. Fincher, T.C. Chung, A.J. Heeger: Phys. Rev. Lett. **50**, 1938 (1983)
39. N. Theophilou, D.B. Swanson, A.G. MacDiarmid, A. Chakraborty, H.H.S. Javadi, R.P. McCall, S.P. Treat, F. Zuo, A.J. Epstein: Synth. Met. **28**, D35 (1989)
40. H.E. Schaffer, A.J. Heeger: Solid State Commun. **59**, 415 (1986)
41. M.J. Rice, S.R. Phillpott, Phys. Rev. Lett. **58**, 937 (1987)
42. T. Holstein, Ann. Phys. **8**, 325 (1959); Ann. Phys. **8**, 343 (1959)
43. R.P. McCall, M.G. Roe, J.M. Ginder, T. Kusumoto, A.J. Epstein, G.E. Asturias, E.M. Scherr, A.G. MacDiarmid: Synth. Met. **29**, E433 (1989)
44. Y.H. Kim, S.D. Phillips, M.J. Nowak, D. Spiegel, C.M. Foster, G. Yu, J.C. Chiang, A.J. Heeger: Synth. Met. **29**, E291 (1989)
45. M. Angelopoulos, G.E. Asturias, S.P. Ermer, A. Ray, E.M. Scherr, A.G. MacDiarmid, M. Akhtar, Z. Kiss, A.J. Epstein: Mol. Cryst. Liq. Cryst. **160**, 15 (1988)
46. See, for example, Proc. ICSM '88 Synth. Met. 28 (1989)
47. H.H.S. Javadi, M. Angelopoulos, A.G. MacDiarmid, A.J. Epstein: Synth. Met. **26**, 1 (1988); H.H.S. Javadi, F. Zuo, M. Angelopoulos, A.G. MacDiarmid: Mol. Cryst. Liq. Cryst. **160**, 225 (1988)
48. A.J. Epstein, J.M. Ginder, M.G. Roe, T.L. Gustafson, M. Angelopoulos, A.G. MacDiarmid: Mat. Res. Soc. Symp. Proc. **109**, 313 (1988)
49. J.M. Ginder, A.J. Epstein, A.G. MacDiarmid: Synth. Met. **29**, E395 (1989)
50. R.P. McCall, J.M. Ginder, A.J. Epstein, A.G. MacDiarmid, to be published

Characterization of Vapor-Deposited Polyaniline

P. Dannetun[1], *K. Uvdal*[1], *R. Lazzaroni*[1], *W.R. Salaneck*[1], *E. Scherr*[2], *A.G. MacDiarmid*[2], *and T. Hjertberg*[2]

[1]Department of Physics, IFM, Linköping University, S-581 83 Linköping, Sweden
[2]Department of Chemistry, University of Pennsylvania, Philadelphia, PA 19104-6323, USA
[3]Department of Polymer Technology, Chalmers University, S-412 98 Göteborg, Sweden

Abstract

High quality films of polyaniline can be prepared by an open boat vapor-deposition process and chemically prepared emeraldine powder as the starting material. These films have been characterized using Optical Absorption Spectroscopy, Fourier Transform Infra Red Spectroscopy, X-Ray Photoelectron Spectroscopy, molecular weight studies and electrical conductivity measurements. The properties of these films, after protonation, are very close to those of conventional emeraldine salt.

1. Introduction

Electrically conducting polymers, including polyaniline, are usually prepared by wet chemical and/or catalytic methods. Many of these methods are incompatible with possible future applications of organic conducting polymers /1/. For example, the possibilities of using polyaniline as the active component in "smart windows" /2/ would be enhanced if a method of thin-film fabrication could be devised, which would allow for a type of "assembly line" device fabrication. In this contribution, we discuss the latest results of a study of thin films of polyaniline fabricated by an open boat vapor deposition technique /3/. The films have been characterized by a variety of methods, including optical absorption spectroscopy, X-Ray Photoelectron Spectroscopy (XPS), Fourier Transform Infra Red Spectroscopy (FTIR) and electrical conductivity.

The general formula for the base (unprotonated) form of polyaniline /4/ is

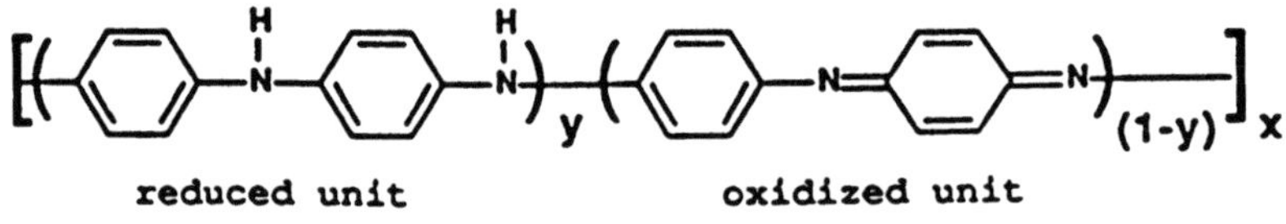

The oxidation state corresponding to y = 0.5 is called emeraldine, and contains equal numbers of oxidized and reduced repeat units. y = 1 corresponds to

Springer Series in Solid-State Sciences, Vol. 91
Electronic Properties of Conjugated Polymers III
Editors: H. Kuzmany · M. Mehring · S. Roth

leucoemeraldine, the completely reduced state, while y = 0 corresponds to pernigrailine, the completely oxidized state. In the emeraldine salt, the electrical conductivity is about 5 S/cm /4/.

2. Experimental

The emeraldine base form of polyaniline was precipitated as a powder from aniline and $(NH_4)_2S_2O_8$, and converted to analytically pure emeraldine base /4/. Vapor deposition was carried out much as described previously /3/. In the present studies, however, the as-made films were exposed immediately to N_2 gas directly after vapor-deposition. In addition, directly following vapor-deposition, the films were exposed immediately to N_2 gas (instead of to laboratory air, as done previously /3/). Thereafter, the films were handled mostly in dry N_2 to almost eliminate exposure to the atmosphere. Electrical conductivity measurements were made by the four-point-probe method, using a commercial Alessi Industries instrument. The thickness of the films was determined using a Sloan Dektak 3030. The deposition times were adjusted to result in typical thickness values of near 5000 Å. For FTIR we used a Brucker IFS 113v spectrometer. Optical absorption studies were carried out using a Perkin Elmer model Lambda 9 spectral analyser, and finally, XPS was carried out in a special instrument of our own design and construction /5/. The molecular weight distribution of the vapor-deposited films was studied by Gel Permiation Chromotography (GPC) using a Perkin-Elmer Series 2 liquid chromatograph employing two Shodex columns (A-80 M/S and A-80 6/S). The solvent was dimethyl formamide (DMF), and correction for the influence of the solvent was made according to Ref. 6.

3. Results and Analysis

The as-deposited films were found to consist of molecular chains with *short conjugation lengths* in a state very close, but not exactly equal to, *leucoemeraldine*. The FTIR spectrum of an as-prepared film, shown in Fig. 1, is *essentially* identical to the spectrum of a phenyl-capped tetramer of leucoemeraldine /7/, but not the same as that of electro-chemically prepared emeraldine base /8/. The emeraldine apparently divides into oligomers, of which the completely reduced portions are selectively deposited, during the vapor-deposition process. One possibility is that the reduction might be due to the higher reactivity of the N=C double bonds. If this is the case, then we have no evidence for where in the process the reduction takes place. Another possibility is that the polymer decomposes in the heating step (in the open boat), and only the more stable portions survive to (re)polymerize on the substrate. The present results are consistent with results of previous studies /9/, however, which showed that is was difficult to evaporate molecules containing more than about five rings.

A comparison of the optical absorption spectrum of an as-deposited film, Fig 2a, with that of Masters /11/ or Cao *et al* /10/ ,Fig 2c, also indicates that the as-prepared films are *essentially* in the leucoemeraldine state. In comparing with spectra of

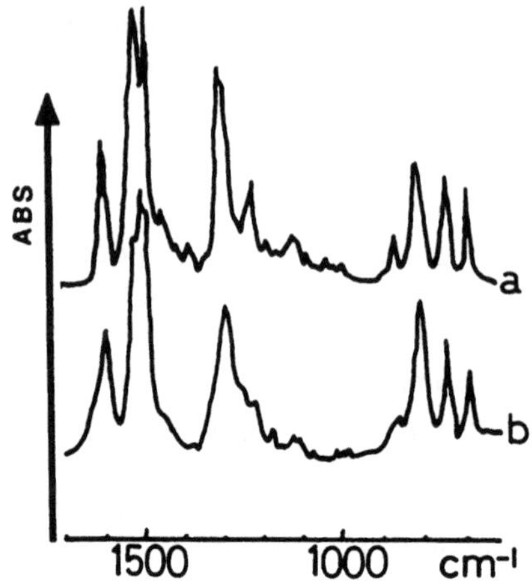

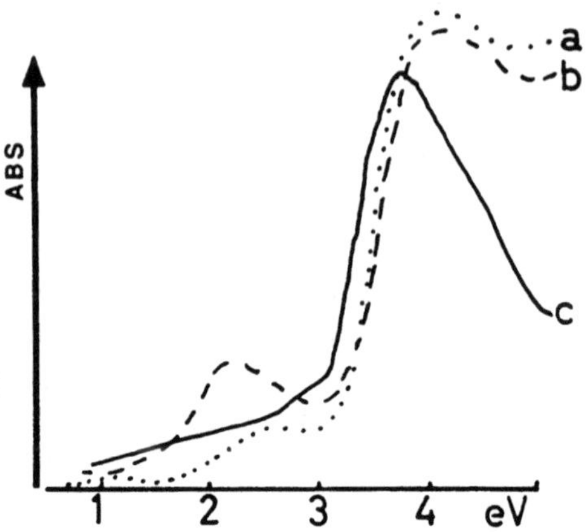

Fig. 1: IR spectra: (a) After Shacklette (Ref. 7), and (b) as-vapor-deposited
Fig. 2. Optical absorption spectra: (a) as-vapor-deposited; (b) air-exposed and (c) leucoemeraldine (Ref.10)

electro-chemically prepared polyanilines, one must take into account the fact that the quinoid rings have a higher optical excitation probability than the benzene rings /12/. Thus, the intensity ratio, between the absorption peak which comes predominantly from the benzenoid rings (4 eV) and the peak corresponding to the quinoid rings (near 2 eV), will vary with film *thickness* /12/.

These vapor-deposited films are completely soluble in dimethyl formamide, DMF. Using GPC, Gel Permiation Chromotography, with polystyrene as a standard, molecular weights up to 10^4 Daltons are detected /3/. This value must be corrected by a factor of 0.08 /6/, however, to take into account the effects of the solvent /6/, yielding values up to about 10^3 Daltons. The molecular weight distribution is bimodal /3/, however, indicating that more than one polymerization mechanism is present /6/. The high molecular weights, together with the results from the IR and optical spectroscopies, indicate that the films could be cross-linked and/or contain conjugation defects, such that the effective conjugation length corresponds to about five aromatic rings.

Upon exposure to air/atmosphere for long periods of time, the as-deposited (mostly) leucoemeraldine films oxidize to a state between leucoemeraldine and emeraldine. During air-exposure, the optical absorption spectra indicate the slow appearence of quinoid rings, for which a new peak around 2.2 eV appears, as seen in Fig 2b. Furthermore, the XPS N(1s) peak broadening (not shown here) can be related to the existence a contribution from both the amine and the imine groups.

Films protonated in 1 M HCl for several weeks, exhibit optical absorption spectra (Fig. 4c) that indicate that they have been converted to a state very near emeraldine. These films have conductivities up to 5 S/cm, which is the same value as obtained for protonated emeraldine prepared by wet chemical and/or electrochemical

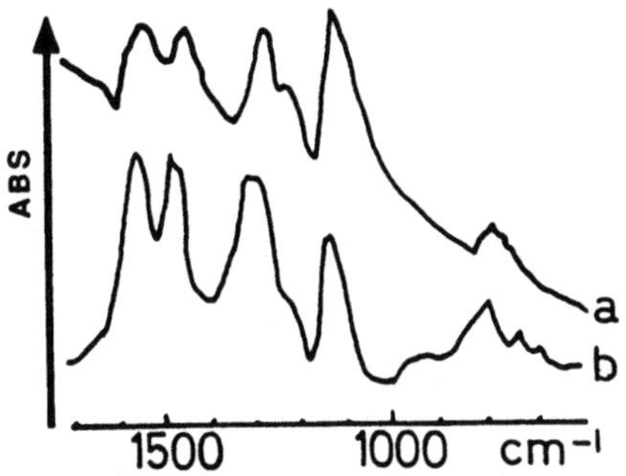

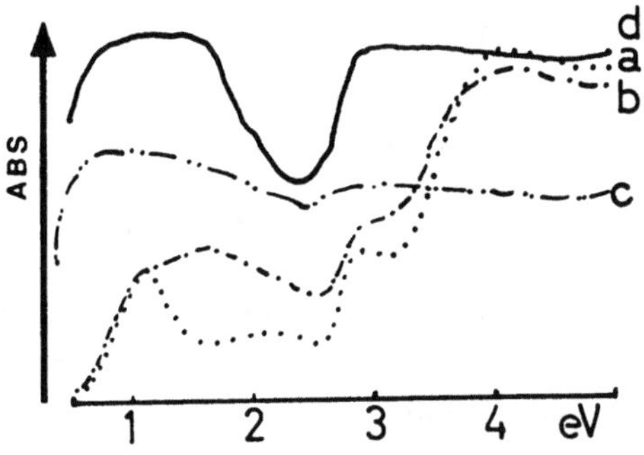

Fig. 3 (left): FTIR: (a) vapor-deposited doped and oxidized, and (b) emeraldine salt (Ref. 7)

Fig. 4 (right): Optical absorption: (a-c), different doping and oxidation levels of polyaniline, and (d) emeraldine salt (Ref. 10)

methods /4/. A combination of exposure to air and treatment with HCl results in intermediate behavior. Freshly-made films, which have been treated in 1 Molar HCl for about 20 minutes, are still near the undoped leucoemeraldine state, indicating that the combined doping-and-conversion process is rather slow.

Such a slow doping/conversion rate is particularly noteworthy, considering the fact that our films are very thin (5000 Å), and that a colour-change is observed visually immediately after contact with the HCl solution. Optical absorption spectra of these same films (20 minutes in 1 M HCl), after the films have been in air for a few days, however, are close to the optical absorption spectra of "conventional" *protoemeraldine* salt ($y = 0.75$) /10/, Fig 4a. After treatment for longer times in the 1 M HCl, the steps of protonation plus the oxidation in the air appear to occur faster. This is shown in Fig 4b (optical absorption) and Fig 3 (FTIR). Further details are under investigation.

4. Conclusions

Thin films of polyaniline of relatively high molecular weight can be prepared using a simple open-boat vapor-deposition process, using chemically prepared emeraldine base as the starting material. The as-deposited films are close to the *leucoemeraldine* form, but can be converted to the conducting *emeraldine salt* by treatment with a protonic acid. The electrical conductivity of these vapor-deposited films, after treatment with HCl, is equal to that reported for electro-chemically made emeraldine salt.

5. Acknowledgements

The authors are indebted to L. Bertilsson for FTIR measurements and L. Gustavsson for sample preparation. Studies of conducting polymers in Linköping are carried out with support from the Nordisk Industrifond (NIF) and the National Swedish Board for Technical Developement (STU).

6. References

1 W.R. Salaneck, C.R. Wu and J.O. Nilsson, *J. Synth. Met.*, **21**, 57 (1987).
2 S.C. Yang, Univ. Rhode Island, USA, private communication.
3 K. Uvdal, M. Lögdlund, P. Dannetun, L. Bertilsson, S. Stafström, W.R. Salaneck, A.G. MacDiarmid, A. Ray, E.M. Scherr and A. J. Epstein, *J. Synth. Met.*, **29**, E451 (1989).
4 J. -C. Chiang and A.G. MacDiarmid, *J. Synth. Met.*, **13**, 193 (1986).
5 W.R Salaneck, R. Bergman, J.-E. Sundgren, A. Rockett and J.E. Greene, *Surf. Sci.*,**198**, 461 (1988).
6 X. Tung, Y. Sun and Y. Wei, *Makromol. Chem., Rapid Commun*, **9**, 829 (1989).
7 L.W. Shacklette, J.F. Wolf, S. Gould and R.H. Baughman, *J. Chem. Phys.* **88**, 3955 (1988).
8 W. R. Salaneck, B. Liedberg, O. Inganäs, R. Erlandsson, I. Lundström, A. G. MacDiarmid, M. Halpern, and N. L. D. Somasiri, *Mol. Cryst. Liq. Cryst.*, **121**, 191 (1985).
9 W. R. Salaneck, C. R. Wu, S. Stafström, M. Lindgren, T. Hjertberg, O. Wennerström, M. Sandberg, C.B. Duke, E. Conwell, and A Paton, *Electrical Properties of Conjugated Polymers*, H. Kuzmany, M. Mehring and S. Roth, Ed's (Springer-Verlag, Berlin, 1987) p-1464.
10 Y. Cao, P. Smith and A.J. Heeger, *Spectroscopic studies of polyaniline in Solution in Sulfuric Acid and of Polyaniline Films Spin-Cast from Sulfuric Acid*, Preprint.
11 J. Masters *et al*, to be published.
12 F. Zuo, R. P. McCall, M. G. Roe, A. J. Epstein, G. Asturias, S. I. Ermer, A. Ray, and A. G. MacDiarmid, *J. Synth. Met.*, **29**, E445 (1989).

XPS Studies of Polyaniline

A.P. Monkman[1], *D. Bloor*[2], *G.C. Stevens*[3], *and J.C.H. Stevens*[4]

[1]Molecular Electronics Group, SEAS, University of Durham, Science Laboratories, South Road, Durham, DH1 3LE, UK
[2]Department of Physics, Queen Mary College, Mile End Road, London, E1 4NS, UK
[3]Central Electricity Research Laboratories, CEGB, Kelvin Avenue, Leatherhead, Surrey, DT22 7SE, UK
[4]Berkeley Nuclear Laboratories, CEGB, Berkeley, Gloucestershire, Gl13 9PB, UK

Abstract. X-Ray photoelectron spectroscopy (XPS) has been used to study thin films of polyaniline (PANi) in various oxidation states. Carbon, nitrogen, oxygen and chlorine core level binding energies have been measured. Comparison to theoretical predictions from a simple potential model show a good fit between experiment and theory. In general the results indicate a change from benzenoid to quinoid nature with increasing oxidation. Only in the semioxidised (emeraldine) form, do positively charged moieties dominate. There is good evidence of metal - chloride-like bonding (dopant ions used in these experiments were chlorine), ring chlorination, and double bonded oxygen.

1. Introduction

The technique of XPS has developed over the last 30 years, but is still firmly based on the pioneering work of Prof. Kai Siegbahn, and his group at the University of Uppsala [1]. The technique has been successfully extended into the field of polymers by Prof. D Clarke and his group at the University of Durham [2]. The basis of XPS originates from Einstein's photoelectric equation. When a material is irradiated with high energy photons ie. x-rays, electrons from the atoms comprising the material will be ejected, with a kinetic energy equal to that of the incident photon, minus an amount of energy E_i. This is referred to as the binding energy of the photoelectron, and is characteristic of each element. There are of course a few correction terms to be considered, but the basis of the technique arises from the characteristic binding energy assoicated with each element. These measured binding energies are perturbed to a small degree, by the atoms local electronic environment. Such perturbations in the case of core level electrons, are the changes in the charge density of the valance electrons due to bonding. Thus, information on atomic bonding, and local environment can be gained by measuring the shifts of the binding energy from some unperturbed reference binding energy.

To first approximation, the perturbations in valence electron charge density can be calculated using electrostatic potential energy arguments. The electrostatic potentials of the core level electrons can be calculated using Pauling electronegativities and group electronegativities. Siegbahn et al.[1] have shown that there exists a linear relationship between these calculated potentials and measured binding energy shifts. We have used this

Springer Series in Solid-State Sciences, Vol. 91
Electronic Properties of Conjugated Polymers III
Editors: H. Kuzmany · M. Mehring · S. Roth

simple model to calculate expected shifts in binding energy for PANi in various oxidation states.

2. Experimental Section

Sample preparation has been described previously [3]. Thin PANi films were deposited on Pt electrodes, and after synthesis, washed, and oxidised to various states, ie. -0.2v, 0.4v and 0.8v with respect to SCE. Samples were then washed in acetone and left to dry (in air). The spectrometer used for these studies was an AEI ES 200B, experimental details can be found in ref. 3. Data acquisition and deconvolution were performed with a DEC PDP 11-03 minicomputer, using the kratos DS800 XPS/AES data system. For peak deconvolution, gaussian line shapes were used with FWMH being fixed. The values of FWHM were taken from a standard reference text [4]. In order to ascertain the degree of sample dedoping whilst in the high vacuum of the spectrometer, optical spectra were recorded after the XPS studies. This indicated that there was still a high degree of doping after the vacuum treatment.

Table 1. C1s core level binding energies

Oxidation Potential	Binding Energy (eV)	Intensity	Area
-0.2 V	284.87 286.50 288.09	93.7 18.1 4.9	80.6 15.6 4.1
0.4 V	284.74 285.64 287.59	72.0 35.9 13.0	58.8 29.3 10.6
0.4 V	284.80 285.54 287.25	68.8 35.6 15.9	56.2 29.1 13.0
0.8 V	284.74 286.03 288.13	86.2 29.0 18.0	64.9 21.8 13.4

Table 2. N1s core level binding energies

Oxidation Potential	Binding Energy (eV)	Intensity	Area
-0.2 V	398.19 399.50 400.54 402.32	26.0 86.1 34.2 2.1	17.4 57.6 22.8 1.3
0.4 V	398.19 399.60 400.75 402.48	10.9 75.3 57.2 13.0	6.8 47.4 36.0 8.1
0.4 V	398.24 399.54 400.75 402.38	24.2 82.8 43.8 9.7	14.8 50.5 26.7 5.9
0.8 V	398.24 399.54 400.85 402.61	44.2 87.0 26.3 9.1	25.9 50.9 15.4 5.3

Table 3. Cl2p core level binding energies

Oxidation Potential	Binding Energy (eV)	Intensity	Area
-0.2 V	197.49 198.54 200.60	16.5 6.1 93.4	14.1 5.2 80.1
0.4 V	197.31 198.63 200.45	92.1 20.2 58.2	54.0 11.8 34.2
0.4 V	197.59 198.89 200.70	83.3 33.6 60.5	46.8 18.8 34.0
0.8 V	197.65 199.20 200.63	17.5 40.0 71.5	13.2 30.4 54.3

Table 4. O1s core level binding energies

Oxidation Potential	Binding Energy (eV)	Intensity	Area
-0.2 V	532.54 533.94	83.8 40.0	66.9 31.8
0.4 V	531.99 533.43	83.2 40.1	66.9 32.2
0.4 V	532.49 534.09	89.8 39.2	68.7 29.8
0.8 V	532.20 533.69	82.6 43.7	65.3 34.5

Table 5. Potential model calculations. ß=quinoid ring, ø=phenyl ring

Group	Core Level	Binding Energy (eV)	Group	Core Level	Binding Energy (eV)
1. ø-N-ø H	N1s	399.3	9. H + N-ø-N- H	C1s	284.6
2. ß=N-ø	N1s	398.8	10. N=ß=N	C1s	285.1
3. H + ø-N-ø H	N1s	401.5	11. + N=ß=N H	C1s	285.3
4. + ß=N-ø H	N1s	402.5	12. N=ß=O	C1s	287.5
5. ø-N-ø Cl	N1s	399.4	13. N=ß=O	O1s	534.6 *
6. + ø-N-ø Cl-	N1s	403.8	14. ø-O-H	C1s	284.6
7. H + ø-N-ø Cl-	N1s	401.8	15. ø-O-H	O1s	531.1 *
8. N-ø-N	C1s	284.5	16. N-ø-Cl	C1s	285.2

Table 6. Assignments of observed binding energy peaks

PANi * B.E. (eV)		Model no. and Cal. B.E. (eV)	Molecule	B.E (eV)
N1s	398.2 399.5 400.7 402.4	2. 398.8 1. 399.3 3. 401.5 4. 402.5	(see below) $EtNH_3Cl$ Me_4NCl	398.8 399.6 401.1 402.1
C1s	284.7 285.5 to 286.5 ≈288	8. 284.6 9. 284.6 16. 285.2 11. 285.3 12. 287.5	Benzene - PVC - $O=(CH_2)_2$	284.5 - 286.4 - 287.6
O1s	532.5 " 533.7	- - 15. 531.1 13. 534.7	water PVOH øOCOOø	532.8 532.8 533.5
Cl 2p	197.6 198.7 200.6	$H_2NC(=NH)NH_2.HCl$ and monovalent metal chlorides divalent metal chlorides PhCl & chloroanil-pyridine		197.8 ca.198 ca.199 200.8

(* an average from all spectra for general comparisons)

Table 7. Effects of delocalisation on the measured binding energies of nitrogen-containing compounds

Compound	Binding Energy (eV)		
	H_2N-ø-	ø-N=N-	ø-NO_2
H_2N-ø-N=N-ø-NO_2 ↕ H_2N^{+}=β=N-N=β=NO_2^{-}	399.6	398.8	405.0

2.1 Results

The measured binding energies of the deconvoluted spectra are given in tables 1 to 4. In table 5 binding energies predicted from model calculations are given. Assignments to the spectra are given in table 6 with the aid of information from model compound spectra and potential model calculations. The effects of delocalisation on binding energy are summarised in table 7.

2.2 Discussion and Conclusions

The Cls spectral feature observed, at ca. 284.7 eV is due to the carbon backbone of the polymer. From the model calculations it is seen that the position of the Cls peaks due to C-Cl, $^{\oplus}$N-ϕ and $^{\oplus}$N=β all lie fairly close together at ca. 285 eV, with the latter showing the greatest shift. Assuming that the experimentally observed peak has components due to all three moieties (unresolvable), and that the C-Cl peak is roughly constant throughout, one can attempt to interpret the spectra. In the reduced film, a small peak is seen at 286.5 eV (highest binding energy position) and so must correspond to a predominance of $^{\oplus}$N=β, indicating that the effect of residual oxidation, such as that caused by oxygen or degradation, is manifest as such moieties. On initial oxidation, the peak is seen at 285.6 eV with increased intensity. This is the lowest position of the peak and as such must be ascribed to a predominance of $^{\oplus}$N-ϕ. In the fully oxidised state, the intensity decreases, though it is still greater than observed initially, and is seen at 286 eV. It is assumed that here all three components contribute nearly equally, with an overall decrease in the amount of each present. The high binding energy peak at ca. 288 eV, is assigned to the presence of C=O bonds, due to polymer degradation. It is seen that this peak increases with increasing oxidation, as expected. In the reduced state, neutral carbon species dominate, with only a small proportion of seemingly charged, oxidised moieties, attributed to the effects of aerial oxidation. In the semioxidised state, neutral carbon content diminishes, and a large increase in charged benzenoid like carbon (rings) is seen, in good correlation to the nitrogen data. In the fully oxidised state, neutral carbon again dominates the spectrum, with some residual charged forms, mainly ascribed to be quinoid in charcter. All of which is self-consistent with the nitrogen spectra.

For the observed Nls spectra, the peaks seen at 398.2 and 399.2 eV, are assigned to neutral amine (1) and imine (2) moieties respectively. The peaks seen at higher binding energy, ca. 400.7 and 402.4 eV are ascribed to the presence of positively charged amine (3) and imine (4) moieties respectively. Note, calculated binding energies for the imines are better than those for the amines. This is put down to the effect of conjugation, which Pauling electronegativities are poor in dealing with. Comparing the measured relative areas gives an insight into the changes of structure during oxidation of the polymer. In the reduced state it is seen that the neutral amine dominates. On oxidation to the partially oxidised form, the level falls sharply and remains fairly constant with further, increasing oxidation. This behaviour is also seen in the netural imine case, although here oxidation to the partially oxidised state causes the peak to all but disappear. This evidence would support the picture whereby imine moieties are preferentially protonated. Such moieties, if unstable with respect to a charge separation, form rings with a dominant benzenoid character. The presence of neutral amine in all spectra, indicate that a proportion of the rings, assumed to be unoxidised segments, remain unprotonated. This may be an experimental artifact since it is probable that to some degree HCl is removed from the samples when placed in the high vacuum of the sample chamber of the spectrometer. With further oxidation, the neutral imine content rises to its highest level, showing that deprotonation must occur during this second oxidation step, even at ca. pH 0 (assuming that not all the HCl is removed by the vacuum)! For the protonated (charged) species, it is clearly seen that the first oxidation step greatly increases the amount of protonated amine, and the second sharply decreases it (to its lowest level). This behaviour is again reflected in the charged imine moiety, although there is a greater amount remaining in the fully oxidised polymer. In all cases the amounts of such quinoid moieties are small. From the potential model calculations, one can conclude that the bulk of the positive charge on the backbone resides on the (amine) nitrogen atoms.

Assignment of the two oxygen species is a little more difficult as the correlations between Pauling charge and binding energy are not as good as those for C and N. However, good agreement to C=O binding energies is seen for the PANi peak at ca. 533.7 eV, ascribed to polymer degradation products. With the ca. 532.5 eV peak being tentatively assigned to either hydrogen bonded water, or C-OH groups. It is seen that the relative intensity of the two components is fairly constant throughout, implying that the latter is more likely.

Finally, the assignment of the Cl2p peaks can be achieved only by comparison to model compounds. It can clearly be seen that the peak at ca. 200.6 eV is due to the presence of C-Cl bonds, whose number increases with increasing oxidation, indicative of ring substitution of the halogen. However, assignment of the lower binding energy peaks is more difficult. The peak of ca. 196.7 eV is assumed to be due to the dopant chloride ions. This peak is seen to increase dramatically after the first oxidation step, as expected, and decreases after the second, characteristic of some degree of deprotonation after the second oxidation. The nature of the dopant interaction with the polymer chain is very similar to that observed for metal chlorides, indicating a large degree of charge transfer and close packing of the Cl ions to the nitrogen sites along the chain. Such an interaction may well give rise to localisation of the positive charge on the nitrogen ie. counter ion pinning. The peak at ca. 198.6 eV is not understood. This could be due to the presence of trapped HCl in the polymer, or due to a reaction product at the end of a broken chain, ie. one containing both O and Cl groups.

2.3 Acknowledgement

I would like to thank Berkeley Nuclear Laboratories for the use of the XPS spectrometer and deconvolution software.

References

[1] K. Siegbahn, "ESCA Applied to Free Molecules", North-Holland, Amsterdam (1971).

[2] D. T. Clark, "Electron Emission Spectroscopy", p.373., 3rd Ed., ed. W. Dekeyser, L. Fiermans, G. Vanderkelen and J. Vennik, D. Reidel Pub. Co., Dordrecht, Holland (1973).

[3] A. P. Monkman, D. Bloor, G. C. Stevens and J. C. H. Stevens, J. Phys D: Appl Phys. 20 (1987) 1337.

[4] "Handbook for XPS Spectroscopy", ed. G. E. Mullenbey, Perkin Elmer.

Photoelectron Spectroscopic Study of Polyaniline as a Function of the Electrochemical Potential

P. Snauwaert[1], *R. Lazzaroni*[1], *J. Riga*[1], *J. Verbist*[1], *and D. Gonbeau*[2]

[1]Laboratoire Interdépartemental de Spectroscopie Electronique, Facultés Universitaires Notre-Dame de la Paix, B-5000 Namur, Belgium
[2]Laboratoire de Physico-Chimie Moléculaire ERA 895, Université de Pau et de l'Adour, F-64000 Pau, France

Abstract. We have studied the electronic structure of electrochemically-prepared polyaniline, by means of X-ray Photoelectron Spectroscopy (XPS). Measurements on a model compound, N,N'-diphenyl guanidine, provided references for the binding energies of the amine and imine groups. Polyaniline N1s spectra have been studied as a function of the electrochemical potential; their evolution has been compared with the results of cyclic voltammetry experiments.

1. Introduction

Among the organic conducting polymers, polyaniline deserves special attention, due to the fact that both oxidation and protonation processes rule its electrical properties. It is by now well-established [1] that the highest values of conductivity are obtained for the pH=0-protonated emeraldine form. Several other oxidation states also exist (Fig. 1), corresponding to different amine (-NH-) to imine (=N-), or benzenic to quinoid rings, ratios.

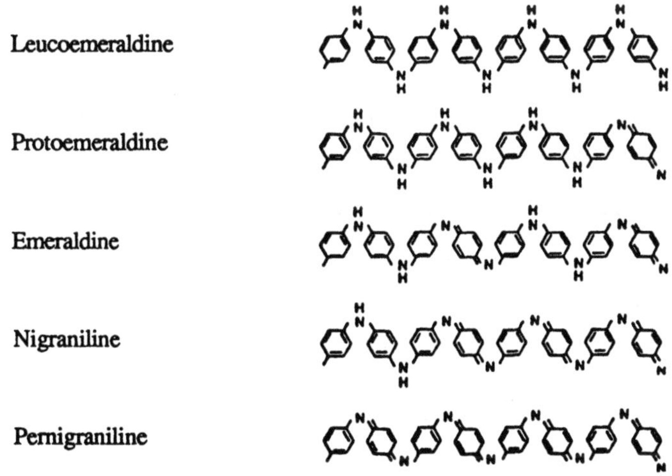

Figure 1: Polyaniline redox states (the chain is represented by an octamer)

Springer Series in Solid-State Sciences, Vol. 91
Electronic Properties of Conjugated Polymers III
Editors: H. Kuzmany · M. Mehring · S. Roth

In this work, we have followed, by means of X-ray photoelectron spectroscopy (XPS), the evolution in the electronic structure of polyaniline as a function of the degree of oxidation, trying to relate this evolution to the electrochemical properties of the polymer.

2. Experimental

Polyaniline is prepared electrochemically from aqueous solutions containing HBF_4 (1M) and aniline (0.1 M). The polymer films are grown onto Indium Tin Oxide (ITO) conducting glass electrodes at a constant potential of 0.8 V vs. SCE; a platinum wire is used as counter-electrode.

The so-obtained compounds are rinsed to remove adsorbed aniline and oligomers, then fully reduced in HBF_4 1M by setting the potential at -0.2 V until the current drops near zero. To obtain the polymer structure typical of a given potential in the -0.2 to +0.8 V range, the ITO electrode is kept at this potential until equilibrium, i.e., until no significant current is observed. NaOH-washing finally leads to the corresponding base form.

X-ray Photoelectron Spectroscopy (XPS) and Cyclic Voltammetry (CV) measurements are performed immediately after synthesis or electrochemical treatment.

3. Results and Discussion

Since benzenic and quinoid rings contribute to the XPS C1s spectrum in a very similar way [2], we focussed our attention upon the N1s region, which usually consists of one broad line, containing contributions from both amine and imine nitrogens. It is well-known [3] that the more basic species, i.e., the imine, appears at lower binding energies relative to the amine, which is a weaker base. In order to obtain more quantitative information, however, we have studied N,N'-diphenyl guanidine, a model compound containing two amine groups and one imine group (Fig. 2).

The two N1s peaks, appearing at 399.1 and 398.0 eV, are easily assigned to the amine and to the imine species, respectively. The experimental spectra of polyaniline oxidized at different electrochemical potentials are then fitted by two contributions, located at these binding energies.

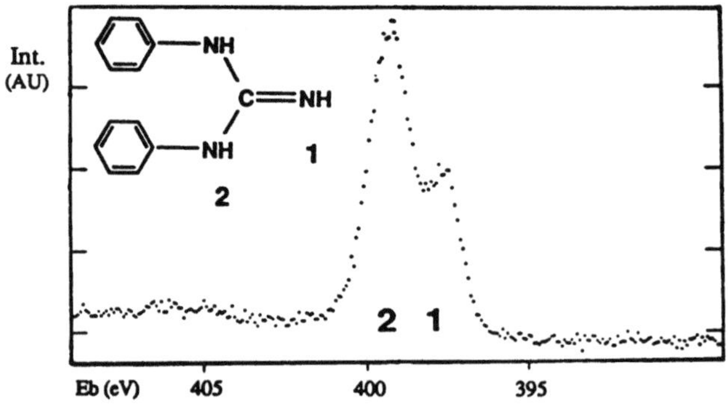

Figure 2: XPS N1s spectrum of N,N'-diphenyl guanidine

The relative areas of these two components yield the amine/imine ratio, hence the oxidation state (Tab. 1).

Table 1: Amine and imine content, and oxidation state in polyaniline as a function of the electrochemical potential, from N1s curve-fitting

Potential (V)	Amine (%)	Imine (%)	Oxidation state
-0.200	97	3	Leucoemeraldine
+0.150	76	24	Protoemeraldine
+0.300	54	46	Emeraldine
+0.600	50	50	Emeraldine
+0.800	30	70	Nigraniline

The standard error on the amine and imine content is ≈ 5%.

At -0.2 V, only amine nitrogen is detected, indicating that the polymer chains are in the leucoemeraldine state. In this case, the N1s binding energy (399.0 eV) is almost identical to that of the amine groups in phenylguanidine. As the potential is increased, the imine concentration rises gradually to around 25% (protoemeraldine), then to 50% (emeraldine) and finally up to 75% (nigraniline). Note that under these conditions, it is not possible to reach the completely oxidized structure, pernigraniline, which contains only imine nitrogen.

As shown in Figure 3, the evolution of the N1s spectrum nicely parallels the cyclic voltammogram of polyaniline. Three plateaus are observed, corresponding to the

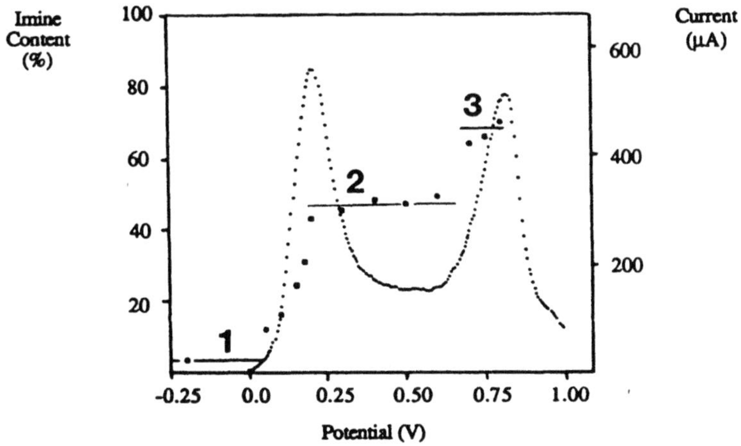

Figure 3: Imine content (•) as a function of the electrochemical potential; comparison with cyclic voltammetry (···)

leucoemeraldine (1), emeraldine (2), and nigraniline (3) forms; the two steps between these plateaus can be related to the two peaks appearing in the CV. The first one (≈ +0.2V), is due to the removal of 4 electrons (and 4 protons) per 8 monomer units, leading to polyemeraldine [4]; the extraction of 2 more electrons (and 2 more protons) gives rise to the second wave at about +0.7V, for which the polymer is in the fully oxidized form [5] (in our case, nigraniline).

We are currently studying related systems, such as substituted polyanilines. The effect of protonation on the electronic structure of unsubstituted polyaniline is also under investigation.

4. Conclusions

From X-ray photoelectron spectroscopy measurements, it is thus possible to determine the oxidation state of polyaniline and to follow its evolution as a function of the electrochemical potential. The changes observed in the electronic structure of the polymer are in nice agreement with its electrochemical properties.

Acknowledgements. We are grateful to P. Dannetun and Prof. W.R. Salaneck for useful comments on the manuscript.

References

1. A.G. MacDiarmid, J.C. Chiang, A.F. Richter, N.L.D. Somasiri: in *Conducting Polymers*, L. Alcacer Ed., (Reidel, Dordrecht, 1986).
2. P. Snauwaert, R. Lazzaroni, J. Riga, J.J. Verbist: Synth. Met. **16**, 245 (1986).
3. T.H. Lee, J.W. Rabalais: J. Electron Spectr. Relat. Phenom. **11**, 123 (1977).
4. E.M. Génies, M. Lapkowski: J. Electroanal. Chem. **220**, 67 (1987).
5. W.W. Focke, G.E. Wnek, Y. Wei: J. Phys. Chem. **91**, 5813 (1987).

Spectroscopic Studies on the Electrochemical Doping of Polyaniline: Quantum Size Effects on the Metallic State

N.S. Sariciftci [1], *H. Kuzmany* [1], *and H. Neugebauer*

[1] Institut für Festkörperphysik, Universität Wien, and Ludwig Boltzmann Institut für Festkörperphysik Strudlhofgasse 4, A-1090 Wien, Austria

[2] Institut für Physikalische Chemie Universität Wien, Währingerstr. 42, A-1090 Wien, Austria

Abstract. In this work we present results of IR studies on polyaniline. A diagram for the dopant concentration as a function of the electrochemical potential is achieved using a spectroscopic technique. Conductivity and enhanced IR absorption correlate linearly with doping. For the metallic state of PANI we discuss the possibility of quantum size effects in small metallic particles.

1. Introduction

Research on conducting polymers in the last decade rediscovered polyaniline (PANI) as an interesting member with unique physical properties. The nature of the metallic state is supposed to be a polaron lattice [1] with a granular morphology where metallic islands are embedded in an insulating matrix [1,2]. From recent results on the spin dynamics, the metallic structures in PANI are suggested to consist of single metallic chains [3].

In this contribution we present for the first time a diagram for the dopant concentration as a function of the electrochemical potential in pH≈0 electrolyte obtained with the ammonia-flush-conversion (AFC) method. The physical properties like conductivity and IR electronic absorption are studied as a function of the dopant concentration. Quantum size effects are discussed for the properties of the microscopic metallic structures in PANI. The broad background IR absorption, the behaviour of the 1 eV absorption band, as well as the magnetic susceptibility data can be qualitatively explained in this model.

Springer Series in Solid-State Sciences, Vol. 91
Electronic Properties of Conjugated Polymers III
Editors: H. Kuzmany · M. Mehring · S. Roth

2. Experimental

Details of the synthesis are discussed in previous works [4]. The electrode was a Au sputtered ZnSe substrate for the IR transmission experiments. The electrolyte consists of 1 M $HClO_4$ + 0.1 M $NaClO_4$. All potentials refer to SCE. The AFC studies were performed by transforming the salt form of the polymer to the base form using NH_3 gas [5]. The produced NH_4ClO_4 is directly correlated with the original anion content of the salt form. This procedure was repeated for several potentials on the same sample, yielding a diagram for the relative anion content during electrochemical processes. This technique allows for the first time the determination of anion content in PANI spectroscopically.

3. Results

Fig. 1 shows the IR response from the anion content in PANI as a function of the electrochemical potential. The middle oxidized metallic form of PANI exhibits the maximum value. During further oxidation a deintercalation of anions is observed. Fig. 2 shows the integrated intensity of the broad background absorption
around 0.5 eV [5] and the conductivity [6] as a function of the dopant concentration using the scaling of Fig. 1. The behaviour of the 1 eV electronic absorption band as a function of the electrochemical potential is displayed in Fig. 3. To visualize the two different branches we plotted

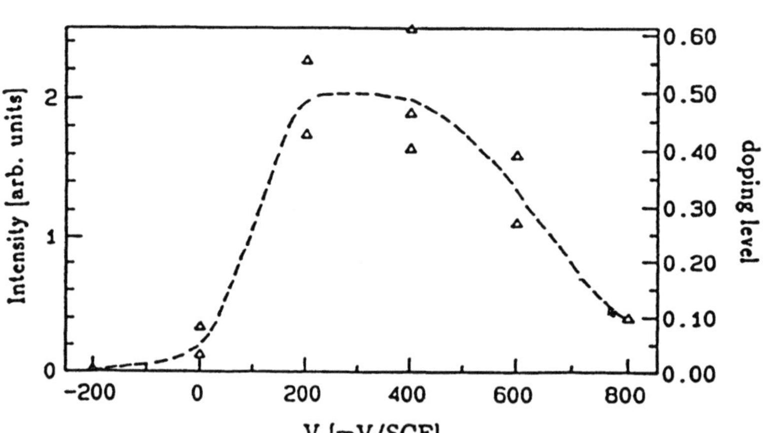

Fig. 1. Integrated intensity of 1100 cm^{-1} absorption band of NH_4ClO_4 for several oxidation levels as obtained with the AFC method. The doping level on the right hand side ordinate is achieved by scaling the data to maximum doping of $y=[A^-/N]=0.5$

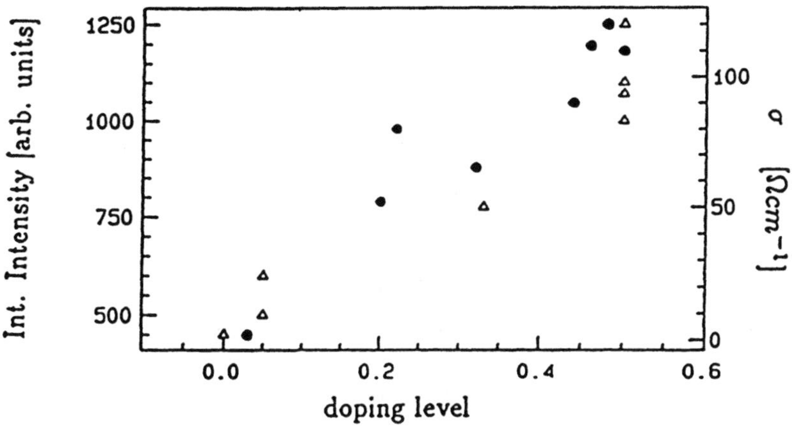

Fig. 2. Integrated intensity of the 0.5 eV background absorption (◁) [5] and dc conductivity (•) [6] as a function of doping level

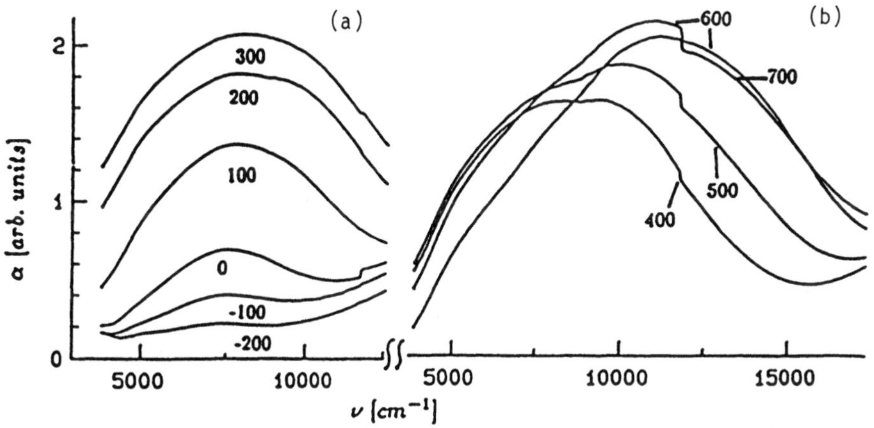

Fig. 3. The behaviour of the 1 eV absorption band for various oxidation levels

in Fig. 3a the potential region where the 1 eV absorption increases in intensity without significant shift of the position of the band and in Fig. 3b the potential region where the absorption band markedly shifts towards higher energies and partially bleaches.

4. Discussion

The scaling diagram in Fig. 1 is in agreement with the observations of Orata *et al.* [7] (obtained with micro balance experiments) giving the maximum doping level in the middle oxidized metallic form of PANI.

The advantage of the AFC method is the direct correlation with the intercalated anions without being affected by the cointercalative changes of the polymer. The scaling of the obtained qualitative data to a fixed value (here y=0.5 for maximum doping level) enables the estimation of the dopant concentration dependence of several physical properties. The linear dependence of the conductivity [6] and the strong IR background absorption around 0.5 eV [5] upon the dopant concentration demonstrates clearly (Fig. 2) the important role of the intercalated anions in these physical properties of PANI. Furthermore, the decrease of the anion content by further oxidation is evidence of a deprotonation connected with the second oxidation process.

For metallic particles with spatial extensions comparable to characteristic lengths (for example mean free path), quantum effects on the band structure are predicted, strongly modifying the bulk properties and resulting in new optical and thermodynamical effects [8,10]. The most striking effect is the discreteness of the electronic energy states in a metallic particle of finite dimensions. Properties of the metallic form of PANI cannot be explained satisfactorily in terms of theories for conventional metals. For the metallic structures within the granular model dimensions ≈ 300Å are reported [1,2]. Thus, we studied some unusual physical properties of metallic PANI in terms of quantum size effects. The predicted enhancement of the far IR absorption of minute metallic particles [8] is qualitatively observed in PANI for the metallic state [5]. The behaviour of the 1 eV absorption band (Fig. 3a) during the electrochemical oxidation processes can be explained by the resonance absorption of a quantum size particle [9] with increasing Fermi energy and volume filling factor. Changes in the sizes of the particles cause shifts of the position of this resonance absorption as shown by the experimental results of Fig. 3b.

The simultaneous increase of Pauli- and Curie-type contributions upon doping [1] is qualitatively explained with the predictions of thermodynamic quantum size effects in a metallic structure with singly occupied highest electronic level (in agreement with polaronic levels) [8,10]. Furthermore, with the assumption for metallic structures [3] consisting of single metallic chains an important connection between effects of the undisturbed conjugation length of polymers and quantum size effects in metallic particles can be derived. A quantitative analysis of experimental data on PANI needs a more refined theoretical analysis of quantum size effects specialized to conducting polymers.

Acknowledgements. Financial support from the Fonds zur Förderung der wissenschaftlichen Forschung in Austria (Proj. No: 5735) and the

Jubiläumsfonds der Österreichischen Nationalbank (Proj. No: 2505) is gratefully acknowledged.

References

1. A. J. Epstein, J. M. Ginder, F. Zuo, R. W. Bigelow, H. S. Woo, D. B. Tanner, A. F. Richter, W. S. Huang and A. G. MacDiarmid, *Synth. Metals* **18**, 303 (1987)
2. H. Kuzmany, N. S. Sariciftci, H. Neugebauer and A. Neckel, *Phys. Rev. Lett.* **60**, 212 (1988)
3. M. Nechtschein, F. Genoud, C. Menardo, K. Mizoguchi, J. P. Travers and B. Villeret, *Synth. Metals* **29**, E211 (1989)
4. N. S. Sariciftci, H. Kuzmany and H. Neugebauer, *J. Mol. Elec.* **3**, 141 (1987)
5. N. S. Sariciftci, M. Bartonek, H. Kuzmany, H. Neugebauer and A. Neckel, *Synth. Metals* **29**, E193 (1989)
6. E. M. Genies, P. Hany, M. Lapkowski, Ch. Santier and L. Olmedo, *Synth. Metals* **25**, 29 (1988)
7. D. Orata and D. A. Buttry, *J. Am. Chem. Soc.* **109**, 3574 (1987)
8. W. P. Halperin, *Rev. Mod. Phys.* **58**, 533 (1986)
9. D. M. Wood and N. W. Ashcroft, *Phys. Rev.* **B 25**, 6255 (1982)
10. R. Kubo, A. Kawabata and S. Kobayashi, *Ann. Rev. Mater. Sci.* **14**, 49 (1984)

In Situ Spectroscopy of Polyaniline

A.P. Monkman

Centre for Molecular Electronics, School of Engineering and Applied Science, University of Durham, South Road, Durham, DH1 3LE, UK

Abstract Data obtained from in situ optical, Resonance Raman (RR) and Subtractive Normalised Interfacial Fourier Transform Infra Red (SNIFTIR) Spectroscopy will be discussed. The effects of both oxidation and pH have been observed and combining the results from these various techniques has given a deeper insight into the complexities of polyaniline (PANi). The results show that oxidatively induced chromaphores are localised and of different structure, contradicting the usual polaronic models for conducting polymers. An absorption band is observed at ca. 0.5 eV which is tentatively ascribed to an internal charge transfer mode between reduced and oxidised units. This would lend agreement to the observed hopping mechanisms for conduction within PANi.

1. Introduction

A number of spectroscopic studies of PANi have appeared in the literature over the last four years [1-6]. However, the effect of both pH and oxidation have not previously been reported. In this brief report, the salient points will be highlighted, and conclusions drawn from the information gathered from all the various techniques mentioned.

The range of spectroscopic techniques used in this study were intended to probe the electronic structure of PANi, the structure of the chain moieties (chromaphores) giving rise to this electronic structure, and the changes which occur to both during oxidation. All of which were studied as a function of pH.

2. Experimental details

The route by which PANi samples were synthesised has been described previously [2]. Samples were grown on ITO glass for optical transmission studies, and Pt electrodes for both the Raman and SNIFTIR studies. In all cases HCl electrolyte was used. During these studies, electrolytes of 2M HCl, 1M HCl and 1/4M HCl, were used. The range of oxidation potentials used were between -0.25V and 0.85V with respect to an SCE reference electrode.

For in situ optical spectroscopy, a special spectro electrochemical cell was constructed, such that the PANi/ITO electrode formed one window, with a quartz disc mounted on the barrel of a syringe (incorporated into the cell) forming the other. The latter could then be pushed up against the sample giving a minimal path length in the electrolyte. Thus, it was possible to probe into the NIR with an aqueous electrolyte. Similarly, for the SNIFTIR work, the Pt/PANi electrode could be pushed tight up against the IR window of the cell. RRS was performed using a 180° backscattering geometry. Here the high reflectance of the Pt electrode allows a double

Springer Series in Solid-State Sciences, Vol. 91
Electronic Properties of Conjugated Polymers III
Editors: H. Kuzmany · M. Mehring · S. Roth

pass of the excitation beam through the sample, increasing sensitivity. Only in the case of the SNIFTIR studies, were the PANi samples maintained oxygen free.

3. Results

A typical in situ optical spectrum (1M HCl electrolyte) is shown in Fig. 1. In the fully reduced state, at all pH's, only one intense absorption is seen, at ca. 4.1 eV. Lower lying absorptions are ascribed to aerial oxidation, etc. On oxidation up to 0.3V, this absorption band loses intensity, and new bands are seen to form at ca. 2.8eV and 1.3eV. The latter is also seen to have an associated low energy tail. There is also evidence of a further underlying absorption at ca. 3.5 eV. The three sharp peaks and rising background in the IR are due to the electrolyte. Further an isosbestic point is observed, indicating that only a single species is being formed on oxidation. The same spectrum is observed with a 2M HCl electrolyte, here the low energy tail of the NIR band is more prominent. In 1/4M HCl, no isosbestic point is observed, and the NIR tail is fairly weak. In all cases the NIR band monotomically blue shifts with increasing oxidation and its intensity increases likewise.

Increasing oxidation (up to 0.85 V) causes further changes to the observed spectra. The intensity of the 4.1 eV band further diminishes. The 2.8 eV band is either lost, or shifts to 2.6 eV, where a new absorption band is observed, also an absorption is seen at ca. 3.5 eV. The broad NIR band is seen to further blue shift with its intensity levelling off. The associated low energy tail is lost. Greater pH effects are observed in this second oxidation regime. In 2M HCl electrolyte the spectra are again similar to that recorded in 1M HCl, however, the NIR band blue shifts

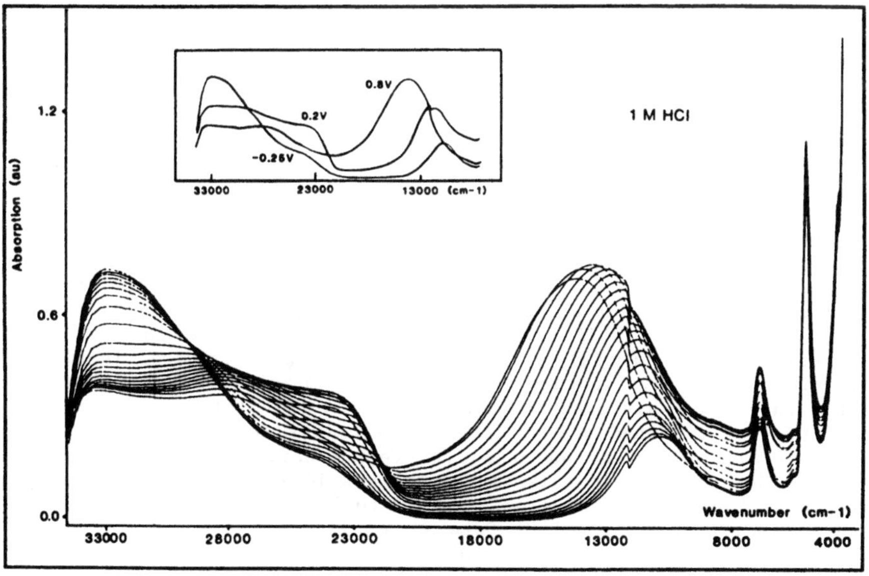

Fig. 1 Insitu optical spectra of PANi in 1M HCl electrolyte. Spectra were recorded at every 0.05 V from -0.25V up to 0.8V vs. SCE. Insert shows the spectra recorded at -0.25, 0.2 and 0.85V respectively

further in the 1M HCl case, and the low energy tail is not lost in the 2M HCl case. With 1/4 M HCl, dramatic changes are seen. The intensity of the 4.1 eV band starts to increase, at the highest applied potentials. Absorptions in the range 3.5 to 2.5 eV are lost. The NIR band shifts up to 2 eV, with its low energy tail being totally lost. The band is considerably broadened, with a small loss of maximum intensity.

The in situ SNIFTIR spectroscopy has given a great deal of information, of which only a few points are noted here. Firstly, in the region 1800 - 1000 cm-1 increasing oxidation causes gain of IR bands at 1570, 1380, 1340 and 1160 cm-1 with associated loss of bands at 1600, 1520, 1210 and 1175 cm-1. Characteristic of loss of benzenoid character and gain of quinoid nature. These changes are independent of pH, although the bands are seen to broaden with low pH electrolytes although they are never washed out. In the spectral range 4500 to 1800 cm-1, a broad band absorption is seen. Its position is pH dependent ranging from ca. 0.5 eV in 1/4 M HCl to 0.4 eV in 2M HCl. This is a true band, not a tail. Further a sharp band is seen, which is independent of pH, at ca. 4400 cm-1. This band is not lost during the second oxidation wave, unlike the broad band which is seen to be lost at high oxidation levels.

RRS were only measured in 2M HCl electrolyte. Three excitation energies were used, 2.71 eV (violet), 2.41 eV (green) and 1.96 eV (red). Using the violet line, maximum resonance enhancement is observed. A characteristic, sharp, Raman spectrum is observed which is invariant on oxidation potential, only resonant enhancement is affected. Strong bands are observed at 1615, 1180, 875, 820 and 670 cm-1. Excitation with the green line, shows Raman spectra similar to this at low oxidation potentials, but above ca. 0.2 v, new bands are seen to form, which become strong at high oxidation levels. Strong Raman bands are seen at 1610, 1565, 1500, 1470, 1315, 1260, 1185, 1160, 875, 630 and 575 cm-1. Using the red excitation line, the same complex spectra are observed, even at low oxidation potentials.

4. Discussions

The 4.1 eV absorption, which is common to all spectra, is also observed in small model compounds including trimmers with all phenyl/amine repeat units [7,8], and hence must be the phenyl ring $\pi \rightarrow \pi^*$ transition. The presence of an isosbestic point at low pH only, indicates that protonation must "lock in" the moiety formed on initial oxidation and its loss at oxidation potentials above ca. 0.3v indicates that a mixed reaction is occurring in this regime. The subtle differences observed between the 2M and 1M spectra can be interpreted simply in terms of oxidation level. It is known that pH strongly affects the potential at which the second oxidation of PANi occurs [9,10]. This potential shifts to higher values with decreasing pH, thus, at the same applied potential, PANi will be more highly oxidised in the 1M HCl electrolyte relative to that in 2M HCl. The same is also true for the 1/4 M HCl case. In this electrolyte, it is seen that the NIR band shifts up to 2 eV, is broadened and suffers a small loss in maximum intensity. This can be interpreted as an effect of deprotonation at high oxidation potentials, the product in this final, fully oxidised state being neutral quinoid, with the absorption being ascribed to formation of localised self trapped excitons as predicted by Duke et al. [11]. Further evidence for deprotonation is seen by the loss of all absorption in the range 3.5 to 2.5 eV, and the increased intensity of the 4.1 eV band. The general shift of the NIR absorption can thus be ascribed to a steady deprotonation of the chromaphore, and increase in its quinoid nature. SNIFTIR data confirms this general increase in quinoid nature with increased oxidation, and further that there is a mixing of benzenoid and quinoid nature, as indicated by the presence of the 1340 cm-1 band, ascribed to a semiquinone moiety. The observed 0.5 eV, electronic band is postulated to be evidence for an internal charge transfer, between reduced and oxidised units. The broadening of the IR bands at low pH indicates an increased amount of charge along the polymer chain, however, at no time are the discrete IR bands washed out, as would be the case if PANi formed a metallic state.

The RRS obtain with the violet excitation line is very characteristic of a p-di-substituted benzene ring. The resonance enhancement arising from the charge transfer between such a protonated moiety and the Cl^- counter ion. The Raman bands are very sharp, indicating a localised system. Hence, the chromaphore giving rise to the 2.8 eV absorption is localised, implying an excitonic transition. The RRS with both green and red excitation show that the chromaphore giving rise to the absorption band in the NIR is different to that yielding absorption at 2.8 eV. This chromaphore is a mixed benzenoid quinoid system, i.e. semiquinoid, as indicated by the observed 1315 cm-1 Raman band. Again the Raman bands are sharp, indicating a localised system. The difference between the spectra observed with red and green excitation lines is explained by the fact that the red line comes into resonance with the NIR optical absorption before the green i.e. it is nearer to the optical absorption in energy terms.

5. Conclusions

From these results it can be seen that both the 4.1 eV and oxidatively induced 2.8 eV absorption can be described in terms of the formation of excitons, which are localised on phenyl rings. At low pH the presence of an isosbestic point indicates that the oxidation reaction produces a single moiety i.e. a simple A → B reaction. However, in 1/4M HCl this isosbestic point is not seen, which indicates that protonation "locks in" the oxidation reaction to produce just one product. In the fully oxidised state, deprotonation occurs, leading to the formation of localised self-trapped excitons on the neutral quinoid units, thus giving rise to absorption at 2 eV. The loss of the isosbestic point may also be attributed to deprotonation. The chromaphores associated with the observed absorptions at 2.8 eV and 1.3 eV are different, the former being a simple di-substituted phenyl moiety, the latter, a mixed benzenoid/quinoid or semiquinoid moiety. From these facts it is difficult to reconcile PANi in the general polaronic picture put forward for other (polyacetylene-like) conducting polymers. As previously reported [12] an ESR signal is induced during the first oxidation wave, which decays at higher oxidation levels (beyond ca. 0.3V). To accommodate this fact, together with the localised nature of all the observed chromaphores, it is postulated that some type of Holstein polaron (localised) is formed. This localised picture does help to explain the facts presented by Wudl et al. [8] that show octomers of PANi to have essentially the same physical properties as the polymer.

The 0.5 eV absorption is ascribed to an internal charge transfer. This would agree with the predicted hopping conduction mechanism indicated by the temperature dependence of the conductivity of PANi [10]. Therefore the picture that emerges for PANi is very different to that of other (more polyacetylene-like) conducting polymers. This must be due largely to the fact that the heteroatom is directly in the conjugation path. It can be seen that there is still a great deal of work to be done in order to fully understand PANi, and perhaps using analogies to molecular crystals will help in this understanding process.

Acknowledgements

I would like to thank Drs. Hamnet and Christenson for undertaking the joint SNIFTIR studies, Professor David Bloor for the many hours of discussion and GEC Marconi Research Laboratories for the use of their Lambda 9 Spectrometer.

References

[1] A. P. Monkman and D. Bloor, "Electronic Properties of Conjugated Polymers", p.232, ed. H. Kuzmany, M. Mehring and S. Roth, Springer Series in Solid State Science 76, (1987).

[2] A. P. Monkman, D. Bloor, G. C. Stevens, J. C. H. Stevens, J. Phys. D : Appl. Phys., 20 (1987) 1337.

[3] R. J. Cushman, P. M. McManus and S. C. Yang, Macromol. Chem. Rapid Comm. 8 (1987) 69.

[4] N. S. Sariciftci and H. Kuzmany, Synth. Met. 21 (1987) 157.

[5] N. S. Sariciftci, H. Kuzmany and H. Neugebauer, J. Mol. Elec 3 (1987) 141.

[6] N. S. Sariciftci, M. Bartonek, H. Kuzmany, H. Neugebauer and A. Neckel, Synth. Met. 29 (1989) E193.

[7] H. Linschitz, J. Rennet and T. M. Korn, J. Am. Chem. Soc., 55 (1933) 1488.

[8] F. Wudl, R. O. Angus Jr., F. L. Lu, P. M. Allemand, D. J. Vachon, M. Nowak, Z. X. Liu and A. J. Heeger, J. Am. Chem. Soc., 109 (1987) 3677.

[9] W. S. Huang, B. D. Humphrey and A. G. MacDiarmid, J. Chem. Soc. Farady Trans. 82 (1986) 2385.

[10] A. P. Monkman, Thesis

[11] C. B. Duke, E. M. Conwell and A. Paton, Chem. Phys. Lett., 131 (1986) 82.

[12] A. P. Monkman, D. Bloor, G. C. Stevens, J. C. H. Stevens and P. Wilson, Synth. Met. 29 (1989) E.277.

In Situ FTIR Spectroscopy of Polyaniline Degradation

H. Neugebauer [1], *N.S. Sariciftci* [2], *H. Kuzmany* [2], *and A. Neckel* [1]

[1] Institut für Physikalische Chemie, Universität Wien, Währinger Str. 42, A-1090 Wien, Austria

[2] Institut für Festkörperphysik, Universität Wien, and Ludwig Boltzmann Institut für Festkörperphysik, Strudlhofgasse 4, A-1090 Wien, Austria

Abstract. The anodic degradation process of polyaniline in aqueous $HClO_4$ and HCl electrolyte was studied using in situ FTIR spectroscopy. As a soluble degradation product p-benzoquinone was found in both electrolytes. The remaining insoluble products are different depending on the composition of the electrolyte solution.

1. Introduction

The stability of polyaniline (PANI), especially in electrochemical systems, plays an important role in any possible practical application of the polymer. In acidic electrolytes PANI tends to degrate forming soluble and insoluble products at high anodic potentials (higher than +700 mV vs. SCE) [1,2]. We used in situ FTIR spectroscopy in the external reflection absorption arrangement in order to get spectral information on this process and to identify the products formed in aqueous $HClO_4$ and HCl electrolytes.

2. Experimental

PANI was polymerized electrochemically on the surface of a Pt mirror electrode in aqueous 1 M $HClO_4$/0.1 M $NaClO_4$/0.2 M aniline electrolyte. The polymer covered electrode was washed, dried and transferred into the spectroelectrochemical cell containing the appropriate electrolyte (1 M $HClO_4$/0.1 M $NaClO_4$ or 1 M HCl). Details of the polymerization process and of the spectroelectrochemical cell arrangement have been published elsewhere [3]. A potential of +800 mV was applied for 30 minutes and FTIR spectra were recorded consecutively as a function of time (all potential values in this paper refer to the SCE electrode). After this degradation process the potential was swept slowly in cathodic direction. Again FTIR spectra were recorded during this process.

Springer Series in Solid-State Sciences, Vol. 91
Electronic Properties of Conjugated Polymers III
Editors: H. Kuzmany · M. Mehring · S. Roth

For comparison of the ex situ spectra of PANI before and after the degradation process spectra of the dry electrode immediately after the polymerization and after the degradation process were measured. These spectra were obtained in the same spectroelectrochemical cell as used for the in situ experiments, but without electrolyte. Before the spectroscopic characterization the polymer electrode was equilibrated at +400 mV, washed with 1 M NH_3 solution (converting the polymer into the base form) and dried in vacuum.

3. Results and Discussion

Figure 1 shows the spectral changes obtained in situ during degradation at +800 mV in 1 M $HClO_4$/0.1 M $NaClO_4$ electrolyte. Bands pointing upwards, belonging to emerging substances, are found at 1655, 1595, 1085, 945 and 890 cm^{-1} (labeled with B in Fig. 1). The bands can be assigned to p-benzoquinone (absorption bands at 1657, 1591, 1084, 943, 896 cm^{-1} [4]). These results prove directly the formation of this substance as a soluble degradation product [2]. A broad band at 1100 cm^{-1} shows the migration of ClO_4^--ions into the thin electrolyte layer due to the electrochemical current [5]. Negative bands (labeled with P in Fig. 1) indicate a decrease in the amount of polymer present on the electrode surface.

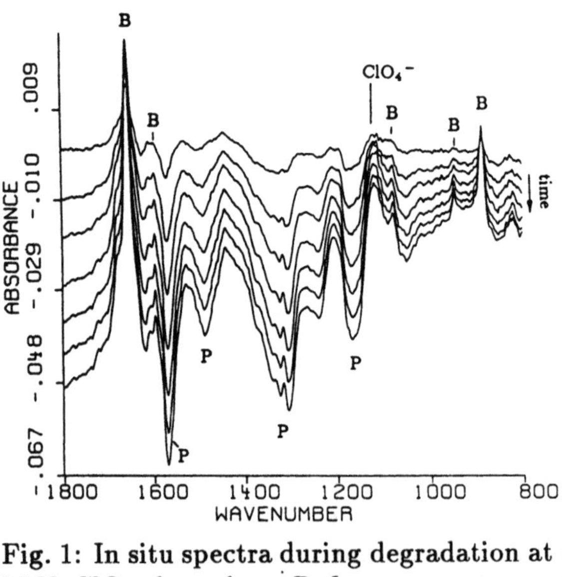

Fig. 1: In situ spectra during degradation at +800 mV in 1 M $HClO_4$/0.1 M $NaClO_4$ electrolyte. Reference spectrum at the beginning of the degradation. The time between the first and the last spectrum is 30 minutes

In 1 M HCl similar spectra are found, but compared with the results in 1 M $HClO_4$/0.1 M $NaClO_4$ the degradation occurs much faster. The inset in Fig. 2 shows the current/potential curve of the slow potential sweep in 1 M HCl after the degradation process. The electrode was kept in place preventing a diffusion of soluble degradation products into the surrounding electrolyte. A cathodic current peak occurs at +450 mV. The in situ spectra during this reduction peak (with a reference spectrum just before the peak) are shown in Fig. 2. Negative bands (labeled with B) and positive bands (labeled with H) indicate the reduction of p-benzoquinone, which was formed during the degradation process, and the formation of hydroquinone.

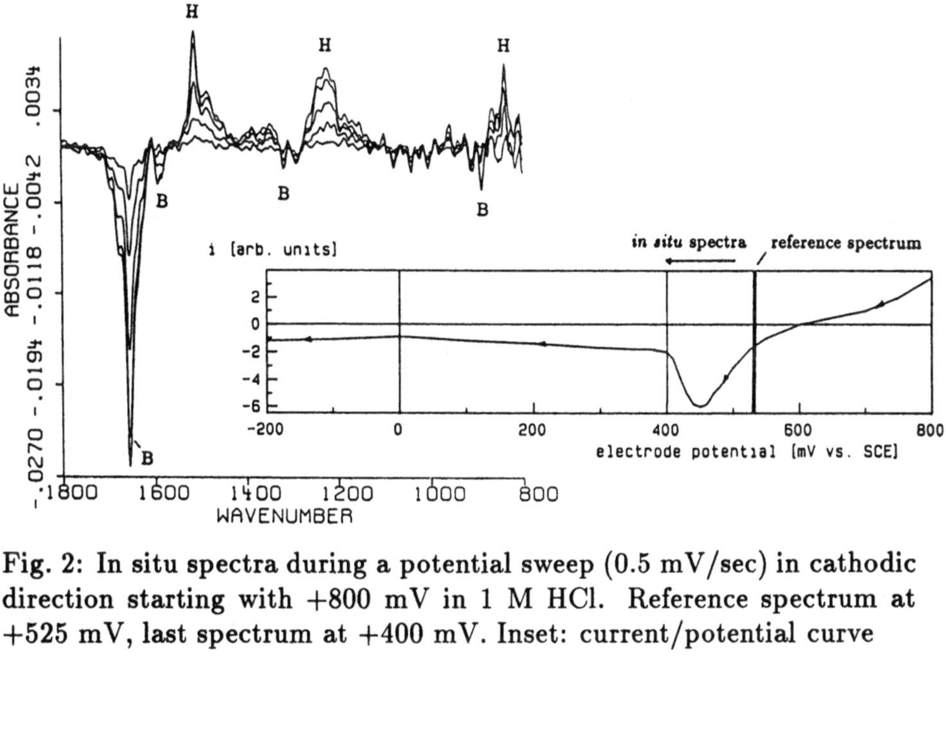

Fig. 2: In situ spectra during a potential sweep (0.5 mV/sec) in cathodic direction starting with +800 mV in 1 M HCl. Reference spectrum at +525 mV, last spectrum at +400 mV. Inset: current/potential curve

The ex situ spectra before and after the degradation process are shown in Fig. 3. In both electrolytes the polymer bands are smaller after the degradation in accordance with the negative polymer bands in Fig. 1. New absorptions occur around 1650 cm^{-1} caused by the absorption of C=O units in the polymer, which were formed during the anodic degradation (probably at the end of the chains). In the spectrum of PANI degradated in 1 M HCl additional bands at 1293, 1041 and 914 cm^{-1} can be seen. A polymer obtained by electrochemical polymerization of

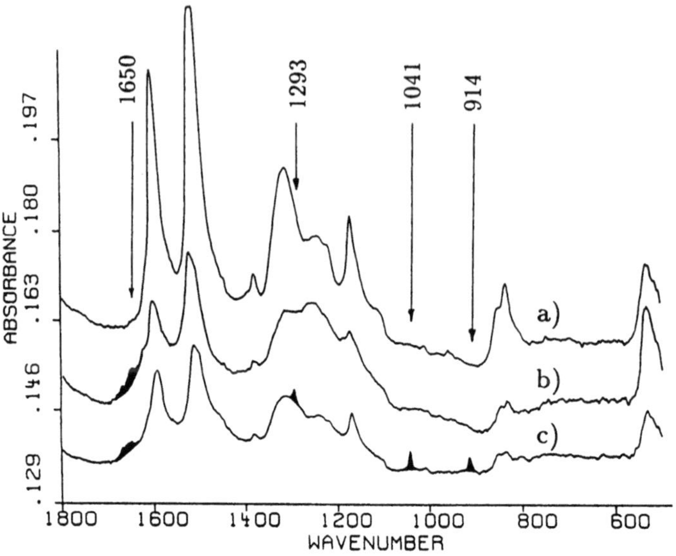

Fig. 3: Ex situ spectra before (curve a) and after the degradation process (curve b: degradation in 1 M $HClO_4$/0.1 M $NaClO_4$, curve c: degradation in 1 M HCl). Significant new bands are coloured black. The spectra are separated

o-cloroaniline showed similar IR bands. These results indicate a chlorination of the aromatic rings during the degradation in 1 M HCl [6].

Acknowledgement. This work was supported by the Jubiläumsfonds der Österreichischen Nationalbank (Proj. No. 2505) and by the Fonds zur Förderung der wissenschaftlichen Forschung of Austria (Proj. No. 5735).

References

1. T.Kobayashi, H.Yoneyama and H.Tamura, J. Electroanal.Chem. **177**, 293 (1984)
2. D.E.Stilwell and S.Park, J. Electrochem. Soc. **135**, 2497 (1988)
3. H.Neugebauer, A.Neckel, N.S.Sariciftci and H.Kuzmany, Synthetic Metals **29**, E185 (1989)
4. The Aldrich Library of FTIR Spectra, C.J.Pouchert (ed.), Aldrich Chemical Company, Inc., Milwaukee, Wisconsin (1985), Edition I, Volume 1, p. 452
5. H.Neugebauer, N.S.Sariciftci, H.Kuzmany and A.Neckel, these proceedings
6. T.Hagiwara, M.Yamaura and K.Iwata, Synthetic Metals **25**, 243 (1988)

Defect States in the Fully Oxidized Form of Polyaniline

M.C. dos Santos and J.L. Brédas*

Service de Chimie des Matériaux Nouveaux,
Département des Matériaux et Procédés,
Université de Mons, Av. Maistriau 19, B-7000 Mons, Belgium

Abstract. The electronic structure of pernigraniline, the fully oxidized polyaniline, is investigated in the framework of Hückel theory with sigma compressibility. This system presents a doubly degenerate ground state. It is shown to accommodate soliton excitations with fractional charges for a pair of either neutral or charged defects. Polarons are however the most stable excitations upon addition or removal of a single electron. Significant differences with what would be predicted from the so-called AB polymer are found.

Polyanilines differ from the usual conjugated polymers in that the connecting units constituting the chains are of different chemical nature: (i) nitrogen atoms which can be in either amine or imine oxidation states and (ii) phenyl rings, which can assume both benzenoid- and quinoid-like geometries. This characteristic has been pointed out as playing an important role on the properties of doped polyanilines [1-2].

A number of theoretical investigations has been carried out on the electronic structure of the simplest heteroatomic chain, the so-called AB polymer [3-5]. Exciting properties, such as charge fractionalization, have been predicted for the diatomic AB chain [3]. Experimental observation of these effects, however, has not been possible to date due to the lack of adequate preparation of well-characterized AB polymer samples [6].

We report here on electronic structure calculations of pernigraniline (PN), the fully oxidized polyaniline, which has recently been synthesized in a pure form [7]. Among the polyaniline family, PN is the only system to present a doubly degenerate ground state. It contains imine nitrogens connecting phenyl rings which adopt alternatively benzenoid- and quinoid-like geometries. The possibility of soliton excitations, together with the heteropolar character of the chemical bond in PN, fulfill the conditions for charge fractionalization to occur. PN is thus a most interesting system to be investigated.

Electronic structure calculations on an infinite PN chain are carried out within a Su-Schrieffer-Heeger-like Hamiltonian. The electron-phonon coupling and the elastic

*Permanent address: Departamento de Química Fundamental, Universidade Federal de Pernambuco, 50739 Recife, PE, Brazil.

Springer Series in Solid-State Sciences, Vol. 91
Electronic Properties of Conjugated Polymers III
Editors: H. Kuzmany · M. Mehring · S. Roth

energy associated to a bond are modeled by a bond-order-bond-length relationship of Coulson type. Hoppings and the energies of the sigma bonds are written as: $\beta(R) = -A \exp(-R/B)$ and $f(R) = C\, \beta(R)(R - R_0 + B)$. In these expressions, R is the actual bond length between neighboring atoms, and A,B,C, and R_0 are adjustable parameters. The model takes into account that site energies for nitrogens, para-carbons, and ortho-carbons assume distinct values. All the adjustable parameters are optimized to reproduce Valence Effective Hamiltonian band structures and an appropriate ground state geometry [8]. The method employed to solve the electronic part of the Hamiltonian is based on a Green's functions technique and a renormalization procedure, which have been described elsewhere [9-10].

In the PN ground state, valence (VB) and conduction (CB) bands appear isolated from the rest of the pi system [10]. These two bands globally extend over 2.9 eV and are separated by a gap of 1.4 eV. In addition to the bond-order density wave present in the «dimerized» ground state, the polymer has also an on-site charge density wave in which nitrogen atoms have a net charge of -0.535 while benzenoid- and quinoid-like rings have the respective charges of 0.121 and 0.948, in units of the magnitude of the electronic charge.

Hyperbolic tangent distortions of the bonds around a nitrogen give rise to two distinct solitons, hereafter referred to as type-1 and type-2 solitons. Their geometries are displayed in Fig. 1. Both defects are centered around a nitrogen atom and have the following properties. Type-1 soliton: (i) has rings around its center which are more aromatic in character; (ii) it is formed by 74% of a state from VB and 26% of a state from CB; (iii) a localized electronic state appears in the gap, 0.11 eV below the zero-energy level; and (iv) the corresponding wavefunction is symmetric with respect to the center. Type-2 soliton: (i) has rings around its center which adopt a more quinoid geometry; (ii) the contributions from VB and CB to form a type-2 defect exactly compensate those for a type-1 soliton; (iii) its associated localized state appears 0.06 eV above the zero-energy level; and (iv) the wavefunction is antisymmetric (node on the central nitrogen).

The creation energy of a soliton pair depends on the occupation of the localized levels. The most stable configuration is calculated to have the lower level doubly occupied and the upper level empty, corresponding to a creation energy of 0.64 eV. This configuration is non-magnetic, see Fig. 2, and the charges around type-1 and type-2 distortions are -0.52 and 0.52, respectively. The energy to excite a magnetic pair (both levels singly occupied) from the ground state is 0.81 eV and the resulting charge values are 0.48 and -0.48.

Strict application of the AB polymer model predicts that a polaron is not stable when the solitons possess charges similar to those calculated for PN [4]. However, stable polarons are found in PN upon single charging. In Fig. 1c, the optimized polaron geometry is shown. Both positive and negative polarons correspond to the same geometric distortion. Two localized levels split in the gap, at 0.38 eV below and 0.30 eV above the zero-energy level, see Fig. 2. The binding energy of a positive polaron is 0.14 eV and 0.21 eV for the negative polaron, which is significantly larger than the 0.05 eV binding energy for a polaron in trans-polyacetylene [11].

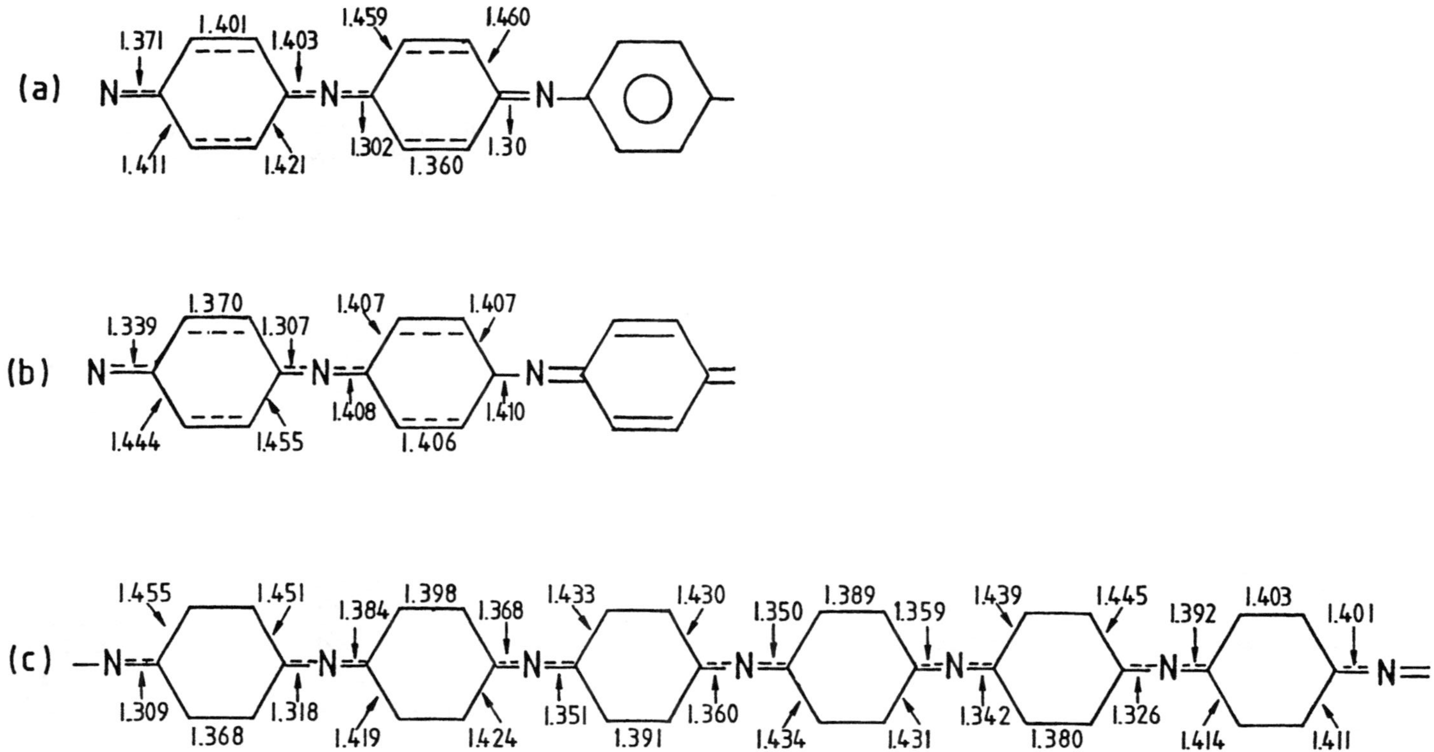

Figure 1. Optimized geometries for: (a) type-1 soliton and (b) type-2 soliton (the N atom to the left corresponds to the distortion center); and (c) a polaron. Bond lengths are given in Angströms

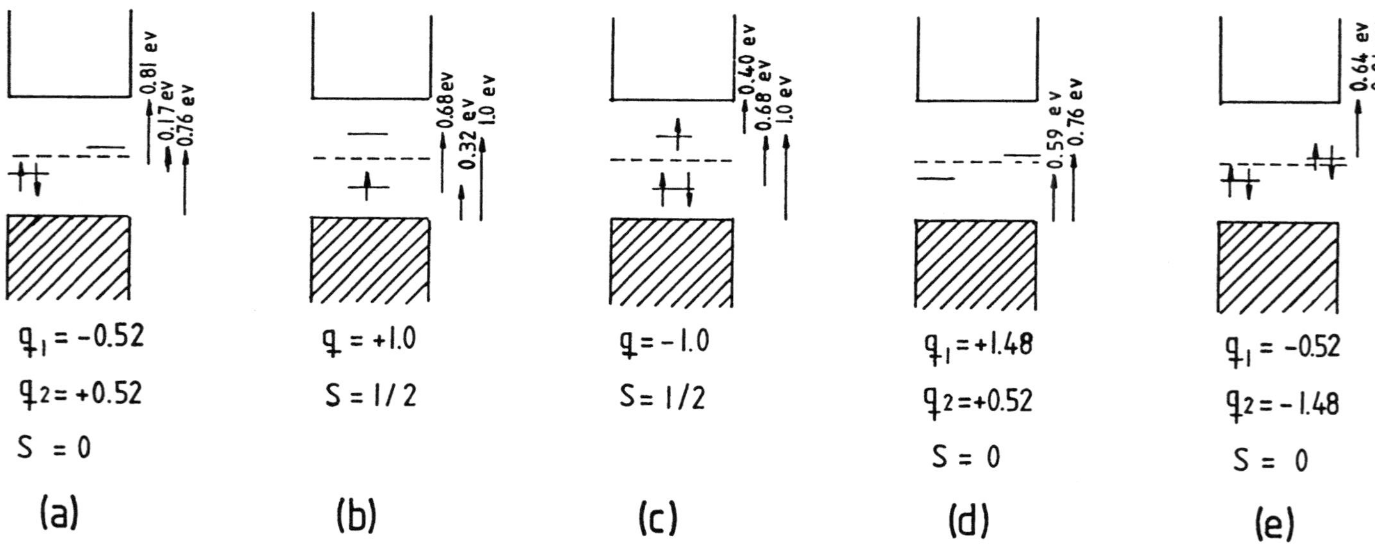

Figure 2. Properties of the low lying elementary excitations in pernigraniline and the corresponding intragap optical absorption energies; q_1 and q_2 are the charges around type-1 and type-2 solitons, respectively, while S is the spin of the excitation. (a) neutral soliton pair; (b) positive polaron; (c) negative polaron; (d) doubly positive soliton pair; (e) doubly negative soliton pair.

Bipolarons are not present in PN. Two polarons spontaneously form a pair of separated charged solitons, more stable by 0.37 eV in the positive case and 0.28 eV in the negative case. Again, the double charge present on the chain is found to fractionalize and separates around type-1 and type-2 solitons, respectively, into 1.48 and 0.52 in the positive charging and into -0.52 and -1.48 in the negative charging. The properties of the low lying elementary excitations of PN are summarized in Fig. 2, where are also shown the intragap optical absorptions arising from these excitations.

In conclusion, our calculations indicate that elementary excitations having fractional charges should occur in PN. The phenomenon of charge fractionalization appears thus to be a general property of a degenerate ground state heteroatomic chain. Experimental efforts to observe these effects in PN are most promising, since PN is becoming available in a pure form [7]. Finally, we note that: (i) the stability region of the polaron solution in a heteroatomic chain can be quite different from that predicted for the AB polymer. (ii) The nonlinear optical response of PN should be most interesting. It has indeed been shown that high hyperpolarizability values are related to efficient mechanisms of charge separation [12]. Such a mechanism is an intrinsec characteristic of PN.

The authors acknowledge IBM-Belgium, the Belgian National Fund for Scientific Research (FNRS) and the Facultés Universitaires Notre-Dame de la Paix for the use of the Namur SCF Computing Facility as well as the CCI Computer Center at the University of Mons. One of us (M.C.S.) has been supported by the Brazilian Agency CAPES.

References

[1] J.M. Ginder, A.F. Richter, A.G. MacDiarmid, and A.J. Epstein, Solid State Commun. 63, 97 (1987); A.J. Epstein et al., these Proceedings.

[2] H.-Y. Choi and E.J. Mele, Phys. Rev. Lett. 59, 2188 (1987).

[3] M.J. Rice and E.J. Mele, Phys. Rev. Lett. 49, 1455 (1982).

[4] D.K. Campbell, Phys. Rev. Lett. 50, 865 (1983).

[5] J.C. Hicks and V.R. Lin-Liu, Phys. Rev. B 29, 5641 (1984); T. Martin and D.K. Campbell, Phys. Rev. B 35, 7732 (1987); A. Saxena and J.D. Gunton, Phys. Rev. B 38, 8459 (1988).

[6] Polymethineimine, $(-CH=N-)_x$, is a prototypical AB polymer which has been synthesized (see D. Wohrle, Adv. Polym. Sci. 10, 35 (1972)), but unfortunately not in a well-characterized form.

[7] A.G. MacDiarmid, private communication.

[8] D.S. Boudreaux, R.R. Chance, J.F. Wolf, L.W. Shacklette, J.L. Brédas, B. Thémans, J.M. André, and R. Silbey, J. Chem. Phys. 85, 4584 (1988); L.W. Shacklette, J.P. Wolf, S. Gould, and R.H. Baughman, J. Chem. Phys. 88, 3955 (1988).

[9] M.C. dos Santos, C.P. de Melo, and H.S. Brandi, Solid State Commun. 52, 99 (1984); Int. J. Quantum Chem. 30, 109 (1986).

[10] M.C. dos Santos and J.L. Brédas, Synth. Metals 29, E321 (1989); Phys. Rev., submitted for publication.

[11] J.L. Brédas, R.R. Chance, and R. Silbey, Phys. Rev. B 26, 5643 (1982).
[12] A.I. Garito, J.R. Heflin, K.Y. Wong, and O. Zamani-Khamiri, Materials Research Society Symposium Proceedings 109, 91 (1988); M. Sinclair, D. Moses, K. Akagi, and A.J. Heeger, ibid., pag. 205; J.L. Brédas, these Proceedings.

Oxidized Polyaniline: Effects of Protonation on the Electronic Structure of Pernigraniline

M.C. dos Santos and J.L. Brédas*

Service de Chimie des Matériaux Nouveaux,
Département des Matériaux et Procédés,
Université de Mons, Av. Maistriau 19, B-7000 Mons, Belgium

Abstract. The electronic structure of a pernigraniline fragment is calculated within a Su-Schrieffer-Heeger-like Hamiltonian. The effect of protonation of the imine sites is investigated. The protonation is modeled by a change in the atomic site energies to account for the transformation of nitrogen imines into amines. Separation between protonated and non-protonated phases in the fragment is obtained. In the protonated phase, protonation adopts a periodicity of order two, commensurate with the underlying charge-density-wave. These results are similar to those predicted earlier for emeraldine.

The protonation of emeraldine base (the half oxidized form of polyaniline) in acidic solutions has been extensively studied in the past few years [1-7]. Upon protonation, emeraldine transforms into a highly conducting salt. Usually, in the conjugated polymers doped via an oxidation or reduction process, the metallic properties appear at high dopant concentration [8]. In emeraldine, however, the protonation results from the very beginning in a continuous growth of the metallic properties [3]. A mechanism, in which protonated phases (forming metallic islands) and non-protonated phases separate in the material, has been proposed to rationalize the peculiar evolution of doped emeraldine to the highly conducting state [3-4].

In a recent theoretical work, Choi and Mele [4] have shown that the phase segregation that occurs in protonated emeraldine originates from the heteroatomic nature of this system. The recent availability from synthesis [9] as well as the prediction of elementary excitations with fractional charges in pernigraniline (PN) [10], the fully oxidized form of polyaniline, has prompted us to study the protonation process of this polymer. The electronic structure of a PN fragment containing up to 30 nitrogens and 30 phenyl rings is calculated at the simple Hückel level including the elastic energy of the sigma framework. Resonance integrals describing the pi system ($\beta(R)$) and the elastic energy associated to a bond ($f(R)$) are written as: $\beta(R) = -A \exp(-R/B)$ and $f(R) = C\,\beta(R)(R - R_0 + B)$. These expressions depend on the bond length R between

*Permanent address: Departamento de Química Fundamental, Universidade Federal de Pernambuco, 50739 Recife, PE, Brazil.

Springer Series in Solid-State Sciences, Vol. 91
Electronic Properties of Conjugated Polymers III
Editors: H. Kuzmany · M. Mehring · S. Roth

neighboring atoms and on adjustable parameters A, B, C and R_0. A representative set of parameters for PN has been taken from a previous work [10].

In order to theoretically simulate the protonation of an imine nitrogen, we have carried out a similar Hückel parameterization for leucoemeraldine (LE), the fully reduced polyaniline. We kept the bond parameters A, B, C, and R_0 found for PN and readjusted the site energies in such a way that the relative positions of the Fermi levels as well as the valence and conduction bands of these systems reproduce Valence Effective Hamiltonian band structure results [5]. Fig. 1a schematically represents the fundamental repeating units for PN and LE and their respective optimized site energies.

It is important to note that, in the neutral state, each imine contributes one electron to the pi system while an amine has two pi electrons. When a proton links to an imine, a bond involving the imine lone pair (which lies in the sigma framework) is formed. An internal relaxation process transforms the protonated imine into an ionized amine, as sketched in Fig. 1b. This does not alter the number of pi electrons in the original PN fragment. At the Hückel level, this process is modeled by a change in the site energy for the chosen protonated unit.

We have started protonating imine sites in the middle portion of the fragment to avoid end effects. The first protonation increases the total pi energy by 0.82 eV. The geometric distortions produced in the chain are confined to the nearest neighborhood of the protonated site. The single and double N-C bonds around the protonated site slightly shorten with respect to their original values (this result is an indication that the electron-phonon coupling of the NH-C bond is somewhat overestimated in our calculation since MNDO geometry optimizations on emeraldine tetramer [7] show an increase of the double bond length upon protonation of the imine site).

The ground state charge density wave (CDW) is locally perturbed by protonation. A charge tranfer of -0.18 e (e is the magnitude of the electronic charge) occurs from the protonated site to the neighboring quinoid ring. We note that, as the Hückel model is exclusively concerned with the pi electrons, the rearrangements on the sigma bonds

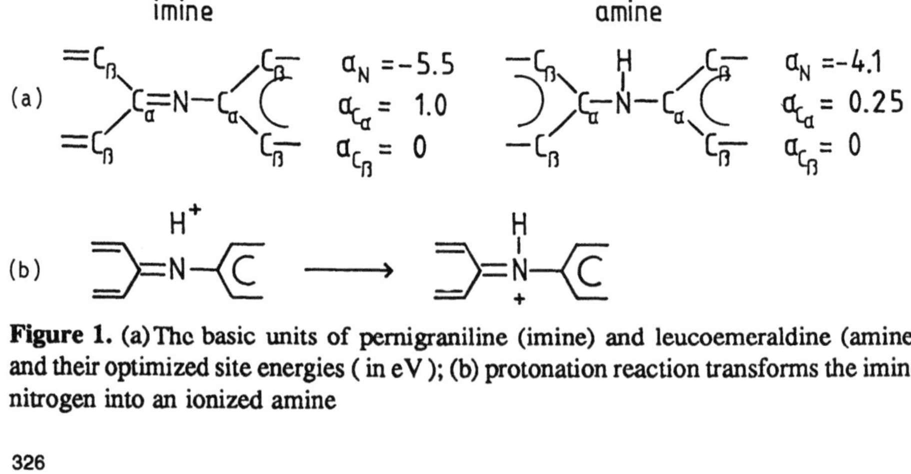

Figure 1. (a) The basic units of pernigraniline (imine) and leucoemeraldine (amine) and their optimized site energies (in eV); (b) protonation reaction transforms the imine nitrogen into an ionized amine

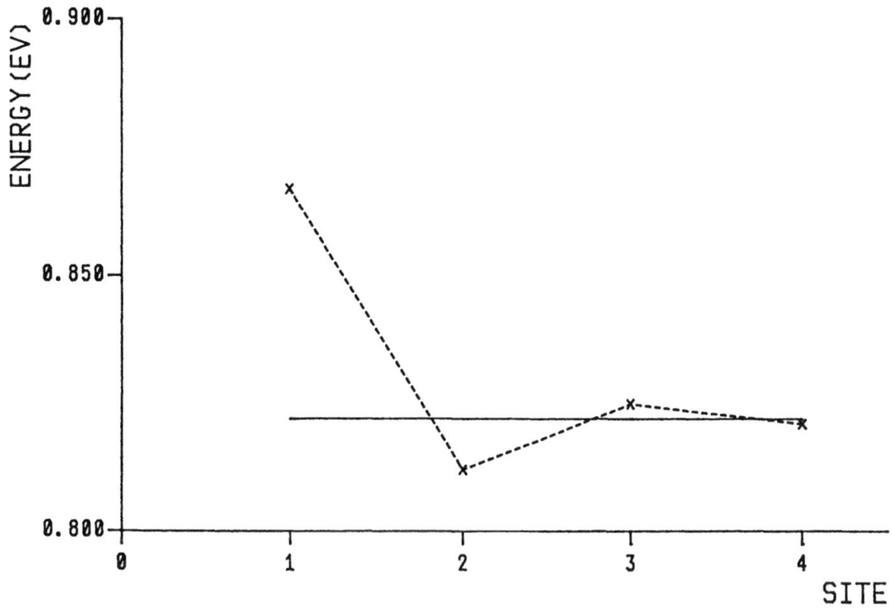

Figure 2. The energy associated to the second protonation when site 0 is protonated as a function of the second site position. The continuous line represents the energy of the first protonation

are not explicitly taken into account; at this level of approximation the distinction between neutral and ionized amines only comes from the band filling.

Fig. 2 shows the energy required by the second protonation as a function of the distance between the protonated sites. A similar evolution is obtained in the subsequent protonations. There is a clear preference for the system to be protonated on every other site. In Fig. 3 the evolution of the atomic pi charges on the polymer is shown. We see that a second CDW develops upon protonation, which has exactly the same period as that of the original CDW. The tendency for the second CDW to keep periodicity of order two (in terms of nitrogen sites) originates from the band filling (one-half) and results in segragation of protonated phases with respect to nonprotonated phases. An equivalent result was obtained for emeraldine, where protonation is shown to follow a periodicity of order four, consistent with a three-quarter band filling in that half-oxidized state [4].

From our study, a comparison can be drawn between the usual redox doping of conjugated polymers and the protonation doping. In the former case, dopants tranfer charges (electrons or holes) to the chains without changing the chemical composition of the polymer. The excess charges modify the band filling and introduce discommensurations which can occur randomly along the conjugated chains.

The protonation doping modifies the chain properties in a different way since it does change the chemical composition (e.g. imines are transformed into amines) without affecting the band filling (the number of pi electrons is kept constant in the process).

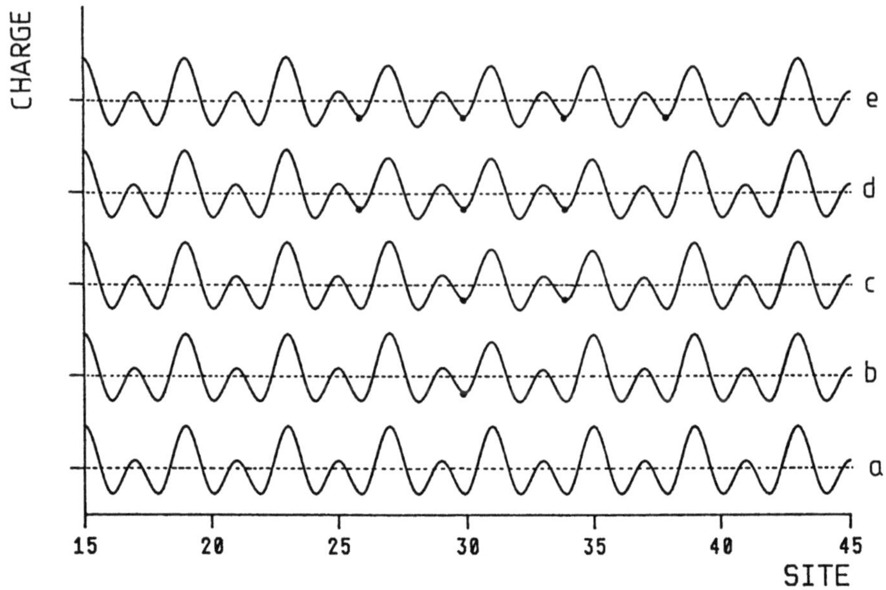

Figure 3. Evolution of the charge density wave pattern upon protonation: (a) ground state CDW; (b) first protonation; (c) second protonation; (d) third protonation; (e) fourth protonation. In these figures, the dashed line represents the zero charge and the amplitudes vary from -0.54 e to +0.95 e. Dots on the curves represent the protonated sites

Furthermore, it is an important aspect of the physics of polyanilines that these polymers are of heteroatomic nature, which results in the presence of a CDW in the ground state; in other words, band gaps in these systems are not only dependent on Peierls instability but also on the CDW amplitude. As a consequence, the appearence of electronic states around the Fermi level upon protonation corresponds to a mechanism which differs from that proposed for the redox doping [8].

The authors are indebted to IBM-Belgium, the Belgian Natinal Fund for Scientific Research (FNRS) and the Facultés Universitaires Notre-Dame de la Paix for the use of the Namur SCF Computing Facility as well as the CCI Computer Center at the University of Mons. One of us (M.C.S.) acknowledges the financial support from CAPES (Brazil).

References

[1] A.G. MacDiarmid, J.C. Chiang, W.S. Huang, B.D. Humphrey, and N.L.D. Somarisi, Mol. Cryst. Liquid Cryst. 125, 309 (1985).
[2] A.G. MacDiarmid, J.C. Chiang, A.F. Richter, and A.J. Epstein, Synth. Metals 18, 285 (1987).

[3] J.M. Ginder, A.F. Richter, A.G. MacDiarmid, and A.J. Epstein, Solid State Commun. 63, 97 (1987).
[4] H.-Y. Choi and E.J. Mele, Phys. Rev. Lett. 59, 2188 (1987).
[5] D.S. Boudreaux, R.R. Chance, J.F. Wolf, L.W. Shacklette, J.L. Brédas, B. Thémans, J.M. André, and R. Silbey, J. Chem. Phys. 85, 4584 (1986).
[6] S. Stafström, J.L. Brédas, A.J. Epstein, H.S. Woo, D.B. Tanner, W.S. Huang, and A.G. MacDiarmid, Phys. Rev. Lett. 59, 1464 (1987).
[7] S. Stafström, Chem. Phys. Lett. 131, 414 (1986).
[8] For a review see A.J. Heeger, S. Kivelson, J.R. Schrieffer, and W.-P. Su, Rev. Mod. Phys. 60, 781 (1988).
[9] A.G. MacDiarmid, private communication.
[10] M.C. dos Santos and J.L. Brédas, Synth. Metals 29, E321 (1989); also in these Proceedings.

Polaron-Induced Resonance Raman Scattering in Polyaniline

M. Bartonek, N.S. Sariciftci, and H. Kuzmany

Institut für Festkörperphysik, Universität Wien, and
Ludwig Boltzmann Institut für Festkörperphysik,
Strudlhofgasse 4, A-1090 Wien, Austria

Abstract. We report on an in situ resonance Raman spectroscopic study of polyaniline during the first oxidation step in an aqueous acidic medium. The resonance excited increase of intensity of benzenoid ring vibrational modes show a sharp maximum at the first oxidation process with a striking resemblence to ESR data. This is explained by resonance enhancement of Raman modes due to polaronic defect states.

1. Introduction

Due to its unique physical properties and technical application potentials, polyaniline (PANI) became a great challenge for research in the field of conducting polymers during the last years [1]. The structure of PANI is a function of both the oxidation level and the pH of the medium where it is equilibrated [2]. In aqueous acidic media PANI exhibits two distinct redox couples during the electrochemical oxidation process. Upon onset of the oxidation spin carrying defects are created during the first and second oxidation process showing a dramatic increase of the ESR signal [3]. Simultaneously electronic in the gap absorptions are observed which are assigned to polaronic defects [4]. However, these features are usually very broad and strongly overlapped by other absorption processes.

In this paper we present a detailed in situ resonance Raman study of PANI in acidic medium covering the first oxidation step. Although resonance Raman spectra were already published previously [5,6,7] these extensive studies revealed that by the means of resonance Raman scattering it is possible to study such defects in more detail than it is possible from absorption experiments. We show that the Raman cross-section of the 1630 cm^{-1} benzene ring vibrational mode is strongly correlated to the amount of polaronic defects (ESR-signal) with a peak of the excitation profile situated at 2.8 eV.

2. Experimental

Aniline was polymerized electrochemically in a three electrode cell using an aqueous acidic electrolyte (1 M $HClO_4$ + 0.1 mol $NaClO_4$ + Aniline). The working electrode was a platinum foil and a coated transparent glass substrate (ITO) for Raman and optical spectroscopic studies, respectively. Synthesis was performed by sweeping the potential on the working electrode

Springer Series in Solid-State Sciences, Vol. 91
Electronic Properties of Conjugated Polymers III
Editors: H. Kuzmany · M. Mehring · S. Roth

with 20 mV/sec between -200 and 800 mV vs SCE. The thickness of the film was about 2000 Å.

For in situ studies the polymer was placed in a special spectroelectrochemical cell including the electrolyte (without aniline monomer). Raman spectra were sampled by using a Spex double mono-chromator and an argon- or krypton ion laser. Optical absorption experiments were performed on a Hitatchi U-3410 spectro-photometer.

3. Results

Fig. 1 shows the Raman cross-section of vibrational modes which correspond to benzenoid and quinoid ring structures (1196 cm^{-1}, 1630 cm^{-1} and 1170 cm^{-1}, 1490 cm^{-1}, respectively) as a function of the electrochemical potential vs. SCE. Quinoid ring vibrational modes are found after the first oxidation step.

A more detailed study of the Raman cross-section of the 1630 cm^{-1} ring vibrational mode around the first maximum using various excitation wavelength is shown in Fig. 2. A pronounced dependence of the Raman cross-section on the laser frequency is observed. The 514 nm excitation causes only a weak increase in Raman scattering, whereas 457 nm laser excitation strongly enhances the cross-section. Excitation with 406 nm laser energy gives no enhancement.

In Fig. 3 we have plotted the Raman cross-section at 120 mV and at 400 mV vs. the excitation energy. For PANI at 120 mV a very pronounced resonance profile around 2.8 eV is observed. The state of PANI at 400 mV shows a different excitation behaviour which reflects a different electronic structure.

The effect of doping concentration [8] on the Raman cross-section is shown in Fig. 4. A linear relationship exists for low doping concentrations up to 20 percent doping (120 mV). For higher doping levels (potentials) this linear relation is lost.

4. Discussion

The Raman signal of the 1630 cm^{-1} mode shows a similar behaviour as the ESR signal [3] of PANI along the electrochemical axis when excited with the blue laser. The ESR-signal [3] exhibits two sharp maxima occurring at the first and second oxidation step (around 200 mV and 750 mV vs. SCE). It is well established that during the first oxidation step radical cations are created at the nitrogen atoms [9]. This species are also responsible for the huge increase in the ESR-signal. In optical absorption spectra a peak at around 3 eV appears and is described as a polaronic defect feature [4]. This radical cation on the nitrogen relaxes due to the electron-lattice coupling to a polaronic defect. The Raman cross-section of the benzene vibrational modes shows a strong enhancement caused by this polaronic lattice defect. From the excitation profile (Fig. 3) one can derive the maximum coupling occurring at 2.8 eV excitation which is in good agreement with band structure calculations [10]. We propose

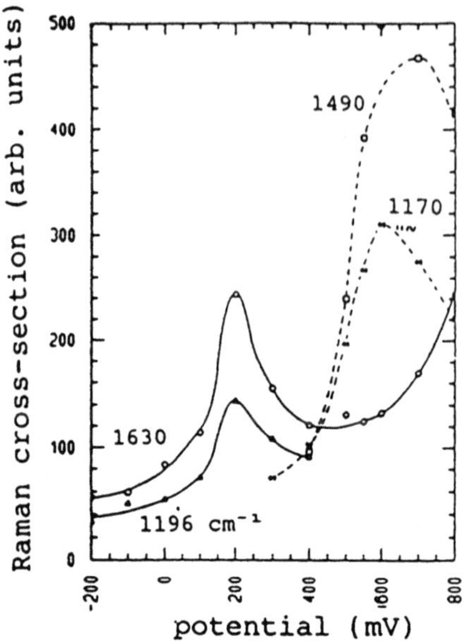

Fig. 1: Raman cross-section of benzenoid (solid) and quinoid (dashed) ring vibrational modes

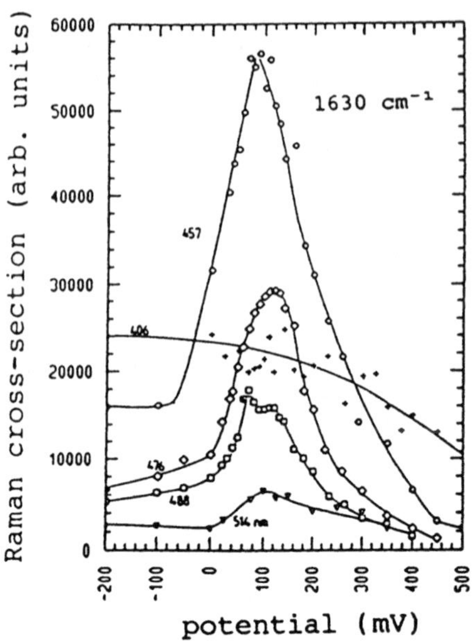

Fig. 2: Raman cross-section measured in situ with various laser lines as a function of potential

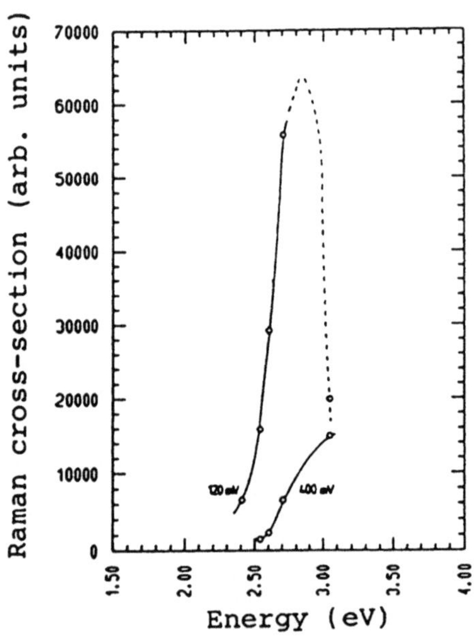

Fig. 3: Excitation profile of PANI at 120 mV and 400 mV

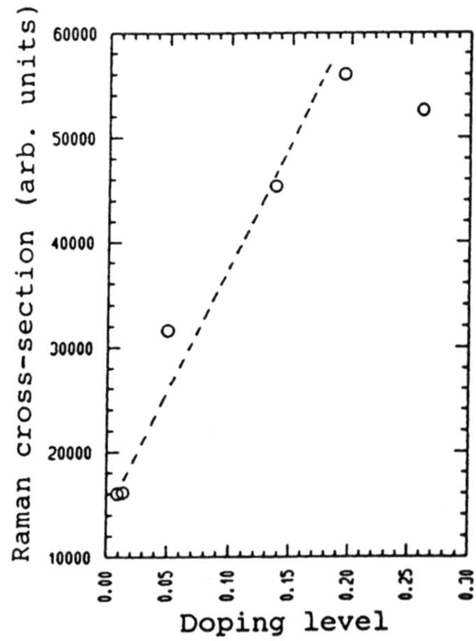

Fig. 4: Relationship between Raman cross-section and doping level

that each isolated polaron causes an additional contribution to the enhancement of the cross-section of the benzenoid ring vibrational mode as can be seen from the linearity between anion concentration and cross-section in Fig. 4. At higher doping levels the interaction between polarons is not negligible and finally leads to a polaron lattice. Then the linearity between doping level and resonance cross-section is lost.

Acknowledgement. Financal support by the Fond zur Förderung der wiss. Forschung in Austria (Proj. Nr. 5735) is gratefully acknowledged. M.B. thanks for financal support by the Kulturamt der Stadt Wien, MA 7, Vienna, Austria.

5. Literature

1 N. S. Sariciftci, M. Bartonek, H. Kuzmany, H. Neugebauer, A. Neckel, Synth. Met. **29**, E193 (1989).
2 J. C. Chiang, A. G. MacDiarmid, Synth. Met. **13**, 193 (1986).
3 S. H. Glarum, J. H. Marshall, J. Phys. Chem. **90**, 6076 (1986).
4 Y. Cao, S. Li, Z. Xue, D. Guo, Synth. Met. **16**, 305 (1986).
5 H. Kuzmany, E. M. Genies, A. Syed in H. Kuzmany, M. Mehring and S. Roth (eds.), Springer Series in Solid State Sci. **63** (Springer-Verlag, Berlin 1985), p. 223.
6 N. S. Sariciftci, H. Kuzmany, Synth. Met. **21**, 157 (1987).
7 Y. Furukawa, T. Hara, Y. Hyodo, I. Harada, Synth. Met. **29**, 189 (1986).
8 N. S. Sariciftci, H. Kuzmany, H.Neugebauer, this proceedings.
9 R. R. Chance, D. S. Boudreaux, J. W. Wolf, L. W. Shaklette, R. Silbey, B. Themans, J. M. Andre, J. L. Bredas, Synth. Met. **15**, 105 (1986).
10 S. Stafström, J. L. Bredas, A. J. Epstein, H. W. Woo, D. B. Tanner, W. S. Huang, A. G. MacDiarmid, Phys. Rev. Lett. **59**, 1464 (1987).

Part X

Polythiophenes

Chromism in the Poly(3-alkylthiophenes)

O. Inganäs, M. Sundberg, G. Gustafsson, J.O. Nilsson, S. Stafström, and B. Sjögren

Department of Physics, University of Linköping,
S-581 83 Linköping, Sweden

Abstract

Thermochromism in the poly(3-alkylthiophenes) is reviewed. The interpretation of these phenomena as an order-disorder transition of the conjugated polymer chain is presented. The nature of the ordered solid state and the minor thermochromic effects occurring at temperatures between 10 and 300 K is discussed. We suggest that the characteristic peaks found in the optical absorption spectra are due to vibronic structure. We also present polarized optical absorption spectra in the undoped and oriented state of the polymers, showing linear dichroism.

1.Introduction

The presence of conformational deformations of conjugated electroactive polymers has been demonstrated in the thermochromic and solvatochromic transformations of the poly(3-alkylthiophenes) (P3AT) [1,2]. These transitions - a continuous and reversible change of colour with temperature in the solid material(Fig.1), and change of colour with solvent quality and temperature in solutions of the polymer - are new to truly conductive polymers, but have been amply observed in the poly(diacetylenes) [3] and in the polysilanes [4]. The ordered phase, found in the

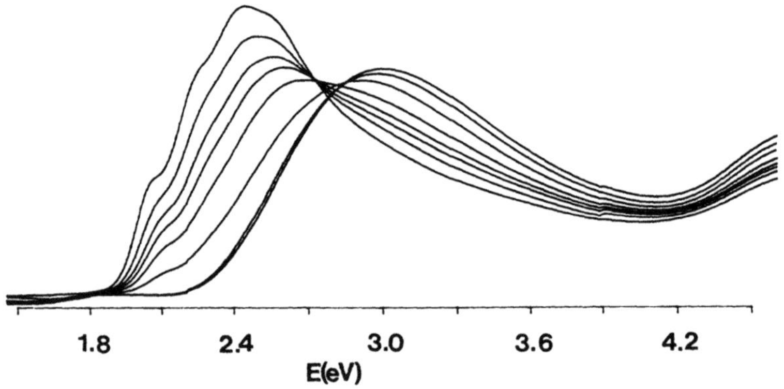

Fig.1 Thermochromism in poly(3-octylthiophene)(POT). The optical absorption spectra are taken at 22, 70, 100, 120, 140, 160, 180 and 200 °C respectively (from left to right)

Springer Series in Solid-State Sciences, Vol. 91
Electronic Properties of Conjugated Polymers III
Editors: H. Kuzmany · M. Mehring · S. Roth

solid at low temperatures and in polymer solutions of poor solvent quality, is characterized by a smaller bandgap than the disordered phase found in the melt and in good solutions. From structural studies [5] it is known that the ordered state in the solid is partially crystalline. The disordered phase has recently [6] been studied in solution and it is found that the Mark-Houwink relation is valid for a range of molecular weights. From these data, it is possible to infer an almost random coil conformation in the disordered phase in good solvents. Similar conclusions have been obtained in neutron scattering studies [7,8] of poly(3-butylthiophene) in some solvents. By the similarity of optical absorption data in the melt and in good solvents, we also infer that the polymer in the melt is highly disordered.

In the ordered phase, most of the thiophene units are coplanar. The disordered phase of high bandgap is characterized by rotational disorder where consecutive thiophene monomers are not coplanar. It has been shown by calculations that this kind of rotational defect can generate a shift of bandgap comparable with that observed in experiments, including both optical absorption and photoemission experiments [9,10]. The driving force behind these order-disorder transformations is not fully understood, but evidence suggests that side chain melting is an important feature. It is observed that the melting transition occurs at lower temperatures with longer alkyl side chains [11]. In the ordered state, diffraction reveals a structure with one periodicity approximately equal to the length of two side chains [5], which disappears at temperatures above the thermochromic transition. All these facts point to the importance of side chain ordering, and possibly crystallization, in the ordered state. The melting transition may then be driven by side chain melting, as concluded for some of the poly(alkylsilanes).

A full theory of order-disorder transitions in conjugated polymers has been proposed by Schweizer [12 - 16] based on the conformation-dependent polarizability of a conjugated polymer, and its interaction with a solvent. This theory predicts phase transitions of a similar kind to those observed in the thermochromic transition, but it is not yet known whether the quantitative predictions of the theory are valid for the P3AT's.

2.The nature of optical absorption in the ordered state

The presence of the characteristic peaks in the optical absorption spectra of poly(3-alkylthiophenes) has been explained as due to vibronic structure [2,11], to different conjugation lengths [1,9]or to excitonic effects[11]. Our studies [17] of the low temperature optical absorption (10 -300 K) (Fig.2) show these structures to be even more enhanced at lower temperatures, in addition to a small shift of the energy position. With different means of sample preparation, we also observe major changes in the overall outline of the optical absorption characteristics. This fact leads us to the conclusion that the vibronic peaks are mainly seen in the ordered low temperature phase. The peaks are spaced equidistant in energy($\approx$ 0.18 eV).

Similar peaks have been found also in poly(paraphenylenevinylene), and attributed to highly crystalline and ordered forms of the material [18]. It is thus clear that ordering is a necessary condition for the presence of vibronic structure in the optical

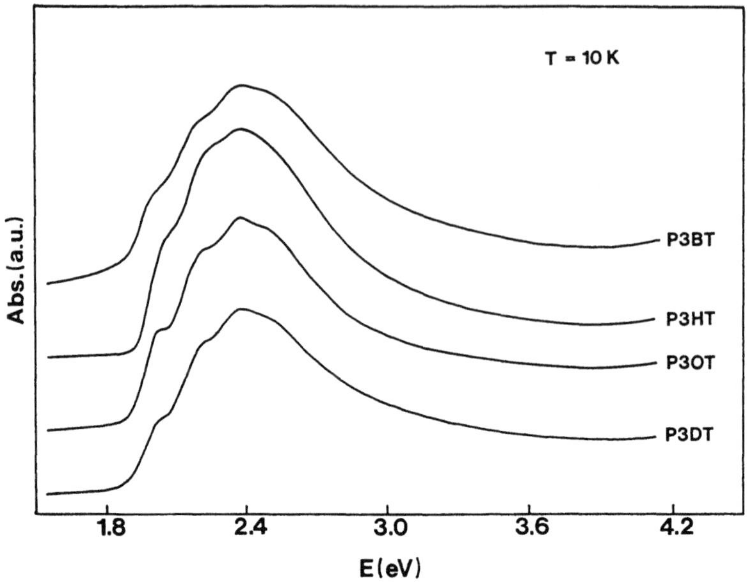

Fig.2 Low temperature optical absorption spectra of poly((3-butyl, -hexyl, -octyl, decyl)thiophene)

absorption spectrum. The vibration that most certainly is the one coupling to the π-electronic system is the C=C stretching vibration at 1460 cm-1 (.18 eV) which is part of the π-electronic path. The presence of alkyl side chains is, in our system, a condition for the attainment of the high order that is a necessary condition for the pronounced optical absorption features due to vibronic structure.

The ordered state of P3AT is also characterized by a major degree of crystallinity [5], as X-ray diffraction on these materials gives well resolved diffraction peaks. The presence of a diffraction peak at low scattering angle is indicative of an order with a periodicity related to the length of the side-chain. But the material is only partially crystalline and we observe an additional amorphous halo related to the non-crystalline phase. It is thus clear that the room temperature optical absorption spectrum is due to the combination of two phases, one crystalline and one amorphous. We suggest that the crystalline phase shows the characteristic vibronic absorption . Additional support for this interpretation is found in the study of polymer blends (Sect.3).

3.Thermochromism in polymer blends

Polymer blends between P3AT's and matrix polymers are easily formed by melt or solution processing. Such blends are of increasing importance for the utilization of conductive polymers, as processing and cost advantages favor the use of small amounts of conductive polymers for attaining the conductivity. It has been shown [19] that the percolation threshold for such polymer blends may be as low as betweeen 5 and 10 %. However, there are still questions on the compatibility of these materials that are left unanswered. We have therefore initiated a study of thermochromism in some of these polymer blends.

Generally, we view this approach as an optical equivalence of the DSC technique used in thermal analysis. In our case, we follow the change of optical absorption with temperature in a polymer blend. The colour is only due to the P3AT component. If we assume that total phase segregation occurs, and that the ratio between the amounts of "bulk" P3AT and "interfacial" P3AT is high, then we would expect the thermochromism in the blend to be exactly the same as in the

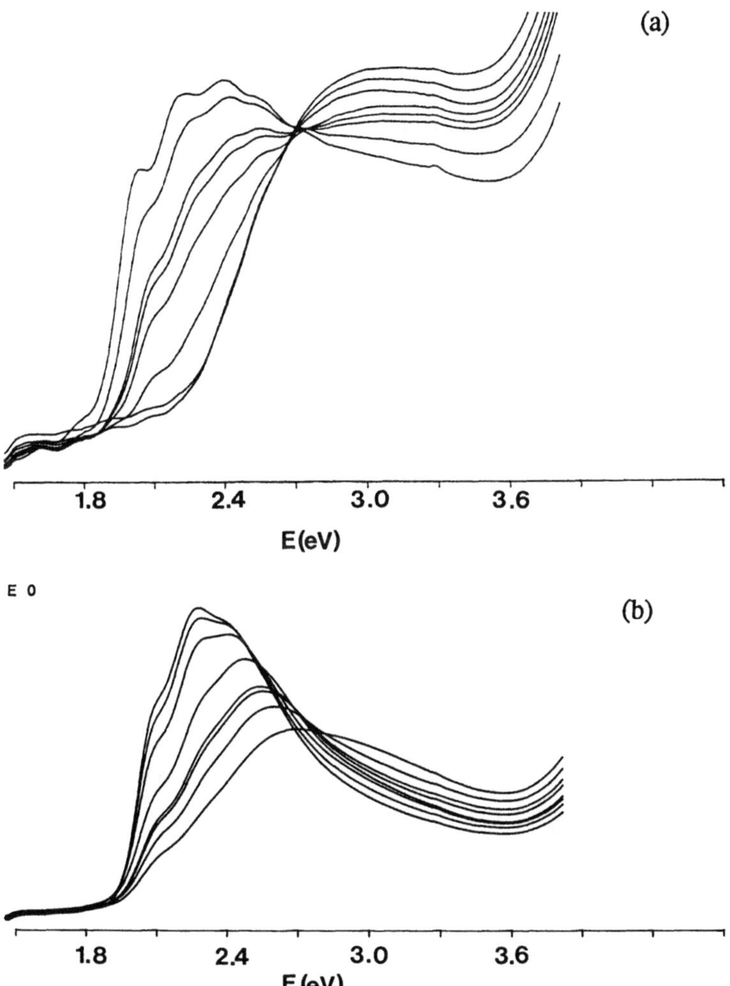

Fig.3 Thermochromism in the polymer blends 10 % POT in polystyrene (PSPOT) and 10 % POT in EVAPOT. Temperatures are in (a) 30, 60, 90, 110, 130, 150, 170, 190 °C, and in (b) 25, 50, 70, 100, 120, 160, 200 °C

homopolymer. If, on the other hand, there is no major phase segregation - i.e. we have a truly molecular blend - then we would expect to see the changes of optical absorption occur at some other temperature.

We report here the thermochromism in blends of poly(3-octylthiophene) with poly(ethylene -*co*-vinyl acetate) (EVA) and poly(styrene) (PS). PS is a totally amorphous polymer with a melting transition at 240 °C, while EVA is partially crystalline with a melting transition at 60-70 °C. POT is in itself a partially crystalline material with a melting transition close to 160 °C. In Fig.3 is shown the thermochromic transitions between 25 and 200 °C for 10 % POT in EVA and in PS. We clearly see an isosbestic point in the PSPOT material, but observe none such for the EVAPOT material. Therefore POT in PSPOT is distributed among two phases, while in EVAPOT more than two (constant) phases are necessary to explain the data. It has been suggested [5] that the homopolymer may exist in one crystalline and one amorphous, nematic phase, undergoing continuous change with temperature. This may be the case also in the EVAPOT blend.

Plotting the energy position of the absorption maximum versus temperature for POT, 10 % EVAPOT and 10 % PSPOT shows quite different behaviour (Fig.4). We see the shift of absorption maximum occur much earlier with EVAPOT than with POT, and also much more rapidly in PSPOT than in EVAPOT. This shift of the optical absorption has been associated with the melting transition in the bulk polymer [11], and we may conclude that the mixing of (partially crystalline) POT with amorphous PS reduces the melting temperature of POT. This is generally expected for mixtures of crystalline and amorphous polymers for the case of negative Flory solvent interaction (polymer/polymer interaction) parameters [20], and indicates that the two polymers form a true blend, i.e. no phase segregation occurs. This conclusion remains to be verified by structural studies and DSC.

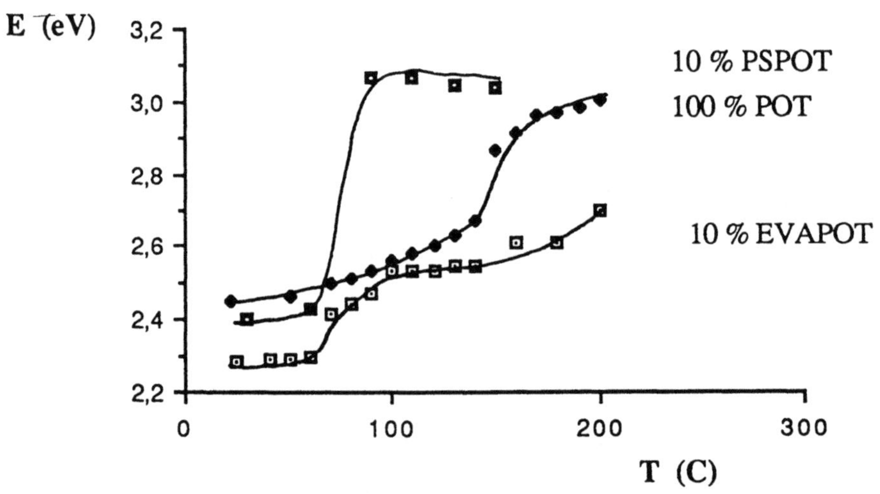

Fig.4 Absorption maximum versus temperature for POT, 10 % PSPOT and 10 % EVAPOT

4. Orientation of P3AT

The study of intrinsic electronic properties of conjugated polymers is greatly assisted by orientation of polymer chains along one direction, where the polymers truly realize their unique low dimensionality. Orientation of conjugated polymers has hitherto mainly been accomplished by orientation of precursor polymers, and subsequent conversion to the conjugated form [21]. With the P3AT, orientation is performed by stretching the polymer in its conjugated form. The orientation level obtained by this technique is so far lower than that obtained by orientation of precursor polymers, but very pronounced linear dichroism is observed. Interestingly, we observe that the colour (in transmission and in polarized light) is deep blue-violet with light polarized along the stretching direction, and faintly yellow for light polarized normal to the stretching direction (Fig. 5). These features are clearly demonstrated in the spectra, where light polarized parallel to the direction of orientation is absorbed for almost all wavelengths in the visible range, except for the far-red and the blue-violet region where absorption is low enough. With light polarized normal to the direction of elongation, the colour is dominated by the transmitted light of red and yellow colour, while blue-violet light is absorbed. This linear dichroism is very pronounced, and we therefore describe it under the heading of chromism.

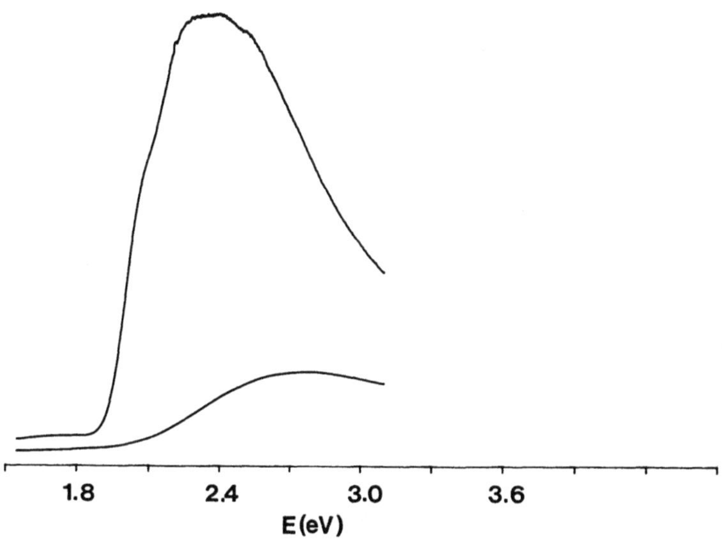

Fig.5 Polarized optical absorption of POT after elongation, parallel versus normal to the stretching direction. The sample was prepared by coating a thin film of POT on polyethylene foil

5.Conclusion

The colour of P3AT's reflect order-disorder transitions of the polymer chain, phase segregation or a true molecular mixing in the polymer blends, and, in polarized light, the orientation of the polymer molecules. It is thus a sensitive probe of many properties believed to be of the utmost importance in conductive polymer transport properties. These properties may also be of importance for applications.

Acknowledgements

We thank Drs. J.-E.Österholm and J.Laakso at Neste Oy, Kullo, Finland for synthesizing and supplying us with P3AT samples. This work was supported by the Nordic Fund for Industrial Development (NI) and the Natural Sciences Research Council of Sweden(NFR).

References

1. O.Inganäs, W.R.Salaneck, J.E.Österholm and J.Laakso: Synth.Metals **22**, 395, (1988)
2. S.D.D.V.Rughooputh, S.Hotta, A.J.Heeger amd F.Wudl: J.Polym.Sci.Polymer Phys.Ed **25**, 1071, (1987)
3. "Polydiacetylenes", ed. D.Bloor and R.R.Chance, (Martinus Nijhoff, Dordrecht 1985)
4. R.D.Miller, D.Hofer, J.F.Rabolt and G.N.Fickes: J.Am.Chem.Soc.**107**, 2172 (1985)
5. M.J.Winokur, D.Spiegel, Y.Kim, S.Hotta and A.J.Heeger: Synthetic Metals **28**, 419, (1989)
6. T.Hjertberg and M.L.Eidrup: these proceedings
7. J.P.Aime, F.Bargain, M.Schott, H.Eckhardt, G.G.Miller and R.LElsenbaumer: Synthetic Metals **28**, 407, (1989)
8. J.P.Aime, F.Bargain, M.Schott, H.Eckhardt, G.G.Miller and R.LElsenbaumer: Phys.Rev.Lett. **62**, 55, (1989)
9. W.R.Salaneck, O.Inganäs, B.Themans, J.O.Nilsson, B.Sjögren, J.E.Österholm, J.L.Bredas and S.Svensson: J.Chem.Phys. **89** , 4613, (1988)
10. B.Themans, W.R.Salaneck and J.LBredas: Synthetic Metals **28**, 359, (1989)
11. O.Inganäs, G,Gustafsson, W.R.Salaneck, J.E.Österholm and J.Laakso, Synth.Metals: **28**, 377, (1989)
12. K.S.Schweizer: J.Chem.Phys. **85**, 1156 (1986)
13. K.S.Schweizer: J.Chem.Phys. **85**, 1176 (1986)
14. K.S.Schweizer: J.Chem.Phys. **85**, 4181 (1986)
15. K.S.Schweizer: Polymer Preprints **27**, 354 (1986)
16. Z.G.Soos and K.S.Schweizer: Chem.Phys.Lett. **139**, 196 (1987)
17. M.Sundberg, O.Inganäs, S.Stafström, G.Gustafsson and B.Sjögren, submitted
18. D.D.C.Bradley and Y.Mori: Jap.J.Appl.Phys. **28**, 174 [1989]
19. J.Laakso, J.E.Österholm and P.Nyholm: Synthetic Metals **28**, 467, (1989)
20. L.H.Sperling:"Introduction to Physical Polymer Science", Wiley-Interscience, N.Y. 1986, Chapt.5.12
21. G.Leising: Polymer -Bulletin **11**, 401, (1984)

Soluble Polyalkylthiophenes Obtained from Substituted Bithiophenes

M. Zagórska[1] *and B. Krische*

Royal Institute of Technology, Department of Organic Chemistry, S-100-44 Stockholm, Sweden

ABSTRACT

4,4'-dibutyl-2,2'-bithiophene and 4,4'-dioctyl-2,2'-bithiophene have been used as starting materials for the synthesis of soluble polythiophenes. Cyclic voltammetry, ^{1}H NMR and UV/VIS spectroscopy give evidence of more regular structure of the polymers obtained from 4,4'-dialkylbithiophenes as compared to those synthesized from 3-alkylthiophenes.

1 Introduction

The development of soluble poly(3-alkylthiophenes) [1][2] was a great improvement in processing technology of conducting polymers. The solubility of these polymers opens new possibilities of characterization of their properties and structure.

3-substituted thiophenes usually used as monomers are not symmetric since their 2 and 5 positions are not geometrically equivalent. Therefore their polymerization may lead to two nonequivalent structures: one originated from "head to tail" and the second from "head to head" coupling. Although "head to head" is a less favoured mode of coupling due to steric reasons, it cannot be totally excluded. Polymerization of symmetrically substituted dimers should result in a more regular chain . In addition, it is known that oligomers exhibit lower oxidation potential than the corresponding monomers. The polymerization can therefore be carried out at milder conditions which in turn lower the risk of the creation of defects due to overoxidation of the polymers [3].

In this paper we report a comparative study on the synthesis and characterization of polydialkylbithiophenes (PDABT-s): namely poly(4,4'-dibutyl-2,2'-bithiophene) (PDBuBT) and poly(4,4'-dioctyl-2,2'-bithiophene) (PDOBT) and the corresponding polyalkylthiophenes (PAT-s): poly(3-butylthiophene)(PBuT)and poly(3-octylthiophene) (POT).

2 Experimental

The polymers have been obtained by both chemical and electrochemical methods.

Electrochemical synthesis and other electrochemical experiments were performed in a two compartment cell using platinum as working electrode and Ag/AgCl electrode as reference. Polymer films were grown at constant current of 1 mA/cm^2 in 0.1 M

[1]Current address: Institute of General Chemistry, Warsaw Agriculture University (SGGW–AR), Rakowiecka 26/30, 02–528 Warsaw, Poland

Springer Series in Solid-State Sciences, Vol. 91
Electronic Properties of Conjugated Polymers III
Editors: H. Kuzmany · M. Mehring · S. Roth

Bu_4NClO_4/acetonitrile. The optimum concentrations of 3-alkylthiophenes were 0.2 M, whereas for dialkylbithiophenes 50 times lower concentrations were sufficient [4].

In chemical synthesis $FeCl_3$ has been used as polymerization agent following slightly modified Sugimoto method [5]. Details of our procedure will be published elsewhere [6].

4,4'-dialkylbithiophenes were synthesized by oxidative coupling of lithiated alkylthiophenes [7]. Monomeric alkylthiophenes were obtained according to the method of Kumeda et al. [8].

3 Results and Discussion

Chemical and electrochemical polymerizations of 4,4'-dibutylbithiophene and 4,4'-dioctylbithiophene give the products which after the reduction to the neutral state are soluble in common solvents and can be easily cast from chloroform solutions giving free standing films. The results of PDABT-s characterization by elemental analyses and FTIR spectroscopy [6] are consistent with the linear chain of 2,5-coupled thiophene rings substituted with one alkyl side group per ring, similarly as is observed for PAT-s. However, ^{1}H NMR spectroscopy reveals the differences between the structure of the polymers obtained from bithiophenes and thiophenes. The representative results are shown in Table 1 and compared with NMR parameters of the monomer and the dimer. The spectra of both polymers are consistent with the presence of a chain of aromatic thiophene rings with aliphatic side groups. However, PDBuBT gives a simple spectrum corresponding to one type of structure, whereas the spectrum of PBuT reveals the coexistence of two magnetically nonequivalent side groups. The shifts of the dominant peaks are different from those of dimer originated polymer. The parameters of the second set of signals are almost the same as those of PDBuBT and close to the ones observed in butyl substituted thiophene and bithiophene spectrum (Table 1). Similar results have been obtained for PDOBT and POT. It is tempting to assigned the two ^{1}H NMR signal sets of PAT-s to polymer regions with different substituents orientation (i.e. "head to head" and "head to tail" coupling) [9]. Therefore the presence of only one set of signals in PDABT-s spectra evidences more regular chain structure.

PDABT-s and PAT-s are very different in colour, the former being bright yellow whereas the latter ones deep red in transmission and greenish in reflection. In UV/VIS spectra a significant blue shift in the absorption maximum is observed for polymers obtained from substituted bithiophenes as compared to the corresponding substituted

Table 1: Chemical shifts of hydrogens of butyl substituted thiophene, bithiophene and corresponding polymers (400 MHz, $CDCl_3$, 25 °C)

	H2	H3	H5	α-CH_2	β-CH_2	γ-CH_2	CH_3
4-BuT*	7.24(dd)	6.95 –	6.92(m)	2.65	1.64	1.40	0.965
4,4'-DBuBT		6.78(d)	7.12(d)	2.60	1.64	1.40	0.965
PBuT		6.98		2.81	1.69	1.46	0.98
		7.02**		2.58**	1.59**	1.35**	0.91**
PDBuBT		7.06		2.55	1.59	1.35	0.90

* - for the consistency of the table, 3–butylthiophene is called 4–butylthiophene,
** - chemical shifts of the lines coresponding to the lower intensity set of signals

Table 2: The main absorption maxima in UV/VIS spectra of PDABT-s and PAT-s

Synthesis method	R	λ_{max} (nm)			
		PAT		PDABT	
		$CHCl_3$ solution	film	$CHCl_3$ solution	film
chemical	butyl	440	496	392	392
electrochemical		432	444	390	390
chemical	octyl	440	484	388	386
electrochemical		430	450	388	388

thiophenes (Table 2). The spectra of PDABT-s are the same for solid films and solutions independently of the solvent used. The corresponding PAT-s display a red shift of the absorption maximum accompanied by a line broadening for solid samples as compared to the one dissolved in chloroform. Much higher energy of $\pi - \pi^*$transition in PDABT-s may reflect the shorter conjugation length than observed in PAT-s. Indeed the maximum of the absorption peak at 390 nm roughly corresponds to the value found for oligomers consisting of four thiophene rings [10]. Since the molecular weight of PDBuBT (M_n = 10390, M_w/M_n = 1.47) is rather high and similar to PBuT (M_n = 10125, M_w/M_n = 1.40) the observed shorter conjugation length in polybithiophenes may arise from steric factors. Close proximity of alkyl groups in adjacent thiophene rings may result in a significant nonplanarity of the system.

Both polymers are electroactive however they exhibit slightly different redox properties (Fig.1). The oxidation and reduction peaks in cyclic voltammetry of PDABT–s are much sharper than in the case of PAT-s. These features may be taken as an additional indication of lower stereoregularity of polyalkylthiophenes as compared to polydialkylbithiophenes.

Electrochemical doping-undoping reaction of PDBuBT have been monitored by UV/VIS spectroscopy. For this purpose a film of chemically prepared polymer has been deposited on ITO electrode. The scanning of the potential to 1.1 V results in a decrease of the 390 nm $\pi - \pi^*$ transition with simultaneous growth of the absorption at ca 800 nm as observed previously for other polythiophenes (Fig. 2).

In addition to electrochemical doping the behaviour of free-standing films towards classical chemical dopants was investigated. Chemical doping carried out in $FeCl_3$/nitro-

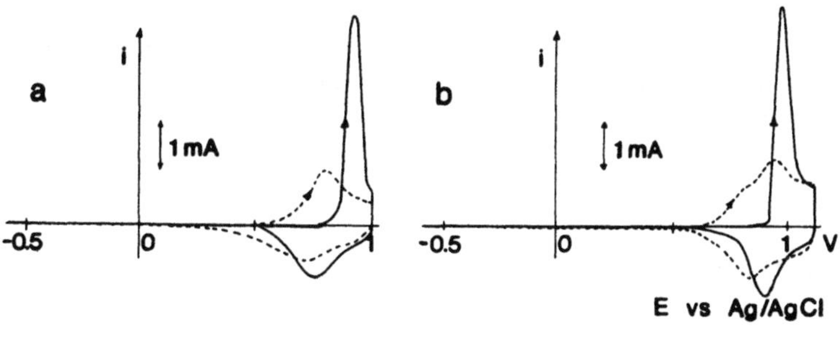

Figure 1: Cyclic voltammograms of PDABT (solid line) and PAT (dashed lines) in Bu_4NClO_4/acetonitrile, 20 mV/s, a–butylsubstituted, b–octylsubstituted

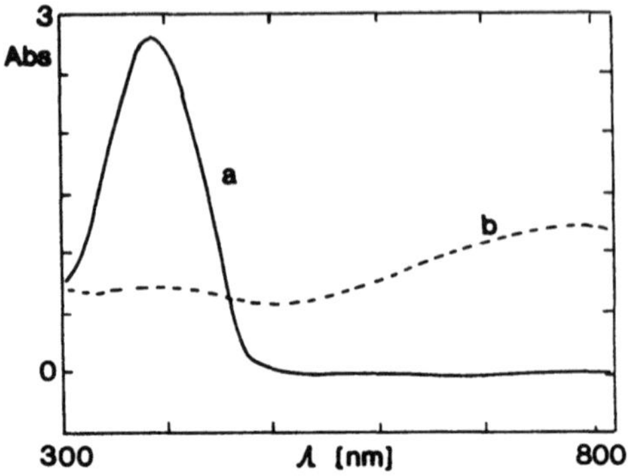

Figure 2: UV/VIS spectrum of PDBuBT film on ITO electrode: a–pristine polymer, b–electrochemically oxidized polymer

methane solution or in iodine vapours results in the transformation of bright yellow transparent material (15 μm thick) into opaque dark blue. The electrical conductivity of the oxidized form of PDABT-s was ca 2-4 S/cm.

4 Conclusions

Electrochemical and chemical polymerization of 4,4'-dialkylbithiophenes results in the formation of soluble polymers which by ^{1}H NMR spectroscopy are more stereoregular than those obtained from monomeric 3-alkylthiophenes. Although, as revealed by UV/VIS spectroscopy PDABT-s apparently have shorter conjugation length than PAT-s, they can be easily doped by electrochemical and chemical methods. It should be underlined that different substituents orientation (i.e. "head to head" and "head to tail" coupling) is only one of the plausible explanations of the observed differences between PDABT-s and PAT-s. Further spectroscopic work is needed to determine the exact geometry of the chain. Such studies (including temperature dependance of NMR spectra) are in progress.

Acknowledgements

We wish to thank Dr. Adam Pron for his critical review of the manuscript. This work was supported by grants from the Swedish National Board for Technical Development (STU) and the Swedish Institute.

References

1. K.Y. Jen, G.G. Miller and R.L. Elsenbaumer: J.Chem. Soc., Chem. Commun., 1346 (1986)

2. M. Sato, S. Tanaka and K. Kaeriyama: J. Chem. Soc., Chem. Commun., 873 (1986)

3. B. Krische and M. Zagórska: Synthetic Metals, 28, C 257 (1989)

4. B. Krische, M. Zagórska and J.Hellberg: to be published

5. R. Sugimoto, S. Takeda, M.B. Gu, K. Yoshino: Chemistry Express, 1, 635 (1986)

6. M. Zagórska and B. Krische: to be published

7. B. Krische, J. Hellberg and C. Lilja: J. Chem. Soc., Chem. Commun., 19, 1476 (1987)

8. K. Tamao, S. Komada, J. Nakajima, M. Kumada: Tetrahedron, 38, 3347 (1982)

9. R.L. Elsenbaumer, Kwan-Yue Jen, G.G. Miller, H. Eckhardt, L.W. Shacklette and R. Jow in "Electronic Poperties of Conjugated Polymers", ed. H. Kuzmany, M. Mehring and S. Roth, Springer Ser. Solid-State Sci., vol. 76 (Springer-Verlag, Berlin 1987), p. 400

10. L. Laguren-Davidson, C. Van Pham, H. Zimmer and H.B. Mark Jr.: J. Electrochem. Soc., 135, 1406 (1988).

Conducting Polymers from 3,4-Cycloalkylthiophenes

R. Rühe[1], *A. Berlin*[2], *and G. Wegner*[1]

[1]Max-Planck-Institut für Polymerforschung, Ackermannweg 10, D-6500 Mainz, Fed. Rep. of Germany
[2]Università Degli Studi Milano, Dipartimento di Chimica Organica e Industriale, via Golgi 19, Milano, Italy

Poly(3,4-cycloalkylthiophene)s are obtained by electrochemical or chemical oxidation of 3,4-cycloalkylthiophenes. With increasing ring size of the substituent both the conjugation length and conductivity decrease. The conduction behavior is in accordance with the predictions by the hopping theory, regardless of changes in the conjugation length. This might indicate that the chain packing is to some extent even more important for the macroscopic conductivity than the electronic structure of the backbone.

1. Introduction

Recently we have shown that the conduction behavior of poly(3,4-cycloalkylpyrrole)perchlorates is controlled by certain topological parameters, which were correlated with the size of the alkyl substituents [1,2]. We were interested in whether similar electrical properties are obtained in poly-(3,4-cycloalkylthiophene)s. Therefore we have synthesized thiophenes with a fused alkyl ring in the 3-and 4-position of the thiophene moiety and polymerized them either electrochemically or by chemical oxidation. In this work the electrochemical, electrical and spectroscopic properties of the obtained poly(thiophene) derivatives are presented. A comparison with the corresponding poly(3,4-cycloalkylpyrrole) salts is made.

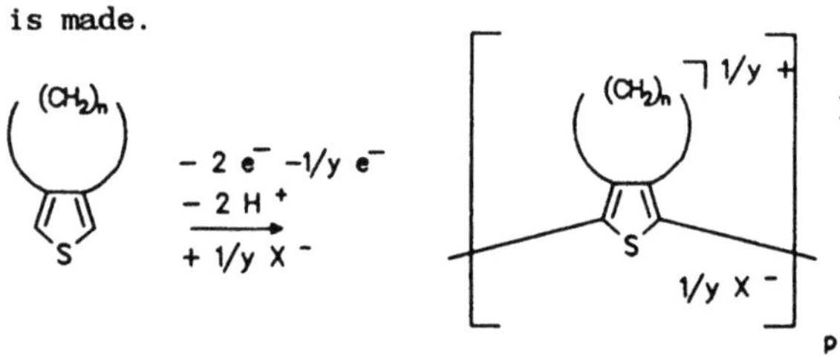

5 - 8   1 - 4

$X^- = ClO_4^-,\ PF_6^-,\ CF_3SO_3^-,\ Cl^-$

n	3	4	5	10
	1	2	3	4
	5	6	7	8

2. Synthesis

Monomers 5-7 were prepared according to the literature [3-5]:

5: cyclopenta[c]thiophene was synthesized by chloromethylation of 2,5-dichlorothiophene followed by condensation with malonic ester, decarboxylation, ring closure, dehalogenation and reduction of the obtained ketone [3]

6: cyclohexa[c]thiophene was obtained by Hinsberg-condensation of 1,2-cyclohexanedione and 2,2'-Thio-bis[methyl acetate], saponification and decarboxylation of the obtained dicarboxylic acid [4]

Springer Series in Solid-State Sciences, Vol. 91
Electronic Properties of Conjugated Polymers III
Editors: H. Kuzmany · M. Mehring · S. Roth

7: cyclohepta[c]thiophene was prepared through reaction of 2-chloromethylene cycloheptanone with methyl thioglycolate, ring closure, saponification of the ester and decarboxylation [5]

8: cyclododeca[c]thiophene was synthesized following the same route as cyclohepta[c]thiophene

Polymeric salts of the poly(3,4-cycloalkylthiophene)s were obtained by electropolymerization in a divided compartment cell at 5 °C, with acetonitrile or nitrobenzene as the solvent. The monomer and electrolyte concentrations were 0.01 and 0.1 mol/l respectively. The optimum current density under galvanostatic conditions was found to be 2 mA/cm^2.

Chemical polymerizations were carried out with iron(III)chloride in acetonitrile or chloroform.

3. Electrochemical properties

The results of cyclovoltammetric investigation of the electrochemical behavior of poly(3,4-cycloalkylthiophene)s and of the corresponding monomers are presented in Fig. 1 and Table 1. In electroanalytical experiments films of poly(3,4-cycloalkylthiophene) salts were obtained by galvanostatic electropolymerization employing a current density of 2 mA/cm^2 until a film

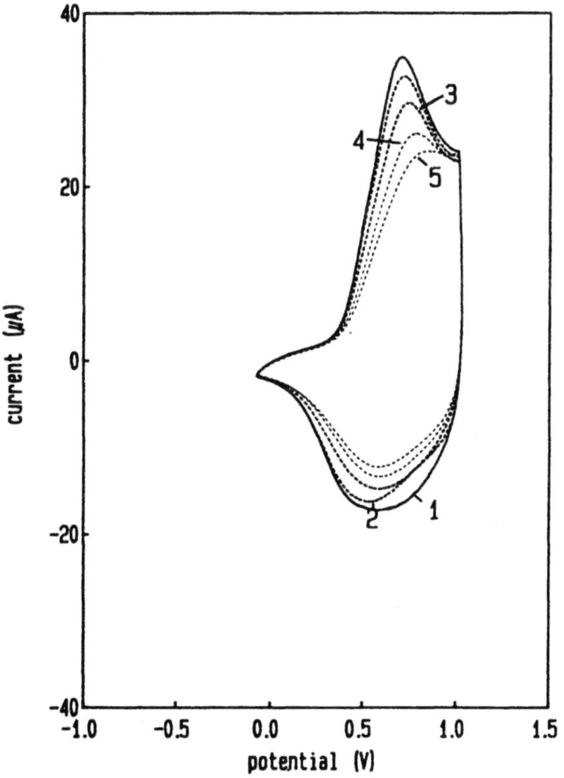

Fig. 1: Cyclovoltammogram of poly(cyclohexyl[c]thiophene); reference Ag/AgCl; sweep rate 100 mV/s; electrode surface 7,85·10^{-3} cm^2; (1)= cycle 1; (2)= cycle 3; (3) = cycle 50, (4) = cycle 100, (5) = cycle 300

Table 1: Electrochemical data of the 3,4-cycloalkylthiophenes 5-8, thiophene 9, 3-methylthiophene 10 and of the corresponding polymers[1,4]

compound	n	E_a^{p}[2] (monomer)	E_c^{p}[2]	E_a^{p}[2]
5	3	1.80 (1.30)[6]	0.47 (0.25)[6]	0.64 (0.74)[6]
6	4	1.70	0.60	0.73
7	5	1.78	1.0 [3]	1.2 [3]
8	10	1.76	1.1 [3]	1.3 [3]
thiophene	-	1.90 (1,60)[7]	0.75 (0.60)[7]	0.90 (0.69)[7]
3-methylthiophene	-	1.65 (1,40)[6]	0.53 (0.45)[7]	0.65 (0.28)[7]

[1] reference Ag/AgCl; sweep rate 100 mV/s; polymers deposited at 23 °C
[2] peak potential of the anodic (a) or cathodic (c) current in volts
[3] only poorly reversible
[4] literature values in brackets (vs. Ag/Ag^+)

of approximatly 1 µm thickness was deposited. The films were removed from the electrolyte, rinsed with acetonitrile, dried and immersed in acetonitrile containing 0.1 mol/l tetrabutylammonium perchlorate as supporting electrolyte.
The oxidation potentials of poly(3,4-cycloalkylthiophene)s (P-3,4-CAT's) increase with increasing ring size of the substituent. The exact position of the peak potentials are strongly dependent on the polymerization conditions. 1 and 2 can be reduced and reoxidized several hundred times, while derivatives with larger substituents (3 and 4) are stable for only a few cycles.

4. Spectroscopical characterization

The oxidized polymers show a broad, featureless absorption over the whole investigated spectral range (330-1100 nm). The wavelength of the absorption maximum of the neutral polymers decreases with increasing ring size of the substituent from 530 to < 300 nm, as shown in fig. 2. The shift shows that the electronic structure of the polymer backbone changes with the size of the substituent. This behavior contrasts to that of poly(3,4-cycloalkylpyrrole)s, where the electronic structure of the polymer is independent from the size of the substituent. A complete analysis of the spectral properties of the obtained polymers will be published elsewhere [8].

5. Electrical properties

Freestanding, conducting films of P-3,4-CAT's are obtained by electrochemical or chemical oxidation. The conductivity of P-3,4-CAT's decreases with increasing ring size of the substituent (Fig. 4).
PF_6^- salts are not very stable against atmospheric conditions; HF is slowly set free. The stability of the oxidized polymer decreases with increasing ring size of the substituent.

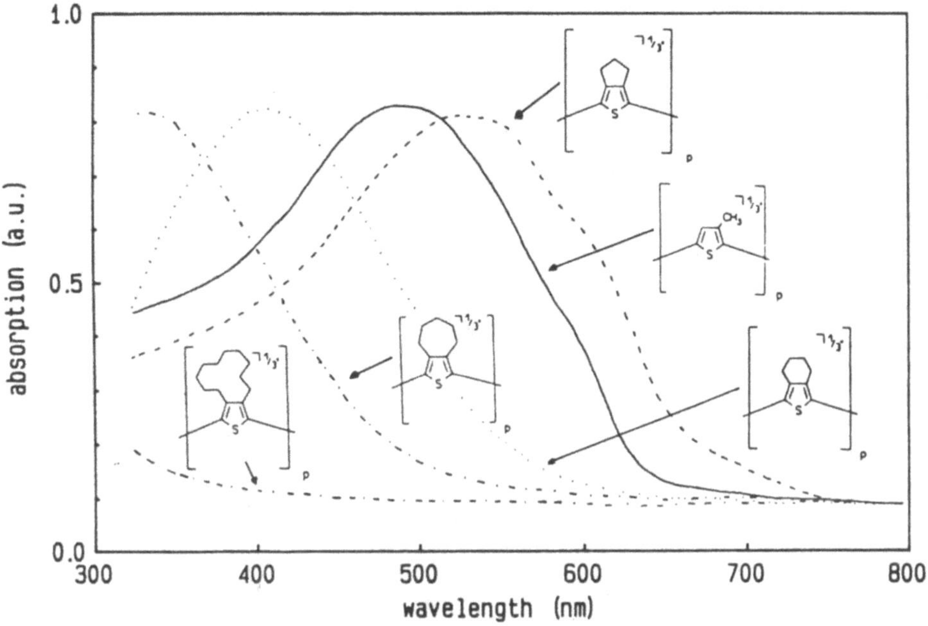

Fig. 2: Optical spectra of neutral poly(3,4-cycloalkylthiophene)s; polymer films on ITO-electrodes; neutral films obtained by electrochemical reduction; 2 μm thickness)

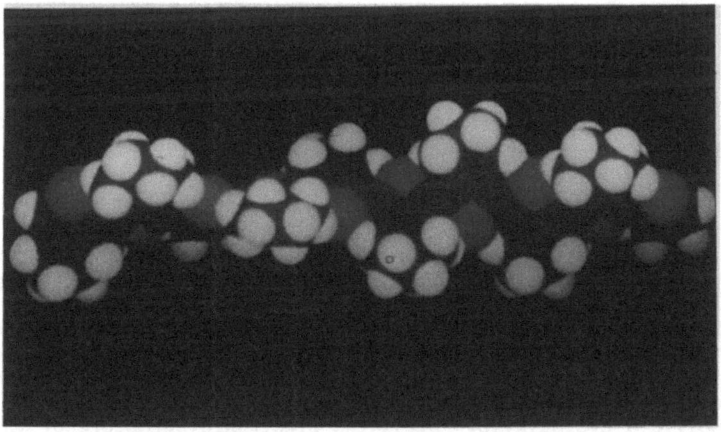

Fig. 3: Space-filling molecular model of poly(cyclohexa[c]thiophene) as obtained by computer simulation (idealized structure)

Molecular models (Fig. 3) show: because of the bulky substituents adjacent thiophene moieties cannot be arranged in plane, but the polymer chain has to be twisted. Consequently the thiophene backbone is surrounded by a layer of methylene groups and two polymer chains cannot get closer than a minimum distance R given by the size of the substituent [2]. According to hopping

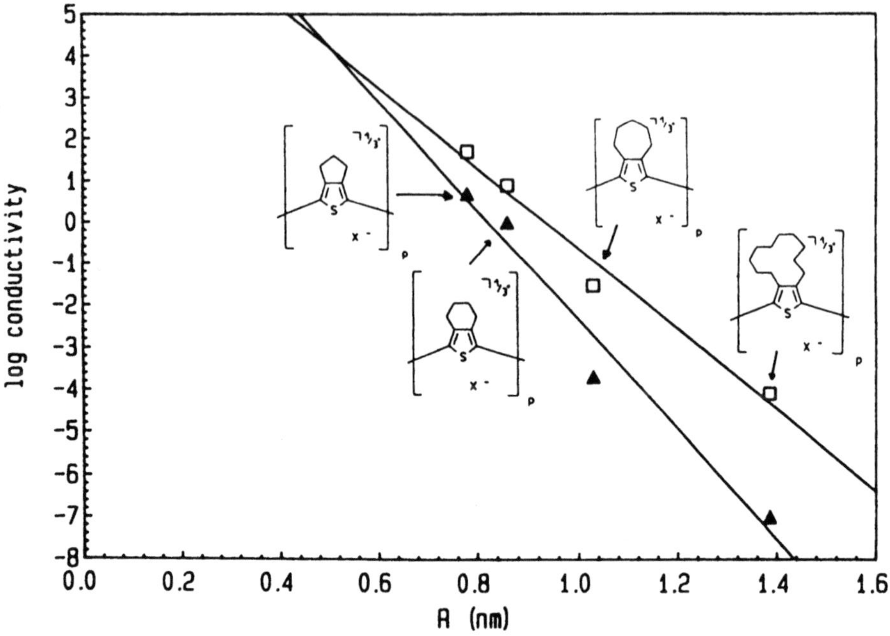

Fig. 4: Logarithm of the conductivity of the poly(3,4-cyclo-alkylthiophene) salts at 22 °C (four probe technique) as a function of the inter-chain distance R ; Cl^- salts (triangles) and PF_6^- salts (squares)

theory [9] of localized charge carriers the conductivity is given by

$$\sigma = \sigma_0 \exp(-2\alpha R) \exp(E_A/kT);$$

σ is a constant, α the inverse localization length of a charge carrier and E_A the activation energy for a thermally activated hopping process.

6. Discussion

1) As concluded from fig. 4 the conduction behavior of P-3,4-CAT's follows qualitatively the behavior which is predicted by the hopping theory considering the topology of the system [2].
2) Changes in the degree of conjugation with increasing ring size of the substituent are probably due to an increase in the steric hindrance against torsion around the bond between adjacent thiophene moieties (atactic copolymer of atropisomers).
3) Contrary to poly(3,4-cycloalkylpyrrole)s the electronic structure of the polymer backbone in P-3,4-CAT's changes with increasing ring size of the substituent. As concluded from molecular models, differences are probably due to a different bonding angle between the aromatic moieties.
4) Due to the fact that both topology and electronic structure of the polymer backbone change with increasing ring size of the substituent, a complete analysis of the conduction behavior cannot be made.
5) The good correlation between chain diameter and conductivity of the polymers seems to indicate that the interchain contacts play a dominant role in the conductivity of these polymers.

Acknowledgements:
J.R. is indebted to the Stiftung Volkswagenwerk for a Kekule scholarship. The valuable technical assistance by Ch. Sieber is gratefully acknowleged. B. Jung is thanked for the computer simulations of the molecular models.

References:

1) J. Rühe, Ch. Kröhnke, T.A. Ezquerra, F. Kremer and G. Wegner, Ber. Bunsengesell. Physik. Chem. 9, 885 (1987)
2) T.A. Ezquerra, J. Rühe, G. Wegner, Chem. Phys. Lett. 144, 194 (1988)
3) D.W.H. Mac Dowell, T.B. Patrick, B.K. Frame and D.L. Ellison, J. Org. Chem. 32, 1227 (1967)
4) R. Meyer et al., J. Prakt. Chem., 20, 244,(1963)
5) S. Hauptmann, E.M. Werner, J. Prakt. Chem. 314, 499 (1972)
6) J. Roncali, F. Garnier, R. Garreau, M. Lemaire, J. Chem. Soc. Chem. Comm, 476 (1987)
7) K. Kaeriyama, M. Sato, S. Tanaka, Synth. Met., 18, 233 (1987)
8) J. Rühe, A. Berlin, G. Wegner, in preparation
9) N.F. Mott and E.A. Davis, Electronic processes in non-crystalline materials (Clarendon Press, Oxford, 1979)

Raman Spectroscopy of Anellated and Alkylated Polythiophenes

*W. Wallnöfer, H. Kuzmany, and E. Faulques**

Institut für Festkörperphysik, Universität Wien, and
Ludwig Boltzmann Institut für Festkörperphysik,
Strudlhofgasse 4, A-1090 Wien, Austria
*Permanent address: Laboratoire de Physique Cristalline, IPCM,
Université de Nantes, F-44072 Nantes, France

Abstract. We report resonance Raman spectroscopy of polythiophene derivatives. The polyisothianaphthene spectra show a series of oligomer lines and give evidence for a quinoid structure. In polyalkylthiophenes luminescence dominates. Disorder upon solving and fusion is seen in the Raman spectra.

1. Introduction

We report resonance Raman (RR) experiments on substituted polythiophenes (PTs): the small gap polymers polyisothianaphthene (PITN, Fig. 1a,b) [1-4] and polyisonaphthothiophene (PINT, Fig. 1c) [5] and the soluble and processable poly(3octylthiophene) (POT, Fig. 1d) [6]. By selectively exciting parts of the sample with different transition energies, e.g. with different conjugation length [7], their vibrations and resonance profiles can be observed separately. In this way order-disorder transitions, e.g. thermochromism, can be studied [8]. Our results also yield information on the molecular structure and changes upon doping.

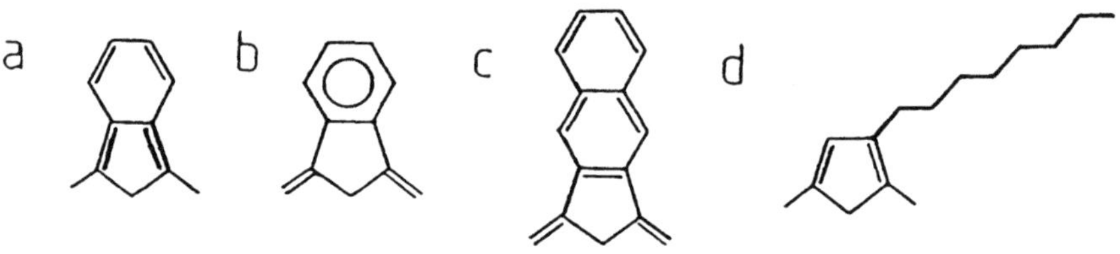

Fig.1 Aromatic (a) and quinoid (b) structure of PITN; (c) PINT; (d) POT

2. Experimental

Argon and krypton ion lasers were used with 11 lines between 406 and 799 nm. The laser light was filtered and line-focused on the sample. The backscattered light was analyzed with a double monochromator and a photon-counting system.

The PITN films were synthesized electrochemically [1] on platinum and ITO-coated glass. Spectra were taken ex situ in a cryostat and in situ in an electrochemical cell [9]. The PAT samples were cast from toluene solution. Spectra were taken in a cryostat in vacuum at temperatures between -196 and +190°C.

Springer Series in Solid-State Sciences, Vol. 91
Electronic Properties of Conjugated Polymers III
Editors: H. Kuzmany · M. Mehring · S. Roth

3. PITN: Results

Raman spectra of undoped (reduced) and doped (oxidized) PITN at various laser wavelengths were measured. They are reported in [9]. The spectrum of undoped PITN at low laser quantum energies (LQE) is dominated by a line at 1460 cm^{-1} which is mainly due to the amplitude mode of the conjugated backbone in which neighbouring C-C bonds are alternatingly stretched [10]. At higher LQEs, this peak evolves into several overlapping lines; the resulting pattern can be deconvoluted to a set of peaks at constant positions (Tab.1).

Table 1 Raman frequencies for PITN oligoenes

# of ITN-rings	calculated frequency [cm^{-1}]	experimental result [cm^{-1}]
2	1539	
3	1514	1520
4	1501	1499
5	1493	1490
6	1487	1484

The resonance excitation profiles of two example lines are shown in Fig. 2. The PITN Raman lines can be divided into two groups. Some are strongly enhanced at low LQE (type 1). Doping reduces the RRS intensity to 1/3. The other lines do not show a clear resonance (type 2); their cross sections are only doubled upon doping.

The depolarisation ratio for all lines is .38 ± .02, except for lines of type 1 at low LQE where it decreases to .28.

Raman spectra of the polymer with one additional fused benzene ring, polyisonaphthothiophene (PINT), were measured at various wavelengths. As Fig. 3 shows, the rich spectrum has similarity with that of PITN. The main peak shifts from 1523 cm^{-1} at 488 nm to 1507 cm^{-1} at 676 nm.

4. PITN: Discussion

Main Raman line. The shift and broadening of the main line with increasing LQE has been seen in related polymers [7] and been attributed to a distribution of conjugation lengths. In PITN we can for the first time distinguish separate peaks.

The conjugational defects in PITN affect mainly the interring bonds, which are easily twisted. Thus the conjugation lengths are mostly multiples of that of the monomer, which is 4 for ITN. A comparison of the expected frequencies, calculated by an empirical formula derived from polyenes [11], and the peak positions, shown in Table 1, shows good agreement for the trimer or quadrimer.

Resonance profiles and ground state structure. The two types of lines can be attributed to different phases of the polymer. We suggest that these are distinguished by their conjugation structure, which is aromatic and quinoid, respectively. By comparison with the absorption spectra [2] the enhancement of type 1 lines can be attributed to resonance with the π-π^*-transition of undisturbed chains. The connected decrease of depolarisation was also observed in polyacetylene [12]. A comparison of measured and

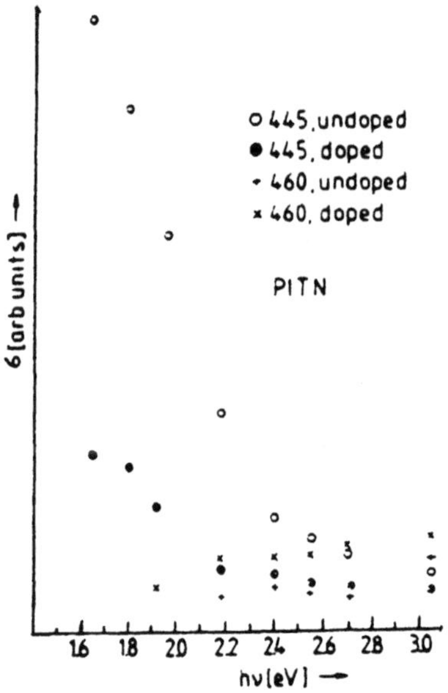

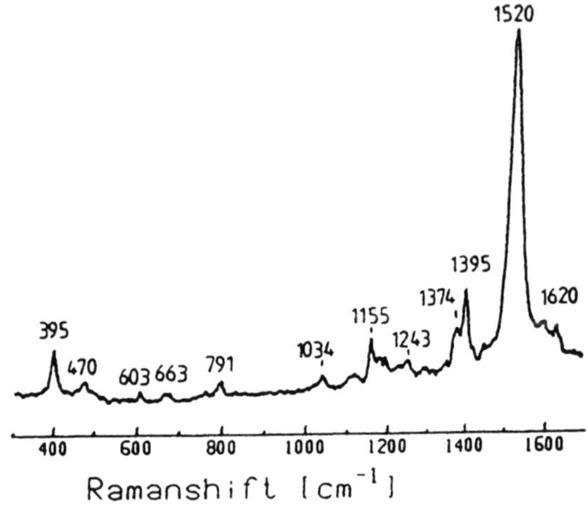

Fig. 2. Raman cross sections of the lines at 445 and 460 cm^{-1}

Fig. 3. Raman Spectrum of PINT at 514 nm

calculated Raman frequencies for aromatic and quinoid geometry [13] gives evidence for a quinoid ground state for the neutral polymer (type 1 lines). This is in agreement with the type 2 line found at 1220 cm^{-1} typical for inter-ring single bond stretching as in PT. From these results, also for PINT a quinoid structure can be postulated.

We attribute the presence of aromatic parts to polarons and bipolarons, where the π structure is out of phase between the charges or radicals, and to very short segments limited by twists or chain ends. A quinoid segment has broken π bonds at its ends. Their energy can make an aromatic state favourable up to a certain length. Our study suggests that the shortest quinoid segment is the quadrimer. Kürti and Surjan [14] predict this switching for slightly longer segments.

5. Poly(3-octyl-thiophene): results

The dominant feature in the light scattering spectra of POT is a strong luminescence, as similarly observed in related polymers [15]. At low temperatures it is centered at 1.7 eV, that is .3 eV below the absorption edge. Up to 4 vibrational peaks are observable with a spacing of 1370 cm^{-1}. Position and spacing are independent of the LQE for the range above 2.4 eV. At 2.2 eV excitation, the spacing is only 1280 cm^{-1}. The luminescence intensity is approximately proportional to the absorption at the LQE. Upon heating the peaks broaden and shift upwards. The shift is gradual from -196 °C to 135 °C by a total of .15 eV. At the thermochromic transition around 150 °C [16] the luminescence pattern shifts upward by .2 eV and its intensity decreases by a factor 2.

Fig. 4 shows Raman spectra of POT and PT. The main peak around 1460 cm^{-1} is similar in both polymers. At higher LQE a side structure of that peak in POT similar to that in PITN is observed. In toluene the red polymer

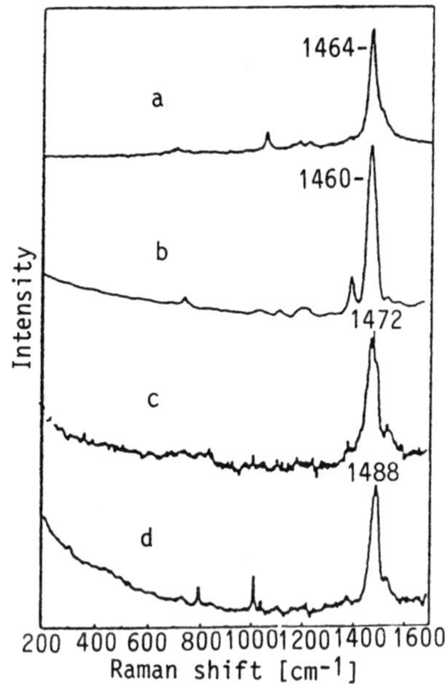

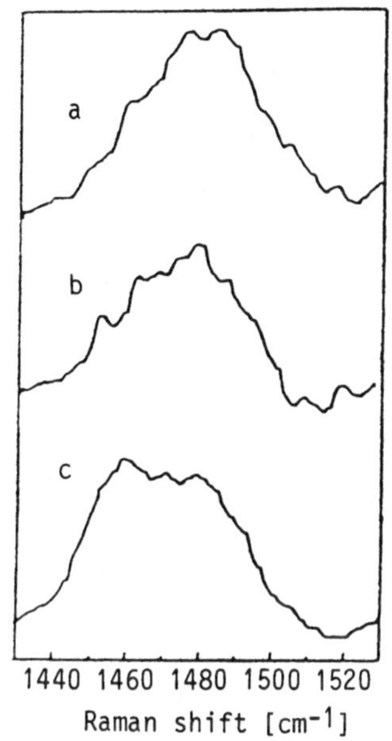

Fig. 4. Raman spectra of (a) PT at 514 nm, (b) POT at 488 nm, (c) POT at 406 nm and (d) POT solved in toluene at 406 nm

Fig. 5. Raman spectra of POT with 406 nm at (a) 180°C, (b) 130°C, (c) 20°C

yields a yellow solution; the main Raman peak is shifted to 1488 cm^{-1}. The effect of heating on the main line is shown in Fig. 5. The main intensity shifts up reversibly and is near the solution line above the transition temperature. Upon electrochemical doping again the peak at 1460 cm^{-1} decreases strongly and the main maximum shifts to higher wavenumbers. This is more pronounced at high LQE, where at 3 eV the intensity at 1485 cm^{-1} is doubled upon doping. At 2.4 eV the peak shifts only to 1470 cm^{-1}.

6. Poly(3-octyl-thiophene): Discussion

The spacing in the luminescence pattern and the structure in the absorption of POT [17] are both close to the wavenumber of the dominant Raman line, as expected from the Frank-Condon theory. This line at 1460 cm^{-1} clearly corresponds to the main peak in the PT spectrum which is attributed to the amplitude mode of the Peierls distortion. Also, the spacing in the luminescence shows a LQE dependency similar to the "dispersive" Raman line. Thus we suggest that the absorption steps and the luminescence peaks are due to backbone vibrational levels.

The upshift of luminescence upon heating follows the change of the absorption edge [16] and thus of the band gap. We thus conçlude that the formation energy of an exciton-polaron out of a hole and a π^* electron is temperature independent.

The appearance of high wavenumber side peaks near 1460 cm^{-1} suggest the presence of disordered parts, in analogy to PITN and related polymers. They appear only at very high LQE because of the wide gap of disturbed POT. In solution, the narrow peak at 1488 cm^{-1} can be explained with shortening of the conjugation length by twisting and nonplanarity of the chain, which also leads to a solvatochromic increase of the band gap [18]. We attribute the temperature dependence to a similar effect. Here the disorder does not come from solution of the side chains, but from their thermally induced defects and melting [19]. The molten, disordered phase shows a very similar backbone frequency to the solution, suggesting a similar defect density. The changes upon doping suggest that, as expected, the long chains are doped more easily than the short conjugated segments. In addition it shows doping-induced disorder, which leads to higher numbers of short segments.

7. Conclusion

In both systems studied,the line around 1460 cm^{-1} shows the most interesting effects. It can be used to study disorder and defects induced e.g. by solution and heating by probing the distribution of conjugation lengths. In PITN, comparison of Raman measurements and calculated frequencies provides evidence for a quinoid ground state structure. In POT, the relation between absorption spectrum, luminescence and Raman spectra shows the importance of the backbone stretching vibration.

We thank K. Eichinger for providing the PINT and the ITN monomer and O. Inganes for providing the POT.

References

[1] F. Wudl, M. Kobayashi, A.J. Heeger, J. Org. Chem. 49 (1984) 3382
[2] N. Colaneri, M. Kobayashi, A.J. Heeger, F. Wudl, Synth. Met. 14 (1986) 45
[3] J.L. Bredas, J. Chem. Phys. 82 (1985) 3808
[4] Y.-S. Lee, M. Kertesz, Int. J. Quantum Chem. Symp. 21 (1987) 163
[5] F. Wudl, Y. Ikenoue, A.O. Patil, in D. Ulrich and P.N. Prasad (eds.), "Nonlinear Optical and Electroactive Polymers", Plenum, New York, in press
[6] R.L. Elsenbaumer, K.Y. Jen, R. Oboodi, Synth. Met. 15 (1986) 169
[7] H. Kuzmany, phys. stat. sol. (b) 97 (1980) 521
[8] H. Kuzmany, J.F. Rabolt, B.L. Farmer, R.D. Miller, J. Chem. Phys. 85(1986) 7413
[9] W. Wallnöfer, E. Faulques, H. Kuzmany, K. Eichinger, Synth. Met. 28 (1989) C533
[10] J. Poplawski, E. Ehrenfreund, H. Schaffer, F. Wudl, A.J. Heeger, Synth. Met. 28 (1989) 539
[11] I. Harada, M. Tasumi, H. Shirakawa, S. Ikeda, Chem. Letters (1978) 1411
[12] H. Kuzmany, P. Knoll, J. Raman Spectr. 17 (1986) 89
[13] E. Faulques, W. Wallnöfer, H. Kuzmany, these proceedings; E. Faulques, W. Wallnöfer, H. Kuzmany, J. Chem. Phys., in press
[14] J. Kürti, P.R. Surjan, these proceedings
[15] S.D.D.V. Rughooputh, M. Novak, S. Hotta, A.J. Heeger, F. Wudl, Synth. Met. 21 (1987) 41
[16] O. Inganäs, W.R. Salaneck, J.-E. Österholm, J. Laakso, Synth. Met. 22 (1988) 395
[17] M. Sundberg, O. Inganäs, S. Stafström, G. Gustavsson, B. Sjögren, these proceedings
[18] K. Yoshino, S. Nakajima, H.B. Gu, R. Sugimoto, Jap. J. Appl. Phys. 26 (1987) L2046
[19] W.R. Salaneck, O. Inganäs, B. Themans, J.O. Nilsson, B. Sjögren, J.E. Österholm, J,.L, Bredas, S. Svensson, J. Chem. Phys. 89 (1988) 4613

Electronic Structure Studies of Undoped and Doped Polyisothianaphthene

E. Pellegrin[1], *N. Nücker*[1], *J. Fink*[1], *W. Wallnöfer*[2], *and K. Kuzmany*[2]

[1]Kernforschungszentrum Karlsruhe,
Institut für Nukleare Festkörperphysik,
Postfach 3640, D-7500 Karlsruhe, Fed. Rep. of Germany
[2]Institut für Festkörperphysik, Universität Wien, and
Ludwig Boltzmann Institut für Festkörperphysik,
Strudlhofgasse 4, A-1090 Wien, Austria

Abstract. The electronic structure of undoped and doped Polyisothianaphtene has been investigated by high-resolution electron energy-loss spectroscopy. Performing a Kramers-Kronig transformation of the loss-function of undoped PITN, the dielectric function could be derived giving information on the π and σ electronic structure. Momentum dependent measurements yield values for the width of the π bands. Further information on the electronic structure is obtained from core-level excitations. The results are compared with those of previous studies of polythiophene and theoretical predictions.

1. Introduction

One way to find polymers which have small or even vanishing band gaps is to consider some of the polymers that are already known and to modify their electronic properties by the addition of substituents. Therefore, it is reasonable to take polythiophene (PT) as parent polymer, because it has already a rather small intrinsic band gap of about 2 eV and because of the ease of substituting a benzene ring along the β-β bond of the thiophene ring. In recent theoretical work of Brédas et al. [1] and Lee and Kertesz [2], the influence of the additional ring on the bond length alternation and on the band gap is considered. Brédas et al. find that the overall characteristics of the polyisothianaphtene (PITN) HOMO and LUMO are such that PITN exhibits an aromatic-type electronic structure. Furthermore, the resulting decrease of the band gap can be rationalized by realizing that the fusion of a benzene ring onto the thiophene ring effectively increases the quinoid character of the electronic structure. For the polymer, this destabilizes the HOMO and stabilizes the LUMO by 0.77 eV and 0.38 eV, respectively. Moreover, this cuts the band gap in half with respect to that of the parent polymer, PT. At variance with these results,

Springer Series in Solid-State Sciences, Vol. 91
Electronic Properties of Conjugated Polymers III
Editors: H. Kuzmany · M. Mehring · S. Roth

Lee and Kertesz find that the quinoid form is more stable than the aromatic one when benzene rings are attached to PT. Furthermore, the gap energy of aromatic (quinoid) form of PT decreases (increases) as the number of rings fused to PT increases. Films of PITN show high conductivity (50 S/cm) and can be reversibly doped with electron acceptors. The oxidized form of PITN is the first highly conducting polymer which is transparent in the visible.

2. Experimental

The monomer benzo(c)thiophene (isothianaphtene, ITN) was prepared from its dihydrogenated sulfoxide by pyrolysis. The polymer was

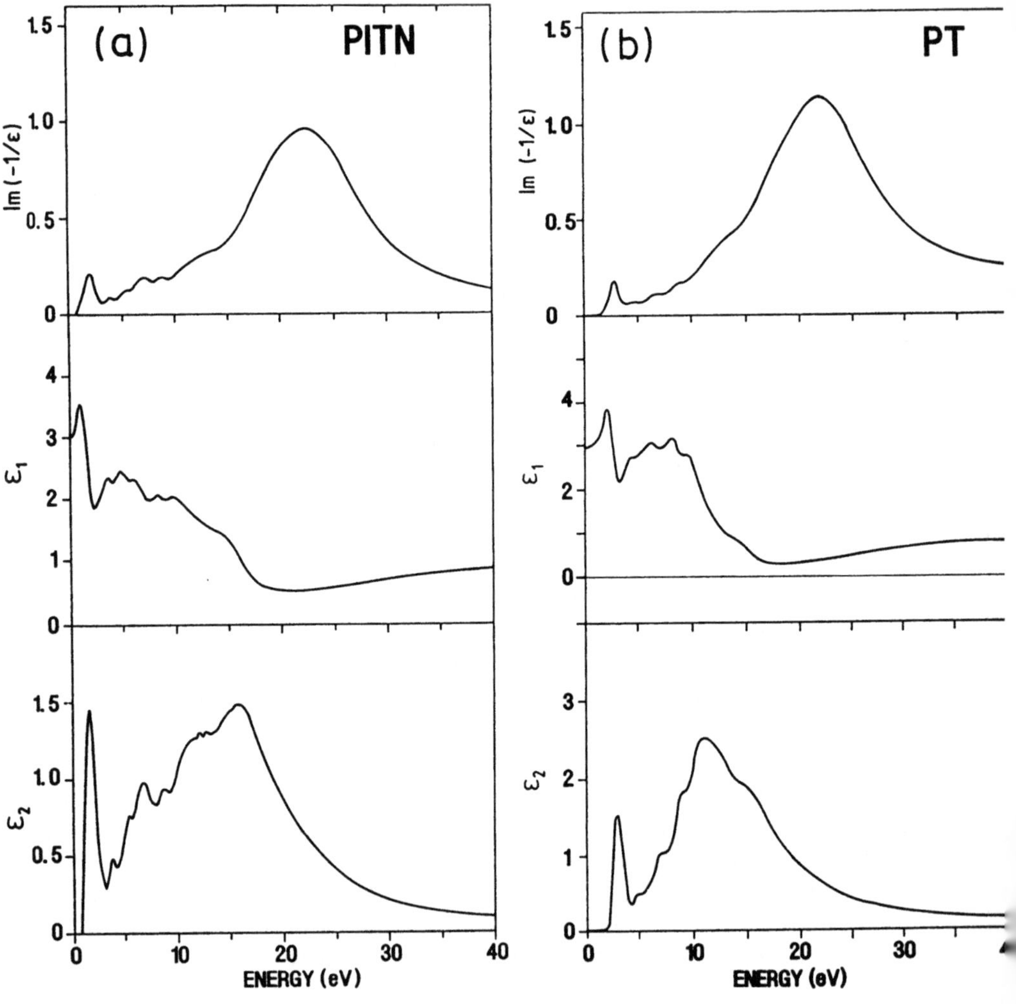

Figure 1a, b

synthesized as thin film on platinum and ITO-coated glass by electrochemical polymerization. Acetonitrile was used as solvant with 0.05 M tetraphenyl phosphonium chloride or lithium fluoroborate as electrolyte; the monomer concentration in the electrolyte was 0.2 M. A platinum wire served as counter electrode and a saturated calomel electrode as reference. The potential was cycled between 400 and 900 mV; all potentials are vs SCE [3].

3. Results

In Fig. 1a we show in the upper part the loss function Im(-1/ε) of PITN. A small momentum transfer $q = 0.1$ Å^{-1} was chosen in order to avoid contributions from surface and radiation losses. The loss function was obtained from the original data by subtracting the elastic line and performing a numerical double-loss correction. The absolute value of the loss function was obtained by using an estimated refractive index of n (ω = 0, q = 0) = 1.7. Im(-1/ε) shows several peaks between 2 and 15 eV which originate from collective excitations related to π-π* and σ-σ* transitions. At about 22.5 eV, the plasmon of all valence electrons (π + σ) appears.

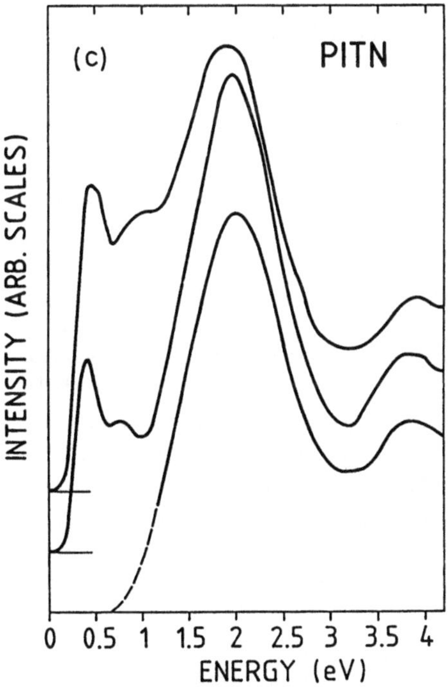

Figure 1c

Fig. 1 Loss function Im(-1/ε), ε_1 and ε_2 for PITN (a) and PT (b); (c) Energy-loss spectra for PITN with different dopant concentrations.

The real and the imaginary part of the dielectric function, ε_1 and ε_2 (also shown in Fig. 1a) where derived from a Kramers-Kronig analysis of the loss functions. At about 1.7 eV, a strong π-π^* oscillator can be realized in ε_2 due to the transition across the fundamental gap. For PT (see Fig. 1b), the same transition is situated at 2.5 eV. The onsets occur at 0.74 eV for PITN and at 1.96 eV for PT. Optical absorption coefficient measurements by Kobayashi et al. [4] on as-grown PITN films (doped with ~5 mol% Cl^- per monomer unit) show an intense interband absorption with a peak at $\hbar\omega \simeq 1.4$ eV, (2.6 eV for PT), a weak shoulder on the low energy side and an indication of a further absorption band at higher energies ($\hbar\omega > 3.5$ eV). The low energy shoulder is not an intrinsic feature of the neutral polymer and will be discussed later on. In these data, the onset of intrinsic absorption occurs at $\hbar\omega \simeq 1$ eV (2 eV for PT), thus being in good agreement with our results and implying that the gap energy of pure PITN is $\simeq 1$ eV, half that of the parent polythiopene. Theoretical predictions for the width of the band gap depend on the methods used in the band-structure calculations. Brédas et al. [1] finds 1.71 eV for PT and 0.54 eV for PITN, whereas Lee and Kertesz [2] yield gap values of 1.83 eV and 1.16 eV for PT and PITN, respectively. Correlation effects between the chains may be the reason for the larger experimental gap energies. Compared to PT, the π-π^* oscillator in PITN is much stronger relative to the oscillator at higher energies. Between 3 and 10 eV, ε_2 shows several peaks which can be assigned to further π-π^* transitions. At about 6.9 eV, an oscillator appears in PITN which is less pronounced in PT. It can be assigned to a π-π^* transition between narrow bands originating from electrons localized in the benzene ring, similar to the situation in PPP [5]. The region above 10 eV consists of a superposition of σ-σ^* transitions and mixed π-σ^* transitions.

We now discuss the effect of p-type doping on the loss function of PITN between 0 and 2 eV. With increasing dopant concentration, excitations in the gap at about 0.4 eV and 0.8 eV are realized, as can be seen in Fig. 1c. They can be assigned to transitions from the highest occupied π band into polaron/bipolaron states in the gap. Furthermore, those two peaks shift slightly towards higher energy with increasing dopant concentration. An additional characteristic third polaron absorption peak could not be observed, indicating the formation of bipolarons at the given dopant concentration; however, the possibility of a small polaron concentration cannot completely be ruled out by our measurements. Kobayashi et al. [6] also find two absorption peaks in the gap, both shifting to higher energies with higher dopant concentration.

Further informations on the electronic structure of PT and PITN are obtained from momentum-transfer dependent measurements of the loss function, shown in Fig. 2. The momentum transfer has been varied from 0.1 to 0.4 $Å^{-1}$ and from 0.1 to 0.6 $Å^{-1}$ for PITN and PT, respectively. In both

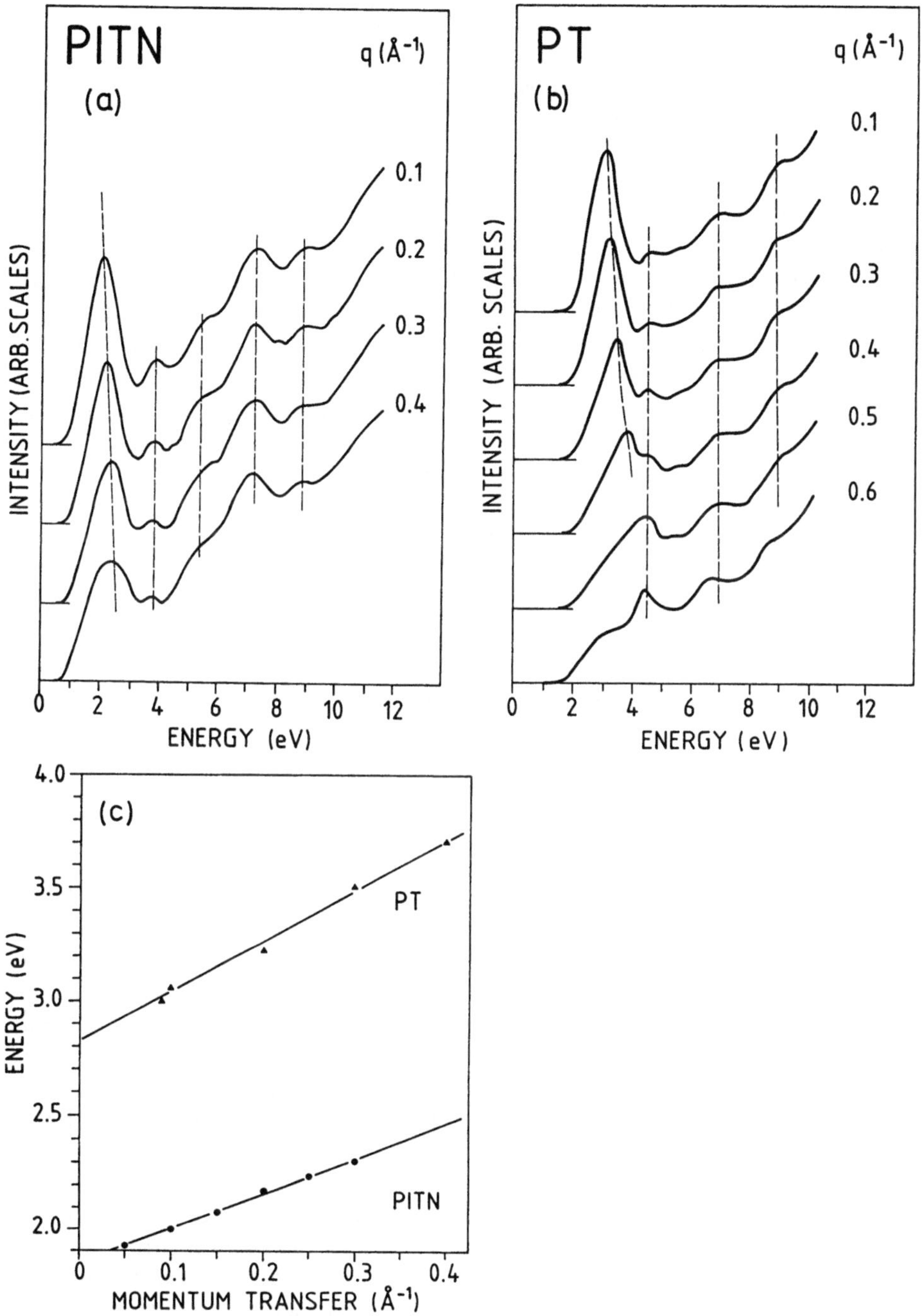

Fig. 2 Momentum-dependence of the energy-loss spectra of PITN (a) and PT (b). The energy of the lowest π-plasmon as function of the momentum transfer is given in (c).

diagrams, only the lowest π plasmon shows a linear dispersion in momentum transfer q indicating a transition between wide π bands. All other excitations in PT as well as in PITN show almost no dispersion. In Fig. 2c the energy of the lowest π plasmon is shown for PT and PITN as a function of the momentum transfer q. Compared to PT, the dispersion coefficient α is considerably lower; PITN:α = 1.5 eVÅ, PT: α = 2.2 eVÅ [7]. This indicates that the bandwidth of the highest occupied and/or the lowest unoccupied band is smaller for PITN relative to that of PT. Band structure calculations by Lee and Kertesz [2] yield a dispersion coefficient of α = 1.5 eVÅ and 1.55 eVÅ for aromatic and quinoid PITN, respectively, both in the momentum transfer range q = 0.1 $Å^{-1}$ to 0.3 $Å^{-1}$ also used in the present measurements on PITN. This is in good agreement with our experimental results, but is not helpful to decide whether the structure of PITN has quinoid or aromatic character. Calculations by Brédas et al. [1] result in a quite different dispersion coefficient of α = 4.1 eVÅ. Similar results are obtained from band structure calculations by Bakhshi et al. [8] (α = 5.9 eVÅ). These discrepancies can probably be explained by the different methods used in these calculations. Another interesting feature in the studies by Bakhshi and Ladik is the comparison between PT and PITN. This band structure yields a dispersion coefficient α = 4.1 eVÅ and 5.9 eVÅ for PT and PITN, respectively. This is in contrast to our measurements on PT, where we obtain a higher α for PT (2.2 eVÅ) than for PITN (1.5 eVÅ).

Core-level excitations

Additional information can be deduced from excitations of core electrons into empty states above the Fermi edge. In Fig. 3, we show carbon K-shell and sulfur $L_{2,3}$-shell spectra from PT and PITN. In Fig. 3a, a carbon 1s-π* transition appears near 285 eV. Probably due to a small chemical shift between carbon atoms having no sulfur neighbours and those having sulfur neighbours a small splitting of ~0.3 eV is observed. The former transition is much stronger in PITN due to the increased number of carbon atoms from the benzene rings. The peak at 287.2 eV in PT (σ shape resonance due to the C-S-bond [9]) is much less pronounced and shifted towards higher energy in PITN. In addition, there are 2s-σ* transitions above 291 eV caused by excitations into σ* states related to C-C bonds.

The S2p edges in PITN and PT are very similar. There is a slight shift to lower energy in PITN as can be seen in Fig. 3b. The four lines between 165 and 169 eV in PT can probably be explained by two spin-orbit doublets caused by unoccupied π* and σ* molecular orbitals associated with the C-S bond in PT.

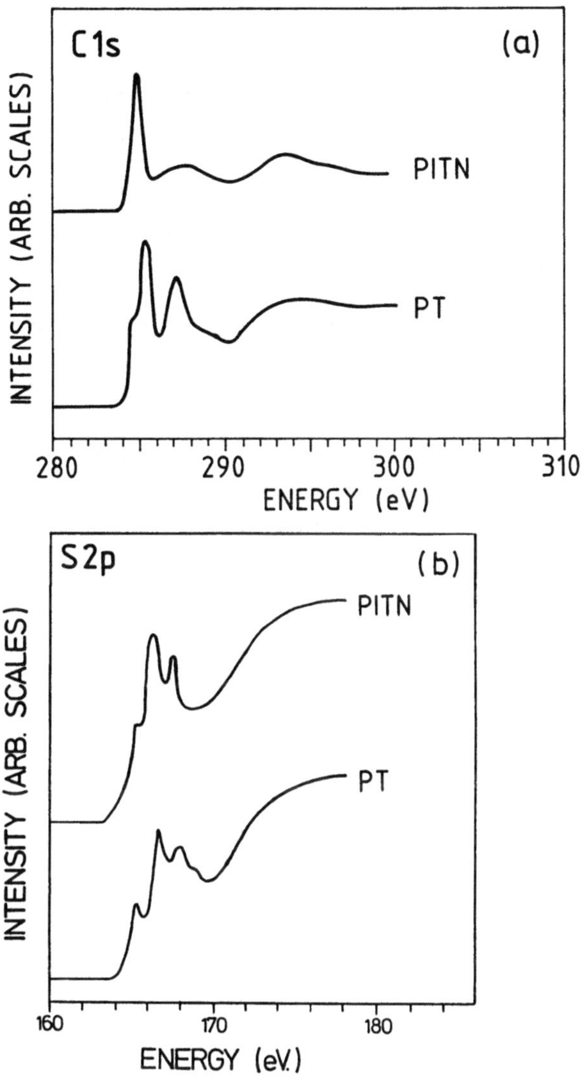

Fig. 3 Core-level excitation spectra of PITN and PT films. (a) Carbon K-shell, (b) Sulfur $L_{2,3}$-shell.

4. Conclusion

To summarize, we confirmed the narrowing of the band gap by about 1 eV when going from PT to PITN. Similarities were found in the spectra of PITN and other polymers such as PT and PPP, indicating similar band structures. The formation of bipolarons by doping could be shown at the given dopant concentration; further investigations are needed to decide whether the ground state has dominantly quinoid or aromatic character.

References

1 J.L. Brédas, A.J. Heeger, and F. Wudl; J. Chem. Phys. **85** (1986) 4673

2. Yong S. Lee and Miklos Kertesz; J. Quantum Chem.; Quantum Chem. Symp. **21** (1987) 163

3. W. Wallnöfer, E. Faulques and H. Kuzmany, Synth. Met. **28** (1988) 553

4. M. Kobayashi, N. Colaneri, M. Boysel, F. Wudl, and A.J. Heeger; J. Chem. Phys. **82** (1985) 5717

5. G. Crecelius, J. Fink, J.J. Ritsko, M. Stamm, H.-J. Freund and H. Gonska; Phys. Rev. B **28** (1983) 1802

6. N. Colaneri, M. Kobayashi, A.J. Heeger, and F. Wudl; Synth. met. **14** (1986) 45

7. W. Czerwinski, N. Nücker, and J. Fink, Synth. met. **29** (1988) 71

8. A.K. Bakhshi and J. Ladik, Solid State comm. **61** (1987) 71

9. J. Fink, N. Nücker, B. Scheerer, and H. Neugebauer, Synth. met. **18** (1987) 163

π-Electronic Structure of Poly(3-hexylthiophene) Studied by Photoelectron Spectroscopy

M. Lögdlund[1], *R. Lazzaroni*[1], *W.R. Salaneck*[1], *S. Stafström*[1], *J.O. Nilsson*[1], *X. Shuang*[1], *J.-E. Österholm*[2], *and J.L. Brédas*

[1]Department of Physics, IFM, Linköping University, S-581 83 Linköping, Sweden
[2]Neste Oy, Corporate Research, Technology Center, SF-06850 Kulloo, Finland
[3]Service de Chimie des Matériaux Nouveaux, Université de Mons, B-7000 Mons, Belgium

Abstract

Ultra violet photoelectron spectroscopy has been used to observe changes in the π-band structure in poly(3-hexylthiophene) induced by doping using $NOPF_6$. Results are consistent with a polaron lattice model of the highly doped state.

1. Introduction

The alkyl-substituted polythiophenes are some of the most studied conducting polymers today because of their high degree of processability. The temperature dependence of the electronic structure of undoped poly(3-hexylthiophene), or P3HT, has been presented previously in the characterization of the thermochromism in P3HT in terms of thermally-induced conjugation defects in the aromatic backbone /1/. The chemical structure of P3HT is shown in Fig. 1.

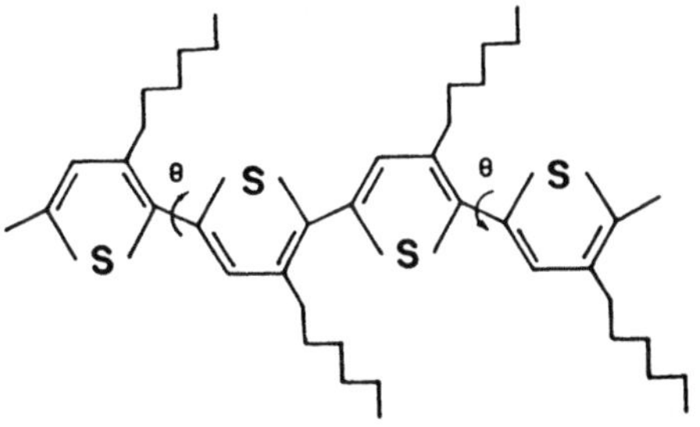

Fig. 1: The chemical structure of poly(3-hexylthiophene)

In this work we have studied the effects on the electronic properties of P3HT caused by doping using $NOPF_6$ dissolved in acetonitrile. We focus upon the structure of the $\pi 3$-electronic band, which lies nearest the Fermi level, E_F, as a function of doping. The results are discussed in terms of the formation of polarons and/or bipolarons, following the recent theoretical work of Stafström and Brédas /2/.

Springer Series in Solid-State Sciences, Vol. 91
Electronic Properties of Conjugated Polymers III
Editors: H. Kuzmany · M. Mehring · S. Roth

2. Experimental

The polymer was prepared as discussed previously /3/. The molecular weight corresponded to polymer chains containing approximately 50 thiophene rings. A great deal of effort was put into preparing very clean thin films. Thin (few hundred Å) solution-cast films were prepared on thin (2000 Å) films of oxidized (15 Å) aluminum, which had been vapor-deposited upon optically flat Si(111) wafers. Following our previous experience with P3HT /1/, thin films were cast from chloroform in a saturated chloroform atmosphere, inserted (through air in transport) directly into a closed N_2 atmosphere in an "introduction chamber" of a specially designed and built photoelectron spectrometer. The films were dried under 10^{-6} Torr, obtained by an oil-free turbo pump, and then moved into the main analysis chamber, at a pressure of less than 10^{-10} Torr, maintained by a special combination of ion- and cryo-pumping. The films were then heated to 180 C to remove hydrocarbon contamination and any remaining solvent /1/. Because of the surface sensitivity of UPS, the heating step was necessary in order to obtain high quality spectra.

Doping was done by moving the sample, at room temperature, into the introduction chamber, substituting the vacuum with a flowing dry N_2 atmosphere, applying a controlled amount of the doping solution, $NOPF_6$ in acetonitrile, on the surface of the sample, which was tilted at 30° to the horizontal. The excess doping solution gathered in a special receptical built into the sample holder. Finally, the chamber was sealed, and the sample dried again under 10^{-6} Torr turbo-pumped vacuum, as above. The N_2 atmosphere is so effective in keeping O_2 and H_2O from the sample that a freshly sputter-cleaned copper surface does not oxidize, even after 5 full minutes in the N_2 flow. Judging from the rate of change of the spectra versus the number of times that a sample was treated with the doping solution, the extent of doping appeared to increase uniformly up to a *saturation* point, beyond which the spectra did not change. For the saturation-doped samples, the samples were rinsed *in situ* with acetonitrile after doping, in order to remove any possible residual salt from the surface.

3. Results and Discussion

Photoelectron spectra of solids are recorded relative to the Fermi energy, E_F, by the manner in which a spectrometer functions. The *vacuum level*, however, is the reference energy of chemical significance. When E_F changes due to some chemical modification of a sample, e.g., in the intercalation of graphite and other layered compounds /4/, and in the doping of conducting polymers /5/, then it is necessary to account for this change, ΔE_F *before* interpreting spectra /4/. The work function, ϕ, can be determined from the width of a UPS spectrum, using the following equation:

$$(E_{CO} - E_F) + \phi = h\upsilon,$$

where $h\upsilon = 21.2$ eV for He I radiation, E_F is the Fermi energy, and E_{CO} is the cut-off energy of the secondary electrons in the UPS spectrum. In this work, we use the standard procedure of estimating ΔE_F from the change in the work function, $\Delta\phi$, which is equal to ΔE_{CO}, and subsequently adjust all binding energy values by this amount. In Fig. 2. is shown the energy level diagram for the XPS C(1s) and S(2p) levels in undoped and saturation-doped samples, including an indication of ΔE_F and the resultant change in ϕ.

The chemical shift observed on the S-atoms is interesting, since calculations of the electronic structure of the polythiophenes show that there is no wavefunction intensity of the highest occupied π-band (referred to as the π_3-band in Ref. 1) on the S-atoms. From these calculations on the neutral (polymer) molecule in the ground state, it might have been expected that the removal of an electron from the polymer chain would not affect the S-atoms. MNDO level calculations /2/ show, however, that in the ionic (charged) state, the formation of polarons and bipolarons result in a local charge redistribution and a corresponding local geometric relaxation, such that a part of the transferred electronic charge does indeed reside on the S-atoms in the vicinity of the polarons or bipolarons.

Using the MNDO charges /2/ and standard estimates of the associated chemical shifts of the C(1s) level (for sp^2 hybridization) and the S(2p) level, we have made a simulation of the expected XPS spectra of undoped and doped P3HT. At the extreme of saturation-doping, only a slight broadening is expected on the C(1s) peak, because the C(1s) spectrum consists of a sum over small chemical shifts in positive and negative directions, both for polaron or bipolaron formation /2/. In addition, the hexyl-groups, which should not exhibit any chemical shifts, will dominate the C(1s) peak, and tend to obscure even the broadening of the C(1s) expected of carbon atoms in the thiophene rings. From the calculated effective charges, a chemical shift

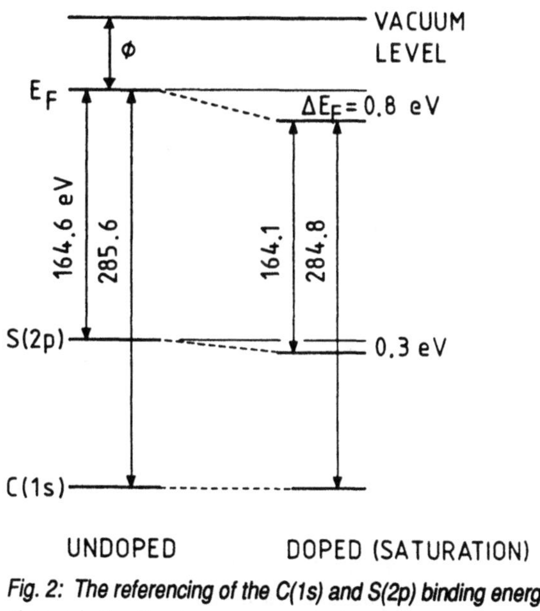

Fig. 2: The referencing of the C(1s) and S(2p) binding energies by consideration of the change in the Fermi energy upon doping, ΔE_F

of a few tenths of an eV is expected for the S(2p) peak upon polaron formation or bipolaron formation. In both cases, however, the shift should be toward *higher* binding energy. This is not what is observed in the *raw data*. When ΔE_F is correctly taken into account, however, the S(2p) peak shifts in the *correct* direction for an electron transfer *from* the polymer chain *to* the dopant species. From the observed shift of about 0.3 eV, however, one cannot really distinguish between polarons or bipolarons.

In Fig.3 are shown the UPS spectra of the π-bands that occur nearest to E_F. Note the evolution of the band structure with doping. In particular, the π_3-band edge is located at about -1 eV relative to E_F in the undoped material, as expected /6,7/ for an intrinsic semiconductor with a band gap of about 2.1 eV. Upon doping, the π_3-band edge moves relative to E_F ***and*** E_F moves relative to the vacuum level, as expected from the VEH calculations of Stafström and Brédas /2/. Most importantly, note that at saturation doping, a finite density of states is observed in the region of the Fermi energy, *consistent with* (but not necessarily *a proof* of the existence of) the presence of a *polaron* lattice in highly doped polythiophenes. This observation is made possible by (i) the use of UPS, which has high cross section for photoionization

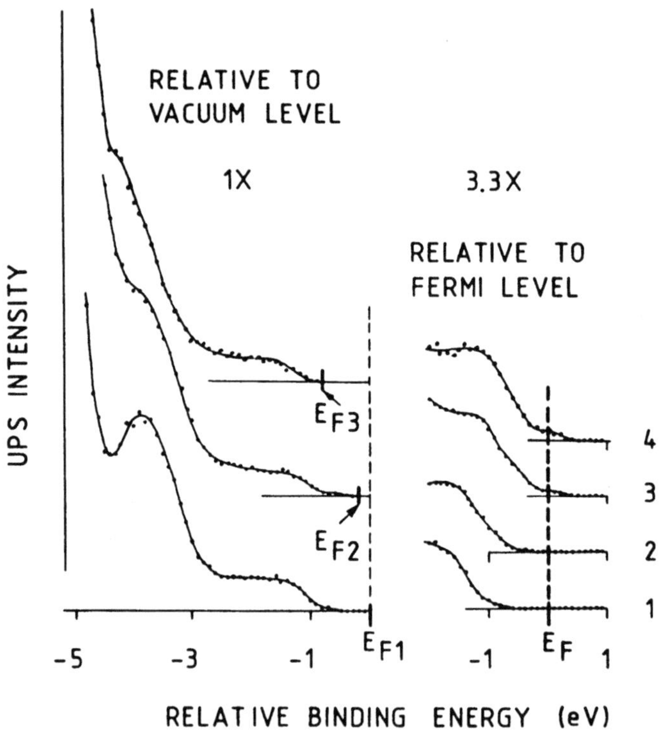

Fig. 3: The UPS spectra of the π-bands are shown aligned relative to E_F (right) and adjusted for ΔE_F (left). 1 refers to undoped, 4 to saturation-doped, and 2 and 3 to intermediate doping

of C(2p)-derived π-orbitals (XPS has notoriously low cross section in this case), and (ii) the special sample preparation and cleaning procedures which can be used on P3HT. To the best of our knowledge, this is the first time that a finite density-of-states has been observed *unambiguously* at the Fermi energy in a doped conducting polymer using photoelectron spectroscopy. There are several other models of the doping of conducting polymers, which predict a finite density-of-states at the Fermi level, e.g., disorder, as discussed by Mele and Rice /8/. The UPS data alone cannot distinguish among the various models.

4. Conclusions

Using special sample preparation techniques, in combination with the high sensitivity of UPS for π-states, we have observed the influence of doping on the π-electronic band structure nearest the Fermi energy in poly(3-hexylthiophene). At saturation doping from $NOPF_6$, a finite density-of-states at the Fermi energy is observed *unambiguously* for the first time in a conducting polymer using photoelectron spectroscopy. These data are consistent with the results of VEH calculations, which predict the existence of states at E_F for the polaronic lattice in highly doped polythiophenes.

5. Acknowledgements

Research on conducting polymers in Linköping is supported by grants from the Swedish Board for Technical Development (STU), the Swedish Natural Sciences Research Council (NFR) and the Nordic Fund for Industrial Research (NI).

6. References

1. Salaneck, W. R., Inganäs, O., Thémans, B., Nilsson, J. -O., Sjögren, B., Österholm, J. E., Brédas, J. -L., and Svensson, S., *J. Chem. Phys.*, **89**, 4613 (1988).
2. Stafström, S. and Brédas, J. -L., *Phys. Rev.*, **B38**, 4180 (1988).
3. Österholm, J. -E., Laakso, J., Nyholm, P., Isotalo, H., Stubb, H., Inganäs, O., and Salaneck, W. R., *Synth. Met.*, **28**, C435 (1989).
4. Wertheim, G. K., van Attekum, P. M. Th., and Basu, G., *Solid State Commun.*, **33**, 1127 (1980).
5. Salaneck, W. R., Erlandsson, R., Prejza, J., Lundström, I., and Inganäs, O., *Synth. Met.*, **5**, 125 (1983).
6. Wu, C. R., Nilsson, J. -O., Inganäs, O., Salaneck, W. R., Österholm, J. -E., and Brédas, J. -L., *Synth. Met.*, **21**, 197 (1987).
7. Jugnet, Y., Tourillon, G. and Tran Minh Duc, *Phys. Rev. Lett.*, **56**, 1862 (1986).
8. Mele, E. J., and Rice, M. J., *Phys. Rev. B*, **23**, 5379 (1981).

Part XI

Polyenes and Oligoenes

Lowest Energy Excited Singlet States of 2,2′:5′,2″-Terthiophene and 2,2′-Thienylpyrrole

D. Birnbaum and B.E. Kohler

Chemistry Department, University of California,
Riverside, CA 92521, USA

Abstract. Fluorescence excitation and emission spectra with full vibrational resolution have been measured for solid solutions of 2,2':5',2"-terthiophene in n-decane and 2,2' thienylpyrrole in n-hexane. Analysis of the vibrational fine-structure for both molecules gives ground and excited state normal mode frequencies and establishes that the S_0 to S_1 transition is symmetry allowed.

1. Introduction

One reason for interest in conjugated polyheterocycles such as polypyrrole (PPy) and polythiophene (PT) is the relatively high electrical conductivities found for the oxidized or reduced (doped) material.[1,2] Mixed species are also of interest: for example, a recently prepared copolymer of alternating thiophene and pyrrole units was found to be 10 (30) times more conducting than PPy (PT).[3] A thorough understanding of the electronic properties of well-defined oligomers of these materials provides the basis for determining such structural properties of the more complex polymers as conformation and conjugation length and is key to the development of a microscopic understanding of charge transport. Trying to understand the electronic structure of a polymer in terms of the electronic structure of a series of better defined and more easily studied oligomers is precedented by our using spectroscopic studies of discrete linear polyenes to better understand the optical and vibrational spectra of polyacetylene.[4-7] Our first results for the polyheterocycles are for the molecules 2,2':5',2"-terthiophene (terthiophene) and 2,2'-thienylpyrrole (thienylpyrrole, Fig. 1).

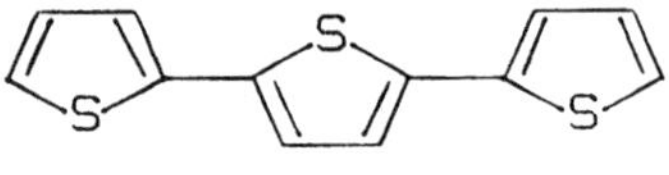

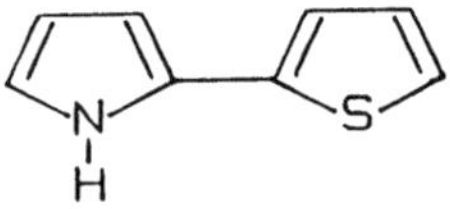

Fig. 1. Terthiophene (left) and thienylpyrrole (right)

2. Experimental

Thienylpyrrole was prepared by B. Jorgensen and M. Aldissi as previously described.[8] Terthiophene is commercially available from Aldrich. Both compounds were purified by vacuum sublimation. Solvents (n-hexane for thienylpyrrole and

Springer Series in Solid-State Sciences, Vol. 91
Electronic Properties of Conjugated Polymers III
Editors: H. Kuzmany · M. Mehring · S. Roth

n-decane for terthiophene) were obtained from Aldrich Chemical Corporation and, after verifying that they showed no detectable absorption or emission under the conditions of our experiments, were used without further purification. For each experiment freshly sublimed samples were dissolved in approximately 10 ml of solvent to make a solution whose absorbance at the lowest energy band maximum was of order 1 O.D./cm. For the high resolution measurements a small alliquot of this solution was put into a quartz tube (approximately 8 mm diameter) then immersed in liquid helium.

Room temperature absorption spectra were measured with a Varian DMS100 spectrophotometer. Fluorescence excitation and fluorescence spectra at 77 K (samples immersed in liquid nitrogen) and 10 K (samples in a CTI Model 21 He closed cycle He refrigerator modified so that the sample and the final cooling stage were under a few torr pressure of He) were measured with a Spex DM1B spectrofluorimeter.

The apparatus for measuring fluorescence excitation and fluorescence emission spectra at 4.2 K has been described previously.[9] Protocols for measuring optical spectra of thienylpyrrole at 4.2 K were, except for the use of DCM and rhodamine B laser dyes, exactly those already reported for terthiophene.[10]

3. Results and Discussion

3.1. Terthiophene

Fluorescence and fluorescence excitation spectra for 10 K solutions of terthiophene in n-decane are shown in Figure 2. These are almost identical to the corresponding spectra measured at 77 K. The close correspondence of the low temperature fluorescence excitation spectrum to the room temperature absorption spectrum supports the idea that our samples contain nearly planar trans-terthiophene. This is the expected conformation since x-ray crystallography has shown that the thiophene dimer is trans planar.[11] The fact that band profiles and the relative intensities of the two most prominent fluorescence features change with excitation

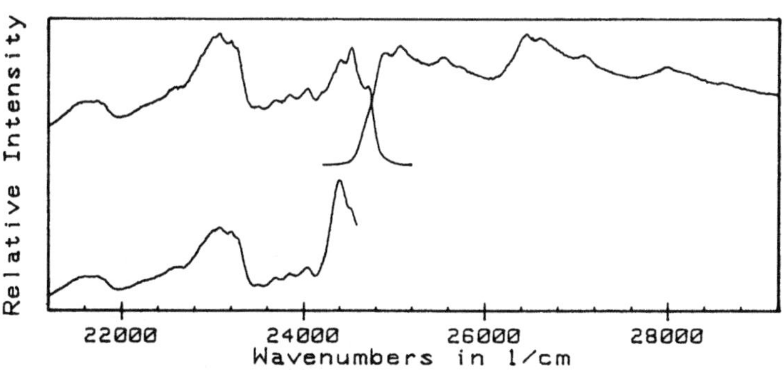

Fig. 2. Fluorescence excitation (upper right), fluorescence for 337 nm excitation (upper left) and 397 nm excitation (lower left) for terthiophene in n-decane at 10 K

wavelength suggests that more than one site is involved. This is clearly seen to be the case when the temperature is lowered further.

Upon cooling to 4.2 K, the fluorescence and fluorescence excitation bandwidths become quite narrow (inhomogeneous FWHM approximately 5 cm^{-1}). Though well resolved, the fluorescence excitation spectrum for broad band detection is very congested, even in the origin region (Figure 3).

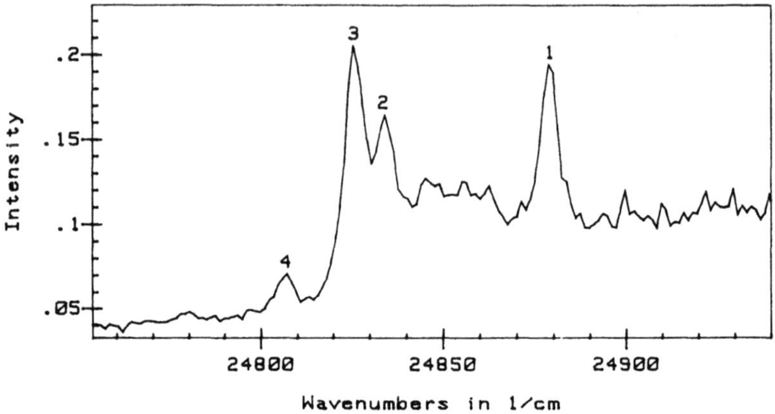

Fig. 3. The fluorescence excitation spectrum for undispersed emission from terthiophene in n-decane at 4.2 K. The four numbered peaks are the S_0 to S_1 origins for the four distinct terthiophene sites

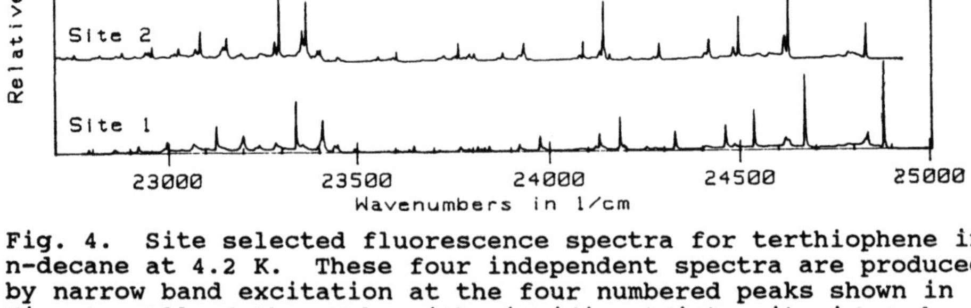

Fig. 4. Site selected fluorescence spectra for terthiophene in n-decane at 4.2 K. These four independent spectra are produced by narrow band excitation at the four numbered peaks shown in Fig. 3. All of the peaks with significant intensity in each of the spectra can be assigned to the same combinations of five fundamental modes of vibration. These assignments are summarized in Table 1. The bandwidths (FWHM) in these spectra are approximately 3 cm^{-1}

Excitation at each of the four numbered bands shown in Figure 3 produces a distinct and independent fluorescence spectrum as shown in Figure 4. Clearly, the sample contains four independent but extremely similar terthiophene species. The origin of the fluorescence excitation spectrum for each of the species overlaps the corresponding fluorescence origin. The fluorescence excitation spectrum of one of the species (peak 1 in Figure 3) is shown in Figure 5. Except for the origins having slightly different energies, the fluorescence excitation and fluorescence spectra of all four species are nearly identical with respect to vibronic development. Origin positions and the vibronic analysis of all four species are summarized in Table 1.

The strongest vibrational modes in emission (ground state) are at 209, 342, 694, 1470, and 1540 cm^{-1}. In fluorescence excitation (excited state) the corresponding modes are at 207, 347, 679, 1220 and 1561 cm^{-1}. The two lowest frequency modes may correspond to C-H bending modes, the 679 cm^{-1} and 1470 cm^{-1} modes are most likely symmetric ring breathing vibrations [12] and the mode at 1540 cm^{-1} in the ground state is reminiscent of the C=C symmetric stretch that dominates the vibronic development of linear polyene optical spectra. Only one of the modes that couple to the electronic transition changes frequency significantly on going to the excited state. This is the mode whose frequency is 1470 cm^{-1} in the ground state. The relatively large change in frequency may be related to the expected transfer of charge to the sulfur atom in the excited state.

The fluorescence decay time for terthiophene in n-decane at 4.2 K is less that our time resolution of about 2 ns. An immeasurably short fluorescence decay time (less than 1 nsec) has been reported for room temperature terthiophene solutions.[13]

In either the planar cis or planar trans conformations terthiophene has C_{2v} symmetry. In a simple molecular orbital

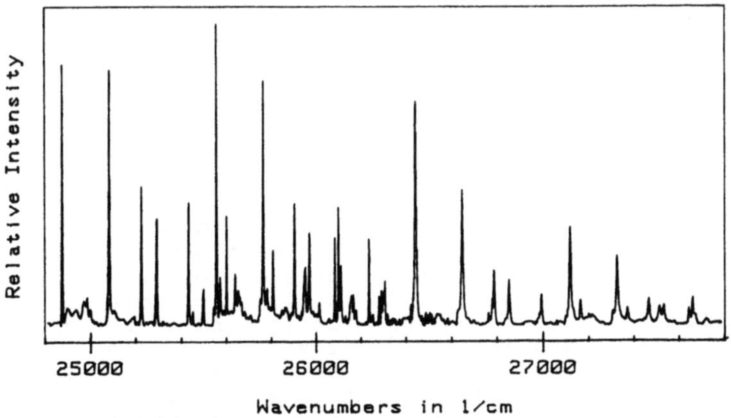

Fig. 5. The fluorescence excitation spectrum for the highest energy site of terthiophene in n-decane at 4.2 K (narrow band detection at 24670 cm^{-1} - origin minus 209 cm^{-1}). The origin for this spectrum is band number 1 of Fig. 2. The complete vibronic analysis is summarized in Table 1

Table 1. Vibronic analysis of terthiophene and thienylpyrrole spectra

	terthiophene				thienyl-pyrrole
Emission	Site 1	Site 2	Site 3	Site 4	
Origin	24879	24835	24827	24806	30006
ν_a	208.99	207.86	209.58	209.77	350.39
ν_b	341.66	338.29	343.31	338.12	408.00
ν_c	694.37	693.11	694.64	694.15	642.88
ν_d	1469.5	1469.1	1474.1	1462.3	1464.9
ν_e	1540.2	1539.7	1544.3	1530.6	1575.9
Excitation					
Origin	24879	24834	24827	24807	30006
ν_a	206.70	206.66	207	207	337.41
ν_b	347.01	344.88	350	344	409.63
ν_c	678.77	676.74	...	...	605.52
ν_d	1220.2	...	...	...	1263.4
ν_e	1560.5	...	...	...	1618.4

treatment all bonding 1-electron orbitals are doubly occupied so S_0 has A_1 symmetry. The promotion of one electron from the highest energy occupied molecular orbital to the lowest energy unoccupied molecular orbital (HOMO to LUMO) generates an excited state of symmetry B_2 where axes are chosen such that the chain axis also transforms as B_2. The coincidence of the fluorescence and fluorescence excitation origins establishes that the observed transition is from the ground state to the lowest energy excited singlet state (S_0 to S_1) and is symmetry allowed, most likely 1^1A_1 to 1^1B_2.

3.2. Thienylpyrrole

The fluorescence excitation and fluorescence (Figure 6) spectra for 4.2 K solutions of thienylpyrrole in n-hexane are fully resolved with inhomogeneous bandwidths of approximately 10 cm^{-1} (FWHM). Our vibronic analysis is given in Table 1. The dominant vibrational modes in emission (ground state) are at 350, 408, 643, 1465, and 1577 cm^{-1}. The corresponding modes in excitation (excited state) are at 337, 409, 606, 1263, and 1619 cm^{-1}. The vibronic development is extremely similar to that of terthiophene suggesting that nearly identical normal modes are involved. This similarity is consistent with the idea that there is strong interaction between the rings (as expected for a highly delocalized pi system) and that the nature of the heteroatom does not greatly affect the vibrational normal modes. This is supported by studies on

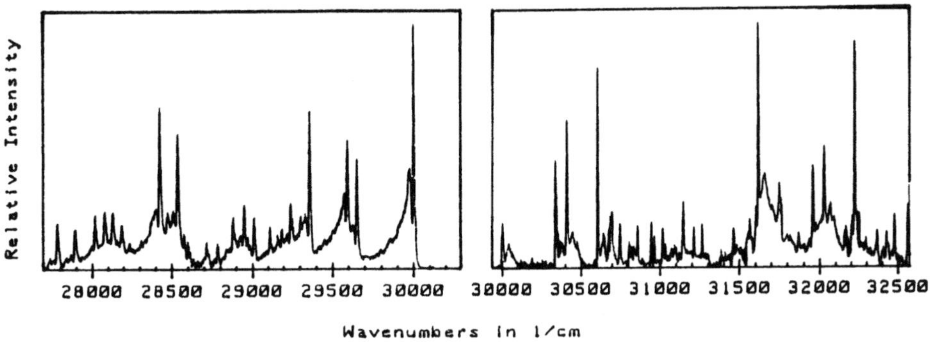

Fig. 6. Fluorescence (left) and fluorescence excitation (right) spectra for thienylpyrrole in n-hexane at 4.2 K

thiophene and pyrrole monomers comparing calculated frequencies with infrared and Raman measurements which show that the ring breathing vibrations of the two molecules are nearly the same.[14,15]

Both the planar cis and trans forms of thienylpyrrole are of low symmetry C_s. Simple MO treatment yields A' as the only possible species for the ground state and all excited states. The coincident origin band energies of 30006 cm^{-1} for both fluorescence and fluorescence excitation establish that the transition is symmetry allowed.

4. Conclusion

It is possible to obtain high resolution optical spectra for oligomers of polyheterocycles. At 4.2 K the fluorescence and fluorescence excitation spectra of n-decane solutions of terthiophene and n-hexane solutions of thienylpyrrole are fully resolved. For both molecules the S_0 to S_1 transition is symmetry allowed as predicted by simple molecular orbital theory. The vibrational development of the spectra is nearly the same for both molecules suggesting strong interaction between the rings and almost no dependence of the normal modes on the specific nature of the heteroatom.

5. References

1. G.B. Street in "Handbook of Conducting Polymers", edited by T.A. Skotheim, Marcel Dekker, New York, 1986, Vol 1 pp. 265-291.

2. G. Tourillon in Ref. 1, pp. 293-350.

3. S. Naitoh, K. Sanui and N. Ogata, J. Chem. Soc., Chem. Commun. 1348 (1986).

4. B.E. Kohler in "Electronic Properties of Polymers and Related Compounds", Springer Series in Solid State Sciences edited by H. Kuzmany, M. Mehring and S. Roth, 1985 pp. 100-106.

5. B.E. Kohler, J. Chem. Phys. 88 (4), 2788-2792 (1988).

6. E. Mulazzi, G.P. Brivo and R. Tiziani, Mol. Cryst. Liq. Cryst. 117, 343 (1985).

7. H. Kuzmany, Pure Appl. Chem. 57, 235 (1985).

8. B.A. Trofinov, A.I. Mikhaleva, R.N. Nesterenko, A.N. Vasil'ev, A.S. Nakhmanovish and M.G. Voromkov, Khim. Geterotsikl, Sowdin, 8, 1136, (1977).

9. B.E. Kohler and J.B. Snow, J. Chem. Phys. 79, 2134 (1983).

10. D. Birnbaum and B.E. Kohler, J. Chem. Phys. 90, 3506 (1989).

11. G. Wisser, G. Heeves, J.Wolters and A. Vow, Acta. Crystallogr. Sect. B 24, 467 (1968).

12. M. Rico, J.M. Orza and J. Morcillo, Spectrochim. Acta 21, 689 (1965).

13. C. Evans, D. Weir, J.C. Scaiano, A. MacEachern, J.T. Arnason, P. Morand, B. Hollebone, L.C. Leitch and B.J.R. Philogene, Photochem. and Photobio. 44, No. 4 441-451 (1986).

14. D.W. Scott, J. Mol. Spect. 31, 451 (1969).

15. D.W. Scott, J. Mol. Spect. 37, 77 (1971).

Studies of Soluble Pyrrole Oligomers

D. Bloor and S. Pugh

Department of Physics, Queen Mary College, Mile End Road, London, E1 4NS, UK

Abstract. The co-electropolymerisation of di-methylpyrrole and tri-methylpyrrole leads to black conductive films that are soluble in a range of common organic solvents. These oligomers have been investigated by a range of spectroscopic techniques. The optical spectra of oxidised and reduced forms display more structure than those of conventional polypyrrole films, particularly at low temperature. The spectra are also sensitive to the counter-ion utilised and the hydrogen ion concentration in methanol solutions. These results together with those of NMR are considered in terms of models of charged states.

1. Introduction

Polypyrrole prepared either chemically or electrochemically has been extensively studied because of its stability under normal environmental conditions [1]. It is, however, a morphologically complex material which in its electropolymerised form is a poly-cationic network incorporating charge compensating anions. The network can expand to incorporate very large anions, which can be removed with relative ease [1-3]. The morphology of samples depend on factors such as the supporting electrolyte, the temperature of electro-polymerisation, the material of the working electrode on which the polymer is deposited and the counter-ion incorporated in the polymer [2,4-8]. This complexity is in part due to the incorporation of monomers bonded at the 3 and 4 positions as well as the 2 and 5 positions. Polymerisation of di-methylpyrrole has been used to produce stereoregular polymer chains. Polymers produced in this way have somewhat more order but are also less conductive than those obtained from unsubstituted monomer. Thus though polypyrrole may be of practical value [10] it is a distinctly non-ideal material with highly variable micro- and macroscopic morphology. Despite this the electronic properties are consistent with a bipolaron (dication) model [11].

Progress in understanding the properties of other conjugated polymers has been aided by the availability of oligomers, e.g. polyenes and thiophenes [12,13] and soluble polymers, e.g. polythiophenes and polydiacetylenes [14,15]. Neither of these possibilities have been realised for polypyrrole; the synthesis of oligomers of reasonable length is difficult and substitution of aliphate and methoxy groups hinders polymerisation. Soluble stereo-regular oligopyrroles can be obtained by the co-polymerisation of 3,4-di-methylpyrrole and 2,3,4- tri-methylpyrrole in an electrochemical cell [16]. The use of di-methylpyrrole ensures a regular chain linkage while the tri-methylpyrrole acts as a chain terminator. These oligomers are soluble in common organic solvents, e.g. methanol, chloroform, etc. The effects of solvent, pH, counter-ion, temperature, etc. on these solutions have been studied by absorption spectroscopy.

Springer Series in Solid-State Sciences, Vol. 91
Electronic Properties of Conjugated Polymers III
Editors: H. Kuzmany · M. Mehring · S. Roth

2. Experimental

The monomers were prepared by literature routes; 3,4-di-methylpyrrole by the method of Ichimira et al. [17] and 2,3,4- tri-methylpyrrole by that of Johnson et al. [18]. Polymerisation of vacuum distilled monomers was accomplished in acetonitrile in a single compartment cell with platinum electrodes. A Thomson potentiostat was used for both constant voltage, at 1V versus a saturated calomel electrode, and constant current, at 2 milliamp per sq. cm, working. There was no detectable difference in material produced by these methods. Counter ions used in the supporting electrolyte were perchlorate, tetrafluoroborate, bromide, toluene sulphanate and dodecyl benzene sulphonate. The oligomers were obtained as black continuous films which could be dissolved in methanol by a combination of manual and ultrasonic agitation.

The state of aggregation of the oligomers was assessed by micro filtration and laser light scattering. More than 90% of the oligomers passed through a 100 nm pore size filter. The sample obtained after this had 'particles' smaller than 5nm, the lower limit of detection of the light scattering experiment. Since the oligomers have sizes in the range 1 to 3nm this suggests that they are not aggregated.

NMR spectra were obtained on samples with bromide counter-ions using a Bruker 80MHz spectrometer. Chemical shifts of 2.025ppm, 2.04ppm and 2.10ppm were observed for 3- and 4-methyls in di-methylpyrrole, the 3-methyl and 2- methyl in trimethylpyrrole respectively. The same shifts were observed for the oligomers with the ratio of terminal to backbone units identical to that of the tri- and di-methylpyrrole components of the electrolyte. This accords with the observation of similar oxidation potentials for the two monomers. Assuming a random incorporation of terminating groups a 1 in 10 initial concentration of them in the electrolyte will produce a distribution of oligomer lengths with a mean value of twenty repeat units.

Spectra of such solutions were recorded on a Lambda 9 Perkin Elmer spectro-photometer. Low temperature experiments were performed dissolving the oligomers in 2-methyl tetrahydrofuran. The solutions were then placed in a specially prepared cell, the temperature reduced using a helium cryostat and the spectra recorded, again using a Lambda 9 Perkin Elmer spectrophotometer.

Spectra of solutions show two main absorption bands with maxima at 1.2 and 2.6 eV and a weaker absorption at 2.25 eV [16]. (The latter is weaker for oligomers of a longer mean chain length.) These can be compared with spectra reported for polypyrrole [1,11,19]. The intense absorptions correspond to those identified as due to bipolarons and the other as HOMO-LUMO (interband) transitions in polypyrrole. In addition there are distinct shoulders at circa 1.4 and 2.2eV in the oligomer spectra, as noted for polypyrrole [20].

Evidence for the true solution state of the oligomers is provided by solvatochromic shifts of the bipolaron bands. In chloroform these occur at 1.3 and 2.3eV [16]. There is also a strong influence of the counter-ion. Dodecyl benzene sulphonate produces sharp bipolaron bonds which move apart when bromide is used and become very weak for toluene sulphonate [16]. In the latter case a sharply rising absorption edge is observed below 3eV. Addition of NaOH to a methonol solution of oligomers with dodecyl sulphonate counter-ions leads to a marked reduction in intensity of the bipolaron bonds (Figure 1). These effects can be analysed using the bipolaron model, section 3.

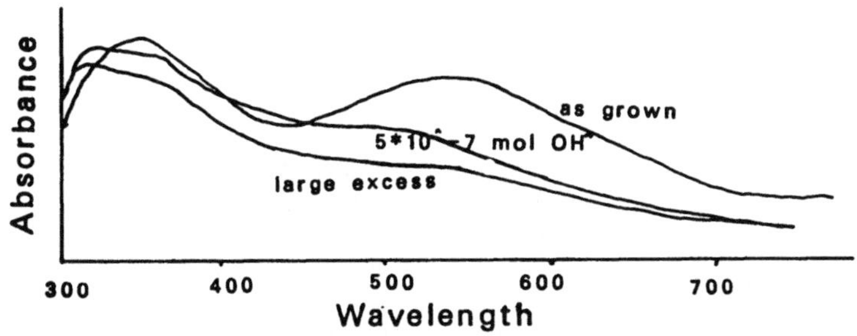

Figure 1. Spectra of dodecylbenzene sulphonate compensated oligomers in methanol (upper) and methanol with 5 x 10^{-7} mol. dm^{-3} OH^- ions (lower)

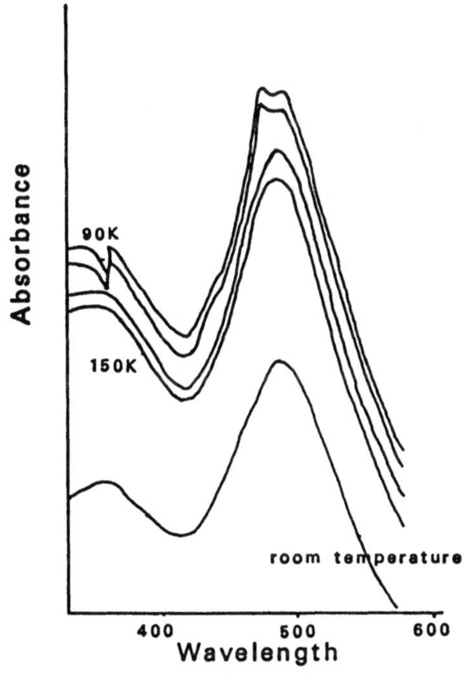

Figure 2. Spectra of dodecylbenzene sulphonate compensated oligomers dissolved in 2-methyl THF recorded at room temperature (bottom), 150, 130, 110 and 90K (top). copyright (C) Controller HMSO London 1989

Cooling of solutions of oligomers in 2-methyl tetrahydrotufan leads to glass formation. For oligomers with dodecyl benzene sulphonate counter-ions the upper bipolaron peak develops fine structure below 120K. The lower bipolaron peak could not be observed in the experimental arrangement employed.

This fine structure, an electronic transition at 2.5eV and vibronic satellites at 2.69 and 2.43eV, is superimposed on a relatively broad band (Figure 2).

3. Discussion and Conclusion

The bipolaron model can be used to explain most of the features of the spectra described above when due account is taken of the interaction of the oligomers with their environment. The appearance of a weak HOMO-LUMO transition and its dependence on mean chain length can be attributed to end effects. The positive charge is delocalised at the ends of the dication oligomers and the bond alternation will relax towards that of the unoxidised chain.

The solvatochrome effects result because the bipolaron levels of isolated oligomers will depend on the polarizability of the solvent in contact with the oligomer. The spectral changes with different counter-ions and with NaOH reflect interactions leading to either partial or complete deprotonation. The latter leads to precipitation of sodium dodecylsulphonate with OH^- and H^+ as water remaining in solution with neutralised (deprotonated) oligomers. A similar effect is produced by the toluene sulphonate counter-ion. This is a strong acid and deprotonates the oligomers, giving the free acid and neutralised oligomers. In neutral solutions the deprotonation is less with dodecyl benzene sulphonate. As the interaction strength increases the bipolaron levels are driven towards the HOMO and LUMO levels until they merge for the neutralised oligomers.

These studies show that for polypyrrole there are a number of factors, e.g. deprotonation that have not previously been generally considered, but see [21], that are important in determining the state of the polymer.

Acknowledgement

This work was supported by the Science and Engineering Research Council by the provision of a studentship for S.P. Collaboration of RSRE Malvern (Dr. K. Harrison) is gratefully acknowledged.

References

1. G.B. Street, in Handbook of Conducting Polymers, Vol. 1. Ed. T.J. Skotheim (Marcel Dekker, N.Y., 1986) p. 256.
2. S. Pugh and D. Bloor this volume.
3. D. Bloor, K.M.C. Cheung and G.C. Stevens, J. Mater. Sci. in press.
4. F. Beck and M. Oberst, Makromol. Chem. Makromol. Symp. 8, 97(1987).
5. R. Qian, J. Qin and D. Shen, Synth. Met. 18, 13(1987).
6. D. Bloor, R.D. Hercliffe, C.G. Galiotis and R.J. Young in Electronic properties of Polymers and Related Compounds, Eds., H. Kuzmany, M. Mehring and S. Roth. Springer Ser. in Solid State Science, 63, 179(1985).
7. M. Osagawa, K. Funahashi, T. Demura, T. Hagiwara and K. Iwata, Synth. Met. 14, 61(1986).
8. K.M. Cheung, D. Bloor and G.C. Stevens, Polymer 29, 1709(1988).
9. A. Nazzal and G.B. Street, J. Chem. Soc. Chem. Commun. 83, (1984).
10. B. Naegele this volume.
11. J.L. Bredas and G.B. Street, Acc. Chem. Res. 18, 309(1985).
12. D. Birnbaum and B.E. Kohler: This volume.

13. D. Fichou, G. Horowitz, B. Xu, F. Garnier : This volume.
14. P. Danneton, K. Uvdal, R. Lazzaroni, W.R. Salaneck : This volume.
15. D.J. Ando, D. Bloor, A. Drake and Udvarhelyi, Polymer in press.
16. S. Pugh and D. Bloor, Synth. Met. 28, C187 (1989).
17. K. Ichimira, S. Ichikawa and K. Immamura, Bull Soc. Chem. Jap. 49, 1157(1975).
18. A.W. Johnson, E. Markman, E. Price and K.B. Shaw, J. Chem. Soc., 4254(1958).
19. K. Yakushi, L.J. Lauchlan, G.B. Street and J.L. Bredas, J. Chem. Phys. 81, 4133(1984).
20. J.H. Kaufman, N. Colaneri, J.C. Scott and G.B. Street, Phys. Rev. Lett. 53, 1005(1984).
21. O. Ingans, R. Erlandsson, C. Nylander, I. Lundstrom, J. Phys. Chem. Solids 45, (4), 427(1984).

Spectroscopic Studies of the Model Oligomer Alpha-Sexithienyl in Its Neutral and Doped States

D. Fichou, G. Horowitz, B. Xu, and F. Garnier

Laboratoire des Matériaux Moléculaires, CNRS,
2, rue Henry Dunant, F-94320 Thiais, France

ABSTRACT - We report spectroscopic studies of the alpha-conjugated thiophene oligomer sexithienyl (6T) in the form of thin solid films and in solution. Optical spectra exhibit well-resolved fine structures characteristic of the purely alpha-conjugated ring sequence. On the other hand, doping experiments of 6T in very dilute solution evidence for the first time the stoechiometric generation of the stable radical cation $6T^{+\cdot}$ in the first step of the process.

1. INTRODUCTION

In the field of conducting polymers, short oligomers of controlled chain length recently received much attention as model compounds towards a better understanding of the electronic properties of their polymeric analogues /1/. The regioselective reconstruction of the chain sequence from the monomer unit by means of organic synthesis affords intermediate size molecules with a well-defined chemical stucture. In addition to their highly-ordered molecular structure, these compounds have low molecular weights which make them processible (solubility, sublimation). Experimental data obtained from such materials may then help to rationalize the progressive appearance of the polymer properties as a function of the chain length and to fit the theoretical predictions of various parameters. Apart from this fundamental interest, these conjugated oligomers may in some cases exhibit electrical and optical properties surpassing those of polymers, which make them potential candidates for device applications.

We recently illustrated this microscopic molecular approach with alpha-conjugated thiophene oligomers (nT with n ranging from 3 to 8). Thus, their third order optical nonlinearities have been studied on thin films as a function of the chain length and the torsion angle between each thiophene ring /2/. Their use as molecular semiconductors in electronic devices has also been reported by us /3-5/. Each oligomer is synthesized independently according to known procedures /6/ or new ones /7/. Due to their relatively low molecular weights, these molecules remain soluble in common solvents up to 6T, while their high stability in air allows them to be sublimed under reduced pressure without decomposing. In addition to a purification process, vacuum sublimation is also a convenient technique for preparing thin films of controlled thickness. Furthermore, vacuum deposition can be made onto any substrate depending upon the sample destination. Thus, optical measurements are performed with films deposited on either quartz or KBr; on the other hand, electronic devices such as Schottky diodes /3-5/ and MISFET /8/ have been built on ITO or n-Si.

From a general point of view, our first studies on oligothiophenes reveal that the most prominent electronic properties of polythiophene (PT) are already existing in short molecules. Nevertheless, out of our five oligomers under study, 6T appears to be the most attractive one for it realizes the best compromise between an intrinsic processability on one side and a PT-like electronic behavior on the other side. For this reason we have recently concen-

Springer Series in Solid-State Sciences, Vol. 91
Electronic Properties of Conjugated Polymers III
Editors: H. Kuzmany · M. Mehring · S. Roth

trated our efforts on this key-compound from a fundamental point of view. We present in this article refined spectroscopic characterizations of 6T thin films together with the first experimental evidence of the production of a stable radical cation (or polaron) in the first step of the doping process of 6T in highly dilute solution.

2. OPTICAL SPECTRA OF 6T

We have already reported the UV-visible absorption spectra of various nT oligomers in thin films at room temperature /2/ and assigned their structured feature to the presence of rotational conformers (rotamers) frozen in the solid state during the sublimation process. Each observed shoulder of the spectra would then correspond to rotamers with different average torsion angles all along the molecule, resulting in different conjugation lengths. This interpretation was supported by CNDO calculations, assuming that the inter-ring bond lengths were not affected by varying the torsion angles. Nevertheless, it does not explain why a discrete distribution of rotamers naturally occurs during the film preparation. We further investigated this behavior by recording the absorption spectrum of a 6T film (1.200 A) at 80°K (see Fig.1). Additional fine structures appear on the long wavelength side, each shoulder of the room temperature spectrum being split into two components (the wavy feature above 600 nm arises from interferences). This observation seems to indicate a vibronic contribution, i.e. absorption transitions from the 6T ground state to various vibrational levels of its first excited state, as suggested by the results of Birnbaum and Köhler on the high resolution fluorescence of 3T in a n-decane matrix at 4.2°K /9/. Although some optical data on higher oligomers have been reported, there have been no high resolution studies .

As a complement to the absorption spectra, Fig.2 shows the fluorescence excitation and emission spectra of 6T in solution in CH_2Cl_2 at room temperature. The excitation spectrum (λ_{max} = 450 nm) corresponds to the fundamental π-π^* absorption band of 6T in solution, with a 20 nm red shift. The emission

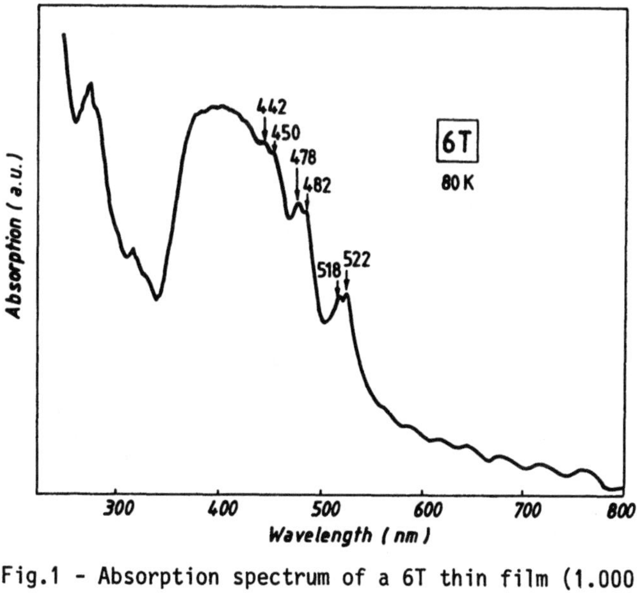

Fig.1 - Absorption spectrum of a 6T thin film (1.000 A) at 80°K

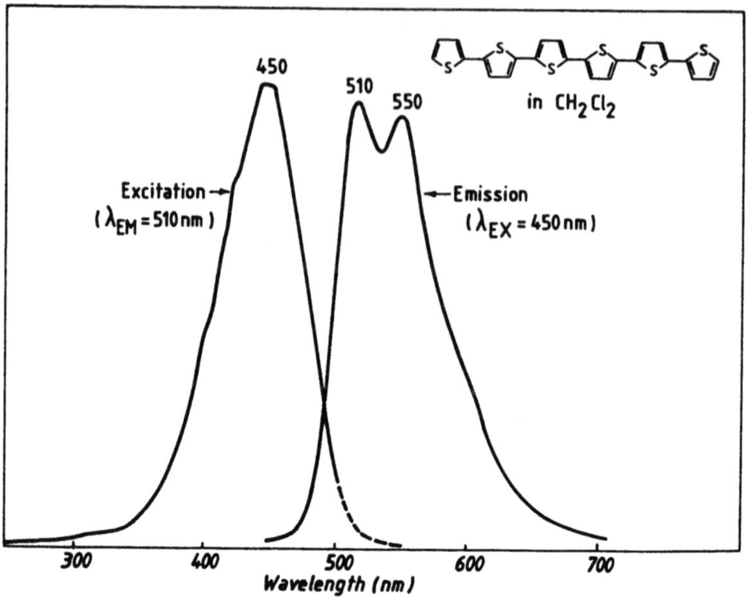

Fig.2 - Fluorescence excitation and emission of 6T in CH_2Cl_2 at room temperature (O.D. = 0.8)

spectrum presents a vibrational structure with two maxima (510 and 550 nm) and a shoulder (∿600 nm); this general feature is identical to that of 3T and 4T in solution /10/ and also surprisingly to that of PT films /11/. This clearly indicates that the three luminescence peaks arise from the alpha-conjugated ring sequence of each of these thiophene derivatives and have nothing to do with structural defects or inter-molecular interactions. A more complete luminescence study of thiophene oligomers is currently under way in our laboratory.

Fig.3 shows the infrared spectrum of a 6T film sublimed on a KBr slice. The good resolution of the bands contrasts with the reported spectra of PT derivatives and allows a refined interpretation of the various bands. If the general feature of the spectrum is close to that of Furukawa et al. /12/ for 6T powders some additional assignments can be made. The two bands at 1045 and 1065 cm^{-1} (numbered 11 and 10 in Fig.3 and shown in the inset) may be assigned to C-H in plane bending vibrations, while the three bands at 1190, 1200 and 1219 cm^{-1} (see inset in Fig.3) would arise from the inter-ring C-C stretching vibrations due to torsion angles between the rings. These last three bands would then be the conformational IR fingerprint of solid state 6T, on the basis of an all-trans alternate ring conformation. The peaks at 3062 and 3100 cm^{-1} (numbered 2 and 1 in Fig.3) correspond to the C-H vibration modes of thiophene in beta and alpha positions respectively.

The effect of a reversible chemical doping ($FeCl_3$ in ethanol) on the IR spectrum of a 6T film is depicted in Fig.4. Intense bands appear in the 950 - 1500 cm^{-1} and 600 - 750 cm^{-1} regions, which can be assigned to translational (T) and ring (R) modes respectively, by analogy with doped PT /14/. The weaker intensities of the R-modes (630 and 720 cm^{-1}) relatively to those of the T-modes (1038, 1136, 1204, 1342 and 1398 cm^{-1} respectively numbered 5, 4, 3, 2 and 1 in Fig.4) evidences the small contribution of the sulfur atom in the electronic structure of doped 6T, which can be viewed as a finite polyene with

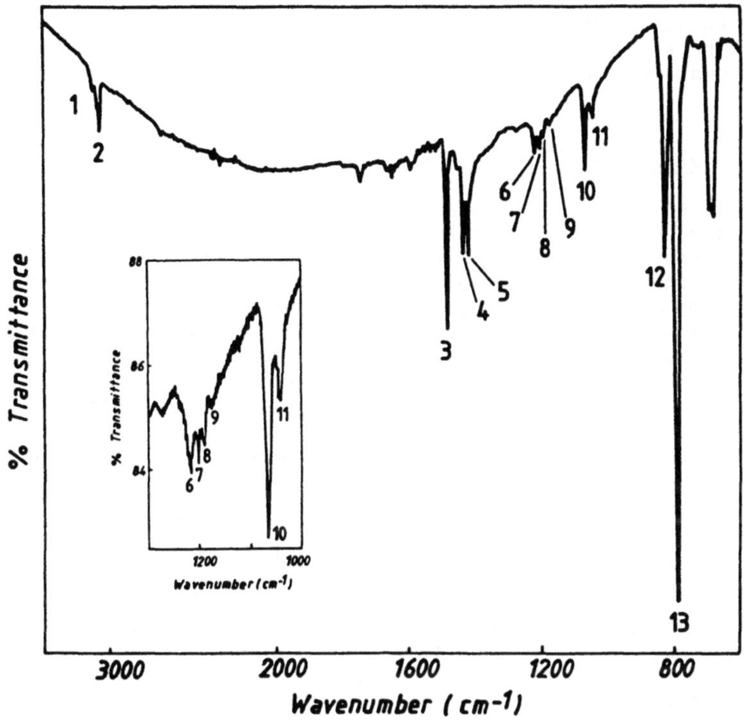

Fig.3 - Infrared spectrum of a neutral 6T thin film (1.200 A) vacuum deposited on a KBr slice

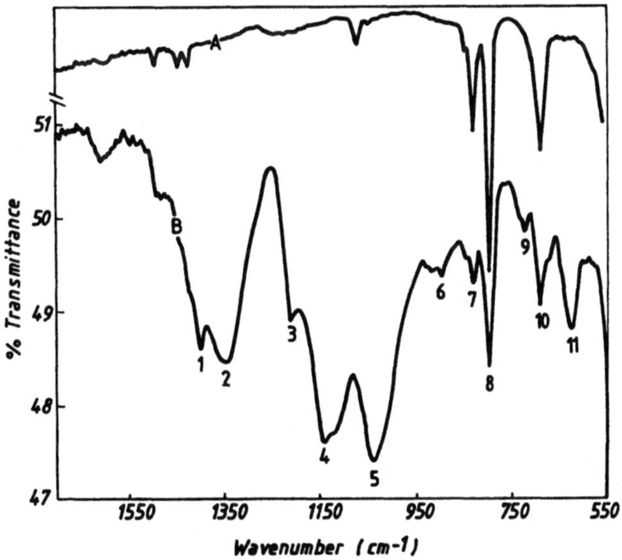

Fig.4 - Infrared spectra of a 6T thin film in (A)its neutral, and (B) oxidized ($FeCl_3$) states

a cis-trans configuration. Undoping of 6T essentially gives back the initial IR (and visible) spectrum of the neutral form; this can be considered as an experimental argument for the reversible character of the doping-undoping cycle of 6T.

3. DOPING OF 6T IN DILUTE SOLUTION - GENERATION OF THE $6T^{+\cdot}$ RADICAL CATION

It has been postulated that the most stable charge storage configuration in doped or photoexcited PT occurs as spinless bipolarons. Nevertheless, an intermediate polaronic regime has been evidenced at low doping levels or in dilute solutions in the case of soluble alkylated PT /15-17/. Despite unambiguous observations on optical and ESR spectra, the complex structure of the polymer does not allow a precise interpretation of the stoechiometry of the doping process. A first attempt to simplify the system has been proposed by Chang and Miller /18/ by using oligomerized 3-methoxythiophene in solution but no polaronic species has been clearly identified. Thus, it was interesting to check the famous polaron-bipolaron concepts with the soluble and highly conjugated 6T. In order to prevent any oligomer-oligomer interaction, we work with concentrations in the range of 10^{-5} mole.l^{-1}.

Doping experiments were realized in methylene chloride with a 6T concentration of 2.4×10^{-5} mole.l^{-1}. The oxidant is $FeCl_3$ in methylene chloride (2.0×10^{-3} mole.l^{-1}). The optical spectrum of neutral 6T in solution (Fig.5, curve A) has a maximum at 432 nm (2.95 eV). Upon progressive doping by a controlled quantity of the $FeCl_3$ solution, the fundamental absorption band disappears while two bands appear in visible region at 680 nm (1.82 eV) and 780 nm (1.59 eV) as shown on curve B. With an excess of $FeCl_3$, these two bands vanish away, while a new one seems to grow beyond 850 nm (curve C), which is our instrumental limitation. Additional modifications appear on the UV side with the emergence of two bands at 350 nm (3.54 eV) and 242 nm (5.12 eV). The analysis of curve B in Fig.5 leads to the following observations:

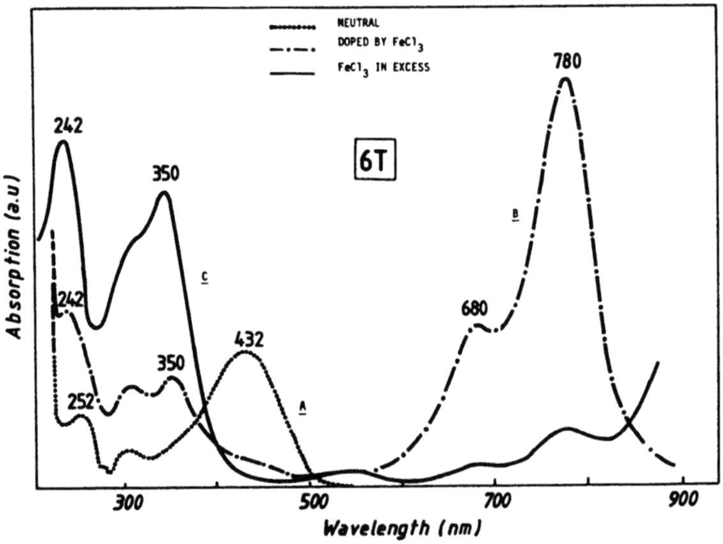

Fig.5 - Optical spectra of 6T in solution in CH_2Cl_2 (2.4×10^{-5} mole.l^{-1}) (A) neutral, (B) after addition of 2 moles of $FeCl_3$ per mole of 6T, and (C) with an excess of $FeCl_3$

a/ The intensity ratio of the two bands at 680 and 780 nm remains constant whatever the $FeCl_3$ quantity added, which shows that they probably arise from the same species in solution.

b/ The maximum intensity of these two bands is reached with exactly two moles of oxidant per mole of 6T. In the reasonable hypothesis of a $FeCl_4^-$ counter-anion, in good agreement with recent works on this dopant /19/, the curve B would correspond to the stoechiometric generation of a mono-oxidized 6T species.

c/ The ESR spectra of this species show an intense signal (despite a very low concentration of starting material) at approximately g = 2.003, with a peak-to-peak linewidth ΔH_{pp} = 8 G (see Fig.6). Furthermore, only 50% of the signal intensity is lost after 24 hours at open air (curve B of Fig.6), which shows the good stability of the paramagnetic species in solution. This narrow ESR signal is superimposed on a broad line also centered around g = 2, and attributed to Fe^{3+} which has another line in the g = 4 region. No hyperfine structure has yet been found on the narrow line.

d/ Identical optical and ESR spectra are produced by using a protic acid like $HClO_4$, H_2SO_4 or HNO_3 instead of the Lewis acid $FeCl_3$. This shows that the counter-anion has no influence on the species in solution which is an ion rather than a charge transfer complex.

From these observations, we propose that the species corresponding to curve B in Fig.5 is the paramagnetic radical cation $6T^{+\cdot}$. Its chemical generation could then be sketched as follows:

$$2\,FeCl_3 \longrightarrow FeCl_4^- + FeCl_2 + h^+ \qquad (1)$$

$$6T + h^+ \longrightarrow 6T^{+\cdot} \qquad (2)$$

$$2\,FeCl_3 + 6T \longrightarrow 6T^{+\cdot} + FeCl_4^- + FeCl_2 \qquad (3)$$

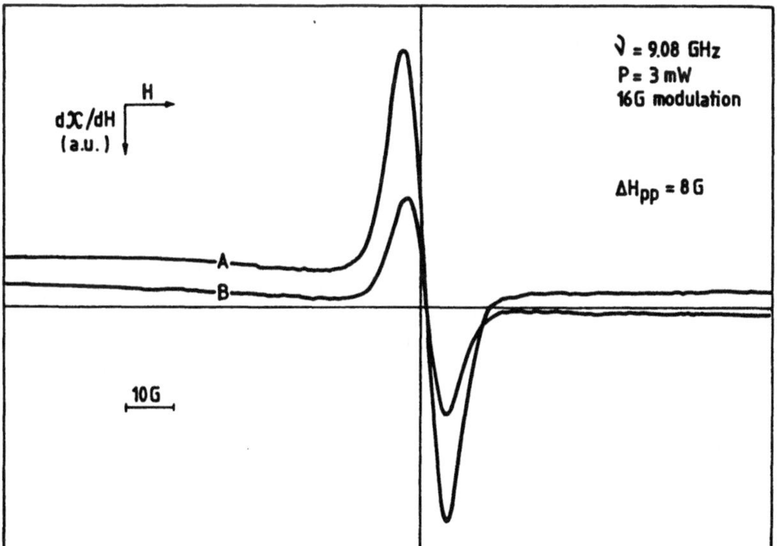

Fig.6 - ESR spectra of a 6T solution in CH_2Cl_2 after addition of 2 moles of $FeCl_3$ per mole of 6T. (A) just after addition, and (B) 24 hours after addition

If now an excess of $FeCl_3$ is added, $6T^{+\cdot}$ evolves towards a more oxidized form, which is highly unsoluble and precipitates as a black solid whose chem· cal composition seems to correspond to the dication salt ($6T^{2+}$, 2 $FeCl_4^-$), as deduced from preliminary elemental analysis. Chemical undoping of this sa· by acetone gives back the neutral 6T, as does also the chemical undoping of $6T^{+\cdot}$ in solution. If no excess of $FeCl_3$ is added, the simple time evolution irreversibly yields an unidentified product, whose absorption corresponds t(the UV part of spectrum C in Fig.5. Since this product is less conjugated th: the neutral 6T for it absorbs at 350 nm, it may come from a free-radical rearrangement of $6T^{+\cdot}$. From a general point of view, the chemical fate of 6T: in dilute solution could thus be summarized:

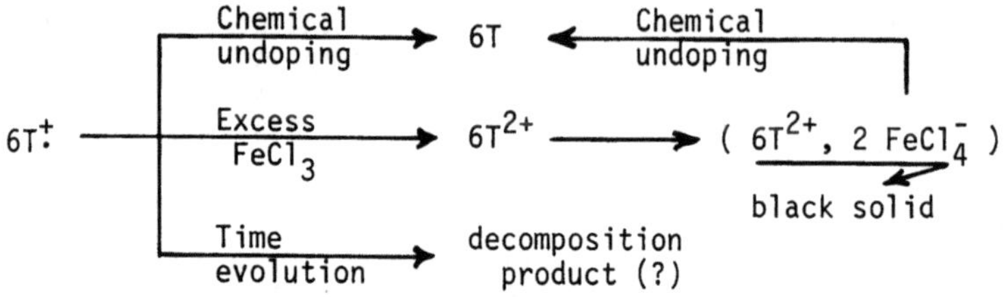

Doping of 3T in the same conditions affords the same spectrum as $6T^{+\cdot}$, showing that $3T^{+\cdot}$ is unstable and dimerizes immediately. Importantly, no 9T i: detected, which would mean that 6T is the shortest oligothiophene that can form a stable radical cation; preliminary results on 4T and 5T also argue in that sense /20/.

It appears then very clearly that the radical cation $6T^{+\cdot}$ possesses the san properties as the postulated polarons of PT. In terms of doped polymers, $6T^{+\cdot}$ would correspond to a 17% doping rate (one positive charge for six thiophene rings, i.e. delocalized on 24 carbon atoms), with a spin to charge ratio of 1 Nevertheless, even if $6T^{+\cdot}$ can be considered as the basic PT polaron, it seem: more realistic to speak about it as a radical cation since these species already form a well-known class of chemical entities /21/.

4. CONCLUSION

We have shown that the simple alpha-conjugated sexi-thienyl molecule possesse electronic properties very similar to those of the very complex polythiophene Its processability allows to prepare films and solutions and to observe clear cut spectroscopic features which help in the understanding of the PT behavior This has been especially the case in the study of the multi-step doping proce in solution, where the stable $6T^{+\cdot}$ radical cation has been isolated and fully characterized by a classical stoechiometric control of the 6T oxidation. It i very likely that the $6T^{+\cdot}$ radical cation and the postulated PT polarons are ve similar species, the instability of the latter ones in soluble PT coming prob bly from intra-chain recombinations which are unavoidable even in dilute solutions.

REFERENCES

1 See for example (a) M. Dory, V.P. Bodart, J. Delhalle, J-M. André, and J-L. Brédas, Mat. Res. Soc. Symp., Proc. vol. 109, p. 239 (1988), (b) L. Laguren Davidson, C.V. Pham, H. Zimmer, H.B. Mark Jr., and D.J. Ondrus, J.Electroch Soc.135, 1406 (1988), (c) C. Kollmar and H. Sixl, J. Chem. Phys. 88, 1343 (1988), (d) S. Pugh and D. Bloor, Synth. Met. 28, C187 (1989), (e) L.W.

Shacklette, J.F. Wolf, S. Gould, and R.H. Baughman, J. Chem. Phys. 88, 3955 (1988).
2 D. Fichou, F. Garnier, F. Charra, F. Kajzar and J. Messier, in "Organic Materials for Nonlinear Optics", eds. R. Hahn and D. Bloor, Royal Society of Chemistry (1989).
3 D. Fichou, G. Horowitz, Y. Nishikitani, and F. Garnier, Chemtronics 3, 176 (1988)
4 F. Garnier, G. Horowitz, and D. Fichou, Synth. Met. 28, C705 (1989)
5 G. Horowitz, D. Fichou, and F. Garnier, Solid State Commun., in the press
6 J. Nakayama, T. Konishi, and M. Hoshino, Heterocycles 27, 1731 (1988) and references herein
7 D. Fichou, G. Horowitz, and F. Garnier, to be published
8 G. Horowitz, D. Fichou, X. Peng, Z. Xu, and F. Garnier, to be published
9 D. Birnbaum and B.E. Köhler, these proceedings
10 J-P. Reyftman, J. Kagan, R. Santus, and P. Morliere, Photochem. Photobiol. 41, 1 (1985)
11 S. Hayashi, K. Kaneto, and K. Yoshino, Solid State Commun. 61, 249 (1987)
12 Y. Furukawa, M. Akimoto, and I. Harada, Synth. Met. 18, 151 (1987)
13 G.J. Visser, G.J. Heeres, J. Wolters, and A. Vos, Acta Cryst. B24, 467 (1968)
14 A.J. Heeger, S. Kivelson, J.R. Schrieffer, and W-P. Su, Rev. Modern Phys. 60, 781 (1988) and references herein
15 K. Kaneto, S. Hayashi, S. Ura, and K. Yoshino, J. Phys. Soc. Jpn. 54, 1146 (1985)
16 N. Colaneri, M. Nowak, D. Spiegel, S. Hotta, A.J. Heeger, and P.A. Pincus, Synth. Met. 28, C399 (1989)
18 A.C. Cheng and L.L. Miller, Synth. Met. 22, 71 (1987)
19 T.E. Jones, W.F. Butler, T.R. Odgen, D.M. Gottfredson, and E.M. Gullikson, J. Chem. Phys. 88, 3338 (1988)and references herein
20 D. Fichou, G. Horowitz, B. Xu, and F. Garnier, to be published
21 G. Vincow, in "Radical Ions", eds. E.T. Kaiser and L. Kevan, Interscience New-York, p. 151 (1968)

Investigation of the Transient Photocurrent in β-Carotene and a Donor–Acceptor Substituted Polyene

Y.Q. Shen, J. Reichenbach, and S. Roth

Max-Planck-Institut für Festkörperforschung,
D-7000 Stuttgart 80, Fed. Rep. of Germany

Abstract Transient photocurrent measurements on β-carotene crystals, films and donor-acceptor substituted polyene films are reported. The conjugated chain in the molecule of donor-acceptor substituted polyenes has the same length as that in the β-carotene molecule. The films are prepared by casting a dichloromethane solution of the compounds on a water surface. The results of field-dependent and intensity-dependent photocurrent measurements are presented and discussed. The photocurrents of the two compounds are compared.

1 Introduction

There has been considerable interest in recent years in the synthesis and study of compounds containing electron donor and acceptor moieties connected by conjugated bridges, since these compounds have been proposed as possible components for molecular electronic devices [1]. The "Soliton Switch" is one of the proposals and is based on the transport of solitons along a conjugated chain. Many studies on conjugated polymers discuss whether the π-electronic system of a polyene chain enables charge transfer through the chain in terms of solitons [2]. The proposal assumes that, when a donor and an acceptor are connected by a polyene chain, charge transfer from donor to acceptor upon excitation will be possible.

In order to study the charge transfer on a short polyene chain, we have carried out investigations of the transient photocurrent in β-carotene and in the donor-acceptor substituted polyene 7-julolidyl-7,8'-diapocarotene-8'-al . Photocurrent investigations are an important method of studying the transport of charge carriers in conjugated systems [3,4]. The photocurrent investigations in conducting polymers show that the photocurrent consists of two components, a fast component, which reflects the intrinsic transport of the charge carriers along the conjugated chains, and a slow component, reflecting the hopping transport between localized states [3,4]. The picosecond photocurrent is dominated by the fast component. Here we report the results of our investigations of the fast component of the photocurrent in various substances with short polyene chains. The transient photocurrent investigation [3] and the chemical preparations of the compounds [5] have been reported elsewhere.

2 Samples

Two compounds have been used in these investigations (see Fig.1), β-carotene and the donor-acceptor substituted polyene 7-julolidyl-7,8'-diapocarotene-8'-al [5]. In the case of β-carotene, single crystals and thin films have been investigated. On single crystal *Au* electrodes can be directly evaporated. The thin films are prepared by casting a dichloro-

Springer Series in Solid-State Sciences, Vol. 91
Electronic Properties of Conjugated Polymers III
Editors: H. Kuzmany · M. Mehring · S. Roth

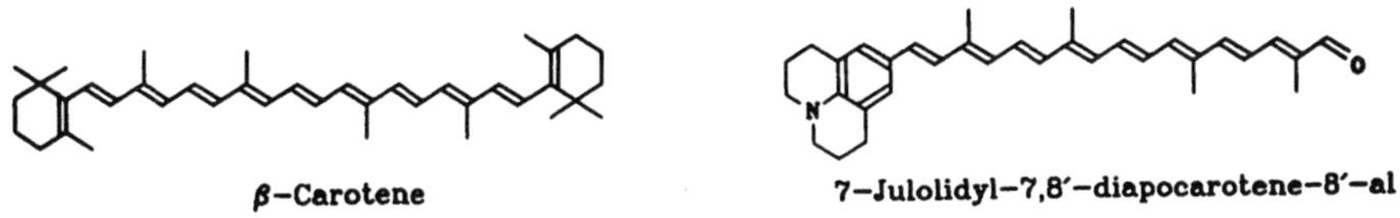

FIG.1: Molecular structure of β-carotene and 7-julolidyl-7,8'-diapocarotene-8'-al

methane solution of β-carotene on a water surface. The dichlormethane evaporates rapidly and leaves a thin, homogeneous film on the water surface. The film can be easily placed on a quartz subtrate with Au electrodes. Since the donor-acceptor substituted polyene cannot be obtained in the form of single crystals only film samples have been investigated. The preparation of the film samples is the same as for the β-carotene films. X-ray diffraction shows that film samples prepared in this way have no crystalline structures [6].

3 Experimental Results

In the transient photocurrent measurements a $500ps$ pulsed N_2 laser is used as the excitation light source. The wavelength of the excitation light has been chosen to be $488nm$, which corresponds to the maximum absorption wavelength of β-carotene and donor-acceptor substituted polyenes [5,7,8]. In the case of the β-carotene single crystal the electric field is applied parallel to the b-axis.

The electric field dependence of the peak photocurrent for three types of samples is shown in Fig.2. For β-carotene crystals the peak photocurrent varies linearly with the electric field E, for $E \geq 1000V/cm$ and sublinearly for smaller fields. For the film samples the peak photocurrent shows an overall sublinear dependence on E up to $5 \times 10^4 V/cm$. Fig.3 shows the field dependence of the photocurrent in a β-carotene crystal measured at

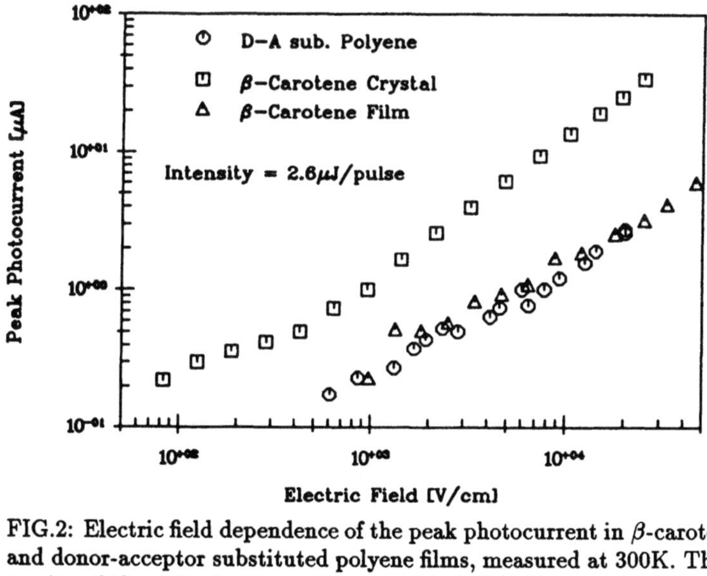

FIG.2: Electric field dependence of the peak photocurrent in β-carotene crystals and films and donor-acceptor substituted polyene films, measured at 300K. The wavelength and intensity of the pulse laser are $488nm$ and $2.6\mu J/pulse$

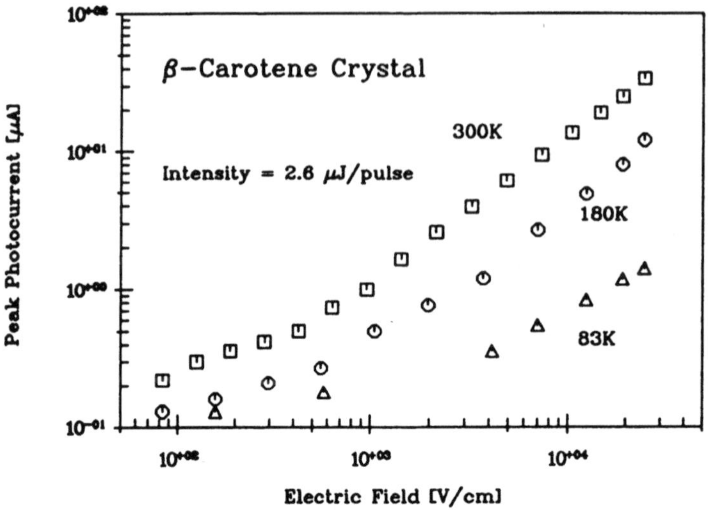

FIG.3: Electric field dependence of the peak photocurrent in β-carotene crystals, measured at 85, 180 and 300K. The wavelength and intensity of the pulse laser are 488nm and 2.6$\mu J/pulse$

different temperatures. As the temperature decreases the photocurrent also decreases and the sublinear E dependence becomes more pronounced. At 83K no linear E dependence can be observed. Furthermore, the decrease of the photocurrent is much stronger in high electric fields than in low fields. The peak photocurrent shows in all samples the same intensity dependence (see Fig.4), in the low intensity region the photocurrent I_p increases linearly with intensity Int and in the high intensity region it follows $I_p \propto Int^{1.54}$.

The E-dependent photocurrent and Int-dependent photocurrent in all samples have also been measured with 366nm excitation light. The E dependence of the photocurrent does not change, whereas the Int dependence of the photocurrent is different from that measured at 488nm. In the low intensity region the photocurrent varies as $I_p \propto Int^{2.15}$ and in the higher intensity region the photocurrent seems to saturate.

4 Discussions

The results of our investigations of the fast component of the photocurrent in β-carotene and donor-acceptor substituted polyenes show that the behavior of the photocurrent is quite similar for these compound. This indicates that the photocurrent is mainly determined by the short polyene chain and is less dependent on the end group and the morphology of these polyene compounds. Compared to the photocurrent in polyacetylene [3] the photocurrents measured in these polyenes are different. First the photocurrent is temperature dependent whereas in polyacetylene the photocurrent is independent of temperature. Second the sublinear field dependence and the superlinear intensity dependence has never been seen in polyacetylene. The sublinear field dependence may be due to the larger distance between the polyene chains, compared to the interchain distance in polyacetylene. The origin of the difference in the intensity and the temperature dependence is not yet clear.

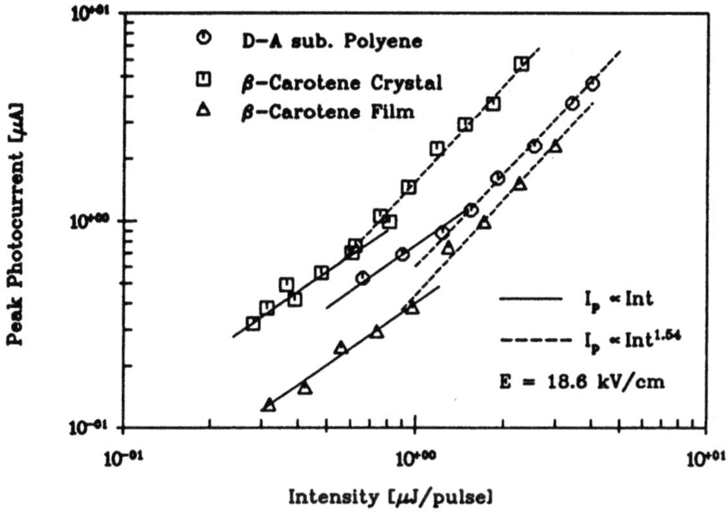

FIG.4: Light intensity dependence of the peak photocurrent in β-carotene crystals and films and donor-acceptor substituted polyene films measured at 300K. The wavelength of the pulse laser is 488nm and the electric field is 16.8kV/cm

5 Conclusions

The transient photocurrent investigations in β-carotene and donor-acceptor substituted polyene show a very different transport behavior compared to polyacetylene. The charge transport seems to occur on the short polyene chain since the photocurrent is independent of the end groups and the morphology. The photocurrent decreases with decreasing temperature. Due to the fact that the molecules of the donor-acceptor substituted polyene are not oriented in the film, the effect of the donor and the acceptor cannot be seen. Therefore the donor-acceptor substituted polyene shows a similar behavior to β-carotene.

6 References

1. F.L.Carter (ed): Molecular Electronic Devices, Dekker, New York(1982).
2. S.Roth and H.Bleier: Advances in Physics, 36(4), 385(1987).
3. H.Bleier, S.Roth, Y.Q.Shen, D.Schäfer-Siebert and G.Leising: Phys.Rev.B, 38, 6031(1988).
4. Y.Q.Shen, H.Lindenberger, H.Bleier and S.Roth: these proceedings
5. Y.Q.Shen, W.Göhring, S.Hagen and S.Roth: to be published
6. K.Carneiro: private communication.
7. D.Chapman, R.J.Cherry and A.Morrison: Proc.Roy.Soc.A., 301 , 173(1967).
8. S.Saito, M.Tasumi and C.H.Eugster: J.Raman Spectrosc., 14, 299(1983).

Langmuir–Blodgett Films from Donor–Acceptor Substituted Polyenes

S. Hagen[1], *H. Schier*[1], *S. Roth*[1], *and M. Hanack*[2]

[1]Max-Planck-Institut für Festkörperforschung, Heisenbergstr. 1, D-7000 Stuttgart 80, Fed. Rep. of Germany
[2]Institut für Organische Chemie der Universität, Tübingen, Fed. Rep. of Germany

Polyene chains are often discussed as possible components for future molecular electronic devices. When a donor and an acceptor are connected by a polyene chain, one assumes that a charge transfer from the donor to the acceptor upon electronic excitation will be possible. To study e.g. the optical nonlinearities of these polyenes we prepared ultra-thin films with the Langmuir-Blodgett technique, which is a well-established method for producing highly ordered structures of organic materials on solid surfaces. To improve the formation of such well-ordered films we synthesized donor-acceptor substituted polyenes with long aliphatic chains.

1. Chemical Preparation

The polyenes (7) and (13) were prepared from the commercially available crocetindialdehyde (1). The synthesis of 7-N,N-dioctylaminophenyl-7,8'-diapocarotene-8'-al (7) is shown in Fig. 1.

For inserting the long aliphatic chains into the donor group, chlorobenzene was treated with di-n-octylamine and n-BuLi in a nucleophilic aromatic substitution reaction. The resulting N,N-dioctylaniline (2) was converted into the phosphonium salt (3) in a reaction with HCHO, $P(C_6H_5)_3$, CH_3COOH, KI, in $CHCl_3$ at room temperature and occasionally stirring for three weeks [1]. The donor was linked to the polyenedialdehyde by the conversion of one carbonyl group into a new C=C double bond in a Wittig reaction [2]. To avoid the formation of the undesired disubstituted product, we protected one of the carbonyl groups of the dialdehyde. In order to do this, crocetindialdehyde was converted into its diacetal (4) with 2,2-dimethyl-1,3-propanediol. Partial hydrolysis of the diacetal led to the monoacetal (5), which was purified by flash chromatography [3]. In the Wittig reaction a solution of the monoacetal in THF was slowly added to the dark-red solution of the donor-triphenylphosphorane, which was generated in a reaction between the phosphonium salt (3) and n-BuLi in THF at 0 °C. After four hours of stirring (control by TLC) at room temperature and final hydrolysis of the acetal group, we applied purification by flash chromatography on silicagel with a mixture of CH_2Cl_2 and ethylacetate 60:1 and obtained monosubstituted polyene with a high yield of 75 %.

Springer Series in Solid-State Sciences, Vol. 91
Electronic Properties of Conjugated Polymers III
Editors: H. Kuzmany · M. Mehring · S. Roth

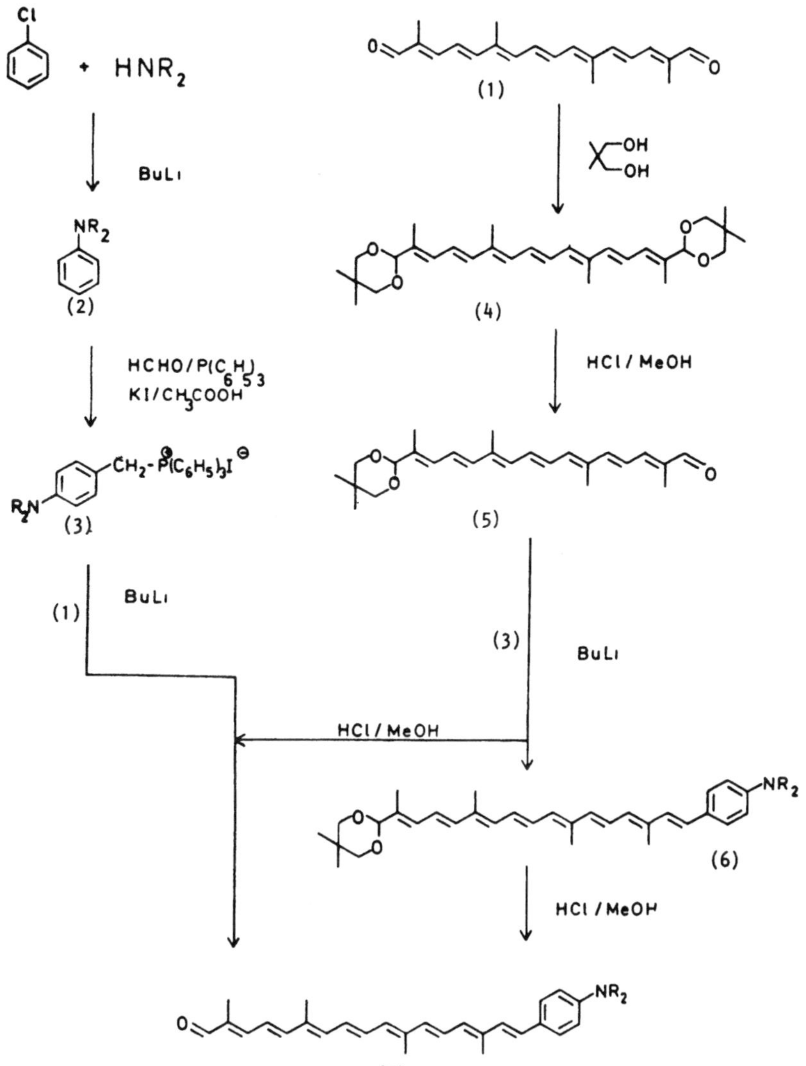

Fig. 1. Synthesis of 7-N,N-dioctylaminophenyl-7,8'-diapo-carotene-8'-al

The synthesis of 7-julolidyl-7,7'-diapocarotene-7'-eicosyl-7'-carboxylic acid (13) is shown in Fig. 2.

In this compound, the long aliphatic chain is connected with the acceptor group. The eicosyl group was introduced in a reaction between trimethylsilylacetic acid and eicosyliodide (9), which was prepared in a reaction between eicosanol (8) and trimethylsilyliodide [4]. Trimethylsilylacetic acid forms a di-anion on treatment with lithiumdiisopropylamide (LDA) in THF at 0 °C. This dianion undergoes alkylation with eicosyliodide at

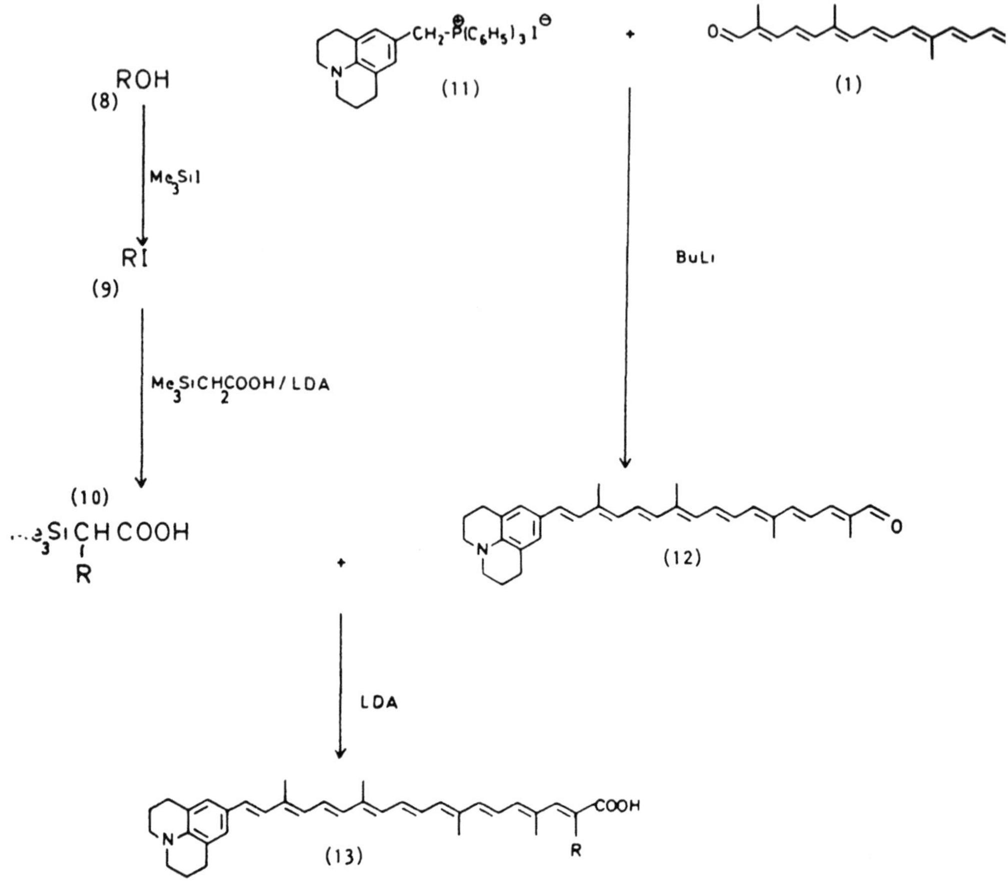

Fig. 2. Synthesis of 7-julolidyl-7,7'-diapocarotene-7'-eicosyl-7'-carboxylic acid

room temperature [5] to the α-substituted carboxylic acid (10). The conversion of the carbonyl group in (12) into a C=C-double bond was done by the Peterson-Olefination [6], the silicon version of the Wittig reaction,. In order to do this α-eicosyltrimethylsilylacetic acid was treated with LDA in THF at 0 °C to form its dianion. The dianion reacts with the polyenealdehyde (12) at 0 °C to the eicosylsubstituted polyene (13). After extractive workup and purification by flash chromatography on silicagel with a mixture of CH_2Cl_2, petroleumether, ethylacetate 10:10:1 we obtained 42 % of the product.

The synthesis of 7-julolidyl-7,8'-diapocarotene-8'al (12) [7] was carried out in the same way as the preparation of (7). The phosphonium salt (11) was prepared from julolidin, which is commercially available, after the same procedure as shown in Fig. 1.

2. Preparation of LB-Films

The LB-Layers were prepared with a commercial Lauda film balance, installed in a class 1000 clean room equipped under yellow light. Chloroform and ethanol used were Merck Uvasol quality. Water was purified in a Milli-Q-System (Millipore) to 18 MX·cm. Commercial Bixin ($C_{25}H_{30}O_4$, Fa. Roth) was purified by chromatographie on ODS-phase (MeOH/Aceton=95:5). Two different substrates, i.e. Au and Si were used. Glass slides (75·10·1 mm) were coated with layers of 200 Å Cr and 2000 Å Au by thermal evaporation. The gold substrates were used immediately after vacuum deposition of the metals. Si substrates (FZ (111), p(B), 1000 X·cm) were etched in 30 % HF and rinsed in Mili-Q water prior to use. For Bixin the subphase contained $5\cdot10^{-4}$ mol/l $CdCl_2$ and was buffered with KOH to pH 7.5. The subphase temperature was 19 °C. For (7) the subphase pure water was used. Bixin was spread from solution containing $2\cdot10^{-3}$ mol/l Bixin in $CHCl_3$/MeOH=10:1 and rested for 5 min before compression. The monolayer was compressed at a constant rate of 1.14 cm^2/s up to the transfer pressure of 32 dyn/cm corresponding to a mean area of 22 $Å^2$/molecule. The dipping speed was 0.83 mm/s for all cycles. The transfer ratio was 0.98. (7) was spread from a $8.9\cdot10^{-4}$ M solution in $CHCl_3$ and rested 5 min before compression. Compression speed was 0.57 cm^2/s up to the transfer pressure of 13 dyn/cm with a mean area of 56 $Å^2$/molecule. Transfer was done at a speed of 0.415 mm/s and yielded a ratio of 0.8.

3. Results

The Langmuir-Blodgett-technique (LB) [8] was used to order the donor-acceptor-substituted polyenes on interface water-air and to transfer these ordered films to solid substrates.

Amphiphilic molecules, consisting of a hydrophilic headgroup and a hydrophobic 'tail' e.g. a saturated long chain fatty acid, are spread on a clean water surface. The molecules lying on the surface form a two-dimensional gas. On reducing the covered surface the mean area per molecule [$Å^2$/mol] is reduced until the interaction of the molecules induces a phase transition into a liquid analogous state. This phase transition is characterized by a change of slope in the surface pressure/area (σ/F)-diagram (Fig. 3). Further reduction of the area induces a second transition into 2D-solid state. This face is comparable to a smectic-b phase, all molecules standing perpendicular on the water surface. In this solid analogous state the ordered film can be transferred to a solid substrate. The film is held under constant pressure and the substrate is dipped into the water, thereby transferring the molecules from the water surface to the subtrate. The compressed solid analogous Langmuir film of Bixin is extremely sensitive to mechanical perturbations. At a surface pressure of 32 dyn/cm the film collapses under the influence of surface waves. We managed this problem by reducing the water level in the trough. The ratio of transferred film area for up- to downstroke motion was almost 1. By comparision of the IR spectra of a multilayer sample on Si (20 LB) to a KBr pellet we could not yet determine if the molecules are stacked in head-to-tail or head-to-head orientation.

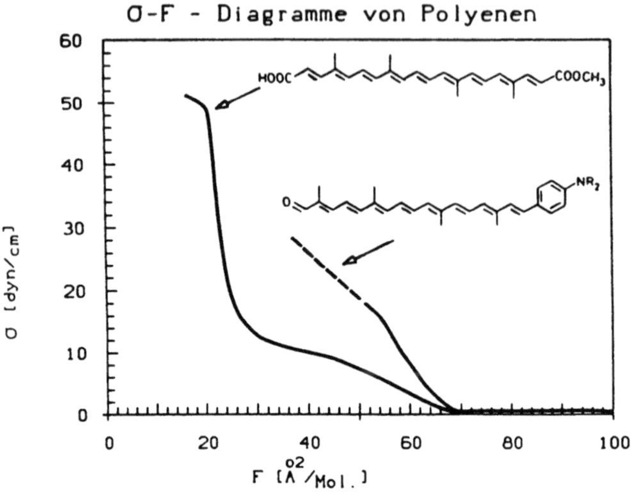

Fig. 3. Surface pressure/area diagram of polyenes

The polyene (7), substitued with two aliphatic chains, showed a lower σ max of 18 dyn/cm. At higher surface pressure the films seems to be stable but consited of, at least partly, folded 'floes' (dashed part of curve in Fig. 3). At the transfer pressure of 13 dyn/cm the film was very insensitive to mechanical vibrations. The transferred film was characterized by IR and UV spectroscopy.

From the IR-spetra we deduce that it is possible to get highly ordered structures from D-A-substituted polyenes with the LB-technique. The main problem is still the high sensitivity against mechanical perturbations during the film transfer and therefore most work on polyenes was done by embedding the active molecules in a inert matrix [9,10]. We could show that it is possible to transfer LB-films from pure polyenes. The handling of polyenes substituted with aliphatic sidechains is more convenient, but the maximum surface pressure and perhaps the ordering is lower than for unsubstituted ones.

Acknowledgement

We want to thank the DFG under SFB 329 for financial support.

References

[1] H. Bredereck: Chem. Ber. 106 (1973), 3732
[2] W.S. Wadsworth, Jr.: Org. React. 25 (1978), 73
[3] K. Bernhard, G. Englert, H. Mayer, R. Müller, A. Rüttimann, N. Vecchi, E. Widmer, R. Zell: Helv. Chim. Acta 64 (1981), 2964
[4] G.A. Olah, S.C. Narang: Tetrahedron 38 (1977), 2659
[5] P.A. Crieco, C.J. Wang, S.D. Burke: J.C.S. Chem. Comm. (1975), 537

[6] D.J. Ager: Synthesis (1984), 384
[7] W. Göhring, PhD Thesis, Tübingen 1987
[8] H. Kuhn, D. Möbius: Angew. Chem. 83 (1971) 672
[9] G. Duda, G. Wegner: Makromol. Chem., Rapid Commun. 9 (1988), 495
[10] R.M. Metzger, R.R. Schumaker, M.P. Cava, R.K. Laidlaw, C.A. Panetta, E. Torres: Langmuir 4 (1988), 298

Part XII

High Temperature Superconductors

Electronic Structure Studies of High-T_c Superconductors

J. Fink[1], *N. Nücker*[1], *H. Romberg*[1], *M. Alexander*[1], *S. Nakai*[1], *R. Manzke*[2], *T. Buslaps*[2], *and R. Claessen*[2]

[1]Kernforschungszentrum Karlsruhe,
Institut für Nukleare Festkörperphysik,
Postfach 36 40, D-7500 Karlsruhe, Fed. Rep. of Germany
[2]Institut für Experimentalphysik, Universität Kiel,
Olshausenstr. 40–60, D-2300 Kiel, Fed. Rep. of Germany

Abstract. An introduction into the electronic structure of high-T_c superconductors is presented. Studies on these compounds by electron energy-loss spectroscopy, angle-resolved photoelectron spectroscopy, and inverse angle-resolved photoelectron spectroscopy are reported. In particular, the systems $La_{2-x}Sr_xCuO_4$, $YBa_2Cu_3O_{7-y}$, $Nd_{2-x}Ce_xCuO_4$ and $Bi_2Sr_2CaCu_2O_8$ have been studied. Information on the dynamics, the character, and the symmetry of the charge carrier is obtained. For $Bi_2Sr_2CaCu_2O_8$ the occupied and unoccupied part of a band crossing the Fermi-level is observed. For $T<T_c$ an opening of a superconducting gap is observed corresponding to $2\Delta(0)/k_BT_c \sim 8$.

1. Introduction

Despite an unprecedented activity since the discovery of the first high-T_c superconductor $La_{2-x}Ba_xCuO_4$ [1] the mechanism for the superconductivity in these compounds remains completely unclear. Even an important prerequisite for the understanding of the mechanism, namely the detailed knowledge of the electronic structure of these materials is not very advanced. In this review we report on our electronic structure studies of typical high-T_c superconductors such as $La_{2-x}Sr_xCuO_4$, $YBa_2Cu_3O_{7-y}$, $Nd_{2-x}Ce_xCuO_4$ and $Bi_2Sr_2CaCu_2O_8$ by electron energy-loss spectroscopy (EELS), by angle-resolved photoelectron spectroscopy (ARUPS) and by angle-resolved inverse photoelectron spectroscopy (ARIPES).

From inspection of the lattice structure of these compounds it follows that the covalent bonds which may lead to a metallic behaviour are the Cu-O bonds in the two-dimensional CuO_2 layers. The rare earth and earth-alkaline ions indicate their fully ionized state, i.e., their electrons are energetically far away from the Fermi level. In Fig. 1 we illustrate the

Springer Series in Solid-State Sciences, Vol. 91
Electronic Properties of Conjugated Polymers III
Editors: H. Kuzmany · M. Mehring · S. Roth

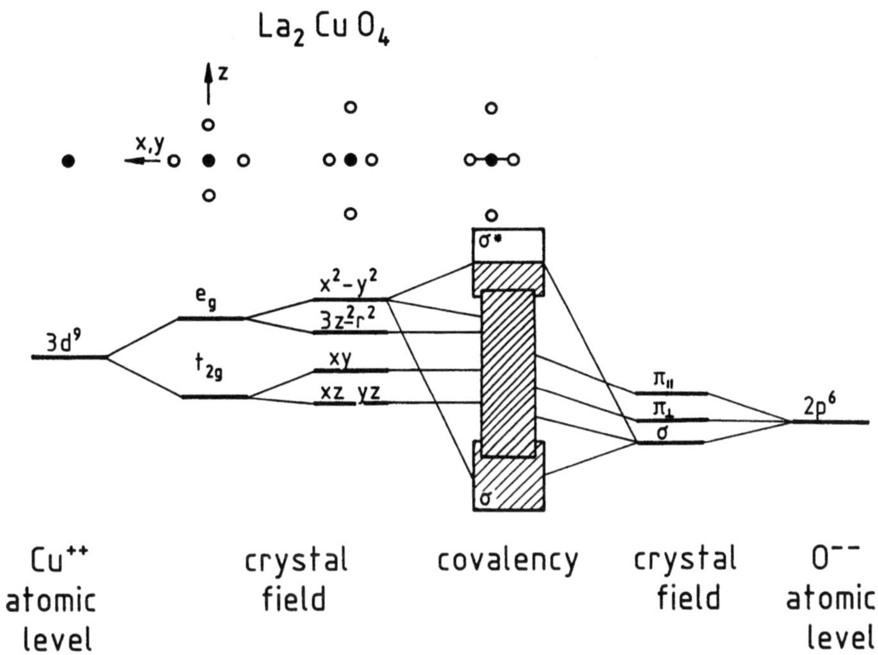

Fig. 1 The energy level diagram of a Cu $3d^9$ ion (Cu^{++}) in a crystal field with atomic positions shown schematically at the top (left side). Crystal field splitting for an O^{-} atom with an almost filled 2p shell (right side). In the center bands formed by hybridization, and specifically bonding and antibonding σ bands, are illustrated

formation of the band structure of the CuO_2 planes in La_2CuO_4. We start with an ionic picture with Cu^{++} in a $3d^9$ configuration (one hole in the 3d shell) and O^{--} in a $2p^6$ configuration (filled 2p shell). The crystal field of the regular O octahedron leads to crystal field splitting of the 3d level into e_g and t_{2g} states. The elongated O octahedron leads to a further splitting of the e_g level into $3d_{x^2-y^2}$ and $3d_{3z^2-r^2}$ states. The hybridization with the O 2p states is strongest for the $3d_{x^2-y^2}$ orbitals in the CuO_2 plane leading to a bonding and an antibonding $3d_{x^2-y^2}$-$2p_{x,y}$ σ and σ* band. In between, there are the Cu-O bands from the other 3d levels. Since there is an odd number of electrons per unit cell, La_2CuO_4 should be a metal as indicated in Fig. 1. Also the present band-structure calculations using the local-density-approximation (LDA) predict a metallic behaviour. In reality, pure La_2CuO_4 is an antiferromagnetic semiconductor with a gap of about 1.7 eV. Contrary to conjugated polymers it is clear that this gap is not caused by electron-phonon interaction. According to W. Weber [2] the gap should then be related to a frozen breathing mode in the CuO_2 layers which was never observed. It is well-known that correlation effects, i.e., local electron-electron interactions play an important role in transition metal compounds. Thus we would better start with localized electrons in an ionic ansatz as shown in Fig. 2a rather than with Bloch states. In the former case we have two types of charge fluctuations: (i) Cu $3d^9$ + Cu $3d^9 \rightarrow$ Cu $3d^{10}$ + Cu $3d^8$ + U were the formation of two holes

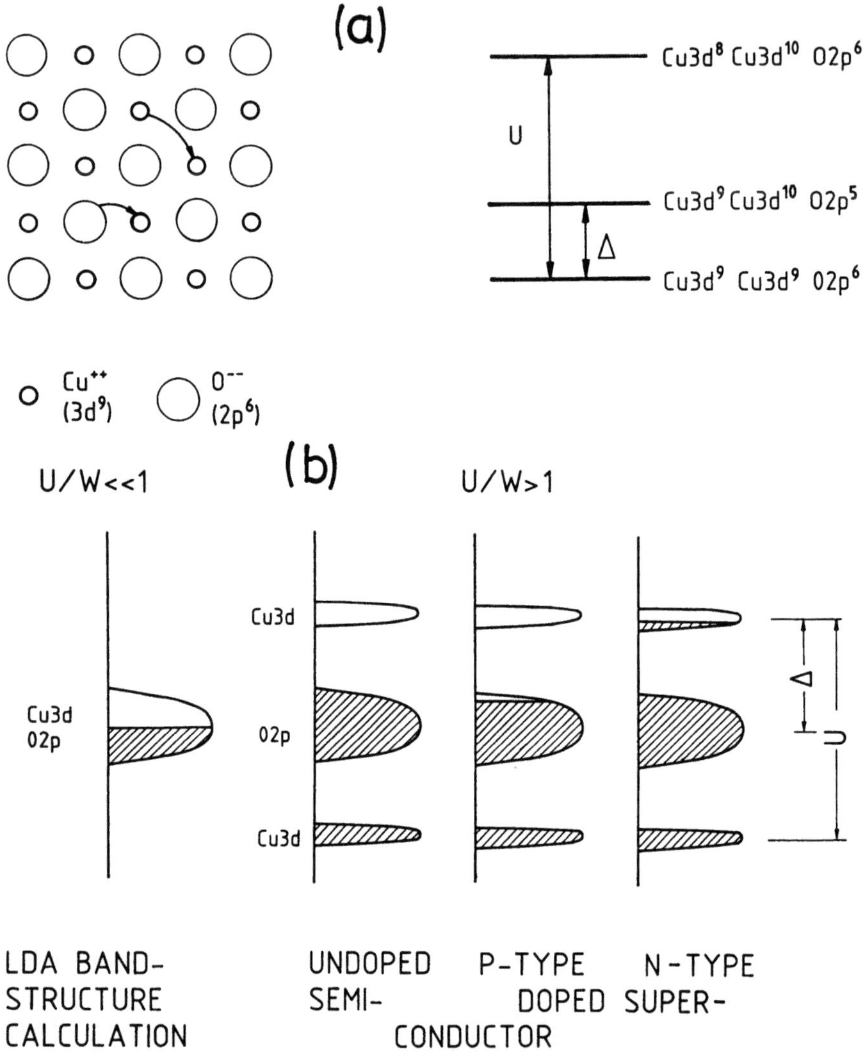

Fig. 2 (a): Charge fluctuations in an ionic CuO layer and total energy diagram for hole excitations. (b): Half filled Cu-O $3d_{x^2-y^2}$-$2p_{x,y}$ σ^* band derived from LDA band-structure calculation. For $U > \Delta$, a charge-transfer insulator is formed. P-type and n-type doping leads to the formation of holes on O and electrons on Cu sites, respectively

on the same Cu site needs the Coulomb energy U; (ii) Cu $3d^9$ + O $2p^6 \rightarrow$ Cu $3d^{10}$ + O $2p^5 + \Delta$ where Δ is the charge transfer energy. For the early transition metal compounds Δ is larger than U and we obtain the conventional Mott-Hubbard picture whereas in the late transition metal compounds and thus also in the high-T_c superconductors $U > \Delta$ and, therefore, the band gap is of a charge transfer type. The theoretical foundation of this model for the electronic structure for transition metal compounds was worked out before the discovery of high-T_c superconductors [3]. While Fig. 2a shows a total energy diagram for hole formation, Fig. 2(b) shows the "band structure" picture for electrons.On the left, the metallic picture derived from LDA band-structure calculations is shown again. Taking fully into account correlation effects on the Cu sites leads to the formation of an upper and a lower Hubbard "band" with dominantly Cu 3d character and a band in between having dominantly O 2p character. The charge transfer gap is formed between the O 2p band and the upper Hubbard "band". The CuO_2 planes can be p-type doped by replacing La^{+++} by Sr^{++} in the LaO layers (in $La_{2-x}Sr_xCuO_4$), by CuO_3 chains (in $YBa_2Cu_3O_7$), by BiO planes (in $Bi_2Sr_2CaCu_2O_8$), and by TlO planes (in $Tl_2Ba_2Ca_2Cu_3O_{10}$). In this case holes will be formed having predominantly O 2p character, i.e., in a local picture O^- will be created. On the other hand, in the recently discovered electron high-T_c superconductor $Nd_{2-x}Ce_xCuO_4$, the replacement of trivalent Nd by tetravalent Ce should lead to n-type doping, i.e., the upper Hubbard band will be filled or in a local picture there should be a formation of monovalent Cu.

The above given model for the electronic structure for the high-T_c superconductors is still strongly discussed and at present it is not clear whether a more localized picture (described above) or a more delocalized picture (described by LDA band-structure calculations) should be applied.

2. Dynamics of Charge Carriers

In Fig. 3a we show the loss function $\mathrm{Im}(-1/\varepsilon)$, the real part of the dielectric function ε_1 and the optical conductivity σ of $Bi_2Sr_2CaCu_2O_8$ in the energy range 0-6 eV (dashed line) [4]. The momentum transfer was chosen to be in the CuO_2 plane and to be small in comparison with the extension of the Brillouin zone. The loss function is dominated by a plasmon at ~1 eV due to the charge carriers, probably holes with predominantly O 2p character. Subtracting the Drude part due to these charge carriers, a gap of about 2 eV is realized in σ, probably the charge-transfer gap. A strong rise in oscillator strength is observed at 3.5 eV due to the other layers exhibiting ionic bonding. In Fig. 3b we show the

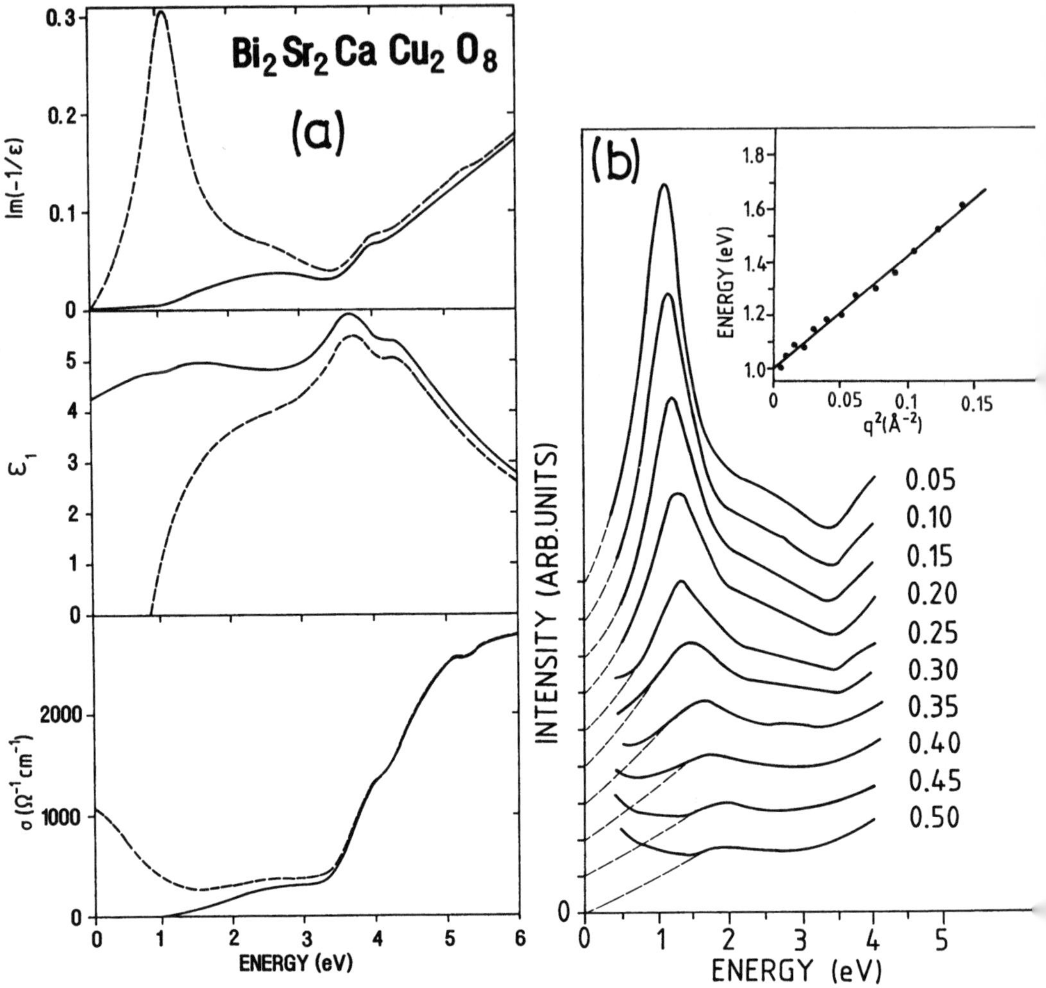

Fig. 3 (a): Loss function Im(-1/ε), ε_1 and σ for $Bi_2Sr_2CaCu_2O_8$ (dashed line). Solid line: The same but Drude part subtracted. (b): Dispersion of the charge carrier plasmon. Inset shows the plasmon energy vs. q^2

dispersion of the plasmon as a function of momentum transfer q. The dispersion is linear in q^2 as shown in the inset. From the dispersoin coefficient $\alpha = 3/10\ (mv_F^2/E_p(0))$ defined (in RPA) by the relation $E_p(q) = E_p(0) + (\hbar^2/m)\alpha q^2$, the Fermi velocity $v_F = 6 \cdot 10^7$ cm/sec could be derived. This value is smaller than values for normal metals and typical for highly doped semiconductors. It is remarkable that these data do not change upon lowering the temperature to 30 K well below $T_c = 83$ K indicating that changes of the electronic structures are well below the energy resolution of 0.15 eV achieved in our EELS measurements. Finally we mention that, up to now, we have not found any evidence for acoustic plasmons which were discussed in relation with possible pairing mechanisms for high-T_c superconductivity.

3. Character and Symmetry of Charge Carriers

In this section we discuss absorption edges which are due to transitions from narrow core levels into unoccupied states. Since the transitions start from a rather localized core level, the transition matrix element is non-zero only for final states overlapping with the core level. Therefore, the local unoccupied density of states is probed. In Fig. 4 we show typical absorption edges of the O 1s level having binding energies of 528-529 eV relative to the Fermi level. For $La_{2-x}Sr_xCuO_4$ no density of states is observed at the Fermi level for the antiferromagnetic semiconductor with $x=0$ (see Fig. 4a). With increasing dopant concentration x, a peak appears at the Fermi level, the intensity of which is roughly proportional to x [5,6]. Thus we directly measure the density of states of the holes on O sites strongly supporting the more localized model given in the introduction for p-type doped La_2CuO_4 (see Fig. 2b). The steep rise of intensity about 3 eV above E_F is due to La 5d and 4f states hybridized with O 2p states.

Information on the symmetry of the hole states can be obtained from orientation dependent measurements of core-electron excitations on single crystals [7]. Typical spectra are shown in Fig. 4b. For momentum transfer q perpendicular to the CuO_2 layers there is no density of states at the Fermi level (~528 eV) while for q in the CuO_2 plane hole states on O can clearly be seen. This indicates that the holes on O are in orbitals parallel to the CuO_2 planes and that there are no holes in $2p_z$ orbitals perpendicular to the CuO_2 and BiO planes. This clearly rules out all models for high-T_c superconductivity based on out-of-plane π-holes in the CuO_2 planes or $2p_z$ holes on the apex O atoms. These measurements, however, cannot differentiate between O σ-holes (as derived from LDA band-structure calculations) or in-plane π-holes (as derived from cluster calculations [8]). Similar measurements on $YBa_2Cu_3O_7$ single crystals together with calculations of the chemical shift of the individual O sites strongly favour σ holes. Measurements on the Cu $2p_{3/2}$ absorption edge reveal that the symmetry of the unoccupied Cu states (upper Hubbard "band") is predominantly $3d_{x^2-y^2}$. A 10% admixture of probably $3d_{3z^2-r^2}$ states could be detected.

For the electron superconductor, the formation of Cu^+ would be expected. Up to now there was no clear evidence of Cu^+ in Cu $2p_{3/2}$ absorption edges. However, as shown in Fig. 4c there are hole states

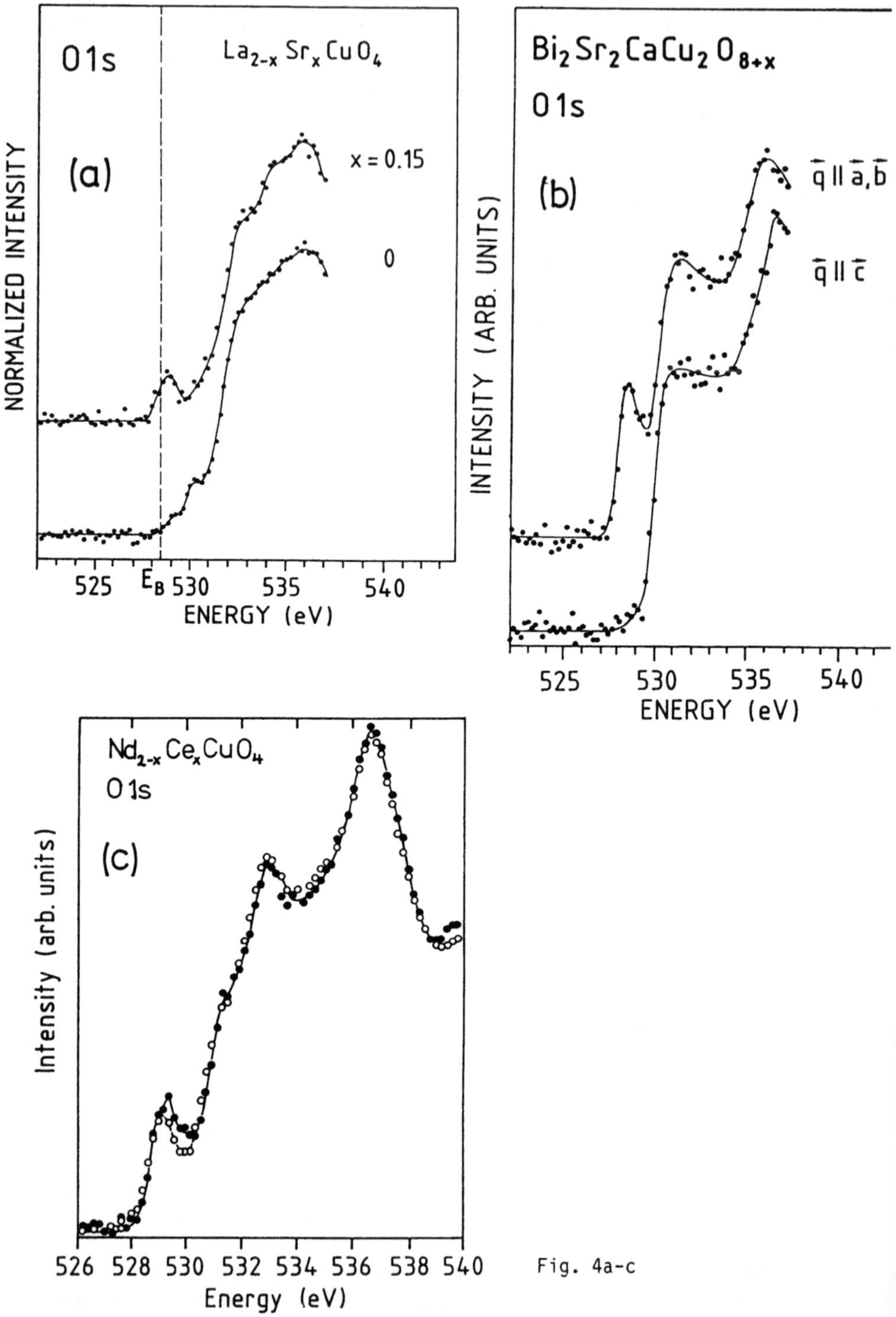
O1s
La$_{2-x}$Sr$_x$CuO$_4$
(a)
x = 0.15
0
NORMALIZED INTENSITY
525
E$_B$ 530
535
540
ENERGY (eV)
Bi$_2$Sr$_2$CaCu$_2$O$_{8+x}$
O1s
(b)
q ∥ a,b
q ∥ c
INTENSITY (ARB. UNITS)
525
530
535
540
ENERGY (eV)
Nd$_{2-x}$Ce$_x$CuO$_4$
O1s
(c)
Intensity (arb. units)
526
528
530
532
534
536
538
540
Energy (eV)

Fig. 4a-c

near the Fermi level (~529 eV) even in undoped Nd_2CuO_4. Upon replacing trivalent Nd by "tetravalent" Ce, i.e., upon n-type doping, the increase of hole states strongly argues against the existence of electron superconductors [9]. Recent experiments on single crystals indicate that the orbitals of the hole states are again predominantly parallel to the CuO_2 planes. More research is necessary to decide, whether electron cuprate superconductors exist or not.

4. Electronic States near the Fermi Level and their Temperature Dependence

In Fig. 5a we show typical angle-resolved photoemission spectra of a $Bi_2Sr_2CaCu_2O_8$ single crystal for various polar angles referred to the surface normal. The measurements were performed using synchrotron radiation with 18 eV photon energy. Similar data have been measured by Takahashi et al. [10]. As will be reported in more detail [11] the bands well below the Fermi level exhibit only a small dispersion. In Fig. 5b we show spectra taken in the vicinity of the Fermi level. There is clear evidence for a band crossing the Fermi level near $\vartheta = 9°$. However, in spite of the rather good energy and angular resolution, 60 meV and 0.5°, respectively, the features indicating the band are rather broad possibly by a reduced life time due to correlation effects. In Fig. 5c similar angle-resolved inverse photoemission data are shown [12]. In the enlarged spectra, one realizes near the Fermi level a sudden increase of the intensity between 30° and 40° indicating again a crossing of a band through the Fermi level. Due to the reduced energy resolution ($\Delta E_{1/2} = 0.6$ eV) in the inverse photoemission data, the maxima related to this band are less pronounced. In Fig. 6a the experimental data of this band (occupied and unoccupied part) are compared with band-structure calculations. The crossing of a flat band through the Fermi level can be realized. This strongly argues against all theories for high-T_c superconductivity not based on Fermi liquids. On the other hand, the measured band (probably a Cu-O band with predominantly O 2p character) exhibits considerable less dispersion than the calculated one. The strong deviation from LDA band-structure calculations may again be related to correlation effects or to strong self-energy corrections of the photoemission process.

Fig. 4 O 1s absorption edges of high-T_c superconductors. (a): La_2CuO_4 undoped and p-type doped with Sr. (b): $Bi_2Sr_2CaCu_2O_8$ single crystal for q parallel and perpendicular to the CuO_2 plane. (c): $Nd_{2-x}Ce_xCuO_4$ "n-type" doped by Ce. Open circles: $x = 0$, closed circles: $x = 0.15$ superconducting sample

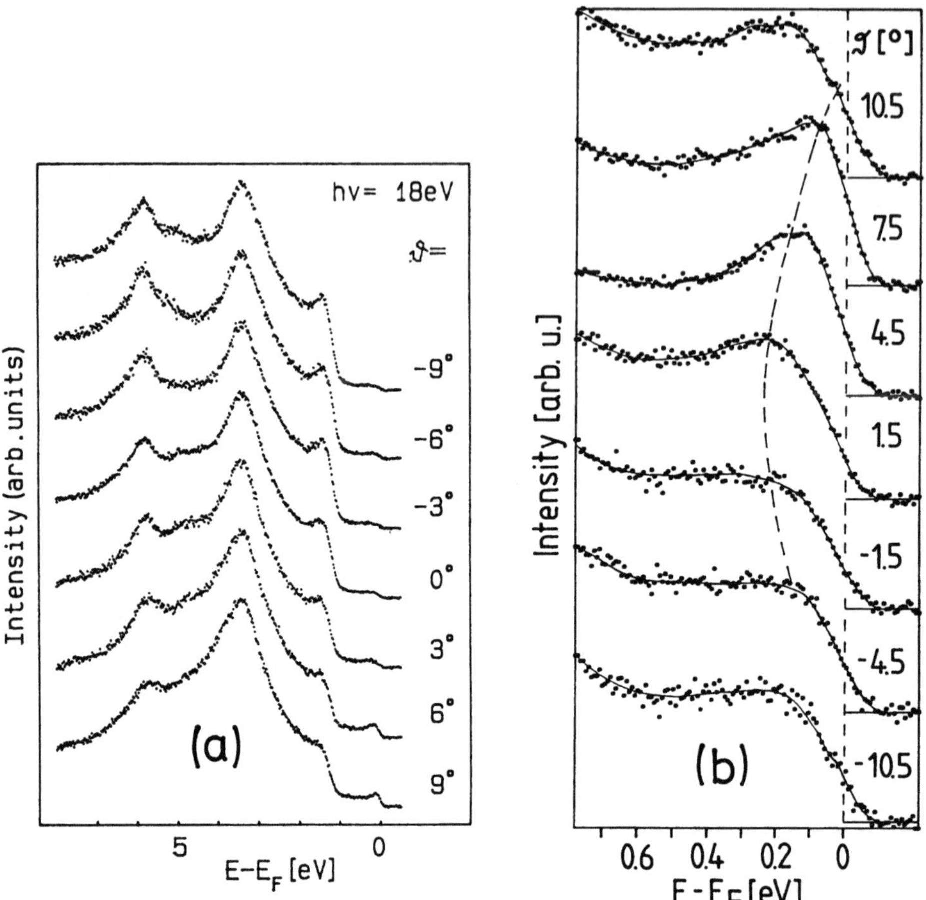

Fig. 5 (a): Angle-resolved photoemission spectra of a $Bi_2Sr_2CaCu_2O_8$ single crystal measured with 18 eV photon energy along the direction ΓX for various polar angles referred to the surface normal. (b): The same but in the vicinity of the Fermi level. (c): Angle-resolved inverse photoemission spectra for various angles of electron incidence along the ΓX direction

In Fig. 6b we show the temperature dependence of an angle-resolved photoemission spectrum of $Bi_2Sr_2CaCu_2O_8$ taken at $\vartheta = 9°$ where the band crosses the Fermi level together with an Au reference spectrum [14]. For $T>T_c$ clear Fermi edges are measured. For $T<T_c$ spectral weight at the Fermi level is shifted to lower energy causing an enhancement of emission at about 80 meV (assigned S). These experimental facts are related with the opening of a superconducting gap $\Delta \sim 30$ meV as derived

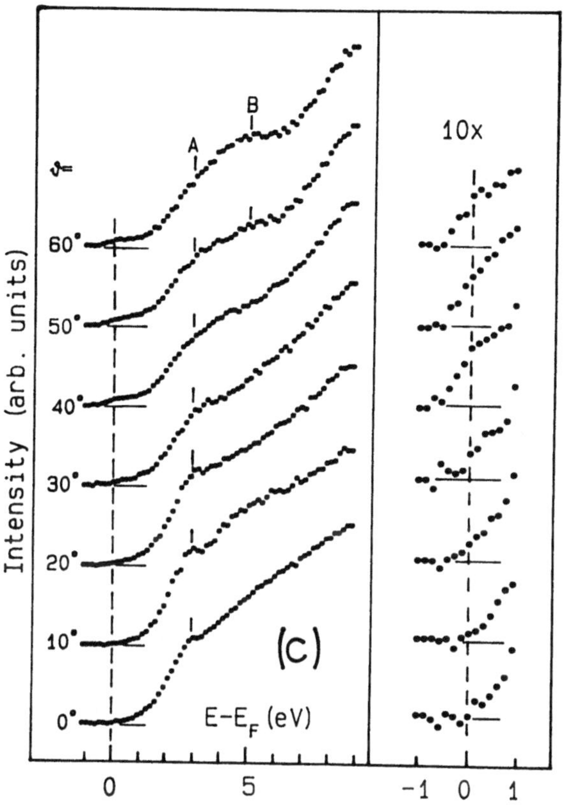

Figure 5c

from the shift of the Fermi level between T = 125 K ($T>T_c$) and T = 63 K ($T<T_c$). The ratio $2\Delta(0)/k_BT_c \sim 8$ is twice as large as the weak coupling value expected from BCS theory indicating that the new high-T_c cuprates are strong-coupling superconductors. Similar values were reported from angle-integrated photoemission measurements [15,16]. The above mentioned measurements [14] have demonstrated for the first time the possibility that changes of the band structure between the normal state and the superconducting state can be measured by angle-resolved photoemission spectroscopy.

References

1. J.G. Bednorz and K.A. Müller, Z. Phys. B **64**, 189 (1986)
2. W. Weber, Phys. Rev. Lett. **58**, 1371 (1987)
3. A. Fujimori and F. Minima, Phys. Rev. B **30**, 967 (1984); J. Zaanen, G.A. Sawatzky, and J.W. Allen, Phys. Rev. Lett. **55**, 418 (1985); S. Hüfner, Z. Phys. B **61**, 135 (1985)

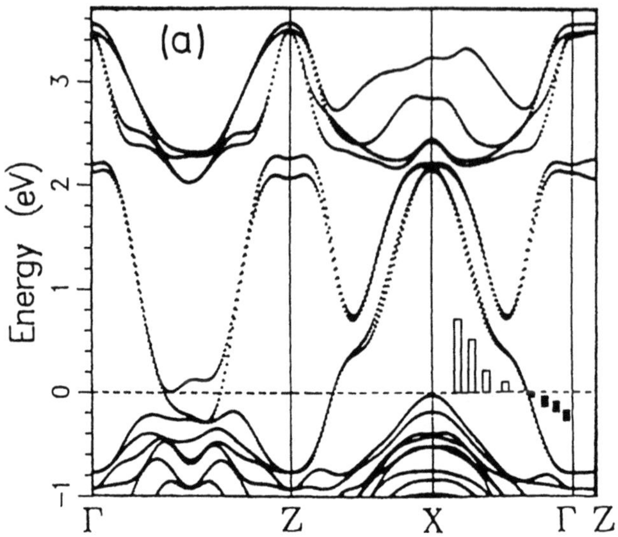

Fig. 6 (a): Band-structure calculation for $Bi_2Sr_2CaCu_2O_8$ (from Pickett and Krakauer [13]). Included are experimental ARUPS and ARIPES values (vertical bars) of a flat band near the Γ point crossing the Fermi level along ΓX. (b): Temperature dependence of the photoemission spectra near E_F of $Bi_2Sr_2CaCu_2O_8$ taken with synchrotron radiation at 18 eV photon energy and 9° emission angle

4. N. Nücker, H. Romberg, S. Nakai, B. Scheerer, J. Fink, Y.F. Yan, and Z.X. Zhao, Phys. Rev. B, in print
5. N. Nücker, J. Fink, B. Renker, D. Ewert, C. Politis, P.J.W. Weijs, and J.C. Fuggle, Z. Phys. B **67**, 9 (1987)
6. N. Nücker, J. Fink, J.C. Fuggle, P.J. Durham, and W.M. Temmerman, Phys. Rev.B **37**, 5158 (1988)
7. N. Nücker, H. Romberg, X.X. Xi, J. Fink, B. Gegenheimer, and Z.X. Zhao, Phys. Rev. B, in print
8. Y. Guo, J.-M. Langlois, and W.A. Goddard III, Science **239**, 896 (1988)
9. N. Nücker, P. Adelmann, M. Alexander, H. Romberg, S. Nakai, J. Fink, H. Rietschel, G. Roth, H. Schmidt, and H. Spille, Z. Phys. B, Rapid Note, submitted
10. T. Takahashi, H. Matsuyama, H. Katayama-Yoshida, Y. Okabe, S. Hosoya, K. Seki, H. Fujimoto, M. Sato, and H. Inokuchi, Phys. Rev. B, in print
11. R. Manzke et al., to be published
12. R. Claessen, R. Manzke, H. Carstensen, B. Burandt, T. Buslaps, M. Skibowski, and J. Fink, Phys. Rev. B, in print

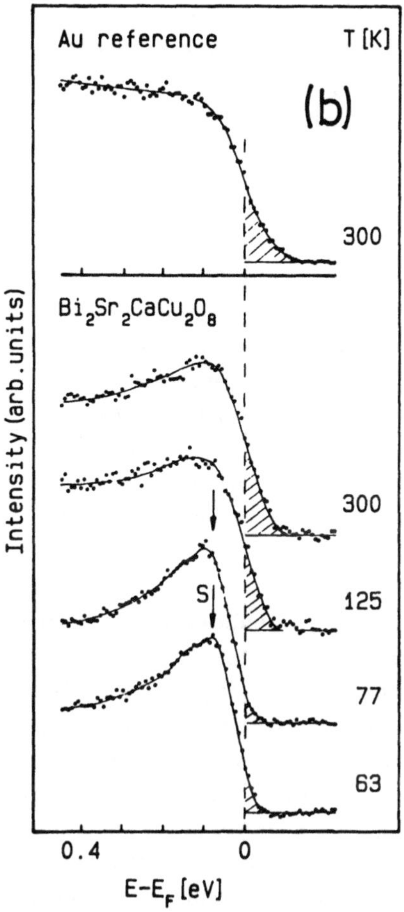

Figure 6b

13. H. Krakauer and W.E. Pickett, Phys. Rev. Lett. **60**, 1665 (1988)
14. R. Manzke, T. Buslaps, R. Claessen, and J. Fink, Europhys. Lett., in print
15. J.-M. Imer, F. Patthey, B. Dardel, W.-D. Schneider, Y. Baer, Y. Petroff, and A. Zettl, Phys. Rev. Lett. **62**, 336 (1989)
16. Y. Chang, Ming Tang, R. Zanoni, M. Onellion, R. Joynt, D.L. Huber, G. Margaritondo, P.A. Morris, W.A. Bonner, J.M. Tarascon, and N.G. Stoffel, Phys. Rev. B **39**, 4740 (1989)

Infrared Properties of High-T_c $YBa_2Cu_3O_7$ Films

K. Kamarás [1;*], *S.L. Herr* [1], *C.D. Porter* [1], *D.B. Tanner* [1], *T. Venkatesan* [2], *E. Chase* [2], *A. Inam* [2], *H.D. Woo* [2], *M.S. Hegde* [2], *B. Datta* [2], *and S. Etemad* [2]

[1]University of Florida, Gainesville, FL 32611, USA
[2]Bell Communication Research, Red Bank, NJ 07701, USA

Abstract: We have measured the reflectance of high-quality ($T_c \approx 88$ K and $\Delta T_c \approx 0.5$ K) $YBa_2Cu_3O_7$ films prepared by pulsed laser deposition. The optical experiments, carried out at temperatures between 15 and 300 K, covered 50–40,000 cm^{-1} (6 meV–5 eV). Analysis of the normal-state spectra in terms of a low frequency Drude contribution and a strong mid-infrared absorption yields Drude parameters corresponding to the "clean limit" ($1/\tau < 2\Delta$ or $\ell > \xi$). At the lowest temperatures and within the experimental error of ±1%, the reflectance is 100% below 150 cm^{-1}. However, the 150 cm^{-1} onset of absorption is *not* the superconducting energy gap.

The energy gap is a key quantity for a superconductor, indicating the minimum energy required for creating an excitation from the ground state. According to BCS theory, the energy gap scales with transition temperature: $2\Delta/k_BT_c = 3.5$. Actual superconductors do not adhere strictly to this relation; however, most elemental superconductors have gap values in a relatively narrow range between the weak-coupling value of $2\Delta/k_BT_c = 3.5$ and the strong-coupling value (e.g., Pb or Hg) of $2\Delta/k_BT_c = 4.5$. Far infrared measurements were important in confirming the presence of the gap in ordinary superconductors and in determining the value of $2\Delta/k_BT_c$. Thus, with the discovery of the high T_c superconductors many attempts were made to measure the energy gap by far infrared means.[1] The results were rather varied, with gap values reported in the range $1.6 < 2\Delta/k_BT_c < 8$.

In studies of $YBa_2Cu_3O_{7-\delta}$ crystals and films, two different features have been identified as the gap. First, a shoulder in the reflectance at 400–500 cm^{-1} (6–8k_BT_c) which is accentuated below T_c has been assigned to the gap by a number of workers.[2,3,4] However, Kramers-Kronig analysis and model calculations[5] have demonstrated that this maximum can be explained by a "plasmon model."[6,7] According to this model, the Drude contribution present in the normal-state $\sigma_1(\omega)$ collapses to a delta function in the superconducting state; the reflectance in this case is governed by dispersive effects as much as by absorption. The key point is that the high T_c superconductors are in the clean limit for superconductivity, with ($1/\tau < 2\Delta$ or $\ell > \xi$). Thus, most of the normal-

* Permanent address: Central Research Institute for Physics, Budapest, Hungary.

Springer Series in Solid-State Sciences, Vol. 91
Electronic Properties of Conjugated Polymers III
Editors: H. Kuzmany · M. Mehring · S. Roth

state oscillator strength is at frequencies below the gap, leaving little or no oscillator strength available for the gap transition.

Perhaps the strongest case for the observation of a superconducting gap by infrared techniques has been the work by Thomas *et al.*[8] who observe a region of 100% reflectance below 150 cm^{-1} in an oxygen deficient ($T_c \approx 68$ K) crystal. The onset of absorption at 150 cm^{-1} ($3.5 k_B T_c$) was identified as the gap. A similar feature in a $T_c \approx 90$ K film at 150 cm^{-1} (and thus at $1.9 k_B T_c$) has recently also been associated with a gap.[4]

In this paper we present results of reflectance studies of $YBa_2Cu_3O_7$ films. An onset of absorption is observed at 150 cm^{-1}. However, in our interpretation, this feature is *not* the superconducting energy gap. Three reasons can be given for this statement: (1) The 150 cm^{-1} feature is present in the normal-state spectra; (2) it does not shift with changes in T_c which result from oxygen concentration changes; and, (3) it occurs at a substantially higher energy (300 cm^{-1}) in $T_c \approx 85$ K $Bi_2Sr_2CaCu_2O_8$ crystals.[9] Instead of the gap, it is the onset of the temperature-independent mid-infrared excitation which is known to occur in these materials.[10,11,5]

The films were produced by pulsed laser deposition from a composite target onto a $SrTiO_3$ substrate. The crystallographic c-axis is normal to the film . Electron microscopy shows an extremely smooth featureless surface and high density. The optical quality of the surface is outstanding. The high density is evidenced by the fact that although the films are only about 2000 Å in thickness, no phonon features associated with the $SrTiO_3$ could be discerned in the spectra.

Although the transition temperature was depressed, $T_c \approx 88$ K, the transition was extremely sharp, with $\Delta T_c \approx 0.5$ K. The temperature dependence of the normal-state resistivity decreased linearly from 200 $\mu\Omega$-cm at 300 K to 70 at 100 K; the extrapolated residual resistivity is approximately zero.

Optical reflectance measurements were made with a Bruker IFS-113v interferometer in the far infrared and infrared and a grating monochromator in the near infrared and visible. Kramers-Kronig analysis gave the frequency-dependent conductivity and dielectric function.

Fig. 1 shows $\sigma_1(\omega)$ for a film at three temperatures: 6, 100, and 300 K. Note the strong low-frequency temperature dependence, in agreement with the dc conductivity. (The magnitude of the far infrared conductivity is only about 60% of the conductivity from dc resistance, however. This is attributed to systematic errors in both measurements.) At higher frequencies there is little or no temperature dependence. The normal state data can be fit with a dielectric function of the form[1,11,5]

$$\epsilon(\omega) = -\frac{\omega_{pD}^2}{\omega^2 + i\omega/\tau} + \frac{\omega_{pe}^2}{\omega_e^2 - \omega^2 - i\omega\gamma_e} + \epsilon_\infty \quad . \tag{1}$$

There are three contributions to this dielectric function: (1) Drude, characterized by a plasma frequency ω_{pD} and a relaxation rate $1/\tau$; (2) mid-infrared, having a strength

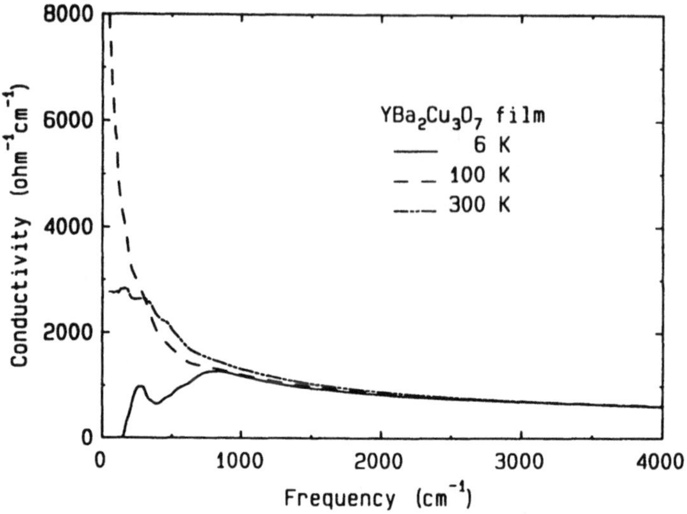

Fig. 1. Frequency-dependent conductivity in the ab-plane for a $YBa_2Cu_3O_7$ film

ω_{pe}, a center frequency ω_e, and a width γ_e; and (3) a high frequency part, with a constant ϵ_∞. Phonon features, strong in ceramic samples [11] are present but very weak in the original data. Fits to the reflectance give parameters similar to those found previously; in particular, $\omega_{pD} = 9200\ \mathrm{cm}^{-1}$ (1.1 eV) and $1/\tau$ decreases linearly from 500 cm^{-1} at 300 K to 160 cm^{-1} at 100 K. Note that using the BCS expression gives $3.5k_BT_c = 220\ \mathrm{cm}^{-1}$, so even at 100 K the scattering rate is below the weak-coupling energy gap.

In the superconducting state a perfect (delta function) dc conductivity develops; in turn there is a large ac inductive response. We model this with the following dielectric function:

$$\epsilon(\omega) = -\frac{\omega_{ps}^2}{\omega^2} + i\frac{\pi\omega_{ps}^2}{2\omega}\delta(\omega) + \frac{\omega_{pe}^2}{\omega_e^2 - \omega^2 - i\omega\gamma_e} + \epsilon_\infty \ . \qquad (2)$$

Here, ω_{ps} is the plasma frequency or oscillator strength of the superfluid electrons, and is proportional to $\sqrt{n_s}$, where $n_s \approx n$ is the superfluid electron density.

The normal-state Drude contribution has been replaced in Eq. (2) by the London-model dielectric function of the superconducting electrons. The first term represents the finite-frequency inertial response while the second is the infinite dc conductivity of the superconductor. The remaining terms are identical to those in Eq. (1). At frequencies low compared to ω_e, the real part of the dielectric function takes on a simple form:

$$\epsilon_1(\omega) = \epsilon_b - \frac{\omega_{ps}^2}{\omega^2} \qquad (3)$$

where $\epsilon_b = \epsilon_\infty + \omega_{pe}^2/\omega_e^2$ is the contribution of bound electrons to the low-frequency dielectric constant.

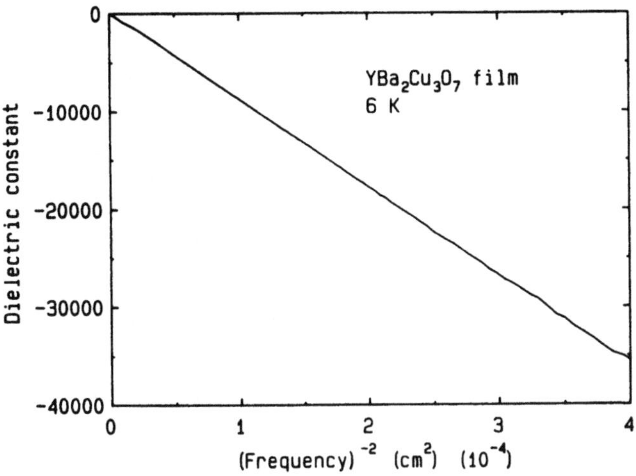

Fig. 2. Real dielectric function at 6 K of a $YBa_2Cu_3O_7$ film, plotted versus $1/\omega^2$. The frequency range spanned by the data is approximately 300 cm^{-1} to 50 cm^{-1}

Eq. (3) suggests that a plot of $\epsilon_1(w)$ versus $1/\omega^2$ should be a straight line with slope ω_{ps}^2 and intercept ϵ_b. Fig. 2 shows such a plot; the slope gives $\omega_{ps} = 9100$ cm^{-1} (in complete agreement with the fit to Drude model) and the intercept gives $\epsilon_b \approx 250$, in accord with the contribution of the mid infrared absorption to ϵ_1.

Returning now to the 6 K $\sigma_1(\omega)$ data of Figure 1, we note the strong onset at 150 cm^{-1}. This is the feature attributed to the gap by Thomas *et al.*[8] and Schützmann *et al.*[4] However, a shoulder is visible in the conductivity in both the 100 K and the 300 K data of Fig. 1. (In the original reflectance it is seen as a small drop in reflectance above 150 cm^{-1}.) Thus it is unlikely to be associated with the *superconducting* energy gap. Further evidence in support of this view comes from the fact that the edge does not shift with T_c, occurring at the same energy in $T_c = 68$ K[8] and $T_c = 88$ K material. If it were the superconducting gap, one would expect a decrease in gap with decrease in transition temperature, keeping $2\Delta/k_BT_c$ constant. Finally in $Bi_2Sr_2CaCu_2O_8$ samples, which have superconducting properties *very* similar to $YBa_2Cu_3O_7$, a similar feature is observed[9] at 300 cm^{-1}, almost exactly twice the energy where it occurs in $YBa_2Cu_3O_{7-\delta}$. It is unreasonable to expect the value of $2\Delta/k_BT_c$ to differ by a factor of two in these two materials which have transition temperatues of 85 K and 88K. Thus, instead of the superconducting gap, the 150 cm^{-1} onset appears to be the beginning of the strong mid-infrared electronic excitation seen in all the high T_c superconductors.[10]

Work at Florida supported by DARPA contract MDA972-88-J-1006.

References

1. For a review of work up to October 1988 see Thomas Timusk and David B. Tanner in *Physical Properties of High Temperature Superconductors,* D.M. Ginsberg, editor. (World Scientific, Singapore, 1989)

2. R.T. Collins, Z. Schlesinger, R.H. Koch, R.B. Laibowitz, T.S. Plaskett, P. Freitas, W.J. Gallagher, R.L. Sandstrom, and T.R. Dinger, *Phys. Rev. Lett.* **59**, 704 (1987).

3. Z. Schlesinger, R.T. Collins, D.L. Kaiser, and F. Holtzberg, *Phys. Rev. Lett.* **59**, 1958 (1987).

4. J. Schützmann, W. Ose, J. Keller, K.F. Renk, B. Roas, L. Schultz, and G. Saemann-Ischenko, *Europhys. Lett.* (1989).

5. T. Timusk, S.L. Herr, K. Kamarás, C.D. Porter, D.B. Tanner, D.A. Bonn, J.D. Garrett, C.V. Stager, J.E. Greedan, and M. Reedyk, *Phys. Rev. B* **38**, 6683 (1988).

6. D.A. Bonn, J.E. Greedan, C.V. Stager, T. Timusk, M.G. Doss, S.L. Herr, K. Kamarás, C.D. Porter, D.B. Tanner, J.M. Tarascon, W.R. McKinnon, and L.H. Greene, *Phys. Rev. B* **35**, 8843 (1987).

7. M.S. Sherwin, P.L. Richards, and A. Zettl, *Phys. Rev. B* **37**, 1587 (1988).

8. G.A. Thomas, J. Orenstein, D.H. Rapkine, M. Capizzi, A.J. Millis, L.F. Schneemeyer, and J.V. Waszczak, *Phys. Rev. Lett.* **61**, 1313 (1988).

9. M. Reedyk, D.A. Bonn, J.D. Garrett, J.E. Greedan, C.V. Stager, T. Timusk, K. Kamarás, and D.B. Tanner, *Phys. Rev. B* **38**, 11981 (1988).

10. K. Kamarás, C.D. Porter, M.G. Doss, S.L. Herr, D.B. Tanner, D.A. Bonn, J.E. Greedan, A.H. O'Reilly, C.V. Stager, and T. Timusk, *Phys. Rev. Lett.* **59**, 919 (1987).

11. D.A. Bonn, A.H. O'Reilly, J.E. Greedan, C.V. Stager, T. Timusk, K. Kamarás, and D.B. Tanner, *Phys. Rev. B* **37**, 1547 (1988).

Coexisting Superconducting Phases in Bi-Sr-Ca-Cu-O Ceramics: New Features of Microwave Absorption

I.I. Khairullin [1], *A.A. Zakhidov* [1], *P.K. Khabibullaev* [1], *Z. Iqbal* [2], *and R.H. Baughman* [2]

[1]Department of Thermal Physics, Uzbek SSR Academy of Sciences, Gazeta Pravda St. 28, Chilanzar C, Tashkent, SU-700135, USSR
[2]Corporate Technology, Allied-Signal Inc., Morristown, NJ 07962, USA

Abstract. The low-field signal (LFS) of microwave absorption is studied in BiSrCaCuO samples in which two superconducting (SC) onset temperatures of 87K and 112K have been found earlier by resistivity. LFS intensity demonstrates features at 60K, 87K, 112K and 118K, indicating existence of two more SC phases: one with higher Tc=118K SC, and one with lower Tc=60K; the last is found inside zero-resistivity percolated SC matrix. The relative content of these four SC phases is estimated. LFS is proved to be highly sensitive probe for discovery of negligible SC fractions with different Tc within insulating and even superconducting multiphase matrices.

1. Introduction

Since the discovery of high temperature superconductivity in LaBaCuO and YBaCuO ceramics, a series of papers have been devoted to quite unusual broad microwave absorption signal appearing in external magnetic field in superconducting (SC) state and known as low-field signal (LFS)[1-6]. Though the nature of LFS is not quite clear yet (number of mechanisms proposed [7,8]) the LFS was found to be a good noncontact probe for determination of SC onset temperature Tc, and also sensitive to external influences [3,6]: e.g. it was recently proposed to use it as a detector of magnetic field [9]. We have found recently that in YBaCuO LFS at zero H-field has a sharp peak called a resonance component [6] that is pronounced at low temperatures T~5 K and is very sensitive to small H-variations and other perturbations [6].

In present paper we analyse the new features that arise in LFS -spectra in multiphased BiSrCaCuO samples in which a small fraction (~1%) of superconducting phase with Tc=112 K was found by resistivity and magnetometry measurements [11] in the background matrix with lower SC Tc=87 K. We demonstrate here that LFS in this sample allows us to detect two more SC phases with Tc=118K and Tc=60K and we compare LFS behaviour with those in standard 1-2-3 YBaCuO.

Springer Series in Solid-State Sciences, Vol. 91
Electronic Properties of Conjugated Polymers III
Editors: H. Kuzmany · M. Mehring · S. Roth

2. Experimental

Preparation of the BiSrCaCuO samples is described in [11], here we used so called B samples [11] that were prepared by high temperature synthesis (at T=845°-870° C). The samples of average size 1 x 0.5 x 0.3 mm were placed in quartz tubes and sealed in the ambient athmosphere. The spectra of microwave absorption were carried out on X-band BRUKER ESP-300 spectrometer at 9.3 GHz, and up to 100 kHz field modulation frequency. The He gas-flow "Oxford instruments ESR-900 cryostat" allowed temperature variation from 5K to 300K with ± 0.2K precision. Additionally equipped external dc-Helmholz coils allowed H-field scanning from -50 to +50 G. The spectra of low-field-signal (LFS) were recorded after cooling the samples in external zero magnetic static field (zero-field-cooling - ZFC) to avoid the magnetization effects due to flux trapping in superconducting phase.

3. Results and discussion

On cooling the BiSrCaCuO samples from 300 to 118 K no changes in resonator adjustment are found. At 118K the broad signal first appears ranging from H=0 to several kG fields with a wide maximum $H_o(T)$ as shown in Fig.1. The position of maximum H (T) shifts to lower fields with decrease of T , while the width of this peak is decreased (Fig.2). This behaviour is opposite to the similar LFS data in YBaCuO ceramics, in which the position of LFS maxima is found to be either temperature independent, or shifts towards higher fields, while the peak is broadened on decrease of T [8].

At Tc =112K the new narrow peak abruptly appears in the LFS spectrum the position of which $H_o(T)$ first shifts to higher fields, and then goes towards zero fields. In the range from

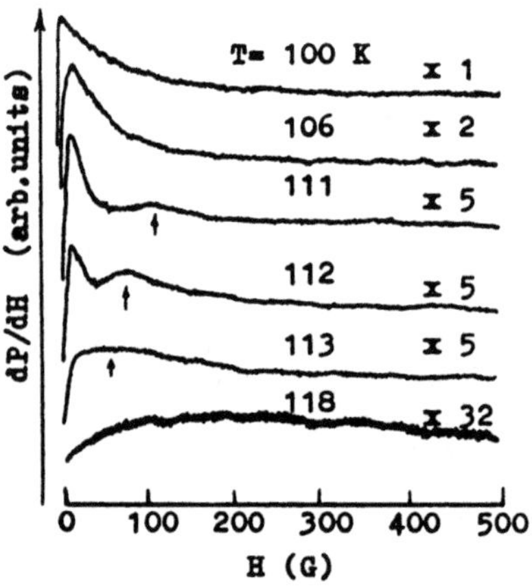

Fig.1 The LFS-dependence on magnetic field at various temperatures for BiSrCaCuO. Arrows show $H_0(T)$ for 118K-LFS. Coexistence of 112K and 118K peaks is clearly seen

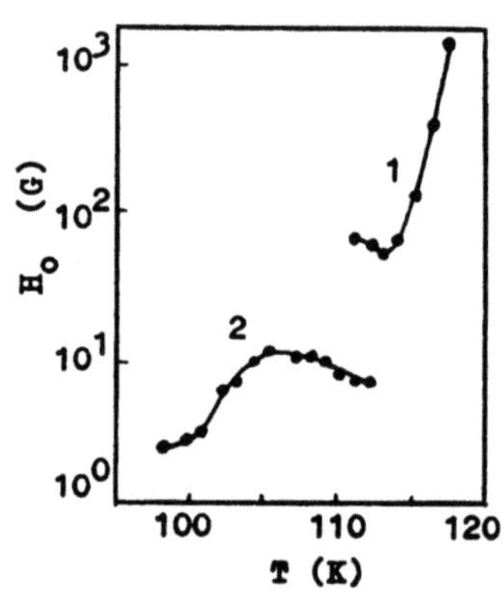

Fig.2 LFS peaks position $H_o(T)$ for 1- wide 118K-LFS and 2- narrow 112K-LFS

110 to 112K the previous broad 118K LFS is still observed rather clearly, so that LFS spectrum consists of superposition of two coexisting signals (Fig. 2). In the inset to Fig.3a the intensity of each signal in this range is shown separately. Further cooling leads to increase of narrow 112K-LFS intensity without any change in its position while the broad 118K-LFS completely disappears as a separate maximum. On cooling till 5K the intensity of LFS remains increasing but at Tc=87K and Tc=60K two local sharp maxima are found at dP/dH(T) curve (Fig.3a). The non-uniform 'behaviour of dP/dH(T), is significantly different both qualitatively and quantitatively from that observed in YBaCuO [1-6] and shown for comparison in Fig.3b.: 1) contrary to YBaCuO with only one SC onset Tc~80K, we have four critical temperatures in BiSrCaCuO samples. Taking into account the results of independent AC resistivity and magnetic moment measurements in the same BiSrCaCuO sample that have identified Tc =112K and Tc =87K superconducting phases we may conclude that two new 60K and 118K SC phases are clearly observed in LFS. Note that 60 K SC was observed in BiSrCaCuO "A" samples prepared by low temperature synthesis[11]; and here we identify this phase also in B-samples. 2) LFS in BiSrCaCuO does not saturate till 5 K, contrary to the situation in YBaCuO (see Fig.3b). 3) The third difference is quantitative: LFS in BiSrCaCuO is significantly less intense then from the same size sample of YBaCuO and its integral value is only about 5-10% of that in YBaCuO. Comparing the form of dP/dH(T) with the microstructural results [11] it could be speculated that the volume fractions of 4 SC phases are distributed in the following order: (40-50%)60K > (20-30%)87K > > (1%)112K > (0,05-0,1%)118K.

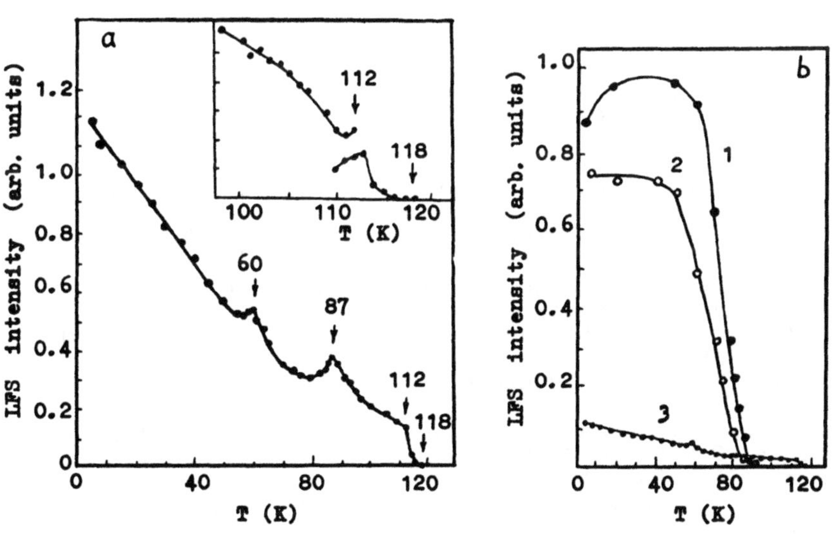

Fig.3 LFS versus T: a) LFS in BiSrCaCuO with four Tc indicated by arrows; b) Comparison of LFS in YBaCuO at air (1) and in vacuum (2), in BiSrCaCuO (3)

4. Conclusion

The study of LFS microwave absorption in multiphase BiSrCaCuO samples allowed us to discover two more superconducting phases with Tc =118K and Tc =60K (in addition to two known phases with Tc=87 K and Tc=112 K) and proved two new useful features:
1) LFS allows one to detect negligibly small fractions of superconducting material with highest Tc inside multidispersed complex matrix , which could not be detected by resistivity (due to absence of percolation). 2) LFS also allows one to detect the fraction of superconducting phase with lower Tc_f inside the superconducting network matrix having higher Tc_m> Tc (due to nonhomogeneous nature and complicated topology of such matrix). Noteworthy that resistivity of such matrix is strictly zero at Tc , so resistivity method is of no use to detect fractions with $Tc_f < Tc_m$ while LFS appeared to work even in this case. More recent microstructural studies [12] have identified the 112K phase as a 3-copper layer compound of ideal composition $Bi_2 Sr_2Ca_2Cu_3O_{10}$ which grows within 87K matrix of ordered crystals of idealized 2-copper composition $Bi_2Sr_2Ca Cu_2O_8$, while the 60K phase corresponds to crystals disordered by intergrowths of a 1-copper layer phase $Bi_2(Sr,Ca)_2 Cu O_6$. The microstructure of 118K phase is to be identified but it is clear from our LFS data (Fig.2) that 112K and 118K grains are separated in spatially independent networks. It could be concluded that LFS-method seems to be the most appropriate probe in the search for new materials with higher Tc, since in the process of synthesis this new superconducting phases are expected to appear first as small fractions (inside nonhomogeneous materials) that could hardly be detected by other methods.

5. References

1. R.Durny, J.Hautala, S.Ducharme, B.Lee, O.G.Symko, P.C.Taylor and D.J.Zheng: Phys.Rev.B 36, 2361 (1987).
2. K.W.Blazey, K.A.Müller, J.G.Bednorz, W.Berlinger, G.Amoretti, E.Buluggie, A. Vera and F.C.Matacotta: Phys.Rev.B. 36, 7241 (1987).
3. S.V.Bhatt, P.Ganguly, T.V.Ramakrishnan and C.N.R.Rao: J.Phys.C 20, L559 (1987).
4. M.D.Sastry, A.G.I.Dalvi, Y.Babu, R.M.Kadam, J.V.Yakhmi and R.M.Iyer: Nature 330, 49 (1987).
5. M.Peric, B.Rakvin, M.Prester, N.Banacavic and A.Dulcic: Phys.Rev.B 37, 522 (1988).
6. I.I.Khairullin, A.A.Zakhidov, P.K.Khabibullaev and L.S.Grigoryan: Synthetic Metals 27, 2/3, F541 (1989).
7. K.W.Blazey, A.M.Portis and J.G.Bednorz: Solid State Commun. 65, 1153 (1988).
8. Y.Maniwa, A.Grupp, F.Hentsch and M.Mehring: Physica C 156 755 (1988)
10. M.D.Sastry, R.M.Kadam, Y.Babu, A.G.I. Dalvi, I.K.Gopalakrishnan, P.V.P.S.S.Sastry, G.M.Phatak and R.M.Iyer: J.Phys.C 21, L607 (1988).
9. S.Tyagi, M.Barsoum, R.V.Rao, Yu.Z.Skumryev and I.L.Costa: Physica C 156, 73 (1988)
11. Z.Iqbal, H.Echardt, F.Reidinger, A.Bose, J.C.Barry and B.L.Ramakrishna: Phys.Rev.B 38 (1988) 859.
12. B.L.Ramakrishna,J.C.Barry, Z.Iqbal,E.W.Ong, A.Bose and H.Echardt, Physica C- in press.

Part XIII

Applications

Polypyrrole Supercapacitors

D. Naegele

Polymer Research Division, BASF Aktiengesellschaft,
D-6700 Ludwigshafen, Fed. Rep. of Germany

Abstract

Conventional DLCs (double layer capacitors) employ the charge storage capability of the electrochemical double layer at the interface between electrolyte and carbons with extremely high specific surface area. A new type of "capacitor" with similar characteristics, but with even higher charge storage capability, could be built using the potential/charge-density characteristics of intrinsically conductive polymers like Polypyrrole.

1. Conductive Polymers as Electrodes for Energy-Storage

Intrinsic conductive polymers like polypyrrole possess a unique combination of properties, which makes them very useful as active electrode materials for rechargeable batteries:

- they are rendered highly electronically conductive by an electrochemical oxidation (or reduction) reaction. This conductivity serves for charge transport from the site of the redox reaction to the current collector.
- in some cases (e.g. polypyrrole) this redox reaction can be performed highly reversible, a prerequisite for a rechargeable electrode of an electrochemical energy storage device.
- during the charging process polycations are formed. For reasons of charge neutrality negative counterions have to diffuse into the electrode material. This is favored by increased ionic conductivity of the charged electrode material, possessing ion exchange character.
- the flexibility of these polymeric electrode materials serves to accomodate internal stresses during the charging/discharging reactions.

2. Performance of Polypyrrole as Electrode Material for Rechargeable Energy Storage Devices

To exclude the influence of the counter electrode, the behavior of polypyrrole electrode materials was studied in half cells, using a Ag/Ag^+ reference electrode [1]. At most, 40% of the pyrrole units in the polymer could carry a unit charge. After 1000 galvanostatic deep discharges the capacity drops by only 20%. Self discharge is around 10% per month in the system studied (Polypyrrole, 0.5 m $LiClO_4$ in propylene

Springer Series in Solid-State Sciences, Vol. 91
Electronic Properties of Conjugated Polymers III
Editors: H. Kuzmany · M. Mehring · S. Roth

carbonate). The kinetic behavior is highly dependent on the morphology of the electrode material being used.

3. Rechargeable Polypyrrole-Lithium Batteries

Such a battery has been developed in cooperation between VARTA-AG and BASF-AG. It contains polypyrrole as the positive electrode and a Lithium alloy as the negative electrode. Charging and discharging reactions are shown schematically in Fig. 1.

The first battery of this kind, that has been presented to the public by VARTA-AG in November 1988, is a button cell of size 2025, a capacity of 5 mAh and a nominal voltage of 3 V.

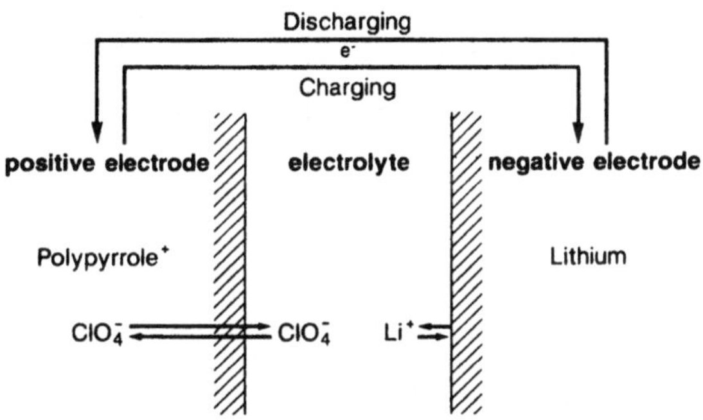

Fig. 1: Electrode reactions of a rechargeable polypyrrole-lithium battery

4. "Polypyrrole Supercapacitors"

This electrochemical cell contains two polymerelectrodes (in contrast to the polypyrrole-lithium battery, which has one), has a charge-voltage relationship typical for capacitors and capacities in the range of 1 Farad. Such capacities are much higher than those for ordinary capacitors and fall in the range of double layer capacitors. The principle of such a cell is shown in Fig. 2.

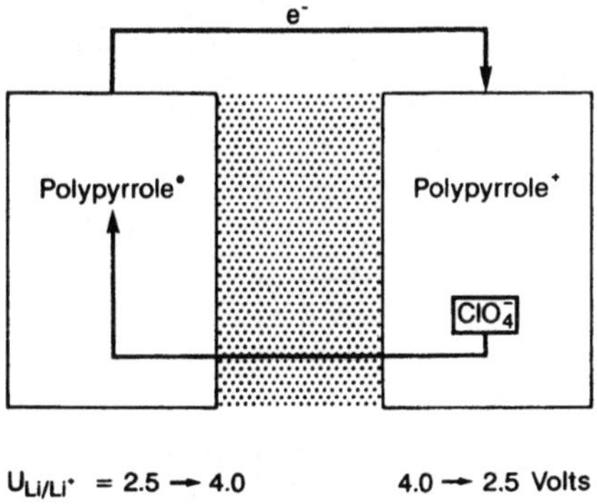

Fig. 2: Electrode reactions of a polypyrrole-supercapacitor

The cell voltage of 1.5 V originates from the fact that one of the electrodes is in the completely discharged state (Potential 2.5 V vers. Li/Li^{+}) and the other in the completely charged state (Potential 4.0 Volts). During the discharge process of the cell the positive charges on the polymer change sides and together with them the negative counter ions.

This means, that there is no change in ion-concentration in the electrolyte during discharge (or charge), and as a consequence this cell system requires - in contrast to the polymer battery - only a minimum amount of electrolyte. For this reason the distance between the electrodes can be extremely small; the cell thickness is virtually twice the thickness of an electrode. The schematics of this voltage-time behavior of such a cell are shown in Fig. 3a.

Fig. 3a represents the change in potential of the two electrodes, whereas Fig. 3b represents the sum of the two, giving the cell voltage.

As soon as one of the electrodes is completely discharged, its potential drops instantly. This means, that the cell-voltage increases instantly. In other words: applying too high a charging voltage to the cell does not result in a too high potential of one of the electrodes, which might damage the polymeric electrode material. One might call this behavior of the cell "overcharge-safe".

To verify these considerations, such a cell was built, the experimental behavior being shown in Fig. 3b'. The shape of the curve is nearly the same as predicted. It resembles very much the behavior of a double layer capacitor, although the charge storage mechanism is different: instead of onthe surface of activated carbon, the charges are stored in the interior of a bulk material. Its structure can be considered as a highly disordered intercalation compound [2]. The charge capacity of such a material [1] is about three times higher than the capacity of the surface of activated carbon materials [3].

Two prototypes of this kind of cell were built as shown in Fig.4: a button cell of 2.5 mm thickness and a

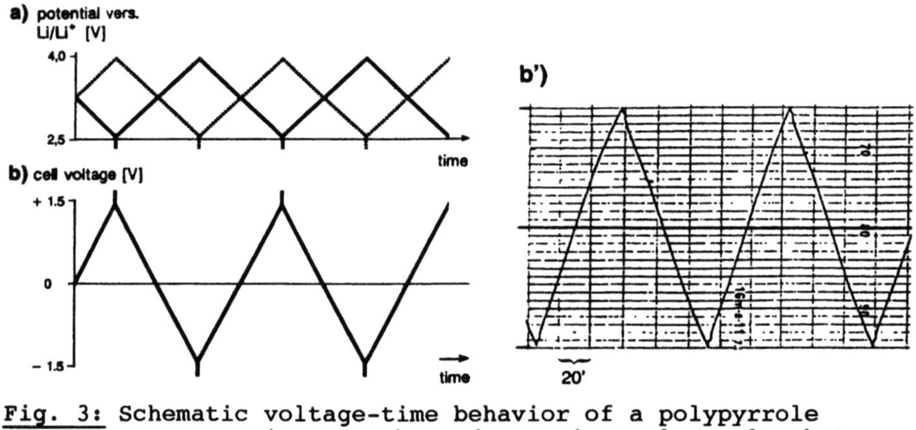

Fig. 3: Schematic voltage-time behavior of a polypyrrole supercapacitor (a)(b) and experimental result (b')

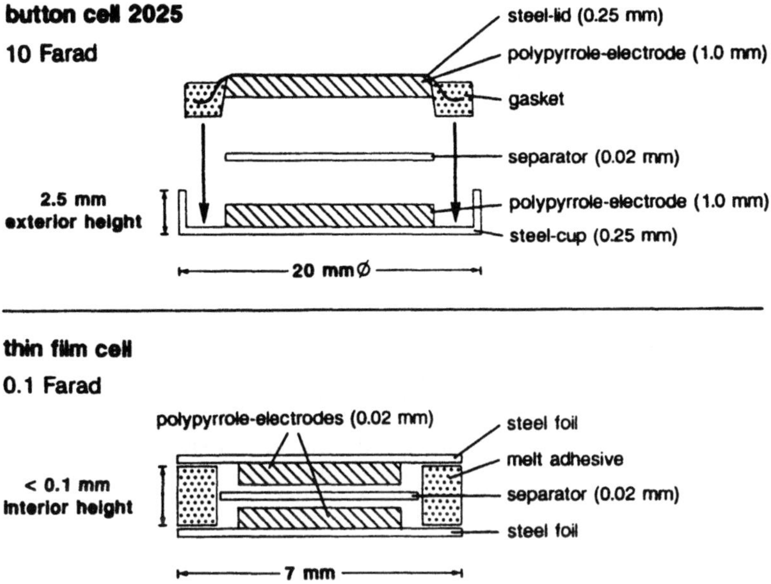

Fig. 4: Cell design of polypyrrole supercapacitors

capacity of 10 Farad and an extremely thin cell using 20 µm thin flexible polypyrrole film. This ultra thin design might be useful for integration of this cell into semiconductor devices.

Acknowledgement

The author is pleased to acknowledge P. Neese for synthesizing the polymers and R. Bittihn of VARTA for stimulating discussions. This work was supported by the Bundesministerium für Forschung und Technologie (03E-8444-A). The author is solely responsible for the content of this publication.

Reference

[1] D. Naegele, R. Bittihn
Solid state Ionics 28-30 (1988) 983-989

[2] L.J. Buckley, D.K. Roylance, G.E. Wnek
J. Polym. Sci., Part B: Polym. Phys., 25 (10), 2179-88 (1987)

[3] M. Nawa, T. Nogami, H. Mikawa
J. Electrochem. Soc. 131 (6), 1457-59 (1984)

Enzyme Entrapment in Conducting Organic Polymers: Application in Amperometric Biosensors

C.R. Lowe [1], *N.C. Foulds, S.E. Evans, and B.F.Y. Yon Hin*

Institute of Biotechnology, University of Cambridge, Downing Street, Cambridge CB2 3EF, UK

Abstract

The entrapment of enzymes in conducting polymers provides a facile means for ensuring proximity between enzyme and electrode with considerable potential for biosensor construction. Preliminary studies have provided evidence that glucose oxidase and a number of other enzymes may be incorporated into conducting polymer films electrochemically deposited on appropriate electrodes. Polypyrrole-entrapped glucose oxidase electrodes respond rapidly to glucose and reach steady state within 20-40s. The technique is particularly amenable to the localisation of enzymes on small or defined electrode geometries such as those fabricated on monolithic devices.

1. Introduction

The last few years have seen unprecedented interest in the development and exploitation of analytical devices for the detection, quantitation and monitoring of specific chemical analytes. The collection of chemical intelligence in real time on the concentrations of ions, gases, metabolites, proteins and hormones is likely to have a significant impact on the health care, agricultural, environmental, security, food, pharmaceutical and fermentation industries [1-4]. Such analytical devices which combine the specificity and sensitivity of biological systems with appropriate electrochemical techniques are often referred to as biosensors [5]. Enzymatic redox reactions are particularly amenable to interfacing with electrochemical transducers since electron exchange is a key step in their natural cycle.

Conventionally, the approach has been to monitor the concentration of one of the products or reactants in the natural enzyme reaction and thereby, indirectly, the concentration of the enzyme substrate [6]. However, the past five years have seen the development of a variety of new strategies in which electron transfer to and from redox enzymes is coupled more closely to the underlying electrode. Such approaches centre on the development of soluble redox mediators as electron shuttles [7], chemically modified electrodes that promote adsorption and orientation of the enzyme on the electrode surface [8] and on electron relay systems covalently bound to the surface of the enzyme [9]. In our work, we have concentrated on techniques to immobilise the enzymes that promote proximity between the enzyme active site and the conducting surface of the electrode.

---

[1] - To whom correspondence should be addressed

Springer Series in Solid-State Sciences, Vol. 91
Electronic Properties of Conjugated Polymers III
Editors: H. Kuzmany · M. Mehring · S. Roth

2. Enzyme Entrapment in Conducting Organic Polymers

The entrapment of enzymes in conducting polymer films provides a facile and controllable method for the deposition of biologically active molecules to defined areas on electrodes [10]. The technique permits the amount of material deposited at the electroactive surface to be readily controlled. Typically, a solution of the electropolymerisable monomer, pyrrole [10], N-methyl pyrrole [11], aniline [12] or phenol [6], in aqueous buffer containing the enzyme is used for the deposition process. The polymer film is then grown either potentiostatically or galvanostatically at a sufficiently oxidising potential to promote formation of the conducting polymer film. Whilst several conducting polymer systems have now been exploited [6, 10-12], polypyrrole [10,13] is probably the most suitable for this application since it is one of the few conducting polymers that is acceptably stable in ambient conditions [14] and can be readily prepared electrochemically from a variety of electrolytes, including aqueous solutions [15]. The polymerisation occurs by electrochemical oxidation of a solution containing pyrrole monomers. An extremely reactive π-radical cation is believed to produce a polymer that is predominantly α,α'-coupled, although some branching of the polymer chains may occur through β-coupling [16]. The resulting polymer displays a net positive charge and incorporates counter-anions from solution during the film growth process. The chemical, physical and morphological properties of the film depend upon factors such as pH, charge passed, electrode potential and the anion incorporated.

Preliminary data has shown that Aspergillus niger glucose oxidase can be incorporated into polypyrrole films deposited on Pt electrodes printed on ceramic substrata [10]. The films were grown from an aqueous solution containing 0-0.13μM enzyme and 0.2M pyrrole in 0.1M potassium phosphate pH 7.0 using a three-electrode cell at a constant potential of +0.8V (vs. Ag/AgCl) for 2.5h. Under these conditions up to 125 mU of glucose oxidase activity could be incorporated per cm^2 of electrode surface. Scanning electron microscopy of the dark brownish-black film of oxidised pyrrole reveals an extremely porous morphology with typical pore sizes up to a few μm. However, when the polypyrrole film is grown in the presence of glucose oxidase the gross morphology of the film alters to a more fibrillar form. It is possible that the topology of the growing surface of the polypyrrole film may be influenced by the incorporation of glucose oxidase as a macromolecular counter-anion [18].

The steady-state responses at +0.7V (vs. Ag/AgCl) in the polypyrrole-immobilised glucose oxidase electrode to the addition of varying amounts of glucose have been employed to construct conventional hyperbolic calibration curves [10]. The maximum current response at infinite substrate concentration for an electrode containing 100mU activity per cm^2 was 6.4μA with an apparent Michaelis-Menten (K_m') constant of 31mM. The latter value is close to that for soluble glucose oxidase from Aspergillus niger. The storage stability of the electrode in 0.1M potassium phosphate buffer pH 7.0 was >21 days at 4° C. This work demonstrates the feasibility of using 'one-step' electrode construction techniques which exploit facile and easily controlled enzyme-immobilisation regimes. These conclusions have subsequently been confirmed by other laboratories [11-13] using several different conducting polymer systems [6], although, interestingly, the detection limits for glucose for the various polymers are markedly different. These observations may reflect differences in the diffusion coefficients and partition coefficients of the reactants and products in the enzyme kinetics [6].

3. Mediated Systems

In the preliminary studies, the enzyme-catalysed oxidation of glucose has been followed by the electrochemical detection of hydrogen peroxide at the surface of the underlying Pt electrode according to the scheme

$$\beta\text{-D-glucose} + \text{GOD(FAD)} \rightarrow \beta\text{-D-gluconolactone} + \text{GOD(FADH}_2\text{)}$$
$$\text{GOD(FADH}_2\text{)} + O_2 \rightarrow \text{GOD(FAD)} + H_2O_2$$

where the flavoprotein (FAD) enzyme glucose oxidase (GOD) becomes reduced in the process and the reduced enzyme (GOD($FADH_2$)) is reoxidised by the natural electron acceptor, oxygen, to generate hydrogen peroxide. However, to measure glucose in whole blood, the test should be independent of oxygen concentration and ideally be performed at less anodic potentials than those required to oxidise hydrogen peroxide in order to obviate interferences from other electrochemically-active contaminants present in whole blood. Mediators such as ferrocene and its derivatives exhibit redox potentials (+100-400 mV vs. SCE) in a range where there is minimal interference from blood components and are well established as electron transfer agents between the reduced form of glucose oxidase and electrode surfaces [7,11,19].

Since it has already been established that glucose oxidase can be successfully entrapped in polypyrrole films [10,13] and that ferrocene analogues can act as effective mediators, it seemed reasonable to propose that the enzyme could be entrapped in ferrocene-containing polypyrrole films to construct a reagentless glucose electrode. Thus, N-substituted ferrocene-pyrrole conjugates containing redox-active side chains designed to accept electrons from the reduced form of the enzyme have been designed and synthesised [20] (Fig.1). Such ferrocene-pyrrole conjugates were efficient oxidants of reduced glucose oxidase and could be anodically polymerised into redox active films either per se or as copolymers with pyrrole in aqueous electrolytes.

Fig.1. The structure of a ferrocene-pyrrole conjugate:[(ferrocenyl) amidopropyl] pyrrole (FAPP)

Furthermore, glucose oxidase could be entrapped in the redox copolymer FAPP/pyrrole from a solution containing sodium phosphate buffer pH 5.6 (10mM), sodium perchlorate (0.1M), pyrrole (1mM), FAPP (1mM) and enzyme (200U/ml) by cycling the working electrode between O and 0.8V at $5mV.s^{-1}$. The redox polymer-entrapped glucose oxidase electrode was operated at +0.3V (vs Ag/AgCl) and the steady state current response recorded following additions of glucose resulting in final cell concentrations of 1-100mM. However, whilst a conventional calibration curve for glucose was noted [20], some sensitivity to oxygen was noted, presumably because of the relatively slow reaction of reduced glucose oxidase with the immobilised mediator compared to the competing natural electron acceptor. Nevertheless, despite problems associated with oxygen sensitivity and long term stability, the technique of co-deposition of enzyme-mediator systems warrants further investigation as a means of producing prototype reagentless sensors in one simple anodic electrodeposition step [20].

4. Multi-Analyte Microelectronic Devices

There is mounting interest in the fabrication of low cost miniature biosensors for clinically relevant analytes which exploit monolithic silicon technology. The novel entrapment procedure described above [10- 20] is ideally suited to the deposition of enzymes on small area metal electrodes. Recent work in our laboratory has demonstrated the feasibility of depositing a number of oxidases on microelectrodes [21]. The microelectronic devices were fabricated by Plessey Research Caswell Ltd. and comprised five pairs of serpentined and interdigitated gold electrodes (1000-3500nm thick) deposited over Ti and Pt metal layers (~ 100nm thick) on a silicon "chip" mounted on a ceramic wafer. The microelectrodes were bonded to pads located at the periphery of the wafer and comprised two large electrodes (~ 500 x 500 μm each) bonded in series and used as a counter electrode and three smaller electrodes (200 x 500 μm). The middle pair of electrodes were converted into an Ag/AgCl reference electrode whilst the two outer electrodes were used to deposit electrochemically glucose oxidase and a second oxidase, galactose oxidase or α-glycerophosphate oxidase respectively. The oxidase-entrapped polypyrrole film was electrochemically synthesised on the gold working electrodes by poising it at a fixed potential of +0.7V (vs.Ag/AgCl) in N_2-purged 0.1M phosphate buffer pH 7.0 containing 10mM potassium perchlorate, 0.1M pyrrole and 2mg/ml enzyme at 25° C. Glucose and α-glycero-phosphate were monitored by following the oxidation of H_2O_2 in phosphate buffer at the respective enzyme electrodes held at a fixed potential of +0.7V (vs.Ag/AgCl). Typically, a steady state current response was obtained within 30-60s following the addition of glucose or α-glycero-phosphate. Fig.2 shows the steady state response of the dual analyte microelectronic sensor. These data show the successful entrapment of flavoprotein oxidases in conducting polymers at a microelectrode surface.

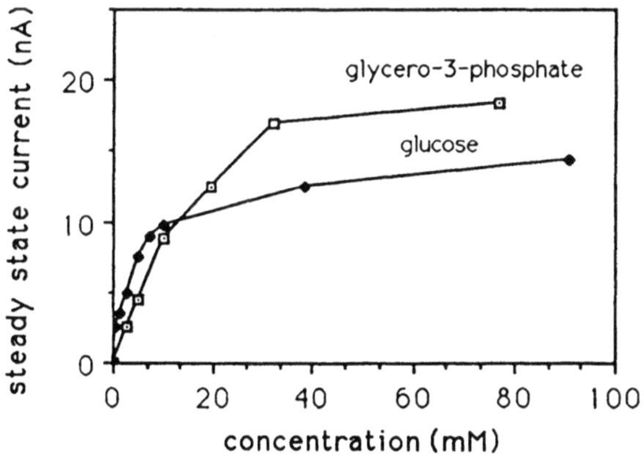

Fig.2 Steady state current responses of the glucose oxidase/α-glycero-posphate oxidase microamperometric sensor to glucose and α-glycerophosphate

5. Conclusions

The electroentrapment of flavoprotein oxidases and mediators in conducting polymer films shows great promise for the construction of amperometric enzyme electrodes. In particular, the technique not only allows the electrodeposition of complete catalytic systems but also offers the precision required to localise enzymes at microelectrodes. The latter can be exploited for the fabrication of multi-analyte miniature biosensors.

6. References

[1] Lowe, C.R.(1984) Trends in Biotechnology, 2, 59-64

[2] Lowe, C.R.(1985) Biosensors 1, 3-16

[3] Foulds, N.C.& Lowe, C.R.(1985) Bioessays 3, 129-132

[4] Higgins, I.J.& Lowe, C.R.(1987) Phil.Trans.R.Soc.Lond. B316, 3-11

[5] Lowe, C.R.(1989) Phil.Trans.R.Soc.Lond., In Press

[6] Bartlett, P.N.& Whitaker, R.G.(1987/88) Biosensors 3, 359-379

[7] Green, M.J.& Hill, H.A.O.(1986) J.Chem.Soc.,Faraday Trans. 82, 1237-1243

[8] Narashimhan, K. & Wingard, L.B.Jr.(1986) Anal.Chem.58, 2984-2989

[9] Degani, Y. & Heller, A.(1987) J.Phys.Chem. 91, 1285-1289

[10] Foulds, N.C. & Lowe, C.R. (1986) J.Chem.Soc., Faraday Trans. 82, 1259-1264

[11] Bartlett, P.N. & Whitaker, R.G. (1987) J.Electroanal.Chem. 224, 37-48

[12] Shinohara, H., Chiba, T. & Aizawa, M. (1988) Sensors & Actuators, 13, 79-86

[13] Umana, M. & Waller, J. (1986) Anal.Chem. 58, 2979-2983

[14] Diaz, A.F. & Hall, B. (1983) IBM J.Res.Dev. 27(4), 342

[15] Asavapiriyanot, S., Chandler, G.K., Gunawardena, G.A. & Pletcher, D. (1984) J.Electroanal.Chem. 177, 229-244

[16] Diaz, A. (1981) Chem.Scripta 17, 145-148

[17] Diaz, A., Castillo, J.I., Logan, J.A. & Lee, W-Y. (1981) J.Electroanal.Chem. 129, 115-132

[18] Salmon, M., Diaz, A.F., Logan, A.J., Krounbi, M. & Bargon, J.(1982) Mol.Cryst.Liq.Cryst. 83, 265-276

[19] Cass, A.E.G., Davis, G., Francis, G.D., Hill, H.A.O., Aston, W.J., Higgins, I.J., Plotkin, E.V., Scott, L.D.L. & Turner A.P.F. (1984) Anal.Chem. 56, 667-671

[20] Foulds, N.C. & Lowe, C.R. (1988) Anal.Chem. 60, 2473-2478

[21] Yon Hin, B.F.Y., Sethi, R.S. & Lowe, C.R. (1989) Sensors & Actuators, submitted for publication

Polymeric Semiconductor Devices

J.H. Burroughes, C.A. Jones, R.A. Lawrence, and R.H. Friend

Cavendish Laboratory, Madingley Road, Cambridge CB3 0HE, UK

Abstract We have been able to fabricate various semiconductor devices using Durham-route polyacetylene as the active semiconductor. The Metal-Insulator-Semiconductor (MIS) structure and the MISFET are of particular interest because, with the formation of charge accumulation and inversion layers, charge can be injected into the semiconductor without the need for chemical doping. Solitons are the lowest energy electronic excitations of polyacetylene, and we show here the various spectroscopic indications that the active charges in these structures have the character of charged solitons.

1. Introduction

Despite the considerable progress in the past decade that has been made in the understanding of the electronic properties of conjugated polymers, there has been relatively little work reported on their use as the active component in semiconductor device structures. There are several reasons for this; the most important of which is that most conjugated polymers cannot be conveniently processed to the forms required in these devices. Most conjugated polymers are not readily soluble in easily-handled solvents, and are infusible. This has severely limited the scope for construction of devices [1,2]. Improvements in the processing of conjugated polymers, through electrochemical deposition during polymerisation, and the use of solution-processible poly(3-alkyl thiophenes) has allowed better device fabrication [3,4], and the use of model oligomers which can be deposited by vacuum sublimation has recently been reported [5]. There are also reports of field-induced conductivity measured in MISFET (Metal-Insulator-Semiconductor Field Effect Transistor) structures [6-9].

We show here some of the results obtained using polyacetylene prepared by the Durham precursor polymer route [10]. The precursor polymer is soluble in solvents such as 2-butanone and can be spin-coated onto the required substrate to produce a very uniform film. As formed the polyacetylene is extrinsically p-doped (10^{16} cm^{-3}). The origin of this doping is not established, though it may result from catalyst residues at chain ends formed during the polymerisation to the precursor polymer. It does, however, provide a convenient level of doping for use in unipolar devices, and we find that the dopants are not mobile on the time scale of the measurements made. We have made an extensive series of measurements on a variety

Springer Series in Solid-State Sciences, Vol. 91
Electronic Properties of Conjugated Polymers III
Editors: H. Kuzmany · M. Mehring · S. Roth

of devices [11,12] and we find that the device performance can be much improved over earlier work with conjugated polymers, for example we obtain rectification ratios for the aluminium/polyacetylene Schottky barrier of up to 500,000.

The fundamental excitations of the polyacetylene chain are not the electrons and holes characteristic of inorganic, three-dimensionally bonded semiconductors, but are self-localised coupled electron-lattice excitations which have the topological characteristics of solitary waves, or solitons [13,14]. Our principal motivation in carrying out this study of the device physics of polyacetylene was to find out how the relaxation of injected charges to form charged solitons controls the device operation and distinguishes this class of polymeric semiconductor devices from traditional structures. Besides the range of electrical measurements that characterise the device performance, we have carried out an extensive range of optical measurements, since these give a clear and unambiguous characterisation of the electronic and vibrational modes of the soliton.

2. Properties of Polyacetylene MIS and MISFET Structures

MIS and MISFET structures have been fabricated by spin-coating a thin film of Durham precursor polymer onto, usually, the insulator layer of a preprepared structure, followed by thermal conversion to polyacetylene. For the devices discussed here, the insulator used is silicon dioxide, which was grown as native oxide on suitably prepared silicon substrates. For the MISFET structure, no further processing is required, as shown in figure 1; for the MIS structure a top, ohmic, layer of semitransparent gold was deposited by evaporation.

The MIS structure allows the possibility of band bending at the insulator-semiconductor interface through the Fermi level to produce a surface charge layer which may be the same carrier sign as the majority carriers (accumulation layer) or as the minority carriers (inversion layer). These formation of accumulation, depletion and inversion layers may be demonstrated through the behaviour of the device

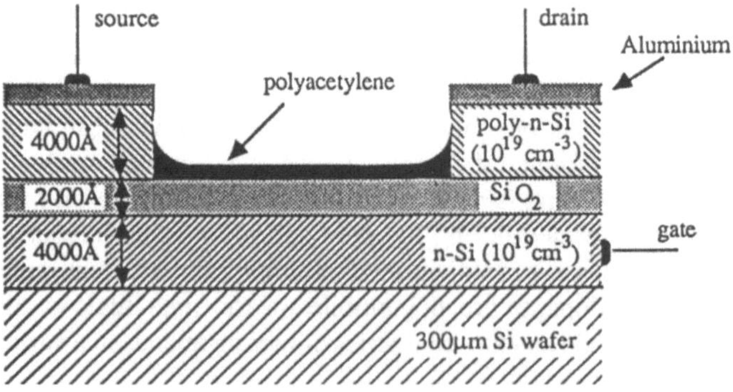

Figure 1 Polyacetylene MISFET. The channel length is 1.5 m

capacitance with respect to the bias voltage. The measured capacitance, C, is that of the series combination of the insulator capacitance, C_i, and the capacitance of the active region of the semiconductor, C_d, and is given by $C = C_iC_d/(C_i+C_d)$. Since C_d is large for the accumulation and inversion layers, C is equal to the geometric capacitance of the insulating layer, but C falls to a lower value for movement of the depletion layer. The capacitance and the conductance versus bias voltage curves for a polyacetylene MIS structure are shown in figure 2. As expected, the capacitance for negative gate voltages flattens out to the geometric capacitance of the insulator, indicating the formation of an accumulation layer. The decrease in the measured capacitance for positive voltages displays the formation of the depletion layer. Saturation of the capacitance for large positive biases sets in when the depletion layer extends across the polyacetylene film (estimated to be about 1200Å for this device).

Demonstration that the charges at the polyacetylene/insulator interface in the MIS structure are mobile is made in a MISFET structure, in which source and drain contacts allow measurement of the conductance of the surface charge layer. The MISFET shown in figure 1 is constructed with n-silicon source and drain contacts, which are expected to make ohmic contacts to the n-polyacetylene layer. We thus expect to find enhanced channel conductance from formation of an n-type inversion layer for positive gate voltages. Figure 3 shows the variation of I_{DS} with V_{GS} at constant V_{DS}. The channel conductance has a minimum at V_{GS} = +10V (full depletion) and rises for gate voltages both more positive (inversion layer) and more negative (accumulation layer), with a maximum on/off ratio for this structure of 100,000 (V_{GS} = -40V to V_{GS} = +10V). We see, therefore, the behaviour expected for positive bias, though the strong enhancement of the conductivity for negative bias indicates that the accumulation layer is still able to make good contact to the source and drain electrodes. Carrier mobilities are low, typically 10^{-4} $cm^2V^{-1}s^{-1}$, and similar to those found for photogenerated carriers and the extrinsic carriers in the

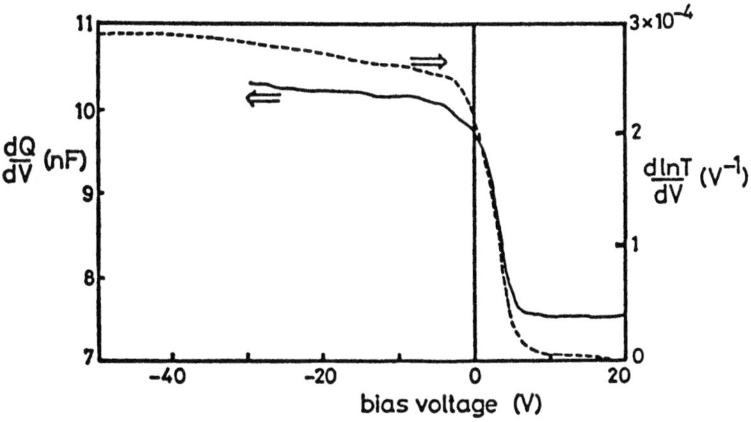

Figure 2 The differential optical transmission, ∂ln(T)/∂V at 0.8 eV (dashed), and differential capacitance, ∂Q/∂V (solid) versus bias voltage for the MIS structure

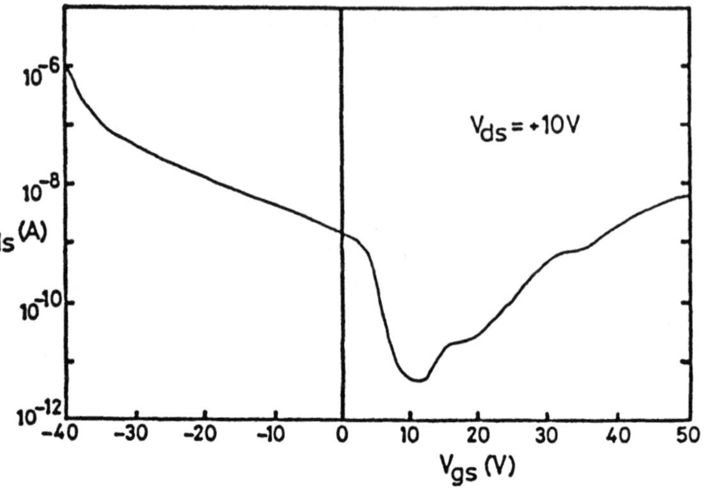

Figure 3 I_{DS} versus V_{GS} at constant V_{DS} (+10V) for a MISFET as shown in figure 1

as-made polyacetylene [15] and we consider that the mobility-limiting process is the transfer of charge between chains.

As discussed in §1, charge is stored in polyacetylene in 'soliton' localised states, which have non-bonding p_z 'mid-gap' states associated with them. We expect to see, therefore, a modulation in the optical properties of the active semiconductor region of these devices with applied bias voltage. For the MIS and MISFET structures we expect a decrease in the device transmission below the band-edge as the device is driven towards accumulation, and new charged solitons are introduced onto the polyacetylene chains at the interface with the insulator. The electromodulation spectrum between 0.4 and 1.2 eV in figure 4 shows that there is a decrease in the

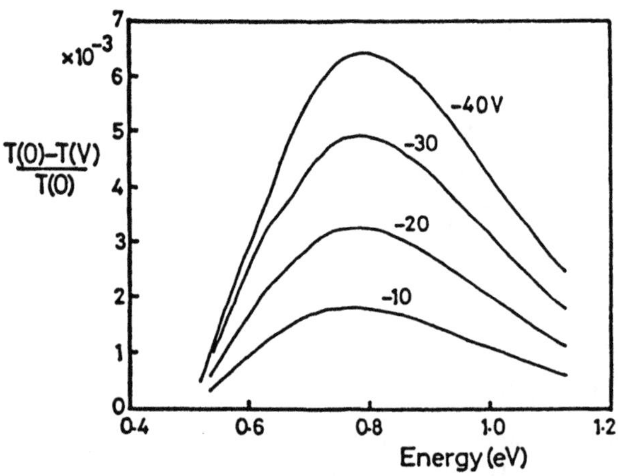

Figure 4 Voltage-modulated transmission for a MIS diode versus photon energy

transmission through the device for negative bias. The peak value of ΔT/T is at 0.8 eV. If all the charge at the polymer-insulator interface is stored in soliton-like states then ∂lnT/∂V should scale with ∂Q/∂V, the differential capacitance, as the bias voltage is varied. This is shown to be the case in figure 2, in which ∂lnT/∂V and the differential capacitance are both plotted against bias voltage. The ratio of ∂lnT/∂V and ∂Q/∂V gives the optical cross-section per charge injected, and in the accumulation regime we find a value at the peak in the 'mid-gap' absorption as 1.2 x 10^{-15} cm^2, in very close agreement with the value measured from dopant-induced absorption [16].

Besides the electronic transitions associated with the soliton state, there is extra IR activity due to new vibrational modes which couple to motion of the charged soliton along the polymer chain. These are the translation modes of the soliton, and are seen in the polyacetylene accumulation layer at 1379 and 1281 cm^{-1} (the lowest mode is obscured by absorption by the silicon substrate) [11,12]. There are additional vibrational modes of the soliton, including the width-modulating amplitude mode which is Raman-active. We report observation of this mode elsewhere in these proceedings [17].

References

1. P. M. Grant, T. Tani, W. D. Gill, M. Krounbi and T. C. Clarke, *J. Appl. Phys.* **52**, 869 (1980)
2. J. Kanicki in "*Handbook on Conducting Polymers*", T. J. Skotheim, ed. (Marcel Dekker, New York) (1986), pp 544-660.
3. F. Garnier and G. Horowitz, *Synthetic Metals* **18**, 693 (1987)
4. H. Tomozawa, D. Braun, S. Phillips, A. J. Heeger and H. Kroemer, *Synthetic Metals* **22**, 63 (1987); *ibid.* **28** 687 (1989)
5. F. Garnier, G. Horovitz and D. Fichou,*Synthetic Metals* **28**, 705 (1989).
6. E. Ebisawa, T Kurokawa and S. Nara, *J. Appl. Phys.* **54**, 3255 (1983)
7. H. Koezuka, A. Tsumura and T. Ando, *Synthetic Metals* **18**, 699 (1987)
8. H. Koezuka and A. Tsumara, *Synthe6ic Metals* **28**, 753 (1989).
9. A. Assadi, C. Svensson, M. Willander and O. Inganäs, *Appl. Phys. Lett.* **53**, 195 (1988).
10. J. H. Edwards and W. J. Feast, *Polymer Commun.* **21**, 595 (1980).
11. J. H. Burroughes, C. A. Jones and R. H. Friend, *Nature* **335**, 137 (1988).
12. J. H. Burroughes, C. A. Jones and R. H. Friend, *Synthetic Metals* **28**, 735 (1989).
13. W. P. Su, J. R. Schrieffer and Heeger, A. J. *Phys. Rev. Lett.* **42**, 1698 (1979); *Phys. Rev.* B**22**, 2099 (1980); Err. B**28**, 1138 (1983).
14. M. J. Rice, *Phys. Lett.* **71**A, 152 (1979).
15. P. D. Townsend and R. H. Friend, Synthetic Metals **28**, 735 (1989).
16. J. Orenstein in "*Handbook on Conducting Polymers*", T. J. Skotheim, ed. (Marcel Dekker, New York) (1986), pp 1297-1396.
17. R. A. Lawrence, J. H. Buroughes and R. H. Friend, these proceedings.

Anisotropically Conducting Polyolefine/CT Complex Composites

A. Tracz[1], *E.El. Shafee*[1], *J. Ulanski*[2], *J.K. Jeszka*[1], *and M. Kryszewski*[1]

[1]Centre of Molecular and Macromolecular Studies, Polish Academy of Sciences, Sienkiewicza 112, PL-90-362, Lodz, Poland

[2]Polymer Institute, Technical University of Lodz, Zwirki 36, PL-90-924 Lodz, Poland

Abstract. Preparation and properties of anisotropic, conductive polyolefine/charge-transfer complex composites consisting of the conductive network of the CT complex crystallites penetrating the polymer matrix are presented and discussed. The investigated composites are prepared using a modification of the reticulate doping technique the so-called zone casting. The influence of the casting parameters on the morphology and electrical properties is discussed. It is shown that the samples obtained under optimum conditions show high anisotropy of morphology and of the electrical conductivity.

1. Introduction

One of the most interesting features of composites is the possibility of a controlled, specific organization of different phases forming the composite. In recent years big progress has been made in the field of conducting polymer composites obtained by using charge-transfer (CT) complexes using so-called reticulate doping [eg.1-3]. Conductive networks in such composites consist of the CT complex microcrystals formed during film casting. These materials when prepared by the standard casting technique show isotropic conductivity.

The technique of zone casting developed in our laboratory allows to prepare anisotropically conductive polymer films with CT complexes [4]. We have shown that using this method one can obtain films of amorphous polymer (polycarbonate) containing nearly parallelly oriented needle-like structures of crystalline tetrathiotetracene-tetracyanoquinodimethane (TTT-TCNQ) complex [5]. Due to anisotropic organization of crystalline structures of the CT complex (conducting phase) these composite materials show anisotropy of conductivity (i.e the ratio of the conductivities in the directions parallel to and perpendicular to the conducting paths) as high as 10^8.

In this paper we present the results of application of the zone casting technique for semicrystalline polyolefines: polyethylene (PE) and polypropylene (PP), containing 1wt.% of TTT-TCNQ CT complex.

Springer Series in Solid-State Sciences, Vol. 91
Electronic Properties of Conjugated Polymers III
Editors: H. Kuzmany · M. Mehring · S. Roth

2. Experimental

The zone casting technique used for obtaining anisotropically conducting films of semicrystalline was partially described in [6]. It consists in casting of a polymer solution containing dissolved CT complex with the zone evaporation of the solvent. The continuity of the film formation and the stability of the CT complex crystallization conditions is assured by continuous supplying of the solution to the evaporation zone, accompanied by simultaneous moving of the support, on which the film is formed, with a constant velocity v. The solution forms a meniscus between a flat nozzle and the moving solidified film. Solvent evaporation from the surface of the meniscus results in a gradient of the polymer and of the CT complex concentrations which determines the CT complex crystals growth. Under appropriate conditions the rate of spontaneous nucleation in the narrow zone Z of the meniscus, where the solution of the CT complex is supersaturated (fig.1).

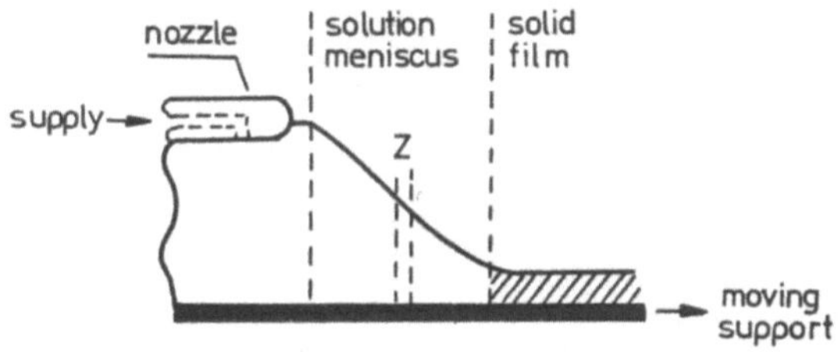

Fig. 1. Schematic presentation of the principle of the zone casting technique

The crystallites grow continuously with the velocity v in the direction opposite to the movement of the support. The main parameters which for a given system (polymer + solvent + CT complex) determine the conditions for oriented growth of the CT complex crystals are: casting temperature, velocity of the support and CT complex concentration.

The films about 25mm thick and 3.5cm wide were obtained from solutions of PE (Lupolen,1810S) or PP (Moplen,C-30G) in o-dichlorobenzene (5g of polymer in 100ml of the solvent) containing 0.05g of the dissolved TTT-TCNQ CT complex (with the donor to acceptor molar ratio 1:2).

The measurements of the DC electrical conductivity of the samples were performed on 2x20mm strips cut out parallel to and perpendicular to the crystals' growth direction.

3. Results and Discussion

In the case of semicrystalline polymers the casting conditions should be carefully selected because also the polymer may crystallize in the solution. The continuous, oriented growth of the CT complex crystals is possible only when the crystallization of the CT complex starts earlier than that of the polymer (i.e. the CT complex crystallization front is before the crystallization front of the polymer). In the opposite case the continuity and orientation of growing CT complex crystals is poor because the solution is not uniform due to randomly growing polymer spherulites. The morphologies of the

PE and of the PP films showing the highest anisotropy of the electrical conductivity obtained for optimalized parameters i.e. casting temperature, velocity and CT complex concentration are shown in fig.2.

Fig.2. Morphology of the crystalline CT complex in polyolefine polymer films showing high anisotropy of conductivity obtained from solution in o-dichlorobenzene: a/ PE +1wt.% of TTT-TCNQ; T= 115°C, v= 0.02 μm/sec, b/ PP +1wt.% of TTT-TCNQ; T= 120°C, v= 0.03 μm/sec. Scale bar 200 μm

Fig.2 shows that the films contain very long crystals (conductive paths) which are nearly parallel to each other separated by the insulating matrix. Such morphology of crystalline CT complex is responsible for high anisotropy of the DC conductivity ($\sigma_{\parallel}/\sigma_{\perp}$) at room temperature: 10^8 for PE+1wt.% of TTT-TCNQ and 10^9 for PP+1wt.% of TTT-TCNQ. The change of casting temperature, velocity or CT complex concentration brings about the changes of the crystallization conditions which results in different morphology and consequently different electrical properties (anisotropy). This effect is illustrated below (taking casting temperature as a variable parameter) for PE and PP films in tables I and II, respectively. The results presented in tables I and II show that the room temperature conductivity in a direction parallel to the direction of the film casting ($\sigma_{\parallel}$) is relatively high and is only slightly affected by the change of casting temperature as compared with the changes of the conductivity in aperpendicular direction ($\sigma_{\perp}$).

Strong changes of the conductivity in perpendicular direction with casting temperature are related to the changes of the morphology of crystalline CT complex. Under non-optimum conditions increased branching of thinner needle-like CT complex crystals and growth of smaller crystalline structures

TABLE I. Electrical properties at room temperature of PE + 1wt.% of TTT*TCNQ films obtained from solution in o-dichlorobenzene with velocity v = 0.02mm/sec at different casting temperatures (T_c)

T_c [°C]	$\sigma_{\parallel}$ [S/cm]	$\sigma_{\perp}$ [S/cm]	$\sigma_{\parallel}$ / $\sigma_{\perp}$
90	1.2×10^{-4}	6.4×10^{-6}	19
105	5.6×10^{-4}	8.3×10^{-11}	7×10^{6}
115	1.2×10^{-3}	2.9×10^{-12}	4×10^{8}
125	2.1×10^{-3}	4.4×10^{-5}	50
140	5.6×10^{-3}	2.4×10^{-4}	24

TABLE II. Electrical properties at room temperature of PP + 1wt.% of TTT*TCNQ films obtained from solution in o-dichlorobenzene with velocity v = 0.03 mm/sec at different casting temperatures (T_c)

T_c [°C]	$\sigma_{\parallel}$ [S/cm]	$\sigma_{\perp}$ [S/cm]	$\sigma_{\parallel}$ / $\sigma_{\perp}$
90	4.7×10^{-3}	4.9×10^{-5}	96
100	2.2×10^{-3}	8.3×10^{-7}	3×10^{3}
110	1.5×10^{-3}	9.0×10^{-11}	2×10^{7}
120	1.7×10^{-2}	2.7×10^{-12}	5×10^{9}
130	3.2×10^{-2}	3.3×10^{-5}	1×10^{3}
140	3.0×10^{-2}	7.2×10^{-4}	41

between oriented needles results in formation of conducting paths in the perpendicular direction. This appears in a strong dependence of anisotropy ($\sigma_{\parallel}/\sigma_{\perp}$) on casting parameters.

The samples with relatively low anisotropy ratio (less than 100) exhibit similar temperature dependencies of the conductivities in both directions (similar also to that observed in the case of the isotropic films [7]).

For highly anisotropic systems the temperature dependence of conductivity in the parallel direction is also similar while in the perpendicular direction is much stronger and the value of anisotropy increases with decreasing temperature. These results show that relatively high conductivity in perpendicular direction for the samples with low anisotropy is due to continuous paths made of the CT complex crystals. In the case of the samples with high anisotropy the conductivity in perpendicular direction is low due to discontinuities of the conductive network in this direction (layers of insulating matrix interrupting these paths).

4. Conclusions

It was shown that using a zone casting technique it is possible to obtain fully organic, anisotropically conducting polymer composites. When inert semicrystalline polymers, like PE or PP, are used as polymer matrices the effectiveness of the technique is high and films with high anisotropy can be prepared using very low amount of the additive. By changing the zone casting parameters the anisotropy of conductivity of the obtained films can be varied in a very broad range.

Acknowledgements

This work was supported by Polish Academy of Sciences under projects CPBP 01.12 and 01.14.

5. References

1. M.Kryszewski, J.K.Jeszka, J.Ulanski and A.Tracz, Pure.Appl.Chem., **56**, 355 (1983).
2. A.Tracz, J.K.Jeszka, J.Ulanski and M.Kryszewski, Chemtronics, 1, **50** (1986).
3. J.Hocker, F.Jonas and H-K.Muller, Die Angewandte Makromolekulare Chemie, **145/146**, 191 (1986)
4. A.Tracz, J.Ulanski, T.Pakula and M.Kryszewski, Polish Patent 131 986.
5. L.Burda, A.Tracz, T.Pakula, J.Ulanski and M.Kryszewski, J.Phys.D;Appl.Phys.,**16**, 1737 (1983)
6. A.Tracz, E.El.Shafee, J.Ulanski, J.K.Jeszka and M.Kryszewski, Materials Science in press.
7. J.Ulanski, A.Tracz, G. Debrue and R.Deltour, J.Phys.D:Appl.Phys., **20**, 1512 (1987)

Post Polymerization Processing of Conductive Polymers (ICP): Integral Part of a New ICP Materials Science

B. Wessling

Zipperling Kessler & Co,
Kornkamp 50, D-2070 Ahrensburg, Fed. Rep. of Germany

Introduction

The approximately 12 year history of intrinsically conducting polymers is characterized, from a materials science viewpoint, by two main features:

- Only since it has been possible to polymerize polyacetylene in the form of a film it has been possible to study this new class of chemical composition in more detail and to invent the principal oxidation or reduction to a conductive state.
- Further progress in developing a consistent theory and also real marketable products was inhibited by the basic problem of nonprocessability.

The research on electrically conducting polymers still consists mainly of isolated approaches for the preparation, understanding and use of ICP.

Processability

Our working hypothesis for our research has been that the globular primary morphology of ICP (which we found to be a general morphological feature of conducting polymers /1/) and their insolubility are basic properties of the ICP materials which are linked to their material properties. The goal of the research therefore has been to take advantage of these basic properties and to develop processing procedures by using these properties. Our results have been described continuously in the literature /2/. This research approach was based on a materials science concept for ICP.

Materials science concept

1. The electrical and other materials properties of ICP are properties of the material as a macroscopic whole and cannot be reduced to molecular properties alone. The materials properties are developed via the configuration and the interactions of the individual molecules in a lattice, morphology, and supermorphology of the end product.
2. Therefore the chemical composition of the individual molecule and the physical structure of the macroscopic whole formed from the whole material together play the dominant role for the materials properties.
3. This is why the study of the primary particles morphology has to be an integral part of an ICP materials science; the spherical form of most of the ICP primary particles has to be taken into account in theoretically and practically oriented investigations.

Springer Series in Solid-State Sciences, Vol. 91
Electronic Properties of Conjugated Polymers III
Editors: H. Kuzmany · M. Mehring · S. Roth

4. Post polymerization processing of conductive polymers dramatically affects the resulting structure of the end product and therefore the materials properties.

Critical comment on an isolated molecular approach

The literature of the conductive polymer research is full of examples which have not been derived from a materials science approach, but have been aimed only at changeing a molecular property. Such examples are:

1. To overcome the infusibility and insolubility of ICP, one has polymerized them in insulating polymer matrices. The resulting polymer blends became conductive, but none of these became processable. This approach did not lead to a defined and reproducible chemical composition and morphological structure /3/.
2. The idea of making soluble conducting polymers has been very attractive in the last 4 years and is actually the broadest area of basic research. Only very few examples have been decribed, in which the complexed ("doped") form was said to be soluble /4/ (e.g. poly-butylthiophene "doped" with $NOSbF_6$), but this is still a matter of controversy /1/. In contrast to the model of these polymers being really soluble, we proposed to consider them also to be dispersions (see below), which have a slightly, but not qualitatively different form of primary particles or their dispersions of "insoluble" conducting, polymers see figs. 1 and 2. Moreover, the properties of the resulting products are not at all different or superior but normally inferior compared to those made from insoluble conducting polymers by other processing techniques. Most of them are less conductive or

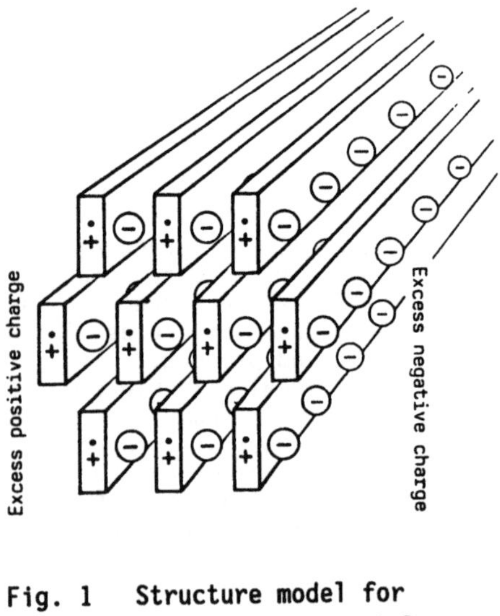

Fig. 1 Structure model for spherical primary ICP particles

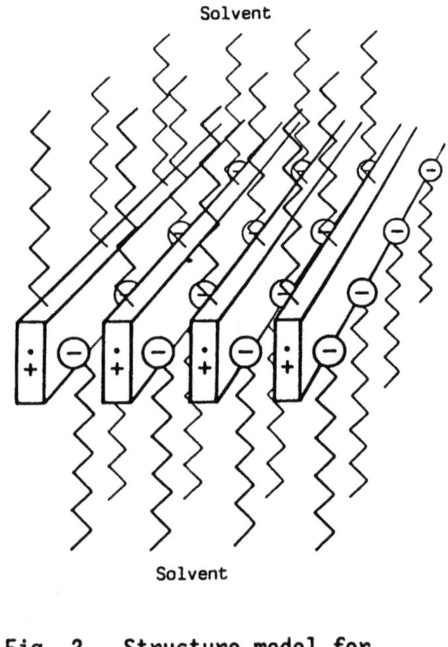

Fig. 2 Structure model for "soluble conducting" polymers

dedope thermally at a temperature of about 80 - 120°C, a phenomenon which might be linked with the thermochromism effect /5/.

3. Polymer blends with "soluble conducting" polymers have been described by Heeger et al., Österholm et al., and Karasz et al. /6/. All of these blends have been produced by first mixing the non-conducting neutral form of the appropriate polymer in the thermoplastic polymer matrix (via solution or melt blending), casting a film and "doping" the resulting mixture In comparison to the technique described by us /2/, we do not see any advantage for such a proposal:
   - the use of solvents in the polymer processing industry from a cost and environmental standpoint is highly disadvantageous
   - the percolation thresholds of all the blends described up to now are significantly higher than those prepared by our routes (fig. 3)
   - the supply of monomers and the polymerization of the soluble polymers needed is highly disadvantageous in terms of cost compared to simple insoluble homopolymers.
4. Recently Heeger et al. described "solutions" of polyanilinehydrochlorides in H_2SO_4 /7/. This work is based on observations made in 1910 by Green and Woodhead. There are two major results claimed by Heeger et al.:
   - "soluble" PAni can be formed into fibers by precipitating the H_2SO_4 solution in water (comparable to the aramide fiber production), therefore it is also the idea to make blends with aramides in this way
   - the product which can be recovered from the solution exhibits XRD reflections, hence is partially crystalline.

   From the materials science standpoint I would like to raise some questions:
   - Why should somebody needing a conductive fiber or the strength of an aramide fiber not go and buy carbon fibers?

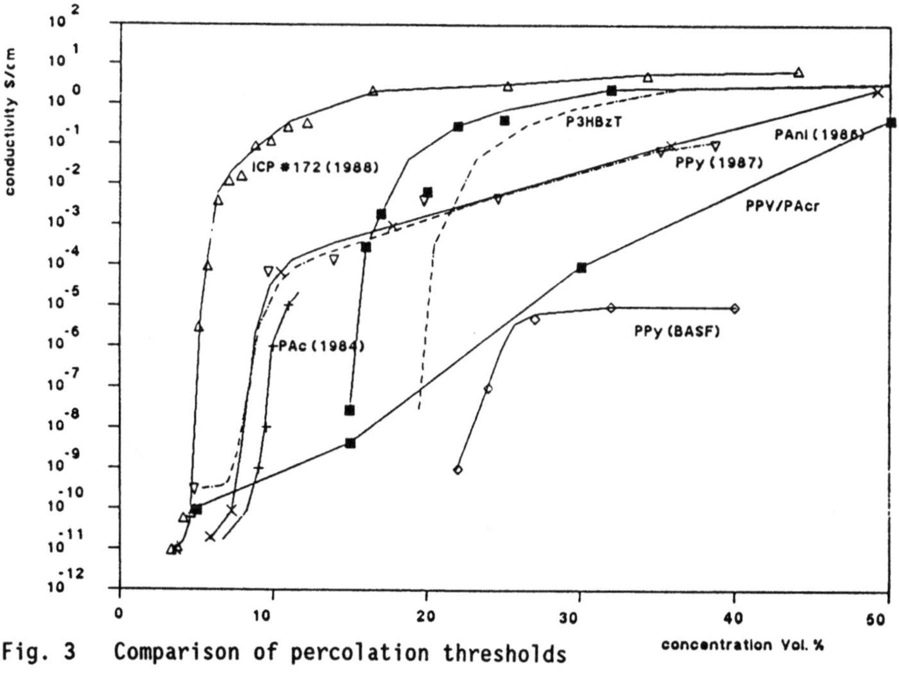

Fig. 3 Comparison of percolation thresholds

- In our experience polyaniline "solvated" in H_2SO_4 cannot becomcensated (neutralized) completely after the "solution" has been made. Could it be that the "soluble" product produced by this method is not a polyaniline any more but a mixture of different ring sulfonated and chlorinated polyaniline derivatives?
- There is no basic difference in the XRD's between those reported by us /2/ and by Heeger, except the precise locations of some of the reflections (due to differences in chemical composition?).

5. The conductivity mechanism for conductive polymers still remains a matter of controversy: although Wegner has pointed out in several publications the effect of the lattice structure and the effect of interactions between the chains on the conductivity, the main work on conductivity mechanisms concentrates on single chain properties. This is why models of solitons and polarons are most popular. The new model of the "granular metal" (Epstein and MacDiarmid /8/) now seems to be a very important step to a conductivity mechanism understanding for the conductive polymers in the direction of a materials science view. Additionally it should be very important for theoreticians to consider the results of Nimtz et al. who have shown that the conductivity of metallic particles dramatically decreases with decreasing particle size of the metals /9/. This is due to the fact that quantum effects play a dominant role in the conductivity mechanism, if the particle size approaches the size of the electron wavelength at the given temperature or frequency used for the measurement. Then only certain states of the electron wave are allowed within a particle of a size of 10 or 100 nm (fig. 4). The alternative to interpreting the ICP conductivity mechanism by looking at a single chain is to combine new insights from conventional materials science research, including new quantum effect theories, with our results on morphology studies.

<u>Advantages of the materials science concept for ICP</u>

The main results which we have obtained over the last 7 years following a materials science concept have been published continuously /2/. Recent results are:

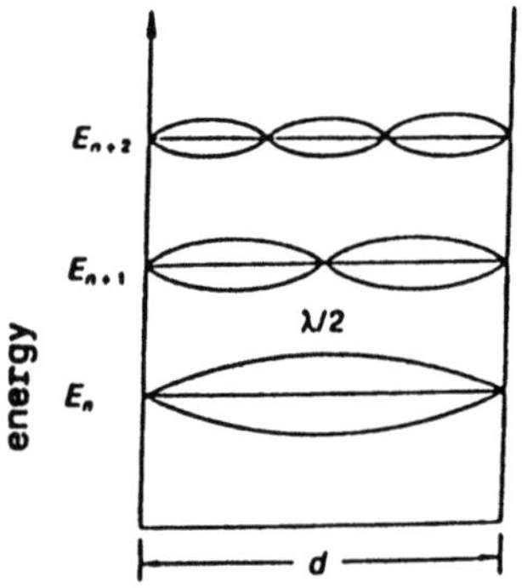

Fig. 4 Electron wave states for a certain particle diameter

Wessling et al.		Heeger et al.	
PAni-HCl	new ICP raw	HCl as synth.	HCl from H_2SO_4 solu.
11°	10°	9°	10°
14°	16°	15.5°	-
19.6°	20°	21°	20°
24.6°	25°	26°	26.5°
26.5°	27°	27°	-
29.3°	?	30°	29°

Fig. 5 X-ray diffractions (2 θ)

1. Where some years ago we only got partially crystalline conducting polymers by processing the raw powder via a high shear and high temperature procedure /2/, a comparable crystallinity is now found in the raw ICP powder, resulting from the improved polymerization procedure (fig. 5). The XRD pattern is comparable to our previous results and those from Heeger et al. /7/, except for differences probably due to differences in the chemical compositions of Heeger's samples.
2. The ductility (of which the model for its mechanism has been described by us /2/) also leads to a significant change of the FT-IR spectra (fig. 6). Those bands which are connected with frequences of the

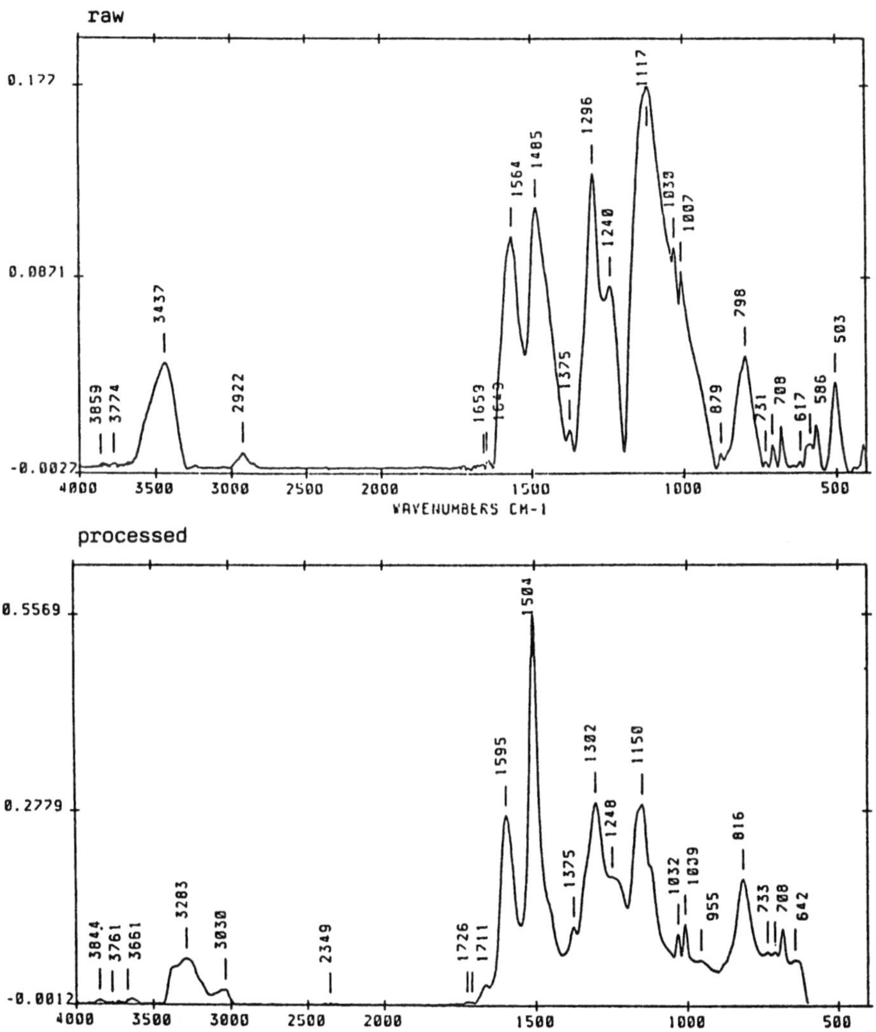

Fig. 6 IR spectrum changes from raw powder to thermal ductilely processed ICP

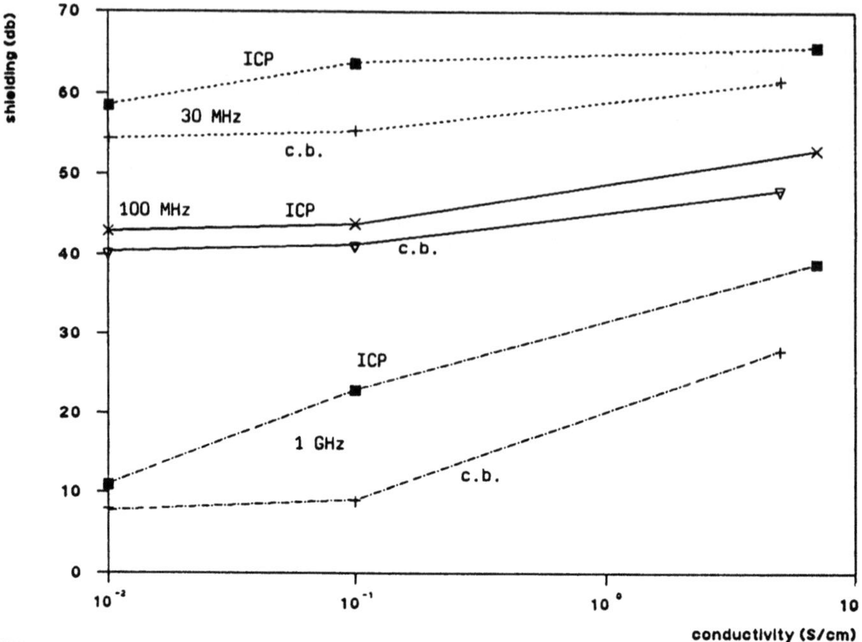

Figure 7

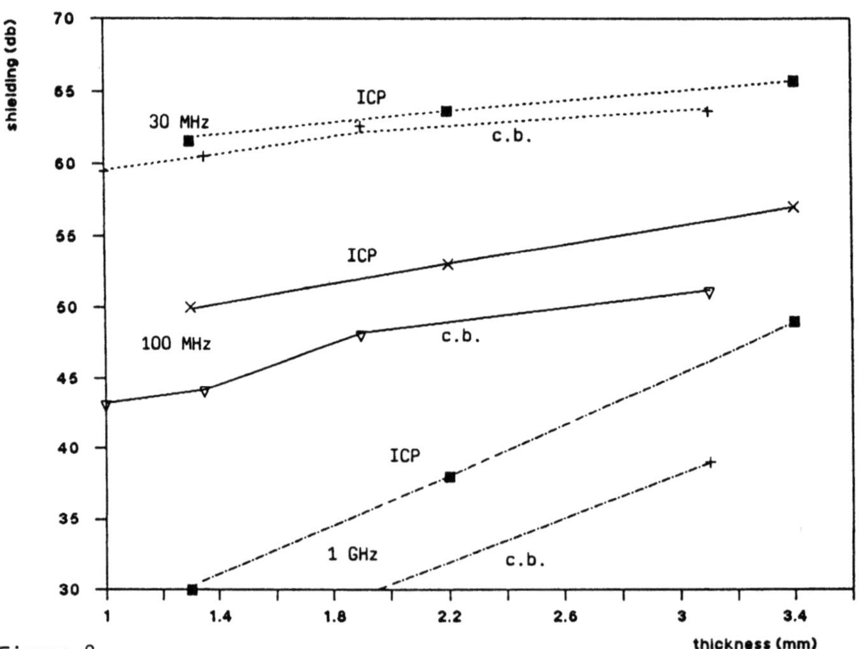

Figure 8

Fig. 7 and 8 Comparison of EMI shielding: ICP blends vs carbon black compounds

polymer backbone are shifted to higher wavelengths. We attribute this to an increase of order.

3. The dispersability of conducting polymers which we invented in 1984 continously leads to improved conductive and processable blends with low critical volume concentrations for the conductivity jump. A new theory for describing the conductivity jump at a critical concentration as being a phase transition has been described elsewhere /10/. We can now announce that the percolation behaviour of conducting polymers in blends could also be reproduced with a 10 times bigger polymerization reactor (10 kg per week).

4. These technologies now allow one to design basic experiments for the evaluation of properties and are interesting for potential application of the ICP as materials. One of the most interesting results is the comparison of an ICP polymer blend with carbon black compounds at comparable conductivity and sample thicknesses (see figs. 7 and 8). In all aspects, conducting polymer blends are superior in EMI-shielding to carbon black compounds due to their superior interactions with electromagnetic waves.

5. It is very interesting to note that ICP blends exhibit a thermopower behaviour comparable to pure ICP's and metals. This needs to be taken into account while discussing the conductivity mechanism in ICP's (fig. 9).

6. The dispersability principle now allowed us also to make stable dispersions of conducting polymers in their complexed ("doped") form. These dispersions when viewed optically look like solutions, but we

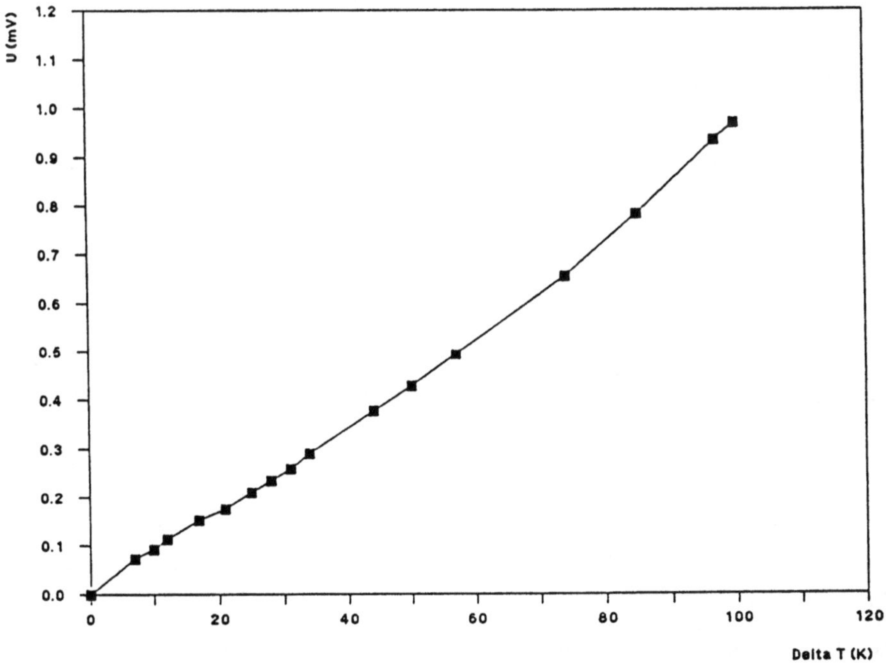

Fig. 9 Thermovoltaic effect of a PAni polymer blend

consider them to be dispersions. This view is supported by some interesting phase separation phenomena, by gelation effects, by their flocculation behaviour and by SEM-evaluations of transparent conductive layers produced from these dispersions. Due to the highly reproducible build-up of these layers from dispersion it has been possible for us to generate transparent ICP layers (50 - 400 nm thin) with a conductivity of up to 200 S/cm (in the case of PAni) (details to be published later).

Conclusions

The materials science concept outlined above contains as 1 of 3 integral parts the post-polymerization processing of conductive polymers to a morphology and structure of the end product. This not only allows one to produce real products which could be used later on an industrial scale, but also serves as a strategically important tool for supplying chemically uniform and pure and morphologically defined structured materials for basic research, in order to elaborate a deeper understanding of the structure and properties of the ICP. The main results of our work following this materials science research concept are:

1. Insoluble ICPs with a globular primary particle structure are on the way to becoming processable and practically useful on an industrial scale.

2. Insolubility and globular morphology, ductility, dispersability, and the conductivity jump phenomenon, interpreted as a phase transition, are the main basic research results based on the integrated view of polymerization and post polymerization processing.

3. The integration of the globular primary particles model as the basic structural unit occurring during polymerization with the quantum effects observed in small metallic particles may lead us to a completely new view of the conductivity mechanism in ICP and to a new concept for the preparation of ICP materials with increased conductivity.

Literature

1. Wessling, B., in: Elektrisch leitende Kunststoffe; H.J. Mair, S. Roth (Hrsg.), Carl Hanser Verlag 1989 (to be published) and literature cited therein
2. a) Wessling, B.: Synth. Met. 28 (1989) C849 and literature cited therein
   b) Wessling, B.: Synth. Met. 18 (1987) 671 and lit. cited therein
3. Baker, G.L.: Adv. Chem. Ser. 218 (1988) 271
4. Aimé, J.-P., Bargain, F., Schott, M., Elsenbaumer, R., Eckhardt, H., Miller, G.G.: Synth. Met. 28 (1,2) (1989) C407
5. Inganäs, O., Gustafsson, G.: Chromism in the poly(alkylthiophenes) Proceedings IWEPP 89
6. a) Hotta, S., Rughooputh, S., Heeger, A.J.: Synth. Met. 22(1) (1987) 79
   b) Österholm, J.E., Laakso, J., Isotalo, H., Stubb, H.: Vortrag auf der ICSM 1988
   c) Machado, J.M., Karasz, F.E., Lenz, R.W.: Polymer 29 (1988) 1412

7. Moon, Y.B., Cao, Y., Smith, P., Heeger, A.J.: X-ray Scattering from Crystalline Polyaniline (preprint), lecture at the ACS-meeting 1989, Dallas/USA
8. Epstein, A.J., MacDiarmid, A.G.: Mol. Cryst. Liq. Cryst. 160 (1988) 165
9. Nimtz, G., Marquardt, P., Mühlschlegel, B., Solid State Commun. 65 (6) (1988) 539
10. Wessling, B.: Synth. Met. 27 (1988) A83

Recent Developments in the Chemistry and Applications of Electroactive Conjugated Polymers

P.C. Allen[1;*], *D.C. Bott*[1;+], *C.S. Brown*[1], *L.M. Connors*[1], *S. Gray*[1], *N.S. Walker*[1], *P.I. Clemenson*[2], and *W.J. Feast*[2]

[1]BP Research, Sunbury Research Centre, Sunbury-on-Thames, Middlesex TW16 7LN, UK
[2]Durham University Chemistry Department, South Road, Durham DH1 3LE, UK

Abstract

Photo-oxidative modification of the Durham polyacetylene precursor polymer followed by heating allows the fabrication of high resolution relief images. The process is described and possible applications are discussed.

1. Introduction

Conjugated polymers have been important targets for synthesis for many years [1], the reason for this interest being their known or anticipated optical, electronic, and more recently electro-optic properties [2]. There are many synthetic routes to such materials but a practical obstruction to progress in studies of the physics and chemistry of conjugated polymers has been their intractability. Two approaches to circumventing this problem have been widely adopted; namely, the synthesis of soluble and consequently processable precursor polymers [3-5] and the modification of the basic polymer structure by addition of solubilizing side chains [6-8].

We have been particularly concerned with the synthesis of well characterized samples of polyacetylene and have previously described several routes to this material via precursor polymers [3,9]. These syntheses are now widely referred to as the Durham route to polyacetylene. The resulting materials have been intensively investigated by the Cavendish-BP-Sussex-Durham collaboration [10], and the method has been widely adopted [11-14]. In the course of studying the dynamics of the transformation from precursor to polyacetylene a novel imaging technique was discovered, the details of this process are the subject of this paper.

2. Results and Discussion

The Durham route, via precursor polymers such as I, has been successfully applied by several different groups of workers; however, in practice this particular precursor polymer tends to undergo partial transformation during normal handling so that the structure is more properly represented as a

*Current address: Thorn EMI, Central Research Laboratory, Dawley Road, Hayes UB3 1HH, UK.
+Current address: Courtaulds Research, PO Box 111, Lockhurst Lane, Coventry CV6 5RS, UK.

Springer Series in Solid-State Sciences, Vol. 91
Electronic Properties of Conjugated Polymers III
Editors: H. Kuzmany · M. Mehring · S. Roth

copolymer of bis(trifluoromethyl) bicyclo [2.2.2] octadiene vinylene units and polyenes with odd numbers of double bonds II. Polyenes are sensitive to oxidation and considerable care to exclude molecular oxygen is necessary if

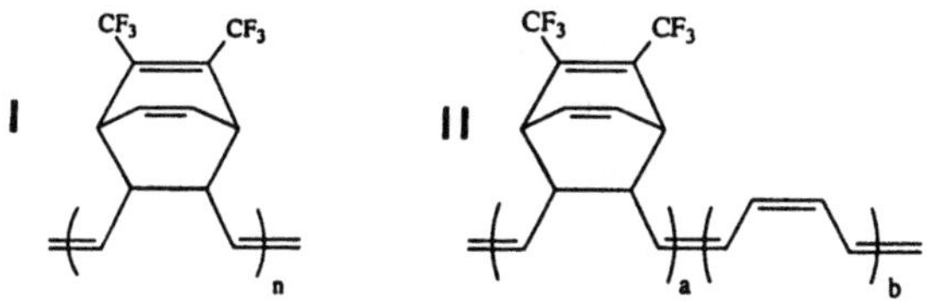

high quality samples of polyacetylene are to be obtained by this route. Some time ago we observed that samples of precursor polymer II exposed to light and air had undergone a radical change in molecular structure and that they were no longer converted to polyene sequences on heating. Photo-oxidized films of precursor polymer II lose only a small proportion of hexafluoroxylene on heating, and give a colourless film which contains carbon, hydrogen, fluorine and oxygen. Figure 1 shows the UV-visible spectra of: a typical sample of pristine precursor polymer II, (a); the product of heating II under reduced pressure in the absence of oxygen (polyacetylene), (b); and the product of heating a photo-bleached film of II under the same conditions, (c). It is clear from spectrum (c) that the fine structure associated with the short polyene sequences present in the precursor has been destroyed, although the product retains some absorption in the region characteristic of unsaturation. Some of the properties of a photo-bleached and heated film of the precursor polymer are compared with Durham polyacetylene in Table 1. A detailed discussion of the chemistry involved in these processes will be presented elsewhere. We now describe how these observations have been developed to provide a technique suitable for production of high resolution surface relief images and discuss some potential technological applications.

The process involved in the production of an image is shown schematically in figure 2. A film of precursor II is spin coated onto a

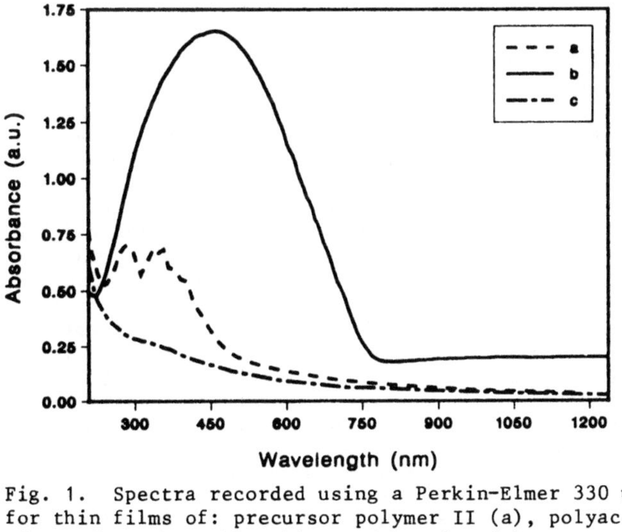

Fig. 1. Spectra recorded using a Perkin-Elmer 330 uv-visible spectrometer for thin films of: precursor polymer II (a), polyacetylene (b), and a film of II after photo-bleaching and heating (c)

Table 1. Comparison of some thin film properties of Durham polyacetylene and photo-bleached polymer II

	PHOTO-BLEACHED AND HEATED II	DURHAM POLYACETYLENE
APPEARANCE	Transparent	Opaque
REFRACTIVE INDEX (@ 633nm)	1.51	3.4
CONDUCTIVITY	$< 10^{-12}$ S cm^{-1}	3 10^{-8} S cm^{-1}
BAND GAP	> 3eV	1.5-1.7eV
THICKNESS (% of original)	70%	33%
I_2 EXPOSURE	No effect	Increased conductivity

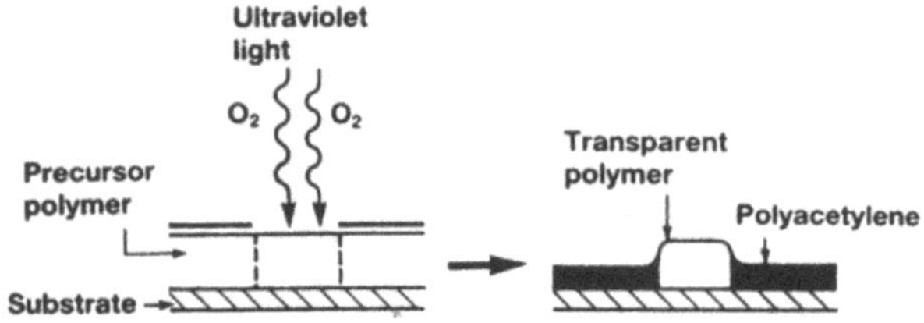

Fig. 2. Schematic for the image generation process in precursor polymer II

polished quartz substrate from a pale yellow solution of precursor polymer II in butanone. Areas of the film are selectively irradiated with uv light in the presence of oxygen via a suitable contact mask. The irradiated film is then heated in a vacuum oven at 100C for 200mins, when the image develops. Images produced by contact copying in this way have been found to possess very high resolution; linewidths of less than 1μm are clearly reproduced. The resolution achievable via contact copying is however limited; higher resolutions may be obtained by interfering two mutually coherent beams from a Helium-Cadmium laser. By varying the angle between the two beams gratings of varying pitch can be defined; the highest resolution grating we have so far studied had a period of 0.6μm, corresponding to a linewidth of 0.3μm. The resolutions achieved here are comparable with those of conventional photolithography and by enhancing the surface reflectivity with a thermally evaporated layer of gold, highly efficient diffraction gratings have been fabricated. Figure 3 shows a micrograph of a cross-section through a diffraction grating produced by this process illustrating the surface relief effect. The efficiency of a grating is determined by the relationship between the height of the relief and the wavelength of the light incident on the grating. Relief heights can be accurately controlled by varying the intensity and/or time of exposure and this has enabled us to obtain diffraction efficiencies approaching the theoretical maximum. Transparent phase gratings can also be produced. If the process is to be used for producing passive optical elements, ie active areas of polyacetylene are not required, then the image need not be developed under vacuum; the irradiated sample is simply heated on a hot plate in air at 100C for 30 seconds in order for the thermal elimination of the volatile hexafluoroxylene to take place. This may be taken a stage further by continuing to heat the sample and hence accelerating the oxidative degradation of polyacetylene. The effects of this degradation are the destruction of the conjugated sequences and the raising of the band-gap, resulting in an air stable transparent film. The irradiated areas of the film are unaffected by this further heating stage

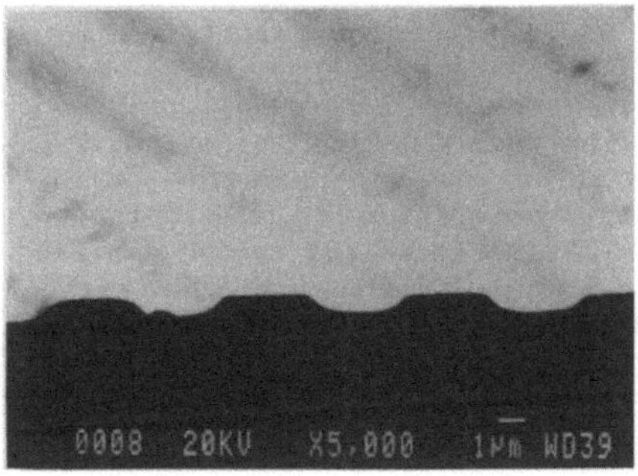

Fig. 3. SE micrograph of a cross-section through a diffraction grating produced by heat-treatment of an irradiated precursor film.

and consequently the relief image is preserved. These films are also characterised by very low scatter, a desirable property of an optical material.

A second area of interest is holography. There is a growing interest in the use of holographic optical elements (HOEs), functional holograms which combine, for example, beam deflection and focussing functions, and are beginning to replace conventional bulky refractive optical components for use in monochromatic systems such as bar-code scanners, head-up displays and compact disc readers. There is also research interest in the use of HOEs for optical interconnects between integrated circuits and optical signal processing. Durham polyacetylene precursor possesses many attributes useful for the fabrication of HOEs; relatively high diffraction efficiencies have been demonstrated and the one-step dry processing offers advantages over competitive materials, such as dichromated gelatin and etched silica.

3. Conclusions

We have described here the properties and applications of surface relief images produced by the uv irradiation of Durham precursor polymer. It should be emphasised however that unexposed areas of the precursor film, when transformed under the right conditions, are converted as normal to polyacetylene. It is easy to see therefore how the uv irradiation process may also be used to define areas of active polyacetylene surrounded by an insulator, in effect a process of direct photolithography.

References

1. J.F.Lennard-Jones,Proc Roy Soc,A158,280 (1937).
2. J.H.Burroughes, C.A.Jones and R.H.Friend, Nature, 335, 137 (1988).
3. J.H..Edwards and W.J.Feast, Polymer, 21, 595 (1980); J.H.Edwards, W.J.Feast and D.C.Bott,Polymer,25,359 (1984).
4. D.R.Gagnon,J.D.Capistran,F.E.Karasz and R.W.Lenz, Polym.Bull, 12(4), 293 (1984); K.Araya, A.Mukoh, T.Narahara and H.Shirakawa, Synthetic Metals, 14,199 (1986).
5. D.G.H.Ballard,A.Courtis,I.M.Shirley,S.C.Taylor,J.Chem.Soc., Chem.Commun., 954(1983).
6. J.M.Bureau, M.Gazard, M.Champagne, J.C.Dubois, G.Tourillon and F.Garnier, Mol. Cryst.and Liq.Cryst.,118(B),235(1985)
7. W.J.Feast, I.S.Millichamp, R.H.Friend, M.E.Horton, D.Phillips, S.D.D.V.Rughooputh and G.Rumbles,Synthetic Metals,10,181 (1985)

8. M.Rehahn,A-D.Schluter,G.Wegner and W.J.Feast,Polymer,accepted for publication, 1989.
9. W.J.Feast and J.N.Winter,J.Chem.Soc.,Chem.Commun.,202 (1985)
10. There are many papers and patents arising from this collaboration. Two of the more exciting recent examples describe new semiconductor device physics (ref 2 above), and third harmonic generation in the oriented polymer(M.R.Drury,Solid State Comm.,68(4),417 (1988)),access to the earlier literature can be had via D.C.Bott,C.S.Brown,C.K.Chai, N.S.Walker,, W.J.Feast, P.J.S.Foot, P.D.Calvert, N.C.Billingham, R.H.Friend, Synthetic Metals,14,245 (1986).
11. G.Leising,PolymerCommun.,25,201 (1984).
12. G.Lieser, G.Wegner, R.Weizenhofer and L.Brombacher, ACS Polymer Preprints, 25(2),221 (1984).
13. S.A.Krouse and R.R.Schrock, Macromolecules,21,1885 (1988).
14. F.L.Klavetter and R.H.Grubbs,Synthetic Metals,26,311 (1988)

Layer-by-Layer Thin Film Deposition of Polypyrrole and Polyaniline

Y.F. Nicolau and M. Nechtschein

DRF/SPh/Dynamique de Spin et Propriétés Electroniques (UPR CNRS 216),
Centre d'Etudes Nucléaires de Grenoble, 85 X,
F-38041 Grenoble Cedex, France

Abstract. Thin films of PPY and PANI are deposited on insulator, semiconductor or conductor substrates. The films are obtained by soaking the substrate alternately in a solution of the monomer and in a solution of a suitable oxidant. The films are synthesized layer-by-layer in the adhering diffusion layer of solution at the interface substrate-solution. The growth rate is in the range of 2-20nm/cycle. Films 20-100nm thick are compact and smooth. Films of PPY and of PANI show a conductivity within 100-200 and 3-15S/cm respectively.

1. Introduction

Both fundamental and applied studies in conducting polymer physics need thin or thick, highly conducting, films having a smooth and compact morphology (showing long and short range order) deposited either on metals, on semiconductors, or on insulators. None of the known deposition methods conveniently fulfil this requirement. Aiming to solve this problem we present here a new deposition method.

2. Film deposition methods, state of the art

The known methods of film deposition are : (i) electrochemical (EC) oxidative polycondensation, (ii) vapor deposition, (iii) casting from solutions and (iv) chemical bath deposition. The EC oxidative polycondensation is the most applied method of concomitant synthesis and deposition of conducting polymers. Thin films of PPY, PANI and PTs are easily deposited on Pt, Au, graphite and on conducting oxide covered glass. These polymers have different short range structures and show different morphologies : crystalline, spheroidal, cauliflower, granular, globular, spongy or fibrillar. These different morphologies are however common for polymers having different short range structures. The morphology strongly depends on the experimental conditions, namely on : (1) the potentiostatic, potentiodynamic or galvanostatic working regime, (2) the oxidation potential or its scanning range, (3) the dopant nature, (4) the temperature, (5) the solvent nature, (6) the monomer concentration, (7) the surface state of the electrode and (8) the number of potential cycles, the total charge or time of deposition.

Important progress in understanding the mechanism of EC oxidative polycondensation of conducting polymers has recently

Springer Series in Solid-State Sciences, Vol. 91
Electronic Properties of Conjugated Polymers III
Editors: H. Kuzmany · M. Mehring · S. Roth

been made using spectroscopic ellipsometry, the rotating electrode technique and chronocoulometry [1]. Some conclusions of these studies are noteworthy : (i) the monomer is not strongly adsorbed, (ii) oligomers are formed near the electrode, (iii) polymer chains grow primordially in solution and are subsequently deposited on an electrode, (iv) the films start to grow by an island-type nucleation, then the growth becomes two-dimensional and finally three-dimensional, (v) compact films grow by a two-dimensional mechanism, porous films grow by a three-dimensional mechanism.

Other studies [2] concern the co-depositon of a compact transparent thin film and of powdery thick black particles. The films show an initial preferred orientation of the polymer chains parallel to the electrode surface, but as the deposition proceeds, the chains begin to stack randomly. Films one or several μm thick are porous. The porous morphology of thick films is explained assuming an enhanced chain branching with increasing thickness.

We consider that the morphology is primordially determined by the growth mechanism. Thus we distinguish three growth mechanisms : (a) deposition of chains, (b) deposition of clusters of chains (globules), and (c) extension of the already deposited chains (heterogeneous oxidative polycondensation). Consequently, the film morphology is determined by the competition among the simultaneously acting growth mechanisms. Only thin films, up to about 150nm thick, deposited on Pt or Au fulfil the above-mentioned requirement. We believe that thicker films become powdery mainly owing to the second mechanism, which becomes dominant as the deposition proceeds.

Compact films of PPY and PANI have been deposited on different substrates by gas phase polycondensation operated under radiofrequency plasma, by CVD and by vacuum evaporation [3]. But the conductivities of these films are low.

Porous films of undoped P3-hexylT have been obtained by solvent evaporation [4]. Doped films show, however, a lower conductivity than insoluble films of P3MT. Many AsF_5 doped conducting polymer films have been obtained from AsF_3 solutions but the films are porous and the conductivities of the films (PPS excepted) are also low [5]. Reduced PANI films have been obtained by solvent evaporation from THF, NMP, and from AcAc solutions. Morphology and conductivity data of doped PANI films are hitherto unknown.

Glass or other insulating polymers have been covered by thin films of PPY or PANI using diluted chemical bath deposition [6]. The morphology of thin films is granular ; the conductivities correspond to those measured on chemically synthesized polymers respectively. Thick films are porous.

3. The layer-by-layer deposition process

We are studying a layer-by-layer (LL) deposition of thin films of PPY and of PANI by soaking an insulator, conductor or semiconductor substrate alternately in a solution of the monomer and in a solution of a suitable oxidant. The liquid Py or solid ANI.HCl layer adhering on the substrate after the evaporation of the solvent is polycondensed on the substrate in contact with a $FeCl_3$ oxidant solution. The excess of oxidant adhering to the substrate emersed from the solution is rinsed carefully before starting a new deposition cycle.

This LL deposition process is akin to the Atomic Layer Epitaxy and to the SILAR (Successive Ionic Layer Adsorptions and Reactions) processes of deposition of thin films of compound semiconductors [7]. The deposition is made using a SILAR machine adapted to LL deposition of conducting polymers. The concomitant synthesis and deposition of the polymer proceed in and from the diffusion solution layer adjoined to the substrate surface as in the EC deposition. As for the EC deposition the growth rate/cycle, the morphology of the films, and their physical properties are strongly dependent on the working conditions that determine the competition among the above-mentioned growth mechanisms (a)-(c), namely :
(1) the oxidation potential of the oxidant solution, (2) the nature of the solvent of the oxidant, (3) the concentration of the oxidant, (4) the thickness of the adhering monomer layer submitted to polycondensation, (5) the dopant nature, (6) the temperature, (7) the nature and surface state of the substrate, (8) the soaking time in the oxidant solution, (9) with or without a supplementary reduction step.

Using a flat (scabbard-like) beaker the oxidant solution can be renewed after each deposition cycle. In this way the deposition of large globules and the agglomeration of globules is hindered. Unlike in the EC deposition, small linear oligomers cannot exist in the oxidant solution because the oxidant is in a very large excess during the polycondensation.

4. Examples of working conditions and preliminary results

PPy deposition : (1) 5s soaking the substrate in an etheric solution of Py 0.5M at 0°C, (2) 5s ether evaporation under 7 ℓ/min N_2 flushing, (3) 30s soaking the substrate in an etheric anhydrous solution of $FeCl_3$ 0.5M at 0-15°C, (4) 4 times 10s rinsing in dry ethanol, (5) 60s ethanol evaporation under N_2 flushing. Depending on the N_2 flushing rate, the growth rate/cycle varies within 2-10nm. The growth rate is of the order of 100nm/h. The conductivity of films up to 1μm thick is within 100-200 S/cm.

PANI deposition I : (1) 5s soaking the substrate in a 1.25M solution of ANI.HCl in dry methanol at 5°C, (2) 50s methanol evaporation under 7 ℓ/min N_2 flushing, (3) 120-300s soaking the substrate in an acqueous solution of $FeCl_3$ 3M (pH=0) at 15-20°C, (4) 2 times 15s rinsing in 80% ethanol-20% HCl 2M (% vol.) and than 2 times 15s rinsing in dry ethanol, (5) 60s ethanol evaporation under N_2 flushing. Depending on the soaking time in the $FeCl_3$ solution the growth rate/cycle varies within 5-20nm. The growth rate is also of the order of 100nm/h. The conductivity of the films up to 1μm thick is within 3-7S/cm. The potential of the films measured vs. SCE in 1M HCl is 0.533 V corresponding to the E_0 Fe^{3+}/Fe^{2+}.

PANI deposition II : (1) 5s soaking the substrate in a 1M solution of ANI.HCl in dry methanol at 5°C, (2) 50s methanol evaporation under 7 ℓ/min N_2 flushing, (3) 300s soaking the substrate in an acqueous solution of $FeCl_3$ 3M (pH=0) at 20°C, (4) 60s rinsing in 0.5 ℓ/min ultrapure flowing water, (5) 60s reduction in a 0.2M $Na_2S_2O_4$ 24% N_2H_4 -aqueous solution (pH = 10.5), (6) 60s rinsing in 0.5 ℓ/min ultrapure flowing water, (7) 15s rinsing in dry ethanol, (8) 60s ethanol evaporation under N_2 flushing. Under these conditions, renewing the $FeCl_3$ solution after each cycle, the growth rate/cycle is 2.5nm. The conductivity of films is about 15S/cm.

Hitherto, we have mostly characterized films 0.15-1μm thick deposited without renewing the oxidant solution and without a reduction step in the cycle. The PPY films are blue, the PANI films are green. The cyclic voltammograms of LL PANI and PPY films are comparable with the best voltammograms recorded on EC films. Also the ESR susceptibility versus pH diagram of LL PANI films is comparable with the diagrams recorded on EC films. The conductivities of LL PPY and PANI films equal the best conductivities measured on EC films.

The morphology of LL thin films up to 200nm thick is smooth and compact, similar to that of EC films. Thicker films keep this morphology but some black globular agglomerates are included in smooth films. Thick LL films (>200nm thick) obtained with oxidant solution renewed after each cycle no longer show black globular agglomerates. Thick LL films obtained with a supplementary reduction step at each cycle keep this morphology, and are still smoother. The compactness of LL PANI films deposited with solution renewing and reduction at each cycle entails an increase in conductivity beyond the conventional 5S/cm of EC films. The wavelength of the radical-cation absorption band determined on LL PANI films as deposited on silica lies at 1225nm and is higher than the intensity of the $\pi \longrightarrow \pi^*$ band lying at 323nm, indicating a longer conjugation length in LL films than in other EC films.

PBT, P3MT, sandwich and composite films can be deposited LL too. More details and results will be published elsewhere.

5. References

[1] a) S. Asavapiriyanont, G.K. Chandler, G.A. Gunawardena and D. Pletcher, J. Electroanal. Chem. **177**, 229 (1984) ; b) P. Lang, F. Chao, M. Costa and F. Garnier, Polymer **28**, 668 (1987) ; c) A.R. Hillman and E. Mallen, J. Electroanal. Chem. **220**, 351 (1987)

[2] a) G. Tourillon and F. Garnier, J. Polym. Sci. Polym. Phys. **22**, 33 (1984) ; b) J. Roncali, A. Yassar et F. Garnier, J. Chim. Phys. **86**, 85 (1989) ; c) G. Tourillon, J. Chim. Phys. **86**, 61 (1989)

[3] A. Mohammadi, I. Lunström, W.R. Salaneck and O. Inganäs, Synth. Metals **21**, 169 (1987)

[4] R.L. Elsenbaumer, K.Y. Jen and R. Obodi, Synth. Metals **15**, 169 (1986)

[5] a) J.E. Frommer, Acc. Chem. Res. **19**, 2 (1986) ; b) M. Aldissi, Synth. Metals **17**, 235 (1987)

[6] D.R. Rueda, C. Arribas, F.J. Balta Calleja, J.L.G. Fierro and J.M. Palacios, Synth. Metals **28**, C77 (1989)

[7] Y.F. Nicolau and J.C. Menard, J. Crystal. Growth **92**, 128 (1988)

Basic Polypyrrylenmethines – Chemical and Physical Properties

R. Becker, G. Blöchl, and H. Bräunling

Wacker-Chemie GmbH, Postfach 1260,
D-8263 Burghausen, Fed. Rep. of Germany

Abstract

Polypyrrylenmethines, a new class of conducting polymers, can be obtained via a polycondensation of pyrrole-2-carbaldehyde. If the polymerization is carried out in an isolating polymer matrix, conductivities up to 10^{-2} S/cm are obtained. Numerous chemical and physical measurements (IR, UV-VIS-NIR, ^{13}C and ^{15}N solid state NMR, XPS and ESR) confirm the proposed structure for the new compounds.

1. Introduction

Polypyrrylenmethines (PPM) are a new class of conducting polymers with promising electrical and photoelectrical properties. Quantumchemical calculations of Brédas et al. indicate, that these systems should have bandgaps around 1 eV, about three times smaller than in polypyrrole [1].

The present paper reports on the synthesis of polypyrrylenmethines [2], a process to produce conducting composites [3], and the physical properties of the new polymers.

2. Syntheses: Preparation of powders, coatings and composits

When pyrrole-2-carbaldehyde 1 is reacted with strong condensating agents (e.g. $SOCl_2$ or $POCl_3$) in inert solvents in the heat, black, insoluble polymers 2 equ. (1) can be isolated in high yields.

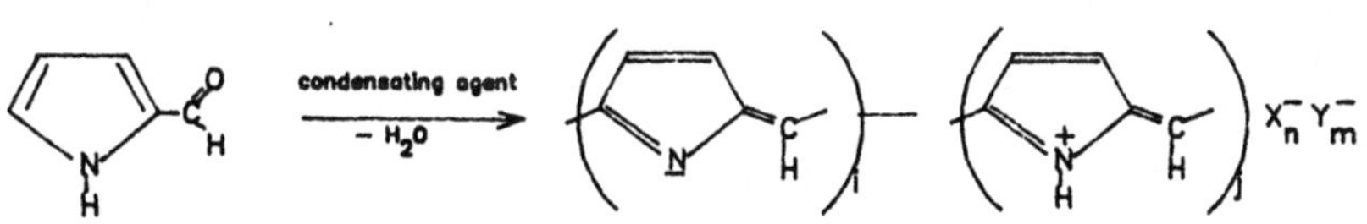

1 2

(1)

$i + j = 1,\ n + m = j;\ X: Cl^-,\ Y: H_2PO_4^-$

It is possible, too, to prepare green-splendid coatings on diverse surfaces (e.g. quartz- glass, silicon wafers) with conductivities up to 10^{-4} S/cm.

Pyrrole-2-carbaldehyde can be polymerized in nonconducting polymer matrices to yield black, free-standing composite

Springer Series in Solid-State Sciences, Vol. 91
Electronic Properties of Conjugated Polymers III
Editors: H. Kuzmany · M. Mehring · S. Roth

films. With PVC as matrix polymer a conductivity of 10^{-2} S/cm is achieved. Besides the good mechanical properties, which are determined by the matrix material, the gain in conductivity is a factor 100 compared to the best coatings.

3. IR- and UV-VIS-Spectroscopy

IR- and UV-VIS-NIR-spectra show that PPM are typical organic semiconductors. The IR-spectrum exhibits a broad absorption above 2000 wavenumbers, caused by the absorption of free charge carriers. The UV-VIS-NIR- spectrum (fig. 1) is dominated by the bandgap-absorption. The onset of the absorption lies in the NIR implying a bandgap value of about 1 eV and being in accordance with the theoretical predictions.

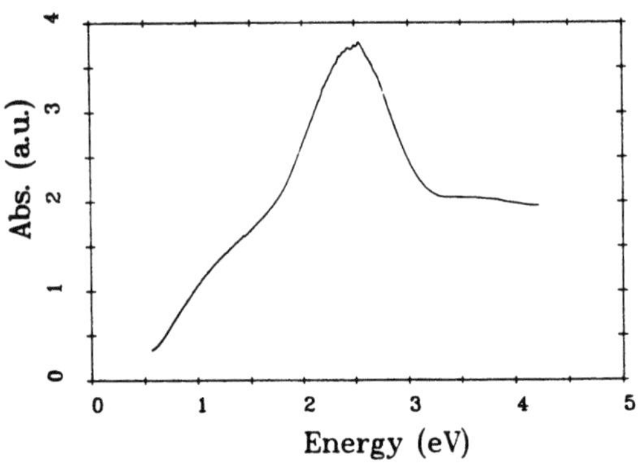

Fig. 1: UV-VIS-NIR spectrum

4. Solid State NMR-Spectroscopy

The ^{13}C-CPMAS spectrum (fig. 2) shows a broad resonance between 100 and 150 ppm, caused by the superposition of the signals of α-, β- and methine carbon atoms. The obvious asymmetry of the resonance is produced by the methine carbon. The absence of signals from saturated carbons (range 0-100 ppm) is remarkable. Only traces of carbonylic defects are visible around 170 ppm. Informations about the chemical nature of the bridging carbon are of particular interest for a proof of structure 2. The spectrum (fig. 3) of a polymer, ^{13}C-labeled at the methine position, shows that the bridging carbon is almost completely present as methine. The methine resonance can be divided into two components, which can be assigned to the nature of the neighbouring ring (protonated-basic).

In accordance to structure 2 it is possible to influence the portion of the basic pyrrole nitrogen via proton acceptors and donators. The degree of protonation can be determined with ^{15}N-CPMAS solid state NMR. If the starting sample is extracted with water, all nitrogen atoms are protonated (fig. 4). After extraction with NH_3 the basic polymer is obtained (fig. 5). The maximum portion of basic nitrogens however is about 30%.

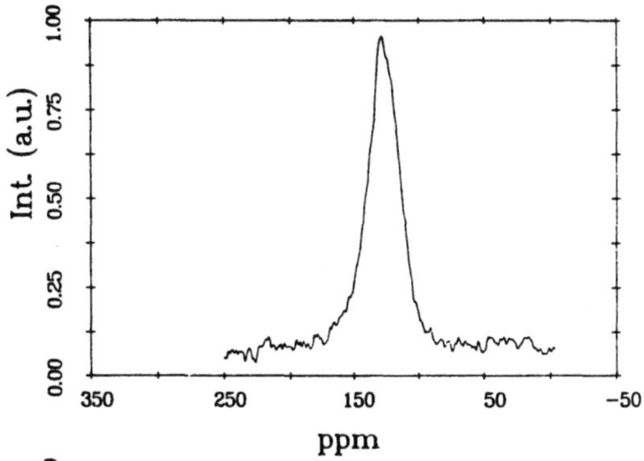

Fig. 2: ^{13}C-CPMAS spectrum (natural abundance)

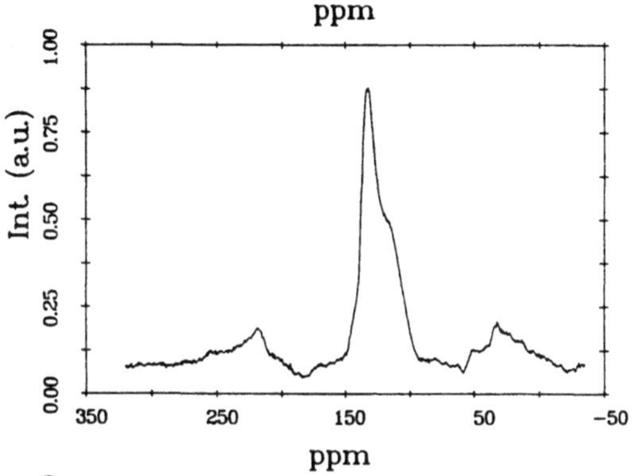

Fig. 3: ^{13}C-CPMAS spectrum (^{13}C-labeled)

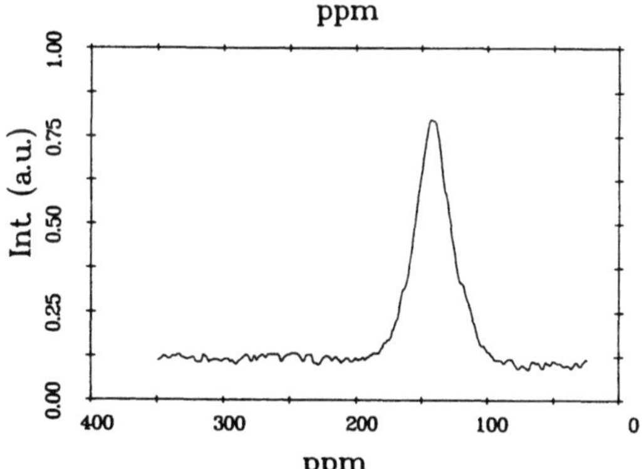

Fig. 4: ^{15}N-solid state NMR: Sample extracted with water

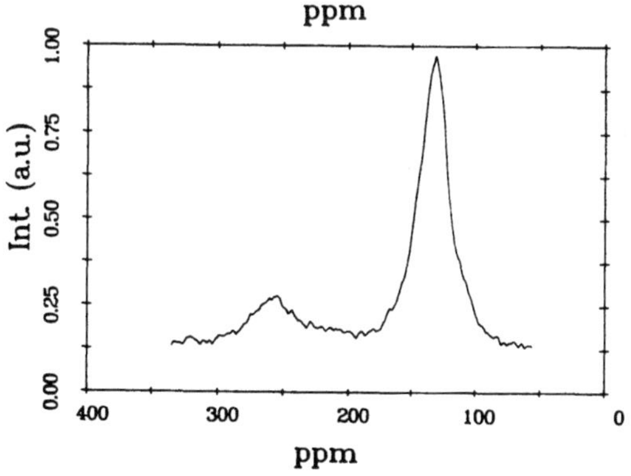

Fig. 5: ^{15}N-solid state NMR: Sample extracted with NH_3

5. Photoelectron Spectroscopy (XPS)

XPS-measurements of not extracted samples show that after the synthesis the sample is completely protonated (fig. 6). In accordance to the ^{15}N-NMR spectrum (fig. 5), there are no basic nitrogens detectable (binding energy 398.5 eV). The positive charge in the ring causes a shift of the N_{1s}- and the C_{1s}-energy levels (fig. 6 and 7) of about 0.5 eV towards higher binding energy compared to polypyrrole. This implies that the positive charge is delocalized over the whole ring. The homogeneous environment of the pyrrol nitrogen is noticable. The width of the N_{1s}-peak is identical with the instrumental resolution. The C_{1s}-peak is separable in nearly equal parts of α- and β- carbons.

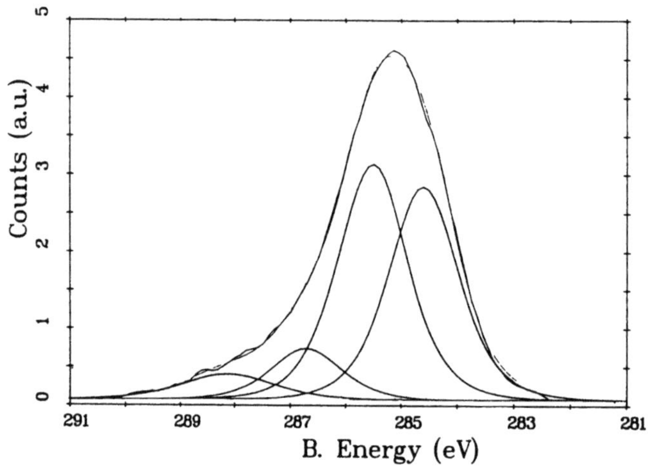

Fig. 6: XPS: N_{1s} spectrum

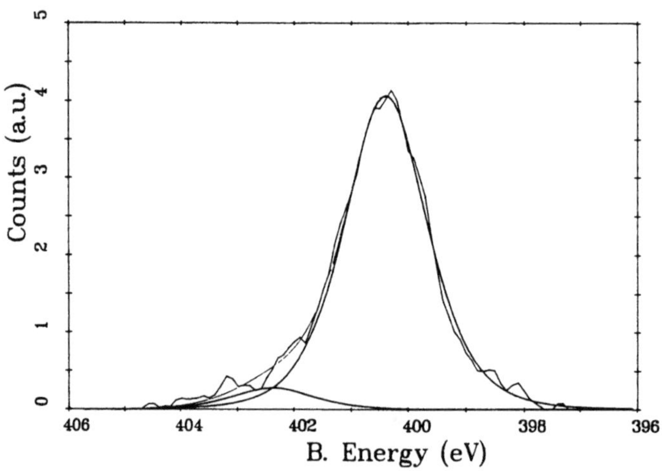

Fig. 7: XPS: C_{1s} spectrum

6. Electron Spin Resonance (ESR)

The values of DC-conductivity (10^{-7}-10^{-5} S/cm), thermopower (S≈1 mV/K) and ESR-measurements show, that after synthesis the polymer is lightly p-doped. In the ESR one resonance with a linewidth of about 7 Gauss and a g-Faktor (g=2.0028) very close to the free electron value is observed. Temperature dependent measurements of the spin concentration allow the separation of free (Pauli-susceptibility) and localized spins (Curie-susceptibility). The obtained values of Curie constant and Pauli spin concentration are $C=2.9\cdot10^{21}$ spins•K/g and $S=1\cdot10^{19}$ spins/g. This corresponds to 1/700 spins/monomer, a value also found in lightly doped polypyrrole and polythiophene. Both extraction of the polymers with H_2O and NH_3 and storing in atmosphere changes these values only slightly.

7. Photoelectrical Properties

As they are semiconductors the PPM are photoelectrical sensitive. The building-up and relaxation of generated charge carriers is a fast process, possessing a time constant $\tau<1$ s. The photogeneration of the charge carriers is controlled by the Onsager mechanism of geminate recombination (Onsager radius ≈ 1.8 nm). The spectral dependence of the photosensitivity essentially follows the absorption spectrum. The quantum yield is still far below 1%.

8. Morphology and Environmental Stability of Composites

Polymerizing the monomer component in the matrix of a nonconducting polymer leads to a fine disperse distribution of the PPM polymer in the host matrix. As shown by REM and TEM pictures, the mean particle diameters are between 50 and 100 nm (1 to 5 µm in the pure PPM polymer). Moreover the REM pictures demonstrate, that the spatial distribution of particles is not statistical homogeneous. The particles coagulate in the shape of a string of pearls and so form a conducting network inside the isolating matrix. This fact explains the higher volume conductivity of the composite compared to the pure PPM.

With regard to technical applications it is of special interest that the composite materials are stable in laboratory atmosphere over long periods of time.

Financial support of the BMFT is gratefully acknowledged.

LITERATURE:

[1] J.M. Toussaint, B. Thémans, J. M. André, J. L. Brédas, ICSM Conference 1988, Santa Fe, Abstract SUN P 61

[2] H. Bräunling et al., Wacker-Chemie GmbH, US-A-4,729,851 (published 8.3.1988)

[3] R. Becker et al., Wacker-Chemie GmbH (Date of application 1.9.1988)

Bridged Macrocyclic Transition Metal Complexes as Semiconducting Materials

M. Hanack, R. Behnisch, A. Lange, A. Leverenz, M. Rein, G. Renz, and P. Vermehren*

Institut für Organische Chemie, Lehrstuhl für Organische Chemie II der Universität Tübingen,
Auf der Morgenstelle 18, D-7400 Tübingen, Fed. Rep. of Germany

Abstract. Bridged quasi-onedimensional macrocyclic transition metal complexes $[MacML]_n$ using phthalocyanine (Pc), tetrabenzoporphyrine (TBP), 1,2- and 2,3-naphthalocyanines (1,2-; 1-,3-Nc) as macrocycles (Mac) with M = e.g. Fe, Ru, Co and L = e.g. pyrazine (pyz), 1,4-diisocyanobenzene (dib), tetrazine (tz) exhibit interesting electrical properties. Regardless of the size of the bridging ligand after either chemical or electrochemical doping stable semiconducting compounds $[MacMLX_y]_n$ (X = e.g. I, BF_4^-, ClO_4^-...) are formed. For the first time the complete characterization of a soluble oligomer $[MacML]_n$ (Mac = R_4Pc, M = Ru, L = dib) by ^{1}H-NMR spectroscopy is reported. Some of these shish kebab complexes e.g. $[PcM(tz)]_n$, (M = Fe, Ru) exhibit good semiconducting properties without additional external doping. Whether or not these compounds are intrinsic semiconductors will be discussed.

Already some years ago bridged quasi-one-dimensional macrocyclic transition metal complexes linked by linear bridging ligands L containing delocalizable π-electrons to form stacked arrangements $[MacM(L)]_n$ have been synthesized by us. As macrocycles Mac phthalocyanine (Pc), tetrabenzoporphyrine (TBP) and more recently 2,3-naphthalocyanine (2,3-Nc) and 2,3-tetranaphthoporphyrine (2,3-TNP) have been used. Transition metals, e.g. Fe, Ru, Co, Rh and others were employed as central metals and as bridging ligands e.g. pyrazine (pyz), diisocyanobenzene (dib), substituted diisocyanobenzenes, tetrazine (tz), substituted tetrazines, cyanide (CN^-) and isothio-cyanate (SCN^-) have been used. Depending on the macrocycle, the central metal and the bridging ligand L these compounds exhibit with and without additional oxydative doping good semiconducting properties [1].

NMR studies to investigate the mechanism of formation of the bridged metal complexes have been carried out with $PcM(Me_4dib)_2$ and $[PcM(Me_4dib)]_n$, M = Fe, Ru [2]. For the first time a soluble oligomer

Springer Series in Solid-State Sciences, Vol. 91
Electronic Properties of Conjugated Polymers III
Editors: H. Kuzmany · M. Mehring · S. Roth

$[MacM(L)]_n$, namely the tetratert.-butyl substituted phthalocyanine oligomers $[t\text{-}bu_4PcRu(dib)]_n$ and $[t\text{-}bu_4PcRu(me_4dib)]_n$ were synthesized. For this purpose $t\text{-}bu_4PcRu$ which was synthesized by us for the first time and which is soluble in quite a few organic solvents was reacted with diisocyanobenzene (dib) and tetramethyldiisocyanobenzene (me_4dib) (in acetone, 24 h reflux). The corresponding oligomers $[t\text{-}bu_4PcRu(dib)]_n$ and $[t\text{-}bu_4PcRu(me_4dib)]_n$ were formed and could be isolated easily. From the ^1H-NMR spectra of both of the compounds dissolved in chloroforme the structure of both oligomers could be deduced unequivocally. The proton-signals of the phthalocyanine rings as well as the ligand-signals were assigned and the chain lengths were estimated from the ratio of the endgroup-signals and the signals of the inner protones of the bridging ligands to be in between 10 - 14 for $[t\text{-}bu_4PcRu(dib)]_n$ and 15 - 19 for $[t\text{-}bu_4PcRu(me_4dib)]_n$ respectively. The chain lengths of both the soluble oligomers could be increased up to 60 by heating them in chloroforme under reflux for 24 h.

The bridging systems $[MacFe(L)]_n$, Mac = Pc, TBP, L = e.g. pyz, dib, have been oxidatively doped with iodine. Incremental doping results in the formation of stoichiometric compounds $[MacM(LI)_y]_n$ which are stable up to 120°C [3-4]. The influence of the length of the bridging ligand L in $[MacM(L)I_y]_n$ on the electrical conductivity has been studied systematically. By changing the bridging ligand L the distance of the cofacial macrocycles in $[MacM(L)I_y]_n$ can be varied ranging from ~700 pm for pyz to ~1150 pm for dib. The powder conductivities of the corresponding polymers $[PcFe(L)I_y]_n$ at the maximum doping level did not change (maximum conductivity $\sigma_{RT} = 10^{-2} - 10^{-1}$ S/cm) using both pyz and dib as the bridging ligand L [3]. This points to a different conductivity mechanism for the transition metal polymers in comparison with e.g. iodine doped $[PcSiO]_n$ [5-6].

Electrochemical doping of $[MacM(L)]_n$ compounds is also possible, e.g. electrochemical doping of $[PcFe(pyz)]_n$ ($\sigma_{RT} = 2 \times 10^{-6}$ S/cm) with a variety of counterions BF_4^-, $B(Ph)_4^-$, PF_6^- and ClO_4^- leads to thermally quite stable semiconducting materials with conductivities $\sigma_{RT} \sim 10^{-2} - 10^{-1}$ S/cm [7]. 2,3-Naphthalocyanine and 2,3-tetranaphthaporphyrine have also been used to prepare the corresponding polymers $[MacFe(dib)]_n$, (Mac = 2,3-Nc, 2,3-TNP) [8]. In a systematic study the redox potentials of a variety of macrocycles shown in Table 1 have been measured by cyclic voltammetry [9]. As can be seen from Table 1 the oxydation potential (E^1 1/2), which is due to the oxydation of the macrocycle is very much dependent upon the nature of the macrocycle. Going from e.g. PcFe to TNPFe a considerable decrease of the oxydation potential is observed. In case of the naphthalocyanines $E^1 1/2$ (see Table 1) is dependent

Table 1

Redoxpotentials[a] of some Pc-, TBP-, NC- and TNP-Fe(+II) and Co(+II)-complexes in pyridine/Bu_4NClO_4 vs. SCE [V]

	E^1 1/2	E^2 1/2	E^3 1/2	E^4 1/2	E^5 1/2
PcFe	1.10	0.69	-1.085	-1.39	-1.93
1,2-NcFe	1.01	0.68	-0.95	-1.21	-1.80
2,3-NcFe	0.81	0.43	-1.09	-1.32	-1.86
TBPFe	0.82	0.34	-0.90	-1.59	-1.87
TNPFe[b]	0.61	0.15	-0.90	-1.49	-1.87
assignment[c]	Mc^{-2}/ Mc^{-1}	Fe(+II)/ Fe(+III)	Fe(+I)/ Fe(+II)	Mc^{-3}/ Mc^{-2}	Mc^{-4}/ Mc^{-3}
PcCo	1.15	0.21	-0.55	-0.91	-1.44
2,3-NcCo	0.77	0.205	-0.575	-0.92	-1.52
TBPCo	0.85	-0.13	-0.88	-1.09	-1.79
TNPCo[b]	0.515	-0.13	-0.90	-	-1.86
assignment[c]	Mc^{-2}/ Mc^{-1}	Co(+II)/ Co(+III)	Co(+I)/ Co(+II)	Mc^{-3}/ Mc^{-2}	Mc^{-4}/ Mc^{-3}

[a]In general, the halfwave potentials E^n 1/2 are reversible (ΔE_p = 58 mV) at lower scan rates (v = 0.02 V/s), but ΔE_p increases with increasing v.
[b]Unstable compound during CV-experiments.
[c]Through comparison with earlier results.

upon the structure of the naphthalocyanine, 1,2-NcFe has a much higher oxydation potential than the symmetrical 2,3-NcFe (see Fig. 1).

Due to the low oxydation potentials of 2,3-NcFe and TNPFe the corresponding diisocyano-bridged polymers (see Fig. 1) show conductivities of ~10^{-3} S/cm already without external iodine doping. The high conductivity of these compounds is caused already by oxygen doping from the air [8]. $[1,2\text{-NcFe(dib)}]_n$ according to the high oxydation potential of the macrocycle given in Table 3 only shows a conductivity of 10^{-10} S/cm without iodine doping.

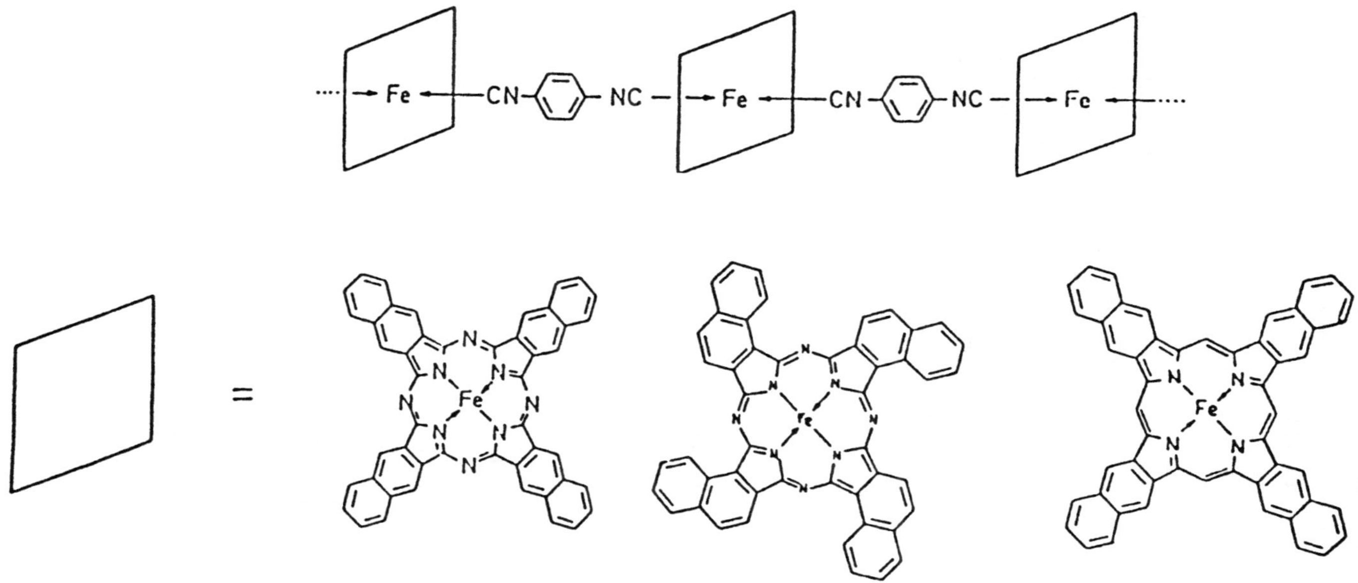

	2,3-NcFe	1,2-NcFe	2,3-TNPFe
conductivity	$[2,3\text{-NcFe(dib)}]_n$	$[1,2\text{-NcFe(dib)}]_n$	$[2,3\text{-TNPFe(dib)}]_n$
[S/cm]	$2 \cdot 10^{-3}$	$6 \cdot 10^{-10}$	$5 \cdot 10^{-3}$

Fig. 1. Schematic drawing of $[2,3\text{-NcFe(dib)}]_n$, $[1,2\text{-NcFe(dib)}]_n$ and $[2,3\text{-TNPFe(dib)}]_n$ and their powder conductivities

1,2-NcFe (Fig. 1) was isolated as a pure isomer for the first time. Separation from the other possible isomers was carried out by column chromatography using the soluble bisaxially substituted cyclohexylisocyanide derivative. The structure of the soluble 1,2-NcFe$(cHxNC)_2$ was determined by NMR spectroscopy. Single crystals of this compound could be grown and the X-ray structure proved that the compound is symmetric as shown in Figure 1 with the macrocycle almost planar [10].

A higher conductivity without additional external doping can also be obtained, if tetrazine or e.g. dimethyltetrazine is used as the bridging ligand in the polymers $[PcFe(L)]_n$ (L = tz, me_2tz). These types of polymers have also been obtained with phthalocyaninatoruthenium leading to $[PcRu(tz)]_n$. They are prepared by treating the corresponding metal macrocycles with tetrazine or dimethyltetrazine in chloroform at 70°C. Table 2 shows some conductivity data of monomeric and s-tetrazine bridged metal macrocycles in comparison with the corresponding pyrazine-bridged systems [11,12]. The monomers, e.g. $PcFe(tz)_2$ show clearly an insulating behavior while the s-tetrazine bridged macrocyclic compounds show, in

Table 2

Conductivity data of monomeric and s-tetrazine bridged metal macrocycles in comparison with the corresponding pyrazine-bridged systems

Compound	σ_{RT} [S/cm]	E_a [eV]
$PcFe(tz)_2$	> 10^{-9} [a]	-
$[PcFe(pyz)]_n$	2 x 10^{-6} [b]	-
$[PcFe(tz)]_n$	2 x 10^{-2} [b]	0.20
$PcRu(tz)_2$	> 10^{-11} [a]	-
$[PcRu(pyz)]_n$	1 x 10^{-7} [b]	-
$[PcRu(tz)]_n$	1 x 10^{-2} [b]	0.10
$[PcRu(me_2tz)]_n$	3 x 10^{-3} [b]	-
$[me_8PcFe(pyz)]_n$	9 x 10^{-6} [b]	-
$[me_8PcFe(tz)]_n$	1 x 10^{-3} [b]	-
$[2,3\text{-}NcFe(pyz)]_n$	5 x 10^{-5} [b]	-
$[2,3\text{-}NcFe(tz)]_n$	3 x 10^{-1} [b]	0.07

[a]Two-probe-technique.
[b]Four-probe-technique.

Table 3

Mößbauer data[a] of tz-bridged macrocyclic iron compounds [12]

Compound	δ [nm/s]	ΔE_Q [nm/s]
$[PcFe(pyz)]_n$	0.24	2.04
$[PcFe(tz)]_n$	0.13	2.27
$[me_8PcFe(tz)]_n$	0.14	2.11
$[(meO)_8PcFe(tz)]_n$	0.15	2.19
$[2,3\text{-}NcFe(tz)]_n$	0.19	1.97

[a]Relative to metallic iron.

general, comparatively high conductivities up to 0.3 S/cm in case of $[2,3\text{-}NcFe(tz)]_n$. The difference between pyrazine and tetrazine as a bridging ligand is especially evident, if one compares the corresponding bridged systems: In all cases an increase in the powder conductivities of 10^3 - 10^4 is observed.

Table 3 shows some Mößbauer data of tz-bridged complexes in comparison with $[PcFe(pyz)]_n$ [13]. In all cases the isomer shift of the tetrazine bridged systems is especially small, which points to a charge transfer from the iron on to the tetrazine in these systems. The electronic structure of $[PcFe(pyz)]_n$ has been studied by means of the tight-binding (LCAO) method [14]. The band gap according to these calculations is mostly determined by the difference in energies between the LUMO of the bridging ligand and the HOMO of the transition metal d_{xy} orbital. The higher conductivities of the s-tetrazine bridged compounds $[PcFe(tz)]_n$, $[PcRu(tz)]_n$ and $[2,3\text{-}NcFe(tz)]_n$ are thereby explainable by the low lying LUMO of this ligand.

References

1. M. Hanack, Chima 37 (1983) 238; M. Hanack, A. Datz, R. Fay, K. Fischer, U. Keppeler, J. Koch, J. Metz, M. Mezger, O. Schneider and H.-J. Schulze in T. Skotheim (ed.), Handbook on Conducting Polymers, Marcel Dekker, New York 1985; M. Hanack, S. Deger and A. Lange, Coord. Chem. Rev., 83 (1988) 115.
2. U. Keppeler and M. Hanack, Chem. Ber., 119 (1986) 3363.

3. M. Hanack, U. Keppeler and H.-J. Schulze, Synth. Met., 20 (1987) 347.
4. B.N. Dicl, T. Inabe, K.K. Jaggi, J.W. Lyding, O. Schneider, M. Hanack, C.R. Kannewurf, T.J. Marks and C.H. Schwartz, J. Am. Chem. Soc., 106 (1984) 3207.
5. M.-H. Whangbo and K.R. Stewart, Isr. J. Chem., 23 (1983) 138.
6. T. Inabe, J.G. Gaudiello, M.K. Moguel, J.W. Lyding, R.L. Burton, W.J. McCarthy, C.R. Kannewurf and T.J. Marks, J. Am. Chem. Soc., 108 (1986) 7595.
7. M. Hanack and A. Leverenz, Synth. Met., 22 (1987) 9.
8. S. Deger and M. Hanack, Isr. J. Chem., 27 (1986) 347; M. Rein and M. Hanack, Chem. Ber., 121 (1988) 1601.
9. M. Hanack and R. Behnisch, Synth. Met., in press.
10. M. Hanack, G. Renz, J. Strähle and S. Schmid, Chem. Ber., 121 (1988) 1479.
11. U. Keppeler, S. Deger, A. Lange and M. Hanack, Angew. Chem., 99 (1987) 349.
12. A. Lange and M. Hanack, unpublished.
13. A. Lange, U. Keppeler and M. Hanack, Inorg. Chem., in press.
14. E. Canadell and S. Alvarez, Inorg. Chem., 23 (1984) 573.

Progress in Science and Industrial Applications of Electroconducting Polymers and Related Compounds

H. Sixl

Hoechst AG, Angewandte Physik, Postfach 80 03 20,
D-6230 Frankfurt am Main 80, Fed. Rep. of Germany

Abstract. This contribution gives a summary of the relevant aspects concerning the present state and recent developments of the physics and chemistry of electroconducting polymers and related compounds with extraordinary electrical and electro-optical properties. A critical assessment of research and development trends with respect to industrial applications will be given. Special attention is drawn to the main topics: electroconducting polymers, organic nonlinear optical materials and new high T_c superconductors. The basic requirements to start industrial research and development towards new market products are extraordinary new material properties such as high conductivity or high nonlinear optical parameters. In addition a series of critical success parameters such as chemical stability or processing capabilities has to be taken into account.

1. Introduction

In the last years scientific and industrial interest has been more and more focussed on electronic and optoelectronic properties of polymers and related compounds such as the recently discovered high T_c-superconducting ceramics. As a consequence of these short term applied research developments the central topics of this winter school have been concentrated on (1) electroconducting polymers, on (2) nonlinear optical polymers, and finally on (3) the new superconductors. Short sessions on molecular electronics and biochips clearly have shown that these activities are long-term scientific research projects which still are on an academic level. Despite some controversial discussions on solitons in trans-polyacetylene the basic theoretical understanding of the electronic properties of polymers is complete. However, concerning the high-T_c superconductors we are far from an understanding of the very basic principal mechanisms, leading to the high T_c values.

2. Electroconducting Polymers

These new materials offer opportunities for various functional properties in addition to electroconductivity. For the future of microelectronics and optoelectronics, polymer systems offer new possibilities for specific material properties, structure and architectures at the molecular level. Especially the combination of classical metal or semiconducting properties with classical polymer properties seems to be most promising. The following main potential advantages over metals or inorganic semiconductors are the driving force for their continued development: plasticity and elasticity, polymer-type processability, low mass densitiy (light-weight) resistance to chemicals and corrosion, tunable optical properties, variable optical transmission, charge storage capacity and rechargeability in battery applications.

There are two classes of electroconducting polymers: **pure materials** and **composites**. **Electroconducting pure materials** are of special interest concerning fundamental science and basic materials properties. They are

Springer Series in Solid-State Sciences, Vol. 91
Electronic Properties of Conjugated Polymers III
Editors: H. Kuzmany · M. Mehring · S. Roth

defined as organic materials with inherent conductivity due to a specific structure (conjugated system) and / or doping. Typical examples are (a) conducting polymers, (b) graphite-like polymers, (c) charge transfer complexes and radical ion salts. **Electroconducting composites** are of special interest concerning applications leading to commercial products. They are defined as processable polymer matrices containing fillers like carbon black, metallic powders or conductive polymers. Commercial products exist already. Applications are (a) pastes, paints, adhesives (b) molding materials (c) fibers and films.

2.1 Characteristic Features of Typical Conducting Polymers

The **polyacetylenes** represent the only conjugated polymer system with a degenerate ground state which allows the existence of "solitons". It is a very popular prototypical polymer with simplest repeat unit. Highly oriented films yield highly conducting films with world record values of 10^5 S/cm /1/. Electronic devices, "MISFET structures", have been fabricated /2/ demonstrating p-and n-type dopability and semiconductor properities.

Important properties of the **polyphenylenes** are mechanical strength, solubility and fusability in the undoped state. **Polyparaphenylene** shows high thermal stability in inert atmosphere (850°C). **Polyphenylene vinylene** has water soluble precursors suitable for processing. These films have been used for electronic devices /3/ and nonlinear optics /4/. Alkoxy derivatives are soluble in organic solvents. Due to their high stability these films represent excellent starting material for highly conducting graphite films.

Owing to its high chemical stability **polyaniline** has become the main polymer of interest for applications. It exists in various molecular structures, which can be reversibly converted electrochemically and chemically. Polyaniline derivatives are soluble in the doped and undoped state. It has been shown /5/ that the polymer "solutions" show the typical features of dispersions.

The characteristics of **heterocyclic polymers** like **polypyrroles** and **polythiophenes** are given by their high stability in the undoped state. As a new result paraffinic polythiophene derivatives turned out to be soluble and fusible /6/ in the undoped state. Progress in this field is given by low melting points (around 100°C) depending on the alkyl chain length and by the thermochromism /7/ similar to that observed earlier in polydiacetylene systems. Other important new results are given by oligomers of polypyrrole /8/ and polythiophenes /9/.

2.2 Applications

In the first commercial application of conducting polymers, polyaniline is used as the positive electrode in a secondary polymer/lithium battery by Bridgestone Corporation /10/. Other important companies active in this field are Showa Denko and Hitachi. Market entry was in 1987 with a small size battery with high energy density and high voltage. Problems of self-discharge and stability have been solved. Typical values of self-discharge are 7 % in 30 days /10/. The high value of polypyrrole (about 50 % in 30 days) now also has been reduced to 10 % using a better morphology of the films /11/.

BASF and Varta Batteries are testing a similar dry cell using a polypyrrole electrode with excellent performance characteristics and structural flexibility. However the energy density and number of charge/discharge cycles will have to at least double before the battery can be commercially successful. Commercialization now starts with a small size button cell; large size batteries will not be sold before 1990. New battery developments now are given by supercapacities using symmetric 50 percent doped polypyrrole electrodes /11/.

Another important new application uses soluble conducting polymers for antistatic coatings. Lockheed and Showa Denko applied for patents on soluble or dispersible polyaniline derivates in the doped state. **Hoechst AG** has developed **fully transparent, soluble conducting** oligomer and polymer composites /12/ which can be coated on a carrier film with high chemical and mechanical stability with surface resistance of $10^6 ... 10^8$ Ohm, which is optimal for antistatic equipment.

Electrochromism of electroconducting polymers is based on the change of the optical absorption by electrochemical doping and undoping. The best **electrochromic display ECD** systems are given by polythiophenes and polyanilines with color changes from blue to red (PT) and green to yellow (PANI). The ECD and "smart windows" (15) have to compete with liquid crystal displays and have still a series of disadvantages (cyclability, large energy consumption, slow, not applicable for colored displays). Several companies applied for patents on ECD (NTT, Hitachi, Maxell /14/). None of the products has yet reached the commercial stage.

3. Nonlinear Optical (NLO) Materials

The commercial interest in NLO materials is driven by the present development of fiber optics, laser diodes for optical data storage and optical processing. Future prospects are given by integrated optics and optical computing. The present interest for application is focussed on NLO polymers or "poled polymers". In these systems an electric field is used to partially orient the nonlinear optically susceptible molecules to break the inversion symmetry of the molecular ensemble. Thus high $\chi^{(2)}$ values are obtained.

The main advantages of NLO polymers over the inorganic highly stable competitor $LiNbO_3$ are given in the following comparison.

	NLO polymers	**$LiNbO_3$**
Response time	subpicoseconds	ns...ps
Nonlinear activity		
for second harmonic generation	$\chi^{(2)}$ = 120 pm/V	10 pm/V
for electrooptics	r = 35 pm/V	30 pm/V
relative dielectric constant	ε = 3	30
figure of merit	r/ε = 10	1
Processability	melt and/or solution	-
Technology	thin films	-

Consequently the relevant properties are approximately a factor of 10 better than in $LiNbO_3$ and in multiple quantum well devices of GaAs/Ga AlAs. In addition polymer technology can be applied to produce poled polymers with NLO active molecules in waveguide devices.

A series of companies are involved in device and material research activities like Kodak, 3M, Rhone Poulenc, Merck, SEL. There are cooperations between Lockheed, Hoechst Celanese, GEC, between Akzo, Alcatel, IMEC /16/, between Philips and Dupont /17/ as well as Thomson CSF, ICI. However progress in nonlinear optical devices is very slow. Using poled polymers, it has been demonstrated in laboratories, that electrooptic modulators now are working (100 MHz ... 1 GHz), with excellent performance. However the key device given by the electrooptic Mach-Zehnder travelling wave modulator still is not available. Concerning laser frequency doubling, all the efforts now are focussed on the method of "spatially modulated poling" /17/. The basic effects have been demonstrated in laboratories, however, there are still no commercial products available. The main problems in all $\chi^{(2)}$ devices are given by monomode waveguide structures and technologies, by the device stabilities and by the optical losses, which still are too large (greater than 1dB/cm). Concerning $\chi^{(3)}$ no relevant progress is observed in the last ten years. The basic problems here are given by the $\chi^{(3)}$ values, which still are by approximately a factor of one hundred far too low for applications. In addition like in the $\chi^{(2)}$ devices the optical absorption and stability represent the major problems.

4. High T_c Superconducting Materials

Since the discovery of superconductivity above liquid nitrogen temperature three classes of oxide superconductors of relevance for applications have been developed. The first class is given by the rare earth oxides superconductors with the most prominent representative, given by $Y_1Ba_2Cu_3O_7$. The superconducting transition temperatures of this class are around 90 K. The second and third classes are given by the Bi-oxide and Tl-oxide superconductors, which show a layer structure. The higher the number of CuO_2-layers, intercalated between the Sr and Ca layers, the higher the transition temperature as shown in the following list:

	dopant,	transition temperature,	number of CuO_2-layers
Bi-Sr-Cu-O	-	40 K	1
Bi-Sr-Ca-Cu-O	-	75 K	1
Bi-Sr-Ca-Cu-O	-	85 K	2
Bi-Sr-Ca-Cu-O	-	110 K	3
Bi-Sr-Ca-Cu-O	Sb,Pb	140 K	4 ?
Tl-Ba-Ca-Cu-O	-	125 K	3
Tl-Ba-Ca-Cu-O	?	160 K	4 ?

All T_c-values are highly dependent on the superconducting phases, compositions, dopants, sintering conditions and on the oxygen content.

The main advantages of the new high T_c materials are the high transition temperature and high critical fields, the prospect of semiconductor-superconductor devices and the simple chemistry. The main disadvantages are their low dimensionality, low coherence length, low critical current density, high sintering temperatures correlated by substrate diffusion problems and the usual deficiencies of ceramic materials.

The best system for applications is the Bi-oxide class. The rare earth class now has the lowest T_c values, the Tl-class is highly toxic. All relevant applications require high critical currents and therefore require "single crystal quality".

The state of the superconducting technology is characterized by high quality c-axis oriented films with critical currents j_c between 2 and 4.10^6 A/cm^2 (zero-field, 77 K). The best preparation methods are pulsed laser or electron beam coevaporation as well as by sputtering or thermal coevaporation of the relevant oxides using additional oxygen ion beams and heated substrates. Thick films, wires and ceramics all show non-satisfactory j_c about 10^3 A/cm^2. At the moment only thin films meet the basic requirmements (T_c far above 77K, high j_c, low surface resistance at high frequency) for applications in **high frequency technology** and **electronics** including **sensors.** However, there is still a long way to go for the most attractive applications with the highest market potential given by **energy technology** and **magnet technology.**

Hoechst AG already offers a product palette on all relevant high T_c superconducting materials especially on the Bi- one, two and three layer systems.

References

1. H. Naarmann, Symposium Materialforschung ed. by BMFT, Vol.1, (1988), 144, H. Naarmann and N. Theophilou, Synth. Met. 22 (1987) 1.
2. J. H. Burroughes, C. A. Jones, R. A. Lawrence and R. H. Friend, this conference.

3. D. D. Bradley and Y. Mori, this conference.
4. C. Bubeck, A. Kaltbeitzel and D. Neher, this conference.
5. B. Wessling, this conference.
6. J.-E. Österholm, J. Laaksso and P. Nyholm, this conference,
M. Zagorska, this conference.
7. O. Inganäs and G. Gustafsson, this conference.
8. D. Bloor and S. Pugh, this conference.
9. D. Fichou, F. Horowitz and F. Garnier, this conference.
10. K. Abe, IEICE Technical Report CPM 87-99 (1988).
11. D. Nägele, this conference.
12. M. Feldhues, G. Kämpf, K. Pampus and P. Wegener, this conference.
13. W. R. Salaneck, P. Dannetum, M. Lögdhund, J. O. Niesson, K. Uvdal,
C. Yang, A. G. MacDiarmid and E. Scherr, this conference.
14. NTT, Japan Kokai Tokkyo 62-34134 (Febr. 14, 1987),
Hitachi Maxell, Japan Kokai Tokkyo 62 - 80631 (April 14, 1987).
15. R. Ulrich, Mol. Cryst. Liq. Cryst. Vol 160 (1988) 1.
16. G. R. Möhlmann, this conference.
17. C. J. E. Seppen, this conference.

Index of Contributors

MIX
Papier aus verantwortungsvollen Quellen
Paper from responsible sources
FSC® C105338

If you have any concerns about our products,
you can contact us on
ProductSafety@springernature.com

In case Publisher is established outside the EU,
the EU authorized representative is:
Springer Nature Customer Service Center GmbH
Europaplatz 3, 69115 Heidelberg, Germany

Printed by Libri Plureos GmbH
in Hamburg, Germany